ENERGY FROM FOSSIL FUELS
& GEOTHERMAL ENERGY

Progress in Energy and Combustion Science

VOLUME 2

Other Pergamon Books and Journals on Energy Sources and Technology

ASHLEY *et al.*	*Energy and the Environment—A Risk-Benefit Approach*
BLAIR *et al.*	*Aspects of Energy Conversion*
BOER	*Sharing the Sun: Solar Technology in the Seventies*
BRATT	*Have You Got the Energy?*
DIAMANT	*Total Energy*
HUNT	*Fission, Fusion and the Energy Crisis*
JONES	*Energy and Housing*
KARAM & MORGAN	*Environmental Impact of Nuclear Power Plants*
KARAM & MORGAN	*Energy and the Environment Cost-Benefit Analysis*
KOVACH	*Technology of Efficient Energy Utilization*
MESSEL & BUTLER	*Solar Energy*
MURRAY	*Nuclear Energy*
REAY	*Industrial Energy Conservation*
SIMON	*Energy Resources*
SMITH	*Efficient Electricity Use*
SPORN	*Energy in an Age of Limited Availability and Delimited Applicability*
MCVEIGH	*Sun Power, An Introduction to the Applications of Solar Energy*
UNITAR	*The Future Supply of Nature Made Petroleum and Gas*

Pergamon International Research Journals

Annals of Nuclear Energy

Energy, The International Journal

Energy Conversion

Geothermics

International Journal of Hydrogen Energy

Solar Energy

Sun World

Thermal Engineering

ENERGY FROM FOSSIL FUELS
& GEOTHERMAL ENERGY

Progress in Energy and Combustion Science

VOLUME 2

Edited by

N. A. CHIGIER

University of Sheffield, England

PERGAMON PRESS

Oxford · New York · Toronto · Sydney · Paris · Frankfurt

U.K.	Pergamon Press Ltd., Headington Hill Hall, Oxford OX3 0BW, England
U.S.A.	Pergamon Press Inc., Maxwell House, Fairview Park, Elmsford, New York 10523, U.S.A.
CANADA	Pergamon of Canada Ltd., 75 The East Mall, Toronto, Ontario M8Z 2L9, Canada
AUSTRALIA	Pergamon Press (Aust.) Pty. Ltd., 19a Boundary Street, Rushcutters Bay, N.S.W. 2011, Australia
FRANCE	Pergamon Press SARL, 24 rue des Ecoles, 75240 Paris, Cedex 05, France
FEDERAL REPUBLIC OF GERMANY	Pergamon Press GmbH, 6242 Kronberg-Taunus, Pfferdstrasse 1, Federal Republic of Germany

Library of Congress Catalog No: 75-24822

0 08 021219 0

Printed in Great Britain by A. Wheaton & Co. Ltd., Exeter

CONTENTS

CONTENTS OF VOLUME 1

POLLUTION FORMATION AND DESTRUCTION IN FLAMES

Prog. Energy Combust. Sci., Vol. 2, pp. 1–25, 1976. Pergamon Press. Printed in Great Britain

NO_x CONTROL FOR STATIONARY COMBUSTION SOURCES

ADEL F. SAROFIM and RICHARD C. FLAGAN*
Massachusetts Institute of Technology, Cambridge, Massachusetts 02139

1. INTRODUCTION

The importance of nitrogen oxides as an atmospheric pollutant has been recognized relatively recently. The past decade has, therefore, seen an intensive effort expended at identifying the major sources of nitrogen oxides, understanding the mechanisms of their formation, and developing technology for abatement of emissions from the dominant sources. The emissions from mobile sources are adequately covered elsewhere in this series. Coverage in this article will therefore be restricted to stationary sources, with extensive reference to the U.S. scene for which the pertinent data are most readily available.

1.1. *Stationary Sources of Nitrogen Oxides*

It is estimated that the U.S. accounts for approximately half the anthropogenic sources of NO_x, and that stationary sources contribute more than half the U.S. emissions.

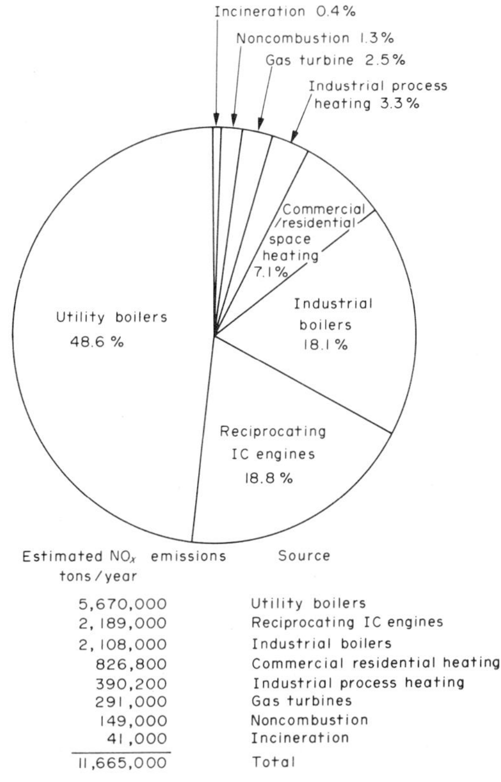

Estimated NO_x emissions tons/year	Source
5,670,000	Utility boilers
2,189,000	Reciprocating IC engines
2,108,000	Industrial boilers
826,800	Commercial residential heating
390,200	Industrial process heating
291,000	Gas turbines
149,000	Noncombustion
41,000	Incineration
11,665,000	Total

FIG. 1. Summary of 1972 stationary source NO_x emissions.

The emissions are dominated by those from combustion sources,[1] primarily utility boilers and stationary engines (see Fig. 1), with small, but sometimes locally significant, contributions from nitric acid and nitrogen fertilizer plants.

The nitrogen oxides emitted by combustion sources are predominantly in the form of nitric oxide (NO) with the residual, usually less than 5%, in the form of nitrogen dioxide (NO_2). The oxides are formed either by the oxidation of atmospheric nitrogen at high temperatures (thermal NO_x) or by the oxidation of nitrogen compounds in the fuel (fuel NO_x). The relative contributions of thermal and fuel NO_x depends on combustor design and operating conditions, and on the nitrogen content of the fuel. As much as half of the total stationary emissions may be contributed by the oxidation of the nitrogen in the fuel, primarily in units burning heavy-oils and coals.

1.2. *Control Options*

The technology for controlling the emission of nitrogen oxides is based on either the modification of the combustion process to prevent formation of the oxides or the treatment of the product gases to destroy or remove the oxides. The latter option is made difficult by the relative inertness and insolubility of NO. Wherever combustion process modification has yielded the needed reduction in NO_x emissions, it has proven to be the most economical method for doing so. Product gas treatment is relatively expensive and is being currently considered[2] only in those cases where the emission standards are set at levels which cannot be met by combustion process modification alone.

2. MECHANISMS FOR NITRIC OXIDE FORMATION

2.1. *Thermal Fixation of Atmospheric Nitrogen*

The formation of thermal NO_x is determined by highly temperature dependent chemical reactions, the so-called Zeldovich reactions.[3] The rate of formation is significant only at high temperatures (greater than 1800 K) and doubles for every increase in flame temperature of about 40 K. The rate of formation increases with increasing oxygen concentration (rate proportional to the square root of the oxygen concentration) except in a small region near the flame zone in which superequilibrium concentrations of oxygen atoms are found.[4,5,6] Although the NO_x emitted by most practical combustors is predominantly in the form of NO, there is evidence from laboratory studies that a significant fraction of the NO_x may be present as NO_2 in localized regions of a flame.[7] Combustion systems, such as gas-turbines or the domestic gas-range, which

* Currently with Department of Environmental Engineering Science, California Institute of Technology, Pasadena, California.

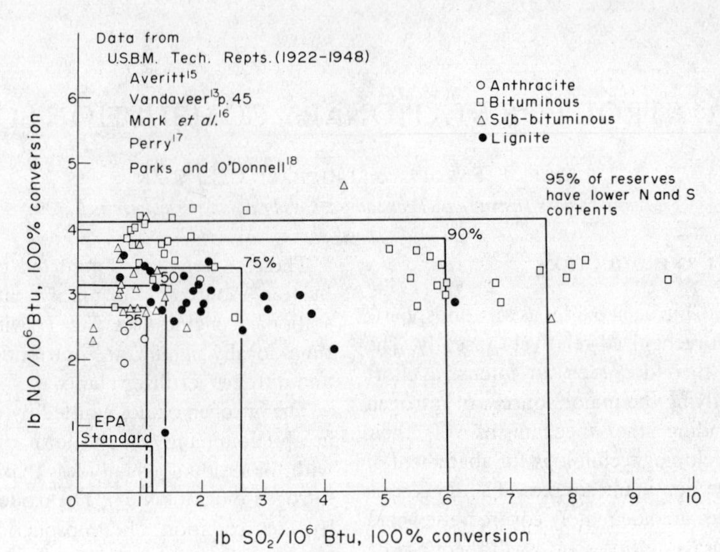

FIG. 2. Nitrogen and sulfur content of U.S. coal reserves.

are characterized by rapid quenching of the primary combustion products may have a relatively higher percentage (as much as 50%) of the NO_x in the form of NO_2.

2.2. *Formation of Fuel NO_x*

An upper bound for the potential contribution of fuel-nitrogen to NO_x formation may be gauged from the nitrogen content of fuels. Natural gas has negligible amounts of organically bound nitrogen, U.S. crude oils have nitrogen concentrations that average about 0.15% by weight,[8,9] and U.S. coals have nitrogen concentrations that average about 1.4% by weight. The nitrogen content in crude oil varies between oil fields (Table 1), and is high for California oils. During the refining of oils, the nitrogen concentrates in the heavy fractions. Residual oils from a California crude may have as much as 1.1% nitrogen, whereas light distillate oils generally have nitrogen concentrations under 0.1%. The spread in nitrogen concentrations for coals is shown in Fig. 2. The concentration is reported as lbs $NO_2/10^6$ B.t.u., assuming 100% conversion of fuel nitrogen to NO_2.* Also shown on the plot are the sulfur concentrations, reported as lbs $SO_2/10^6$ B.t.u., and the percentage of U.S. coal reserves that have nitrogen and sulfur concentrations below certain levels. Although use of low-sulfur coals has been proposed for controlling SO_x emissions it is apparent that there is much less opportunity to control NO_x emissions by use of low-nitrogen coals.

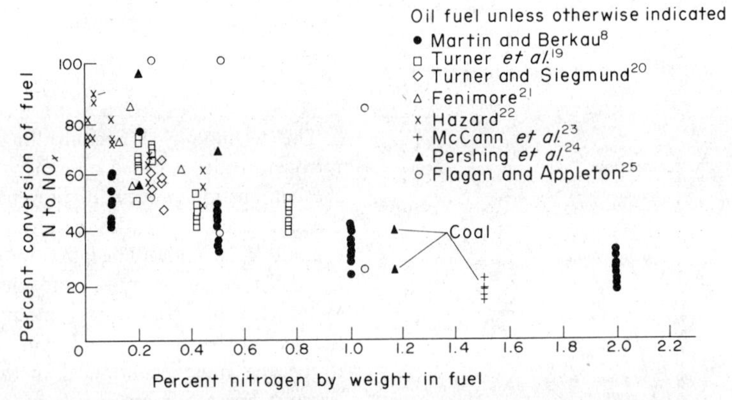

FIG. 3. Percent conversion of fuel N to NO_x in laboratory scale combustors.

TABLE 1. Nitrogen content in U.S. crude oils[10] (From Refs. 11, 12, 13, 14) In Million Barrels

	Proved reserves (1972)		Production (1972)		Average weight % nitrogen
	MM bbl	%	MM bbl	%	
Texas	12,144	33.4	1,258	38.3	0.074
Alaska	10,096	27.8	73	2.2	—
Louisiana	5,028	13.8	780	23.8	0.056
California	3,554	9.8	346	10.5	0.49
Oklahoma	1,303	3.6	198	6.0	0.151
Wyoming	950	2.6	139	4.2	0.183
New Mexico	582	1.6	106	3.2	0.082
Other	2,682	7.4	331	11.8	—
U.S. Total	36,339	100	3,231	100	—

* Conversion factors to other units are presented in an Appendix. Multiplication by 1.8 yields g/10⁶ cal.

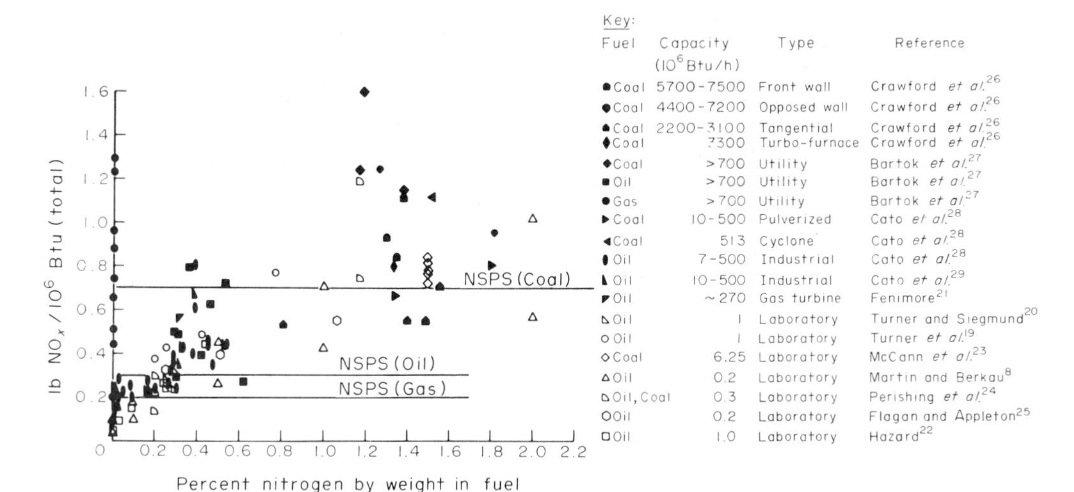

FIG. 4. NO_x emissions from uncontrolled combustors as a function of fuel nitrogen content. (NSPS refers to New Source Performance Standards in the U.S. for steam-generating equipment with capacities in excess of 250,000 B.t.u./hr.)

Only a fraction of the fuel nitrogen is converted to nitric oxide in practical combustors. Tests on carefully controlled laboratory-scale units have typically shown that 15–100% of the fuel nitrogen is converted to NO_x (Fig. 3), with the higher conversion efficiencies obtained when the nitrogen content is low or when the combustor is operated lean.

It is not easy to separate the contribution of fuel NO_x from thermal NO_x; consequently the degree of conversion of fuel nitrogen to NO_x in large scale units is not known with certainty, but estimates obtained from field tests show trends similar to those obtained in the laboratory. Consideration of the total uncontrolled NO_x emissions from large and small scale units in Fig. 4 suggests that emissions increase with increases in fuel nitrogen content for units ranging in size from laboratory burners to utility boilers. Uncontrolled emissions from large gas-fired units account for the high emission data at zero nitrogen content.

The mechanism by which the fuel nitrogen is converted to nitrogen oxides is imperfectly understood. The conversion efficiency of fuel nitrogen to NO_x is thought to depend on the competition between reactions of fuel nitrogen fragments leading to NO and N_2. Such a competition can qualitatively explain why the observed conversion efficiency of fuel nitrogen to NO_x increases markedly with increased oxidizing contributions but is relatively insensitive to changes in temperature.[8,21]

3. FACTORS INFLUENCING NO_x EMISSIONS

Increases in air/fuel ratio in practical combustors usually result in increases in both fuel and thermal NO_x; increases in flame temperature, however, result mainly in increases in thermal NO_x. Any factor which influences temperature and concentration profiles in combustors may, therefore, have an influence on NO_x emission. This makes generalizations for practical systems difficult since fuel/air ratios, fuel/air mixing patterns, fuel type, interaction of different burners, and the placement of heat transfer surfaces all influence emissions. Additional complications are introduced by the observations that NO may be reduced by reactions with hydrocarbons in flames[31] and that hydrocarbons can react with atmospheric nitrogen to form nitrogen-containing compounds that are in turn oxidized to nitric oxide.[32] Factors that influence emissions include those listed below.

3.1. *Air Preheat*

Increases in air preheat, result in increases in flame temperature and, correspondingly, in the contribution of thermal NO_x. The observed effects of changes in air preheat on NO_x emissions from premixed flames are found to be in agreement with theoretical predictions.[33]

Although reducing air preheat has been shown to decrease emissions on full-scale utility boilers[34] it carries with it an associated penalty in thermal efficiency. On the other hand, reducing the inlet air temperature increases the thermal efficiency of reciprocating engines and is a very attractive NO_x control technique for these systems.

3.2. *Water or Steam Injection or Water-in-Oil Emulsions*

Flame temperatures can be reduced by use of steam or water injection. Injection of water in an amount equal to 15% by weight of a fuel/air mixture has resulted in a 90% reduction in emissions.[35] The water may be introduced by forming a water-in-oil emulsion which has the additional benefit of providing better fuel atomization and lower soot and particulate emission.[36] Although the effectiveness of water injection for reducing NO_x emission has been demonstrated on a full-scale utility boiler,[34] the associated loss in thermal efficiency was high. Water injection, as expected, has little effect in controlling the emission of fuel—NO_x.

The major application of water injection for NO_x control is in engines, where it provides substantial NO_x reductions without incurring large losses in fuel economy.

3.3. *Flue Gas Recirculation*

The recirculation of flue gases has the two-fold effect of reducing temperature and of diluting the oxygen in the air, resulting in reductions of thermal NO_x by amounts that are consistent with theoretical predictions. Flue gas recirculation, however, has little influence on fuel NO_x. In order to achieve the maximum reduction in thermal NO_x the flue gas has to be injected into the flame zone. Tests on a small scale furnace[35] yielded a 60% lowering of thermal NO_x with a recirculation of 10% (by weight) of the flue gases and a 75% reduction with a recirculation of 20%. In those furnaces utilizing flue-gas recycle for temprature control, the cost of implementing NO control is small; otherwise the cost of installation of ducting makes flue-gas recycle economically unattractive as a control method for existing units. Flue gas recycle, unlike water injection, does not have an adverse effect on the thermal efficiency of a furnace, but does on an engine.

3.4. *Furnace Load and Size*

As the size of a furnace of a given type increases, the emissions per unit weight of fuel increases slightly.[37] This is consistent with the postulate that flames radiate less effectively to the walls in larger furnaces and, therefore, run somewhat hotter.* The level of emissions per unit weight of fuel varies much more dramatically with furnace load, being a little more than proportional to load for gas-fired field units and a little less than proportional to load for coal and oil-fired units.[27] This difference in behaviour between fuels can be explained by the influence of fuel NO_x which is expected to be relatively load insensitive. Reduction of NO_x emissions by use of reduced loads or the oversizing of combustion chambers carries with it a capital cost associated with underutilized equipment.

3.5. *Excess Air*

As the amount of excess air is increased in a combustor, the oxygen content in the flame zone generally increases and the temperature decreases. As a consequence of these two opposing effects, the emissions of thermal NO_x pass through a maximum[38–40] as the excess air is increased with the position of the maximum depending upon the mixing patterns in the combustion zone. Most, but not all, furnaces operate in a range where NO_x emissions decrease with a reduction in excess air. Fuel NO_x, being temperature insensitive, increases monotonically with increased excess air.[8,19,21] Low-excess-air firing can reduce NO_x emissions, but care must be taken not to replace those emissions with CO and soot emissions. When operating a multi-burner system near stoichiometric, care must be taken to regulate the fuel/air ratio of each burner so as to prevent some burners from running fuel rich. Low excess air firing of furnaces, although necessitating more control equipment and instrumentation to monitor emissions, provides fuel savings through increased thermal efficiency. In contrast, fuel lean combustion with high-excess air can reduce the NO_x emissions from engines, but improved carburation, fuel injection, or stratification may be necessary for efficient operation.

3.6. *Staged Combustion*

Staged combustion takes advantage of both low temperatures and low oxygen concentration by operating some burners fuel rich, allowing the partially combusted products to cool between stages, and completing the combustion by addition of the residual air. Several variations on staged combustion are used. In a multi-burner system the lower burners may be run fuel rich and the upper burners lean; this is known as biassed or off-stoichiometric combustion. Alternatively all of the fuel can be injected through the lower burners and the air introduced through the upper burners or higher in the furnace through "NO" ports; this is referred to as overfire-air[41] or two-stage combustion.[42,43] Staged combustion reduces the emission of both thermal and fuel NO_x. By operating the first stage with as little as 70% of stoichiometric air, the emission from a 0.25% nitrogen oil could be reduced to under 100 ppm.[44] As much as a 90% reduction in the fuel nitrogen has been observed by use of fuel-rich operation in laboratory burners.[21] The potential for NO_x control in practical systems will depend on many factors. In retrofitting units the number of burners, the ability to change fuel rates to individual burners, and the availability of "NO" ports for air addition above the burners are physical constraints. In addition, in changing operation or design conditions to reduce NO_x, care must be taken that other pollutants are not substituted for NO_x. Low excess air operation increases the potential for emissions of unburned hydrocarbons, carbon monoxide and soot. Staged combustion may result in the formation of hydrogen cyanide or ammonia in the first stage, which will be partially oxidized to NO in the second stage.[32] Changes of fuel/air ratio on burners may create problems of fuel stability and noise.[45] These complicating factors and the very wide variety of combustion systems in use make it difficult to estimate the potential reduction in emissions that may be achieved by combustion process modification. Staged combustion, however, provides the most successful method currently in use for the control of NO_x emission from combustion sources.

3.7. *Burner Design*

Variation in the momentum of the fuel and air streams, in the swirl or rotation of the air stream, in the

* The effect of firing rate on load is greatest when the flames occupy only a small fraction of the combustion chamber volume, as is often the case in utility boilers.

shape of a burner, in the positioning of the fuel nozzle in a burner, in the atomization of liquid fuels, in the amount of air that is premixed with the fuel, or in the positioning of the flame stabilizer, if any, can make large differences in mixing and combustion patterns and hence NO_x emissions from different burners. The potential for major reductions in NO_x emission by modification of burner design has been established on experimental burners.[46] Although a complete characterization of the processes is not possible, some general trends are identifiable.

High intensity combustion favors the formation of thermal NO_x. Evidence for this is provided by an inverse correlation of NO_x emission with residence time in the combustion zone in a laboratory burner as the swirl is varied.[47] Other, less quantitative, evidence is provided by the observation that burners which entrain air gradually and produce relatively long flames, such as in a tangential boiler, produce low thermal NO_x whereas high intensity burners, for example cyclones, yield high emissions.

Low emissions of thermal NO_x also result when a burner produces entrainment of relatively cold combustion products into the flame, a form of internal flue gas recirculation.[48]

High emissions of fuel NO_x are found when air is premixed into a burner or when a flame is lifted so that air is entrained before ignition.[46] For oil and coals containing fuel nitrogen it is preferable that the fuel nitrogen be released into an oxygen-deficient ambient atmosphere.[46] Long diffusion flames therefore favor low fuel NO_x in addition to low thermal NO_x.

Through the use of multiple concentric fuel and air ports, apparently providing delayed mixing or a form of staging, emissions as low as 150 ppm have been reported for an experimental coal-fired burner.[46]

The above results suggest that with developmental effort significant reduction in NO_x emissions may be derived from changes in burner design.

3.8. *Fuel Atomization*

Under normal operating conditions in practical systems, the technique of fuel atomization and atomizer pressure have little effect on either thermal or fuel NO_x. Normal atomization procedures produce fuel droplets which are sufficiently large and which have a velocity relative to the air flow sufficiently low that the droplets burn with an attached flame.[49] This diffusion type flame results in near stoichiometric combustion of much of the fuel vapor and results in both thermal and fuel NO_x levels which are relatively insensitive to the overall fuel/air ratio of the combustor.[20,25] By using more efficient atomizers which produce much smaller droplets with a high velocity relative to the air flow, distillate fuels may evaporate and partially premix with air prior to burning. Thus, in a burner operating fuel-lean, thermal NO_x may be significantly reduced by increasing the atomizer efficiency.[20] Fuel NO_x, on the other hand, tends to increase with increasing atomizer efficiency.[25]

3.9. *Burner Interaction*

Interaction between burners either on the same wall[50] or on opposed walls[27] leads to higher emissions, probably as a consequence of the reduced heat transfer from the central flames to the walls.

3.10. *Reduction of NO by Reaction with Hydrocarbons*

Laboratory experiments in which hydrocarbons and ammonia were injected into the post-flame zone have shown significant reduction in NO_x levels,[52] and may suggest new methods for NO_x control. Other studies demonstrate that NO is partially (30–95%) reduced by the injection of the NO into a burner with the combustion air of a burner,[20] into a fluidized bed coal combustor,[53] and into a diffusion flame.[5] In practical turbulent combustors, some NO may be destroyed by these processes as the turbulent eddies bring fuel rich pockets together with NO.

The factors that influence NO formation are many and are imperfectly understood. The results reported above suggest some directions for the development of control strategies but what is practically achievable must be determined from field studies for the different classes of stationary sources, as described in the subsequent sections. The large variation in peak temperatures, pressures, sizes, and residence time in different combustion systems explains the wide range of emission factors encountered with residential heating units having the lowest emissions per unit fuel consumption and reciprocating engines the highest. These differences necessitate different control methods for the different classes of units and may limit the degree of control which can be achieved by combustion modification alone.

4. UTILITY BOILERS

4.1. *Background*

Emissions of NO_x by utility boilers rank second to those from mobile sources in magnitude. In contrast to mobile sources, however, the NO_x from utility boilers is discharged from a relatively small number of tall stacks. The boilers are fired by coal (54.3% in 1972[1]), gas (27%) and oil (18.6%), sometimes in combination. Environmental constraints on sulfur emissions had resulted in a substitution of low-sulfur oils and gas for coals, with a short term decrease in percentage use of coal, but the trend has been reversed and projections[54] indicate that in the U.S. there will be a continued reliance on coal, a decrease in the rate of installation of oil-fired units, and a gradual phasing out of gas-fired boilers. Emissions are contributed by over 3,000 boilers in use for steam-electricity generation, varying widely in size, design, and age. The most common boiler types, designated by the location of the burners in the combustion chamber, are tangential (T), horizontally opposed (HO), front wall (FW), cyclone (Cyc), vertical (V), turbo-fired (turbo) and all wall (AW).[55]

4.2. *Control Methods for Gas-Fired Units*

The emissions from gas fired units are entirely due to thermal NO_x and fall typically in the range 0.30–1.31 lbs/10^6 B.t.u. for uncontrolled units at full-load (Table 2, taken from a field study[27]). The lowest emissions are expected for units with delayed mixing of fuel and air and in which the flames can radiate effectively to the walls; the highest emissions are expected for large units with high combustion intensities. Generally, tangentially-fired boilers yield the lowest emissions. Reduction in load has a major influence on emissions (Table 2); to a first approximation there is a proportionality between emissions and furnace load.

tangentially-fired boilers designed with flue-gas recirculation to the windbox (the compartment confining the air to the burners). Recirculation of flue gases at a rate of 15% by weight of the sum of the fuel and air reduced the emissions by 30–60%; recirculation of 30% yielded emission levels below 100 ppm (0.131 lb $NO_2/10^6$ B.t.u.).[57]

A field study[27] of NO_x emission control methods determined the control achievable on representative front wall, horizontally opposed, and tangentially fired boilers at three load levels, by use of various combinations of low-excess-air firing, staging, and flue gas recirculation (see Table 3). Use of low excess air achieved modest reductions in emissions (13–32%). The

TABLE 2. Uncontrolled emissions from gas, oil, and coal-fired utility boilers operated at full, intermediate and low load (From Ref. 27)

Fuel	Size (MW)	Type of firing	Full load (MW)	ppm @ 3%	lb NO_2 / 10^6 B.t.u.	Intermediate load (MW)	ppm @ 3% O_2	lb NO_2 / 10^6 B.t.u.	Low load (MW)	ppm @ 3% O_2	lb NO_2 / 10^6 B.t.u.	% N in fuel
Gas	180	FW	180	390	0.51	120	230	0.30	70	116	0.15	—
Gas	80	FW	82	497	0.65	50	240	0.31	20	90	0.12	—
Gas	315	FW	315	992	1.29	223	768	1.00	186	515	0.67	—
Gas	350	HO	350	946	1.23	—	—	—	150	341	0.44	—
Gas	480	HO	480	736	0.96	360	610	0.79	250	363	0.47	—
Gas	600	HO	559	570	0.74	410	335	0.44	325	253	0.33	—
Gas	220	AW	220	675	0.88	190	550	0.72	125	313	0.41	—
Gas	320	T	320	340	0.44	240	230	0.30	—	—	—	—
Gas	66	V	66	155	0.20	—	—	—	—	—	—	—
Oil	180	FW	180	367	0.50	120	322	0.44	80	266	0.36	0.29
Oil	80	FW	80	580	0.79	50	361	0.49	21	258	0.35	0.36
Oil	250	FW	250	360	0.49	172	306	0.41	—	—	—	0.31
Oil	350	HO	350	457	0.62	—	—	—	150	264	0.36	0.46
Oil	450	HO	455	246	0.33	365	219	0.30	228	186	0.25	—
Oil	220	AW	220	291	0.39	170	267	0.36	120	324	0.44	0.42
Oil	320	T	320	215	0.29	220	220	0.30	—	—	—	0.30
Oil	66	T	66	203	0.27	—	—	—	—	—	—	0.62
Oil	400	CY	415	530	0.72	258	205	0.28	—	—	—	0.53
Coal	175	FW	—	—	—	140	660	0.90	—	—	—	1.36
Coal	315	FW	275	1490	2.04	190	1280	1.75	160	1200	1.64	1.36
Coal	600	HO	563	838	1.15	462	781	1.07	363	643	0.88	1.38
Coal	800	HO	778	905	1.24	580	741	1.01	—	—	—	1.17
Coal	575	T	—	—	—	470	405	0.55	310	264	0.36	—
Coal	300	T	300	568	0.78	240	418	0.57	—	—	—	—
Coal	700	CY	665	1170	1.60	545	882	1.21	—	—	—	1.19

The control techniques that have been successfully demonstrated on field units are low-excess-air firing, staged combustion, flue-gas recirculation, water injection, and reduced air preheat. The concept of staged combustion was pioneered on gas units in the late 1950's and 1960's by Babcock and Wilcox and Southern California Edison.[42,43,56,57] A 33–75% reduction in the emissions level was attained on twelve units ranging in size from 175 MW to 480 MW, by the use of staged combustion (two-stage and/or off-stoichiometric combustion). Off-stoichiometric combustion, in which the fuel from up to 25% of the burners was diverted to the remaining burners, yielded better results than the two-stage combustion, in which part of the air to all of the burners was diverted to "NO" ports higher in the furnace.

Southern California Edison also tested the potential for control of NO_x by flue-gas recirculation on

minimum value of the excess air that could be used was determined by problems of CO emissions which increase dramatically below a critical excess air level. The minimum practical level of excess air is a function of burner and furnace design; low excess air firing therefore requires a fairly careful control of the fuel and air rate to individual burners and a careful monitoring of products of incomplete combustion in the stack. These disadvantages, however, are partially offset by the increase in the thermal efficiency of a boiler with a reduction in the excess air level. The reduction in NO_x achieved by staging and flue gas recirculation in Table 3 are similar to those observed in the earlier Southern California tests. A combination of control techniques, as expected, yielded a smaller gain than that computed from the gains observed when each technique was applied individually.

The final three columns in Table 3 show that a

TABLE 3. Percentage reduction in NO_x emissions achieved by staging low excess air firing and flue gas recirculation (From Ref. 27)
Combustion operating modification and furnace load(1)

Fuel fired	Type(2) of firing	% Reduction in NO_x emission														
		Low exc. air			Staging			LEA + staging			Flue gas rec.			"Full"(3)		
		Full	Int.	Low	Full	Int.	Low	Full	Int.	Low	Full	Int.	Low	Full	Int.	Low
Gas	FW	13	24	7	37	30	30	48	42	36	—	—	—	48	42	36
	HO	17	15	32	54	35	59	61	48	68	—	—	20	73	52	72
	T	—	—	—	—	—	—	—	—	—	—	60	—	66	65	—
	ALL (Average)	16	19	26	45	31	52	54	44	52	—	60	20	64	51	60
Oil	FW	27	20	28	29	20	20	39	32	21	46	31	—	50	41	21
	HO	10	16	12	34	34	47	35	44	42	—	—	—	38	35	55
	T	28	22	—	—	17	—	—	45	—	10	13	—	—	59	—
	ALL (Average)	19	19	18	30	22	34	38	37	32	28	23	—	47	42	38
Coal	FW	—	14	—	—	40	—	—	55	—	—	—	—	—	60	—
	HO	—	—	—	—	—	—	—	—	—	—	—	—	—	—	—
	T	27	18	—	—	39	—	—	50	42	—	—	—	—	50	42
	ALL (Average)	27	17	—	—	39	—	52	52	42	—	—	—	—	55	42

(1) Furnace load: "Full" = 85%–105%, "Intermediate" = 60%–85%, Low = 50%–60% of rating.
(2) Type of firing: FW = front wall, HO = horizontally opposed, T = tangential.
(3) "Full control": combination of techniques achievable on boilers tested.

50–60% reduction in emissions was achievable when combinations of all the techniques available on each of the boilers in the field study were utilized to their fullest.

Injection of water and reducing air preheat are all less desirable methods of NO_x emissions control as they impose a penalty in thermal performance. Water injection tests carried out by Combustion Engineering on a 150 MW tangentially fired unit showed a maximum reduction in NO_x emissions of 50% at a water injection rate of 45 lb/10^6 B.t.u. fired and an associated 5% decrease in furnace efficiency.[34] Significant reductions in NO_x emission and furnace efficiency were also observed when the air preheat was reduced from 490 to 81°F at three-quarter and half loads.[34]

In summary, the techniques most readily adapted to reducing emission from existing gas-fired units are low excess air firing and staged combustion, particularly in units where a multiplicity of burners permits the changes to be made without any equipment modification. Flue gas recirculation provides an attractive control option for existing units only when these are already equipped with a recirculation system. Water injection or reduction in air preheat are less desirable control techniques in view of the penalty of increased fuel consumption entailed with their use. For gas units, a reduction in emissions of 50–60% can be expected with the use of low excess air and staged combustion alone, and greater reductions are possible when these can be combined with flue-gas recirculation. It should be noted, however, that units differ significantly in design and individual units may show much lower potential for emission control.

4.3. *Control Methods for Oil-Fired Units*

The emissions from oil-fired units are due to a combination of thermal and fuel NO_x. Control methods which are based on a temperature reduction, such as flue gas recirculation, water injection, and load reduction, are effective in controlling the thermal NO_x contribution but have little effect on the fuel NO_x. Methods which reduce the oxygen availability in the primary combustion zone, such as low-excess-air firing, staged combustion, and long flames with delayed oxygen-uptake are effective in controlling both sources of NO_x. The emissions from nine oil-fired utility boilers included in a systematic field survey[27] range from 0.27 to 0.79 lbs/10^6 B.t.u. at full load (Table 2). Tangentially-fired units were found to yield the lowest NO_x emission for a given size unit, as had been previously found for gas-fired boilers. It was estimated from results obtained using oils with different nitrogen contents that 30% of the fuel nitrogen in the nitrogen concentration range of 0.3–0.6% by weight was converted to NO_x, corresponding approximately to an incremental emission of 0.06 lb/10^6 B.t.u. per 0.1% increment in fuel nitrogen. Independent tests[34] on tangentially-fired boilers indicate a range of conversions of fuel nitrogen to NO_x from 43% for 0.2% nitrogen oils to 30% for 1% nitrogen oils when the boilers where operated with 3% O_2 in the stack gases.

Reduction in load for oil-fired boilers resulted in a reduction in emissions but the dependence of emissions on load is less marked than for the case of gas-fired boilers (Table 2).

The control methods tested in the field study included

low-excess-air firing, staging, flue-gas recirculation and combinations thereof (Table 3). The trends are similar to those obtained with gas but the fractional reduction in NO_x emissions is lower. The results show an average reduction in NO_x for the best combination of control techniques ranging from 48% at full load to 38% at low load. Although the uncontrolled emissions of gas-fired units were higher than the emissions from gas units of the same design and size, the emissions with the best control achievable were generally lower for the gas-fired units. Difficulty in controlling the emissions from oil fired units is due in part to the fuel nitrogen content of oils and in part to the difficulty of controlling the very complicated processes that define the atomization and combustion of liquid fuels.

4.4. *Control Methods for Coal-Fired Utility Boilers*

The uncontrolled emissions from coal-fired boilers are higher than those for gas and oil (Tables 2 and 4), ranging from 0.53 to 2.04 lbs/10^6 B.t.u. at full load. Tangentially-fired units again show the lowest emissions. The emissions from coal-fired units show a less than proportional dependence on load (Table 2). From the dependence of the emissions on load it is inferred that about 20% of the fuel nitrogen content, typically 1.3% by weight, is converted to NO_x and that the fuel nitrogen contributes 50% of the emission at full-load.[26] The relative proportions of fuel NO_x and thermal NO_x will depend upon many variables including coal moisture content, nitrogen content, burner design, furnace design and operating conditions.

TABLE 4. Uncontrolled emissions from coal-fired utility boilers and percent reduction in emission achievable by staged combustion (From Ref. 26)

Fuel	Size (MW)	Type of firing	Load (MW)	ppm @ 3% O_2	lb NO_2/10^6 B.t.u.	% N in fuel	% Reduction at full load
Coal	105	FW	101	454	0.60	—	53
Coal	125	FW	125	634	0.84	1.35	40
Coal	256	FW	253	703	0.93	1.31	48
Coal	320	FW	350	832	1.11	1.38	34
Coal	218	HO	219	569	0.76	—	34
Coal	480	HO	490	711	0.95	1.81	35
Coal	800	HO	800	935	1.24	1.27	48
Coal	250	T	250	410	0.55	1.49	25
Coal	330	T	334	531	0.71	1.56	59
Coal	350	T	350	415	0.55	1.4	35
Coal	348	T	306	434	0.53	0.80	4
Coal	850	Turbo	370	600	0.80	1.34	34

More conclusive evidence on the contribution of fuel nitrogen in coal to NO_x emissions is provided by laboratory studies in which pulverized coal was burned in argon-oxygen mixtures.[24] In these tests 25–40% of the 1.2% nitrogen content of the coal was converted to NO_x.

Based on the importance of both thermal NO_x and fuel NO_x it is expected that techniques that reduce temperature in the flame zone will be less effective in controlling NO_x emission than those that reduce the oxygen content. Low-excess-air firing and staged-combustion have proved to be effective in reducing NO_x emissions from coal-fired units, by an average of 37% for twelve units included in field tests.[26] The carbon content of the ash and the carbon monoxide in the stack gas increased slightly during the tests.[26] Accelerated corrosion studies[26] for the 300-h duration of the trials with staged combustion showed negligible wear on the corrosion strips but additional long range trials will be needed to confirm these preliminary findings.

4.5. *Design Modifications and Costs*

The success of low-excess-air and staged combustion has enabled all major U.S. boiler manufacturers to develop units that will meet the U.S. emission standards for new, large steam-generating plants.* To guarantee performance some manufacturers specify a maximum nitrogen content on the oil to be fired. Further reduction in emissions can be achieved at considerably increased costs by use of flue gas recycle. Water injection may be used but with a penalty in thermal efficiency.

Severe constraints may be imposed on the ability to make modifications on existing units by the design of the unit and the space available for adding ducting. The modifications easiest to apply are low-excess-air firing and off-stoichiometric combustion. Addition of NO ports may be feasible in many units but the addition of flue-gas recycle would require a major reconstruction.

Costs for modifying new and existing coal-fired tangentially-fired units have been prepared by Combustion Engineering[58] and are presented in Figs. 5 and 6. The costs were prepared for the introduction of 20% of the total combustion air over the fuel firing

* The standards are 0.2 lbs $NO_2/10^6$ B.t.u. (0.36 g/10^6 cal) for gas, 0.3 lbs $NO_2/10^6$ B.t.u. (0.54 g/10^6 cal) for oil, and 0.7 lbs $NO_2/10^6$ B.t.u. (1.26 g/10^6 cal) for coal where all the NO_x is reported as NO_2 equivalent.

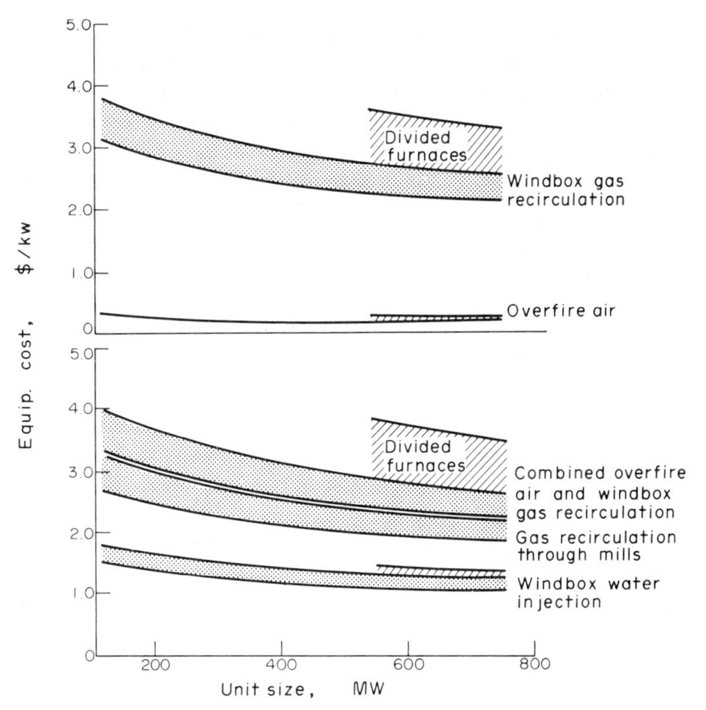

FIG. 5. Costs of NO_x control methods for new coal fired units (controls included in initial design).[58]

zone as overfire air, recirculating 30% of the flue gas to the secondary air ducts and windbox, a combination of overfire air and flue-gas recirculation, recirculation of 17% of the flue-gas through the coal pulverizers (mills), and for water injection into the fuel-firing zone at 5% of the steam rate for the boiler. The practical limit on the size of boilers constructed with a single combustion combustion chamber is about 600 MW;

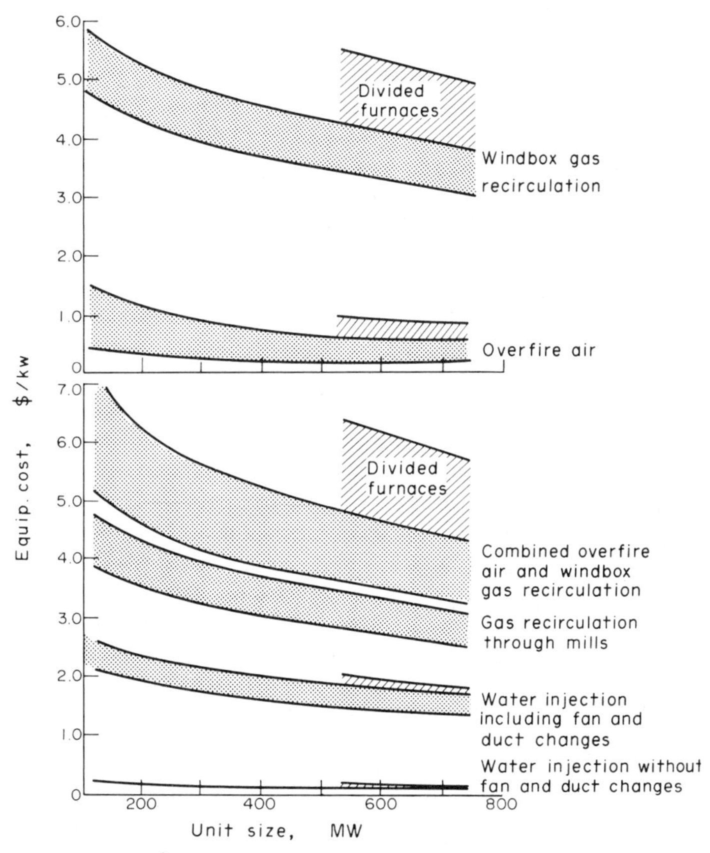

FIG. 6. Costs of NO_x control methods for existing coal fired units (heating surface changes not included).[58]

the higher costs for the large units in Figs. 5 and 6 are associated with divided furnaces.[58] Cost figures are expected to vary considerably between units particularly for the case of retrofit. The costs in Figs. 5 and 6 are only capital costs. In addition to capital costs, operating[55] and testing costs will be incurred.

emissions of sulfur oxides, recent trends have been toward the use of natural gas and low sulfur oils in all but the largest units. Due to increasing prices and reduced availability of these fuels, however, the industrial sector will probably be forced to rely increasingly upon coal and residual oil.

TABLE 5. Uncontrolled-NO_x and low-NO_x emissions from industrial boilers (From Ref. 28)

Fuel	Size MBH	Burners		Test load MBH	Baseline		Low NO_x		% Reduction
		Type	Number		ppm @ 3% O_2	lb NO_2 / 10^6 B.t.u.	ppm @ 3% O_2	lb NO_2 / 10^6 B.t.u.	
Gas	25	Ring	1	20	72	0.086	65	0.077	10.0
Gas	29	Ring	1	22	70	0.083	—	—	—
Gas	29	Ring	1	11	97	0.115	68	0.081	29.6
Gas	59	Ring	6	48	116	0.138	—	—	—
Gas	60	Ring	2	48	101	0.124	86	0.102	17.7
Gas	60	Ring	4	46	242	0.288	138	0.164	43.1
Gas	160	Ring	1	136	374	0.445	355	0.423	4.9
Gas	158	Ring	4	125	299	0.343	110	0.131	61.8
Gas	300	Ring	4	259	199	0.237	169	0.201	15.2
Gas	10	Ring	1	8	55	0.065	38	0.045	30.8
Gas	20	Ring	1	14	107	0.127	75	0.089	29.9
Gas	11	Ring	1	10	87	0.104	83	0.099	4.8
Gas	7	Ring	1	6	71	0.085	67	0.080	5.8
Gas	10	Ring	1	7	91	0.108	90	0.107	0.9
Gas	13	Ring	1	12	57	0.068	—	0.067	1.5
Gas	18	Ring	1	17	56	0.067	53	0.063	6.0
Gas	20	Ring	1	13	79	0.094	67	0.080	14.9
Gas	8	Ring	1	6	70	0.083	45	0.054	35.0
Gas	20	Ring	1	10	101	0.120	—	—	—
Gas	33	Ring	1	24	90	0.107	81	0.096	10.3
Gas	65	Ring	6	53	98	0.117	—	—	—
Gas	110	Ring	1	85	94	0.112	—	—	—
Gas	225	Nozzle	8	180	212	0.252	147	0.175	30.6
Gas	325	Nozzle	8	260	320	0.381	230	0.274	28.1
#2 Oil	110	Steam	2	88	177	0.231	—	—	—
#2 Oil	10	Air	1	7	169	0.220	149	0.194	11.8
#2 Oil	18	Steam	1	14	65	0.085	63	0.082	3.5
#2 Oil	18	Air	1	14	97	0.127	86	0.112	11.8
#2 Oil	18	Mech	1	12	80	0.104	80	0.104	—
#2 Oil	11	Air	1	11	128	0.167	—	—	—
#2 Oil	18	Air	1	16	116	0.151	—	—	—
#2 Oil	18	Steam	1	16	118	0.154	—	—	—
#2 Oil	20	Air	1	16	193	0.252	141	0.184	27.0
#2 Oil	29	Steam	1	10	103	0.134	—	—	—
#2 Oil	7	Air	1	7	127	0.166	—	—	—
#2 Oil	158	Steam	4	115	181	0.236	120	0.157	33.5
#2 Oil	33	Steam	1	23	123	0.160	104	0.136	15.0
#2 Oil	13	Air	1	11	84	0.110	—	—	—
#5 Oil	19	Cup	1	12	200	0.261	139	0.181	30.7
#5 Oil	85	Steam	4	59	329	0.429	243	0.317	26.1
#5 Oil	11	Air	1	12	183	0.239	162	0.211	11.7

5. INDUSTRIAL BOILERS

5.1. *Background*

Boilers with capacities in the range 10–500 million B.t.u./h emit an estimated 18.1% of the U.S. stationary source NO_x. The industrial boiler population is made up of about 75 000 units of various types, ages, and applications which burn gas (57% of the energy supplied in 1968), coal (23%) and oil (20%).[45] The oil usage includes both distillate and residual oils, but primarily residuals. As a result of regulations on the

The smaller units, 10–16 million B.t.u./h capacity, account for about 35% of the units and 10% of the industrial capacity.[45] These are primarily packaged firetube boilers which burn oil or gas. The 17–100 million B.t.u./h units, which are primarily packaged watertube boilers, account for 53% of the units and 45% of the capacity. The larger units are watertube boilers and are mostly field erected. Currently only 1% of the units and 9% of the capacity are greater than 250 million B.t.u./h capacity, the size range for which new units are subject to regulation.

The data on emission levels and combustion modification effectiveness for NO_x control are limited, mostly derived from a field study by KVB Engineering[28] in which about eighty units were tested for baseline emission levels and minimum NO_x achievable without hardware modifications. The results of this study are presented in Table 5. Due to the limited flexibility in combustion operating conditions for these units, only moderate reductions were obtained in most cases.

When combustion modification strategies are considered, it is necessary to consider the many problems which may arise: corrosion and deposits on boiler tubes, flame instability, blow-off, flashback, combustion driven oscillations, and combustion noise or roar. The need for tuning individual units to minimize these problems will place constraints on retrofit programs due to the large number of units and their diverse characteristics.[45]

5.2. *Gas-Fired Boiler Emissions*

The gas-fired units without air preheat, firetube boilers and some of the watertube boilers, had NO_x emission levels less than 0.14 lb $NO_2/10^6$ B.t.u.[28] The watertube units with air preheat had generally higher emissions, from 0.08 to 0.45 lb $NO_2/10^6$ B.t.u. By comparing units with different degrees of air preheat, it was estimated that the preheat effects vary from 0.02 lb $NO_2/10^6$ B.t.u./100°F for units of less than 30 million B.t.u. per hour capacity to 0.15 lb $NO_2/10^6$ B.t.u./100°F for much larger units. This appeared to be a function of the burner firing rate (B.t.u./h/burner). Reducing the oxygen level had only a small effect on emissions from units without air preheat, but did decrease the NO_x

TABLE 5. (continued)

Fuel	Size MBH	Burners		Test load MBH	Baseline		Low NO_x		% Reduction
		Type	Number		ppm @ 3% O_2	lb NO_2/10^6 B.t.u.	ppm @ 3% O_2	lb NO_2/10^6 B.t.u.	
#5 Oil	18	Air	1	18	177	0.231	154	0.201	13.0
#5 Oil	18	Steam	1	17	161	0.210	152	0.198	5.7
#5 Oil	13	Air	1	11	181	0.236	171	0.223	5.5
#5 Oil	7	Air	1	7	275	0.359	—	—	—
#5 Oil	59	Steam	6	46	619	0.807	516	0.673	16.6
#5 Oil	65	Steam	6	50	466	0.608	431	0.562	7.6
#5 Oil	125	Steam	1	100	337	0.440	322	0.420	4.5
#5 Oil +RG	110	Steam	4	88	172	0.224	166	0.217	3.1
#5 Oil +RG	110	Steam	4	90	215	0.280	211	0.275	1.8
NSF Oil	17	Cup	1	15	184	0.240	—	—	—
#6 Oil	18	Steam	1	14	350	0.457	331	0.432	5.5
#6 Oil	18	Air	1	15	334	0.436	277	0.361	17.2
#6 Oil	80	Steam	1	51	305	0.398	268	0.350	12.1
#6 Oil	90	Steam	3	71	346	0.321	175	0.228	29.0
#6 Oil	65	Steam	2	54	186	0.243	174	0.227	6.6
#6 Oil	105	Steam	4	80	251	0.327	222	0.290	11.3
#6 Oil	260	Steam	4	130	240	0.313	201	0.262	16.3
#6 Oil	500	Steam	9	400	267	0.348	180	0.235	32.5
#6 Oil	7	Air	1	7	298	0.389	249	0.325	16.5
O/C	513	Cyc	2	320	716	0.934	—	—	—
O/C	513	Cyc	2	320	710	0.926	—	—	—
Coal	60	UFS	7	48	266	0.327	188	0.263	29.3
Coal	60	UFS	7	46	224	0.314	198	0.277	11.8
Coal	135	Sprd	2	110	370	0.518	335	0.469	9.5
Coal	50	Sprd	1	40	465	0.651	330	0.462	29.0
Coal	75	Sprd	1	63	465	0.651	387	0.542	16.7
Coal	225	Pulv	8	181	378	0.529	360	0.504	4.7
Coal	210	Sprd	5	120	553	0.774	471	0.659	14.9
Coal	230	Sprd	6	162	547	0.766	360	0.504	34.2
Coal	500	Pulv	6	400	580	0.812	—	—	—
Coal	513	Cyc	2	320	800	1.120	742	1.039	7.2
Coal	10	UFS	1	8	273	0.382	—	—	—
Coal	10	UFS	1	8	346	0.484	—	—	—
Coal	325	Pulv	8	260	484	0.678	—	—	—

Abbreviations: NSF = Navy standard fuel; Cup = rotary cup fuel atomizer; Air = air assist fuel atomizer; Steam = steam-fuel atomizer; Cyc = cyclone furnace coal combustor; UFS = underfed stoker coal burning equipment; Sprd = spreader stoker coal burning equipment; Pulv = pulverized coal burning equipment; MBH = million B.t.u. per hour.

levels from preheated boilers. Off-stoichiometric firing, achieved by taking one burner out of service, resulted in NO_x reductions of 16–42%. By combustion modification, the percentage of units with emissions below 0.2 lb $NO_2/10^6$ B.t.u. was increased from 75% to 82%.

5.3. *Oil-Fired Burner Emissions*

The baseline emission levels for those units fired with No. 2 fuel oil were 0.13–0.26 lb $NO_2/10^6$ B.t.u.[28] Increasing combustion intensity resulted in increased NO_x, but other changes in operating parameters had little effect.

The heavy fuel oil emissions were higher, 0.20–0.81 lb $NO_2/10^6$ B.t.u., and more sensitive to the operating conditions. The nitrogen content of the fuel had a particularly strong effect since an estimated 44% of the fuel nitrogen was converted to NO_x. The NO_x emissions decreased with decreasing oxygen levels, burner firing rate, and combustion intensity. Emission reductions from 6 to 29% were obtained by using off-stoichiometric or staged combustion. The technique of fuel atomization did not strongly influence the emissions so long as good atomization was achieved. The two highest emission levels for No. 5 oil were the result of preheating the oil to only 130°F, rather than 160–180°F as in most other tests. Increasing the oil temperature improved the atomizer effectiveness and reduced the NO_x emission levels.

5.4. *Coal-Fired Boiler Emissions*

Due primarily to the high nitrogen content of coals, 1.29–1.80% by weight in the KVB study, the emissions from coal fired units were generally higher than for oil or gas, i.e., 0.31–1.12 lb $NO_2/10^6$ B.t.u.[28] The lowest emissions from coal-fired units were from underfed-stoker coal-burning equipment. In these units the air fed up through the grating is insufficient for complete combustion, so additional air must be introduced above the grating through overfire air ports. The combustion is, therefore, effectively staged, and the NO_x emissions are quite low, i.e., 0.31–0.48 lb $NO_2/10^6$ B.t.u.

Spreader stokers, in which the fuel is introduced with the air flow above the grate, have intermediate emission characteristics, 0.52–0.77 lb $NO_2/10^6$ B.t.u. Some of the fuel is burned in the fuel spray; the remaining fuel is burned on the grate as in the underfed stoker. The resultant combustion is only partially staged. The combustion intensities were also higher than for underfed stokers, possibly increasing thermal NO formation.

Pulverized coal units, in which all of the fuel is burned in suspension, have higher emissions, 0.53–0.81 lb $NO_2/10^6$ B.t.u. A unit equipped with two cyclone-type coal combustors produced 1.12 lb $NO_2/10^6$ B.t.u., the highest of all the units tested due to the very high combustion intensity.

Reductions of about 0.07 lb $NO_2/10^6$ B.t.u./% reduction in oxygen were obtained by reducing the excess air. A decrease in underfire air-rate with a compensating introduction of air through openings higher in the furnace for a spreader stoker-fired boiler resulted in about 46% NO_x reduction, but the grate temperature increased beyond its allowable limits. The most acceptable operating conditions reduced NO_x by 20–25%. The controlled NO_x emissions were below 0.68 lb $NO_2/10^6$ B.t.u. for all but the cyclone fired boiler.

6. RESIDENTIAL AND COMMERCIAL SPACE HEATING

6.1. *Background*

Space heating equipment of less than 10 million B.t.u./h capacity emits about 7.1% of the U.S. stationary source NO_x largely near ground level in areas of high population densities.[59] The problem may be more severe than this number would indicate since the emissions are confined to the heating season.

The commercial units, which generally fall in the range of 0.3–10 million B.t.u./h capacity, are a mix of packaged firetube, cast iron, and watertube boilers.[51] These units are equipped to burn natural gas (56%), number 2 oil (38%), number 4 and 5 oils (8%), and number 6 oil (4%). About 6% of the units have dual fuel capabilities. Commercial boilers are designed to operate with a minimum of manual control and maintenance for a life of from 10 to 45 years depending upon the type of unit.

The residential units are much smaller, i.e., about 70% have capacities of less than 0.2 million B.t.u./h, and 96% have capacities of less than 0.42 million B.t.u./h.[51] The units include warm-air furnaces, hot-water and steam boilers, and water heaters, and are fueled primarily by distillate oil (No. 2) or natural gas. Operation is generally controlled by a thermostat and maintenance is provided on an annual schedule, at most.

6.2. *Emissions from Residential Space Heating Equipment*

The specific NO_x emission levels from residential units average about 0.14 lb $NO_2/10^6$ B.t.u.,[51] which is low primarily as a consequence of low combustion intensities in these units. About 90% of the residential units tested in a recent field survey[51] of oil-fired space heating equipment had emissions below 0.2 lb $NO_2/10^6$ B.t.u.; the highest emission level measured in that survey was about 0.29 lb $NO_2/10^6$ B.t.u. Normal service and adjustment practice had very little effect on the emission level of NO_x or any pollutant other than smoke, which was minimized during adjustment.

A recent EPA study[59] provides some more detailed information on the effects of operating conditions and burner design on residential heating equipment emissions. Combustion improving devices, which utilize flame retention to produce a more intense flame region, tend to increase both efficiency and NO_x emissions. The spark ignition system was found to produce between 7 and 10% of the NO_x due to continuous operation of the ignition arc during the burning cycle. Turning the ignition system off after startup, or using low power output ignition could eliminate this small source. Other modifications to existing equipment designs were found to have very little effect. However, significantly lower levels of NO_x emissions have been demonstrated without increases in other pollutant emissions or loss of efficiency, on both optimized operation of conventional systems and new combustor concepts.[59,60,61] Low emission burners based on conventional design practices have been developed.[61] One unit tested for 128 h was a 20 000 B.t.u./h, residential size burner which produced about 0.048 lb $NO_2/10^6$ B.t.u. A commercial size 180 000 B.t.u./h unit, which was tested for a total of 112 h produced only 0.025 lb $NO_2/10^6$ B.t.u. Burners which use hot combustion products to prevaporize the liquid fuel can achieve very low levels of all pollutants with 0.02–0.07 lb $NO_2/10^6$ B.t.u. during steady state operation. During startup, however, the fuel vaporization is not effective in some systems and the CO and smoke emissions are high. Internal recirculation of combustion products also very effectively reduces thermal NO formation by decreasing flame temperatures. NO_x emissions as low as 0.015 lb $NO_2/10^6$ B.t.u. have been reported. The lowest emission levels reported, less than 0.005 lb $NO_2/10^6$ B.t.u., have been achieved using catalytic combustion;[60,61] however, the hydrocarbon emissions were very high. Further development will be required before the very low NO_x emission levels of these demonstration units are achieved in commercial and residential heating units.

6.3. *Emissions from Commercial Boilers*

The specific NO_x emissions from commercial boilers fired with natural gas or No. 2 oil are comparable with the emissions from residential units, i.e., 0.07–0.20 lb $NO_2/10^6$ B.t.u.[51] Combustion of residual oils, No. 4, 5, or 6 oil, resulted in emissions as high as 0.47 lb $NO_2/10^6$ B.t.u. in a Battelle field study,[51] primarily due to the nitrogen content of these fuels, about 60% of which was converted to NO_x.

For heavy oil combustion, reduced excess air resulted in reduced NO_x emissions, and high smoke and carbon monoxide emission levels. Techniques such as staged and low-excess-air combustion which have been applied for fuel NO_x control in utility boilers might be adaptable to large commercial boilers, but have not been applied to date. Switching to low nitrogen content fuels, e.g., natural gas and No. 2 oil, if they are available, would be one means of NO_x control for commercial units for which modification of the combustion process is not feasible. New combustion concepts such as prevaporization of the fuel or surface combustion also have potential for commercial units, but have not been demonstrated.

7. INTERNAL COMBUSTION ENGINES

7.1. *Types and Locations*

Internal combustion engines produce about 21.3% of the total stationary source NO_x emissions, second only to utility boilers.[62] The installed capacity of stationary engines is distributed about equally between reciprocating engines (34.7 million horsepower) and gas turbines (35.5 million horsepower).[63] The applications and NO_x emissions of these two classes of engines are, however, quite different. Estimates of power generation by engines indicate that reciprocating engines generate about 71.8% of the power while gas turbines generate the remaining power, corresponding to utilization of 58 and 22%, respectively, of their installed capacities. Emission estimates for these two sources are somewhat uncertain, particularly for gas turbines whose total emissions have been estimated at 131[62] to 291[63] thousand tons NO_x per year. Nonetheless, it is clear that reciprocating engines produce the vast majority, i.e., 88–94%, of the stationary engine emissions.

Recent survey estimates[63] of the installed power, fuel usage, emissions, and power generation for stationary engine applications are tabulated in Table 6. In terms of installed horsepower, the major uses of engines are electric power generation (54.8%), oil and gas pipelines (22.4%), agriculture (10.7%), natural gas processing (5.6%) and production (4.6%). The major sources of stationary-engine NO_x emissions are oil and gas pipelines (41.6%), natural gas processing (19.2%), and natural gas production (13.8%) which use primarily spark ignition gas engines, and agriculture (14.2%), which uses primarily diesel engines. Electric power generation, which accounts for most of the installed horsepower, produces only about 5–10% of the stationary engine NO_x due to two factors: (1) the overall load factor for engines used in electric power production is only 12% since the primary application is for peak and stand-by power generation; (2) most of the installed horsepower is in the form of gas turbines which produce much lower NO_x emissions than reciprocating engines. Reciprocating engines are used for peak power generation in older installations and continue to find application for electricity generation in municipalities, hospitals, schools, and shopping centers which are too small to use gas turbine units. In recent years, both large electric utilities and natural gas pipelines have been favoring large gas turbines.

Recent projections from the electric utility industry[54] for the period 1974–1983 indicate a 75% increase in the gas turbine generating capacity and a four-fold increase in the electric power generated by gas turbines for continuous power generation in addition to peak power generation. This will be accompanied by slightly reduced usage of natural gas and increased usage of both distillate and residual oils as gas turbine fuels.

Large concentrations of stationary engine horsepower, e.g., 10 000–60 000 B.h.p. reciprocating engines at some pipeline compressor stations, may represent significant local pollution sources. Most pipeline compressor stations are remote from population centers; however, there may be some human exposure to local high NO_x levels at compressor stations.

Stationary reciprocating engines can be classified into several categories. The fuel/air mixture may be ignited in reciprocating engines either by an electrical

TABLE 6. Stationary engines in the United States
(From Ref. 63)

	Load factor	Installed horsepower—10^3 Bhp					Fuel consumption[a] 10^{12} B.t.u.		Power generation 10^6 Bhp hr		Annual NO_x emissions Tons			Emission factors gm NO_x/Bhp hr	
	%	Diesel	Dual fuel	Gas engine	Gas turbine	Total	Natural gas	No. 2 oil	Recip	Gas turbine	Recip	Gas turbine	Total	Recip	Gas turbine
Electric power	12	1,570	3,710	90	30,440	35,810	127.24	280.14	5,900	33,240	62,440	62,920	125,360	9.61	1.72
Oil and gas pipeline	69	830	390	10,990	3,520	15,730	802.06	40.77	73,700	21,260	930,200	39,800	970,000	11.46	1.70
Natural gas processing	—	—	—	2,410	1,530	3,940	432.60	—	31,280	15,010	429,690	28,130	457,820	12.47	1.70
Oil and gas exploration	15	1,500	—	500	—	2,000	5.09	14.67	2,580	—	31,720	—	31,720	11.16	—
Crude oil production	—	—	—	852	—	852	42.66	—	5,410	—	62,370	—	62,370	10.47	—
Natural gas production	—	—	—	3,237	—	3,237	190.04	—	24,100	—	308,200	—	308,200	11.61	—
Agricultural	40	7,500	—	—	—	7,500	—	198.94	26,280	—	318,700	—	318,700	11.01	—
Industrial process	—	—	—	230	—	230	11.88	—	1,510	—	19,300	—	19,300	11.61	—
Municipal water and sewage	75	465	—	465	—	930	24.08	23.14	6,110	—	76,100	—	76,100	11.31	—
TOTAL	—	11,865	4.100	18,774	35,490	70,229	1635.75	557.66	176,870	69,510	2,238,720	130,850	2,369,570	—	—
											Per cent 58.12%	Capacity 22.36			

[a] Calculated assuming higher heating values of 5.8×10^6 B.t.u./Bbl for oil and 1070 B.t.u./SCF for natural gas.

spark discharge (spark ignition engines) or by compression heating (diesel engines). Natural gas engines are almost always spark ignited and may be either two or four cycle. In diesel engines combustion is controlled by injecting fuel into the cylinder through a spray nozzle at the proper time during the compression stroke. Dual fuel engines can operate on 100% fuel oil or natural gas to which fuel oil has been added (about 5% on the basis of heating value) to serve as a pilot for ignition.

Stationary gas turbines have evolved from aircraft jet engines. In its simplest form, the gas turbine engine (simple cycle) consists of a compressor in which the air is compressed to pressures ranging from 100 to 200 psig, followed by a combustor in which fuel (liquid or gaseous) is added. Combustion occurs at mixtures of fuel and air which are overall very fuel-lean. The combustion gases are then expanded through a turbine or turbines to near atmospheric pressure, converting their energy to power. The simple cycle gas turbine thermal efficiency is 24–31%. The efficiency of the gas turbine can be increased by further extraction of heat from the exhaust gases which normally leave the turbine at 800–1100°F. In a regenerative cycle this is done by passing the air leaving the compressor through a heat exchanger in which some of the heat of the exhaust gases leaving the turbine is transferred to the pressurized air. For comparable turbine inlet temperatures, less fuel is required, increasing the thermal efficiency to about 34–38%. Another way of utilizing the exhaust waste heat is in a combined cycle. In this case the exhaust gases are used to produce steam which can then be used to drive a steam turbine to produce additional power. Due to its high thermal efficiency, currently 40–42%, the combined cycle gas turbine is being favored by electric utilities in new installations.

7.2. *Emission Levels*

Table 7 summarizes the emission factors used in the estimates of exhaust emissions from stationary engines, described above. While there is considerable variation in the emissions of each of the engine types, indicated by the numbers in parentheses, these figures provide a useful basis for comparison of the uncontrolled emission levels of the various engines.

Except for gas turbines and prechamber diesel engines, the NO_x levels, in the range 2.8–4.2 lb $NO_x/10^6$ B.t.u., exhibit only minor differences among the different engine types.[63] Turbocharging results in slight increases in NO_x emissions over naturally aspirated engines due to increased temperatures. Prechamber diesel engines, in which combustion is initiated in a fuel rich environment, emit about half as much NO_x as direct injection diesels.

Gas turbines yield one-tenth the emissions of reciprocating engines, primarily as a result of lower peak temperatures and shorter residence times in the

TABLE 7. Stationary engine emission factors
(From Ref. 63)

	Engine type	# Cycle	Charging: Air	Charging: Fuel	BSFC (B.t.u./Bhp-h)	NO_x (lb/10^6 B.t.u.)	CO (lb/10^6 B.t.u.)	HC_t (lb/10^6 B.t.u.)
13 Mode test	Diesel	4	TC	DI	7252	4.2 (3.5–4.0)	1.2	0.43
13 Mode test			NA	DI	8428	2.9 (1.9–5.6)	1.4	1.4
13 Mode test			TC	PC	7252	1.6 (1.1–2.2)	0.49	0.10
13 Mode test			NA	PC	8428	1.5	0.65	0.08
13 Mode test		2	SC	DI	7840	4.1	1.7	0.22
Constant conditions		4	TC	DI	7252	3.3	1.2	0.04
Constant conditions			TC	PC	7252	2.3	0.28	0.03
Constant conditions			NA	PC	8428	1.6	0.25	0.05
	Dual fuel	4	TC	DI	5970	(2.9–3.1)	0.74	1.1
	Natural gas	4	TC	DI	6830	3.9 (3.5–4.9)	0.32	0.65
			NA	DI	7150	3.6 (2.4–4.9)	0.43	0.62
			TC	High speed	7000	4.0 (3.5–4.9)	1.8	0.66
		2	TC	DI	6635	3.5 (2.3–6.7)	0.90	1.5
			NA	DI	7100	3.3 (1.4–5.1)	0.09	3.3
	Gas turbine (natural gas)				11185	0.34 (0.17–0.51)		

Abbreviations: TC = turbocharged; NA = naturally aspirated; SC = supercharged; DI = direct injection; PC = prechamber engine; Oil: HHV = 19600 B.t.u./lb.

combustion chamber. Emissions from distillate-oil fired gas turbines are about a factor of 2 higher than emissions from gas turbines fueled with natural gas.[64,65] Since the electric utility projections indicate increasing reliance on oil as gas turbine fuel, including use of residual oils to supply a substantial fraction of gas turbine fuel by the mid-1980's,[54] the uncontrolled specific emissions for gas turbines may be expected to increase. Fuel-nitrogen may play an important role in determining the achievable emission levels, since the limited data available on fuel-NO_x formation in gas turbines [21,66] suggests a high conversion efficiency of fuel-N to NO_x.

7.3. NO_x *Emission Control Technology*

Many emission control techniques which have been developed for mobile sources are potentially applicable to stationary engines; however, different design constraints may make techniques either more or less desirable for stationary engines than for mobile engines. Size and weight constraints are severe for mobile engines, and durability is an important characteristic of an engine used in stationary applications. Techniques to reduce NO_x include engine derating, fuel injection and ignition timing retard, exhaust gas recirculation, catalytic converters, water or steam injection, and engine component and operating condition modifications. Methods developed or evaluated for mobile-source NO_x-control may be applicable to stationary engines, but the emission control achievable, and operating parameters in stationary applications cannot be directly inferred from similar effects on smaller mobile engines. Since the carbon monoxide and unburned hydrocarbon emissions of large, stationary, reciprocating engines tend to be quite low, moderate increases in these emissions due to NO_x control may be acceptable.

(a) *Diesel engines*

NO_x emissions from diesel engines can be reduced significantly by retarding the fuel injection timing a few crank angle degrees.[63,65,67] Injection retard is probably limited to about 6° or less for most engines because fuel consumption increases rapidly beyond that point. Typically 6° retard reduces NO_x emissions by about 40% while increasing specific fuel consumption and carbon monoxide emissions by about 4 and 50%, respectively. This technique may require minor hardware modifications, or none at all, and can be incorporated in both new and existing engines. The increased temperature of the exhaust gases which result from injecting the fuel later in the cycle may reduce the durability of the exhaust valves and the turbocharger.

The fuel injection system may, in some cases, be optimized to reduce NO_x without increasing fuel consumption. NO_x reductions of about 20% have been achieved by injecting the fuel in a shorter period.[65] The use of higher injection pressures may reduce the engine durability as a result of higher pushrod and cam loadings.

Water injection into the intake system (water induction) is an effective method to reduce NO_x emissions.[63,65,67] Typically, a water induction rate about equal to the fuel injection rate reduces NO_x emissions by about 50% with very little change in the fuel consumption. Water induction could be applied fairly easily to both new and existing engines, but problems of corrosion and wear of the intake system, intake valves, and water injection nozzles, and degradation of the lubrication oil would have to be resolved before these techniques could be applied to stationary engines. These problems might be alleviated by using distilled or demineralized water, which would increase cost, or by water injection in the form of emulsified fuel, which might increase fuel system corrosion.

Exhaust gas recirculation has been shown to be a very effective NO_x abatement technique, particularly where the EGR is cooled before admission into the intake manifold.[63,65] Incorporation of 10% cooled EGR typically results in about 50% reduction in NO_x, 1.5% increase in fuel consumption, and 100–150% increase in CO and smoke.[65] Corrosion and deposit build-up in the EGR circuit, particularly in the EGR cooler, excessive engine wear and fuel oil contamination are potential problem areas.

Reduction of the air intake temperature reduces NO_x emissions by about 20%/100°F temperature drop, with a slight reduction in fuel consumption.[65] In turbocharged diesels, intercooling (cooling the compressed intake air) is a proven technique to increase the power output capability of the engine and to lower its specific fuel consumption. Additional improvements in engine performance and NO_x might be achieved by further cooling the compressed intake air. By combining turbocharging (which generally increases NO_x emissions), retarded injection timing, and intercooling, NO_x reductions of up to 35% have been demonstrated without any loss in fuel consumption over the equivalent naturally aspirated engine.

Combustion chamber modifications may provide small reductions in NO_x emissions without increasing the fuel consumption. Diesel engines in which the fuel is injected into a precombustion chamber containing a fraction of the combustion air, illustrated in Fig. 7, generally produce less NO_x than comparable direct injection diesels, but the reduced emissions may be accompanied by a small increase in fuel consumption.

Nitric oxide can be removed from the exhaust gases with high efficiency (30–70%) by catalytic reduction by carbon monoxide, hydrogen, or methane if the oxygen concentration is sufficiently low or by ammonia even in the presence of oxygen. This technique is promising for the long term control of NO_x from stationary engines, but much development will be required before the technique is applied on practical systems. Moreover, the cost will be relatively high, about $3/B.h.p. for a 1000 B.h.p. engine.[63]

It is difficult to assess the initial and maintenance costs for the various control techniques accurately in view of the very limited emission control work con-

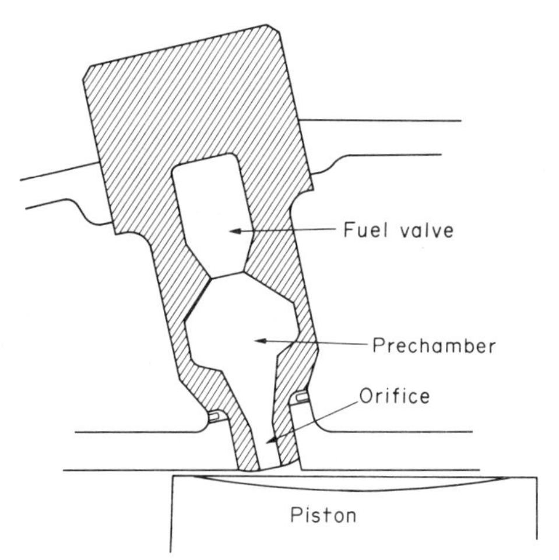

FIG. 7. Precombustion chamber system for a diesel engine.[63]

ducted to date. Incorporation of a turbocharger on a medium size diesel has been estimated to increase the cost by about 10%, or \$2.50–\$3.00/h.p. The percentage cost increase in large stationary diesels is somewhat lower.[65] Addition of an intercooler would add about \$0.30–\$0.50/h.p. while increasing the operating efficiency. It is estimated that the initial cost of a water injection system would be comparable to that of a fuel injection system, with possible additional costs for water purification or distillation equipment. The cost of a typical emission control system has been estimated at \$1.25–\$3.00/h.p. with a related maintenance cost increase of about 10–15%.

Estimates of the operating cost of NO_x reduction in terms of fuel cost alone[65] indicate that timing retard is the least cost-effective technique. Water induction produces small changes in the operating cost. Intake cooling, either alone or combined with 10% EGR, reduces the specific fuel consumption. Exhaust gas recirculation or increased fuel injection rate combined with timing retard may achieve moderate NO_x reductions with small fuel consumption increases.

Only fuel injection timing retard requires no hardware changes for NO_x control, but it is the least cost effective technique considered. Injection timing retard combined with an optimized fuel injector improves the fuel consumption and is applicable for both new and retrofit application. Intake cooling when combined with EGR, water injection, or timing retard appears to be a very attractive technique. Water injection and prechamber design diesels are other techniques which show promise for short term control.

Over the longer term, prechamber diesel engines utilizing a combination of techniques such as intake cooling, EGR, or water injection and catalytic reduction of NO_x in the exhaust may yield lower NO_x emissions with low specific fuel consumption.[63,65]

(b) *Spark ignition engines*

Due to the efforts over recent years to develop emission control techniques for automobile engines, many NO_x control strategies are well defined. Use of fuel rich mixtures result in low NO_x emissions at the expense of increased fuel consumption and emissions of CO and hydrocarbon; the CO and hydrocarbons can subsequently be oxidized in catalytic converters or thermal reactors.[65] Very fuel lean combustion which would result in the lowest emissions of all three pollutants, appears feasible only by means of fuel stratification or improved carburation or fuel injection. Stratified charge offers the possibility of substantially reduced emissions as well as reduced fuel consumption.[63,65] Since application of this technique would require redesign of the engine, further development work is required.

Reducing the temperature of the fuel/air mixture results in a decrease of the NO_x emissions and, generally, also in the specific fuel consumption.[68] The temperature reduction can be accomplished by eliminating inlet manifold heating in gasoline engines and by passing the air through an evaporative cooler for gas engines.[65] In the latter case NO_x would be even further reduced because of the increased moisture content of the air. Reducing the coolant temperature also results in moderate NO_x reductions, e.g., 20% has been achieved by decreasing the coolant temperature from 370 to 212°F.

Water injection can reduce the NO_x emissions by as much as 80%, but this may be accompanied by more than a 10% increase in fuel consumption based on tests on a large gas engine. Results obtained on a single cylinder test engine have shown comparable reductions in NO_x with a 2% reduction in fuel consumption suggesting that the fuel consumption of the larger engine might be improved.[65]

Exhaust gas recirculation can reduce NO_x emissions by as much as 80%, but the fuel economy loss is considerable.[63,65] If, however, only 5–10% of the exhaust gas is recycled, advancing the spark timing or altering the mixture fuel–air ratio can be used to minimize the loss, making 40% NO_x reductions possible. As with all techniques for NO_x control for spark ignition engines, the effects of the control technique on engine durability must be established before application to stationary engines, many of which have an expected useful life of 30 years.

Increasing the engine speed at constant power results in a substantial reduction in NO_x, but the effects on engine life might be significant.[63,67,68] Valve timing can be modified to increase the amount of combustion products retained in the cylinder from cycle to cycle (in effect, internal EGR), reducing NO_x with little or no penalty in fuel consumption.[65]

Ignition timing retard can produce moderate reductions in NO_x, accompanied by a loss in engine power and fuel economy, and possibly at high loads, overheating or burn-out of exhaust valves.[63,65,68]

A meaningful assessment of the cost of emission control for large stationary gas engines has not yet been performed.

(c) *Gas turbine engines*

Emission control technology for gas turbines has been developed primarily in studies of automotive and aircraft gas turbine engines.[65] While the methods developed for mobile sources may be applicable to stationary engines, strategies for stationary gas turbine NO_x control should not be limited to these techniques since stationary engines are not subject to the severe size and weight limitations of mobile engines. In addition, the problems of emission control may be complicated by the combustion of residual oils which will, according to a recent electric utility projection,[54] account for 40% of the fuel burned in gas turbine generating sets by 1983. The heavier fuels require increased residence times of the hot gases for complete combustion, resulting in increased thermal NO formation.[64] Further, the heavier fuels may contain significant fuel nitrogen. These problems have not been resolved in the studies to date. It is clear that fuel nitrogen may substantially increase the NO_x emissions,[64] making limitations on the nitrogen content of the fuel necessary if the proposed Environmental Protection Agency emission standards for oil combustion (75 ppm NO_x at 15% oxygen) are to be met.

Several combustion modifications, which are applicable to present combustor designs, have achieved modest NO_x reductions. These include air-blast or air assist fuel atomization (which can reduce the fuel droplet size by more than 50%), operating the primary zone fuel lean, quenching the flame early in the combustor, and internal flow recirculation.[65] NO_x reductions achievable using these techniques are small, but the costs are low.

Water or steam injection has been shown to be a very effective technique for NO_x control, capable of meeting current standards for stationary gas turbine NO_x emissions.[65,70–78] Siting flexibility is somewhat reduced by the requirements for water in quantities comparable to the fuel supply. Furthermore, to avoid detrimental effects on turbine durability the water has to be purified to a maximum of 2–5 ppm of dissolved solids. The added costs of the water injection system (6–10% of the baseline cost of a large gas turbine), water treatment, and the water (5–7% of the baseline operating cost) must be considered in the application of this technique.[65] For small gas turbines (less than 2 MW) the costs may be prohibitive. A decrease in the cycle efficiency is observed with water injection since the heat of vaporization is not recovered in the cycle. Steam injection increases the mass flow through the turbine without requiring additional compressor work, thereby increasing the thermal efficiency. NO_x reductions of 50–75% have been achieved with water injection with only slight increases in the CO emissions.[65] Water injection is currently being used by some utilities, without observed deleterious effects as long as the water purity is maintained at a high level.

Several advanced combustor designs have been proposed which may meet the NO_x standards with "dry" (no water injection) operation.[65] Internal or external recirculation combined with fuel prevaporization and lean primary zone operation has demonstrated this potential. Catalytic combustion, in which prevaporized fuel and air are injected into a catalyst bed, looks promising for long term emission control.[60]

8. INDUSTRIAL PROCESSES

8.1. *Industrial Process Heating*

It is estimated that industrial process heating contributed 3.3% of total stationary source emissions in 1972.[1] The sources include open-hearth furnaces with high air-preheat used for glass-melting and to a decreasing extent in steel making; rotary, vertical, and tunnel kilns used in cement, lime, and ceramic industries; pyrolysis process furnaces for ethylene and propylene production; petroleum heaters, cat-cracking regenerators, and CO boilers in the petroleum industry billet-reheating and ingot-soaking furnaces, coke-ovens and blast furnaces in the steel industry; and a large variety of heat-treatment furnaces in the metallurgical industry. Some measurements of the emissions from these sources have been provided by a recent survey of the NO_x emissions from stationary sources in the south coast air basin of California.[79] This data is summarized in Table 8.

The overall emission factor for fuel combustion in oil refinery operations was about 0.28 lb $NO_x/10^6$ B.t.u.[79] The predominant fuel consumed was refinery gas which varies widely in composition from heater to heater within a refinery. Heating values may vary from less than 800 B.t.u./scf to more than 1900 B.t.u./scf. Due to the higher flame temperature, the high heating value gas may be expected to produce more NO than the gas with a lower heating value. The gas burned in carbon monoxide boilers contains some ammonia which is readily oxidized to form NO in a manner similar to other fuel nitrogen compounds. The NO emissions are, therefore, high in spite of the low heating value of the CO gas. The variations in fuel composition, burner design, and operating cycles result in a larger range of emissions, e.g., 0.09–1.01 lb $NO_x/10^6$ B.t.u. for process heaters, 0.04–0.17 lb $NO_x/10^6$ B.t.u. for crude oil heaters, and 0.35–0.81 lb $NO_x/10^6$ B.t.u. for carbon monoxide boilers.[79] NO_x emissions from these systems may be reduced through the use of combustion modifications previously described, i.e., low excess air, staged combustion, and flue gas recirculation.

Industrial processes which require a high degree of air preheat produce higher levels of NO_x than do most other combustion sources. The combustion air flow into cement kilns is typically heated to temperatures of 800 K or hotter. In addition, the hot gas residence times are long due to the relatively inefficient heat transfer. Emissions of 2.38–4.0 lb $NO_x/10^6$ B.t.u. were measured in the flue gases of units fueled with natural gas.[79] Oil fired kilns produced somewhat lower emissions, e.g., 1.08–2.0 lb $NO_x/10^6$ B.t.u., probably as a result of radiative heat transfer cooling the flame more quickly for oil than for gas.

The combustion air flows into glass melters through regenerative air preheaters which preheat the air to temperatures of 1100 K or higher. The extensive air preheat results in high flame temperatures and high NO_x emissions, e.g., 0.79 lb $NO_x/10^6$ B.t.u. in a glass melter burning natural gas.[79]

Open hearth furnaces are a primary source of NO_x

TABLE 8. NO_x emissions from industrial process heating (From Ref. 79)

Industry	Unit	Fuel	Firing rate (10^6 B.t.u./h)	Emissions (lb $NO_x/10^6$ B.t.u.)
Petroleum refinery process heater	20	FG	74	0.096
	24	FG	62	0.405
	25	FG	39	1.01
	31–34	FG	55	0.463
	44	FG	58	0.148
	45	FG & Oil	109	0.257
	46	FG & Oil	109	0.237
	54	FG	167	0.093
	57	FG	256	0.136
	58, 59	FG	113	0.347
	63	FG	62	0.133
	64	FG	312	0.159
	66, 67	FG	130	0.341
	69	FG	16	0.114
	70	FG	24	0.171
	71	FG	18	0.149
	99	FG	153	0.088
	100	FG	365	0.179
	103	FG	68	0.110
Crude oil heater	23	Oil	60	0.177
	38	FG & Oil	82	0.173
	47	FG	236	0.065
	47a	FG	228	0.031
	48	FG	180	0.144
	49	FG	105	0.040
	56	FG	66	0.144
	65	FG	74	0.182
	68	FG	241	0.146
	96	FG	394	0.158
	97	FG	141	0.168
	98	FG & Oil	114	0.116
Carbon monoxide boiler	13	FG & CO	506	0.58
	14	FG & CO	521	0.53
	37	FG & CO	61	0.70
	72	FG & CO	291	0.35
	102	FG & CO	321	0.81
Cement kiln	A	NG	220	2.32
	B	NG	220	3.67
	C	NG	350	4.2
	D	NG	350	4.2
	E	NG	75	4.2
	F	NG	75	4.2
	A	Oil	220	1.08
	B	Oil	220	1.08
	C	Oil	350	2.0
	D	Oil	350	2.0
	E	Oil	75	2.0
	F	Oil	75	2.0
Glass melter		NG	50	0.788
Steel open hearth furnace	PI	Coal	55–72	1.55–1.61
	Scrap	Coal		2.40–2.85
Sinter machine			116	0.40
Coke ovens			112	0.21
				0.21
Boiler		BFG	28–314	0.12
		BFG & COG		0.076
Reheat mills		NG		0.109
				0.138
Soak pits				0.113

Abbreviations: BFG = blast furnace gas; CO = carbon monoxide gas; COG = coke oven gas; FG = fuel gas; NG = natural gas.

emissions from steel production. The combustion air is preheated to very high temperatures, typically 1640 K, by flow through regenerative air preheaters. The flame temperature is further increased to as much as 2700 K by replacing part of the combustion air with oxygen. In addition to the high thermal NO_x levels produced by high temperature combustion, nitrogen contained in the coal which is burned may result in significant NO formation. The NO_x emissions are high, 1.55–1.61 lb $NO_x/10^6$ B.t.u. in furnaces where molten pig iron is added to the melt and 2.40–2.85 lb $NO_x/10^6$ B.t.u. in open hearth furnaces which produce steel only from scrap iron.[79] NO_x emissions from glass melters and open hearth furnaces can be reduced about 30% by reducing the level of excess oxygen in the flue gases. Further reductions may be achieved by modifications to the operating schedule of the equipment, e.g., adding the hot metal at an earlier time in the cycle.

Other elements of a steel production facility produce lower, but not insignificant NO_x emissions. Sinter machines produce about 0.40 lb $NO_x/10^6$ B.t.u. and coke ovens produce about 0.21 lb $NO_x/10^6$ B.t.u. Boilers burning blast furnace gas, reheat mills, and soak pits emit 0.08–0.13 lb $NO_x/10^6$ B.t.u.,[79] as summarized in Table 8.

8.2. *Non-Combustion Sources*

The main non-combustion sources of NO_x are the manufacture and use of nitric acid which were estimated to contribute 1.3% of the stationary source emissions in 1972.[1] The techniques for controlling NO_x emission from non combustion-sources are based on removal of the NO_x from the tail gas by selective or non-selective reduction, adsorption, and absorption. Uncontrolled emissions from nitric acid plants were typically in the range of 2000–3000 ppm (approx. 28–41 lbs per ton of acid) with approximately equal concentrations of NO and NO_2. Inasmuch as nitric acid is produced by the absorption of nitrogen oxides, the level of uncontrolled emissions is determined by economic considerations. With the introduction of emission standards of 3.0 lbs NO_2/ton of acid produced in new plants, some manufacturers have reduced the emissions levels to meet the standards entirely by process modification. For older units, several control techniques are available, the most common of which is the catalytic combustion of the NO_x using a fuel, generally natural gas, and a platinum or palladium catalyst.[30,55] In this process sufficient fuel must be supplied to consume the oxygen present in the tail-gas and part of the energy released is recovered in a waste heat boiler and/or a power recovery turbine. A selective reduction technique has been developed using ammonia to reduce the NO_x selectively without consuming the oxygen. For this technique to work, the catalyst bed temperature must be carefully controlled and maintained in the range of 410–520°F. Several selective abatement techniques have been installed on commercial units.[55]

A commercial adsorption process has been developed utilizing a molecular sieve bed to catalytically convert NO to NO_2 and to adsorb the NO_2 product. Continuous operation is achieved by use of two beds in an adsorption–desorption cycle. The NO_2 is recycled to the nitric acid plant. High recovery efficiencies are claimed for this process at costs competitive with catalytic reduction.[80] Two units have been installed on commercial plants.

Absorption of NO_x is hindered by the relatively low solubility of NO. A process enhancing the adsorption of NO_x by catalyzing the oxidation of NO to NO_2 has been commercialized.[9]

For typical NO_x concentrations in a tail gas of about 3000 ppm, the methods described above can achieve over 90% removal.

9. FLUIDIZED BED COMBUSTORS

9.1. *State of Development*

The combustion of coal in fluidized beds operated in the temperature range 1400–2000°F appears to have a number of advantages over conventional utility pulverized-coal boilers with flue gas desulfurization. These advantages include:

(a) high heat release rates per unit volume permitting the construction of relatively compact boilers without penalizing the over-all combustion efficiency;
(b) high rates of heat transfer to tubes immersed in the bed and, therefore, a relatively small tube surface area;
(c) SO_2 removal capabilities of 80–95% by injection of limestone or dolomite in the bed;
(d) operation below the ash softening temperatures and, hence, elimination of slagging problems and a reduction in emissions of hazardous trace metals (operation at temperatures above 2050°F may be used, however, to promote self-agglomeration of the ash);[81]
(e) use of coarsely pulverized coal in contrast to under 200 mesh for a pulverized-coal boiler;
(f) low nitrogen oxide emissions;
(g) potential for increased cycle efficiency by use of pressurized boilers in a combined steam/gas-turbine cycle;
(h) ability to handle low-grade solid fuels.

Studies of fluidized bed combustion for coal burning were initiated in 1961 by the National Coal Board in England on small-scale units operated at pressures from atmospheric up to 6 atmospheres.[82] In 1965, work was initiated in the U.S. at Pope, Evans, and Robbins under contract to the Office of Coal Research.[83] With the recognition of the potential of fluidized bed combustion for reducing the emissions of sulfur and nitrogen oxides, the U.S. program has expanded to include a number of organizations working under the sponsorship of the Office of Coal Research and the Environmental Protection Agency.

Fluidized bed combustion is believed to offer the greatest potential in competition with conventional fossil-fuel utility boilers burning a low-sulfur fuel or a high sulfur fuel in combination with flue gas desulfurization. The economic analyses indicate that the pressurized units used in combined cycles have greater potential in the long-range than the atmospheric pressure units. The development of atmospheric pressure combustors is, however, more advanced and a 30 MW unit is scheduled for completion in 1975 in Rivesville, Virginia. Depending on the successful operation of this unit it is expected that multicellular 200 MW_e units will be installed before 1980 and 800 MW_e units shortly thereafter. It is anticipated that the installation of pressurized units will follow the atmospheric-pressure units. It is clear that the acceptance by the utility industries of the fluidized-bed combustors will depend greatly on whether or not these units will be able to match the relatively trouble-free operation of conventional pulverized-coal units.

9.2. NO_x *Emissions from Fluidized-Bed Combustors*

The emission of NO_x from fluidized-bed combustors has the potential of meeting the existing standards for coal-fired steam-generating plants by a considerable margin. The factors that control these emissions are, however, imperfectly understood. At the temperatures of fluidized bed negligible amounts of thermal NO_x are expected. The major source of NO_x emission is the fuel nitrogen. This has been proved conclusively by tests at the Argonne National Laboratories which showed no change in NO_x concentrations from a fluidized-bed coal combustor when the nitrogen in the combustion air was replaced by argon.[84] The emissions from fluidized bed combustors is therefore a function of both the fuel nitrogen content and the fraction of fuel nitrogen converted to NO_x during combustion. Many factors influence the conversion of fuel nitrogen to NO_x. Tests on a batch laboratory reactor have shown that part of the NO_x is contributed by oxidation of the volatiles emitted by the coal and the remainder by oxidation of the char, with the fractional contribution of the volatiles increasing with increasing temperature.[85] The emissions from laboratory and pilot units have been found to increase with increasing temperature and increasing excess air.[53,84,86,87] Increased pressure in some cases yields lower emissions.[88] The emissions of NO_x from the beds are also found to be a function of the composition of the solids in the bed. Calcium sulfate is found to catalyze the reduction of nitrogen oxide with carbon monoxide and the emissions from a bed using calcium sulfate as a bed material are lower than the corresponding emissions with an alundum bed material.[53] In addition a partially sulfated lime is found to reduce nitric oxide.[53] Some additives, such as cobaltic oxide, however, cause the emissions to increase[86,89] possibly as a consequence of catalysis of the oxidation of the fuel nitrogen. Representative values of emissions for a range of conditions are shown in Table 9.

TABLE 9. NO_x emissions from fluidized bed combustors (From Refs. 90 and 91)

T (°F)	O_2 (%)	Pressure (atm)	NO_x (lb $NO_2/10^6$ B.t.u.)	% N in fuel	Comments
1500	6–7	1	0.11		
1500	4	1	0.11–0.17		
1400	4	1	0.45	1.4	$CaSO_4$ bed material
1500	4	1	0.80	1.4	$CaSO_4$ bed material
1800	4	1	0.99	1.4	$CaSO_4$ bed material
1600	1	1	0.81	1.4	$CaSO_4$ bed material
1600	4	1	0.95	1.4	$CaSO_4$ bed material
1600	6	1	1.05	1.4	$CaSO_4$ bed material
1600	8	1	1.12	1.4	$CaSO_4$ bed material
1600	1	1	1.05	1.4	Alundum bed material
1600	4	1	1.12	1.4	Alundum bed material
1600	6	1	1.19	1.4	Alundum bed material
1600	8	1	1.26	1.4	Alundum bed material
1445	2.7	8	0.27	1.1	
1565	3.0	8	0.29	1.1	
1575	3.0	8	0.20	1.1	
1650	3.0	8	0.25	1.1	
1460	3.0	8	0.21	1.1	
1565	3.0	8	0.27	1.1	
1550	3.0	8	0.22	1.1	
1460	2.7	8	0.21	1.1	
1630	3.0	8	0.38	1.1	
1665	2.9	8	0.17	1.1	

Emission levels of 0.10–0.17 lbs $NO_2/10^6$ B.t.u. have been attained in both atmospheric and pressurized combustors operated slightly fuel lean (15% excess air or 3% oxygen).[88,92] Other atmospheric pressure tests have yielded higher emissions than the above but still lower than the emissions standard for coal units.[53] If needed, significant additional reduction in NO_x emissions may be attained with fluidized bed combustors by the use of staged combustion.[53]

10. FUTURE TRENDS

10.1. *Fuel Usage*

It is projected that in the near future the U.S. will need to rely increasingly on fuels derived from its major domestic reserves of coal and possibly shale. As the nitrogen content of both coal and shale are high, increased utilization of fuels from these sources will result in increased emission unless the nitrogen content is reduced prior to combustion or a capability for burning high-nitrogen content fuels with acceptable NO_x emission levels is developed.

The forecast for changes in fossil fuel requirements by the utility industry[54] in the U.S. for the period 1974–1983 is that coal usage will increase from 417 to 742 million tons/year (approx. $10.8–19.3 \times 10^{15}$ B.t.u./year), oil will increase from 685 to 968 million barrels/year (approx. $4.0–5.6 \times 10^{15}$ B.t.u./year), and natural gas will decrease from 2.8 to 1.9×10^{12} ft^3/year (approx. $2.9–2.03 \times 10^{15}$ B.t.u./year). These projections lead to the conclusion that if all utility boilers by 1983 emitted at levels equal to the standards for new sources, the total emissions of NO_x would increase from the estimated 5.7×10^6 tons/year in 1972 to 7.8×10^6 tons/year in 1983.

It is expected that synthetic gaseous and liquid fuels derived from coal will be used to meet the deficit between the U.S. production and demands for gaseous and liquid fuels. Little is known about the NO_x emission potential of new fuels, other than that thermal NO_x will be affected by their adiabatic flame temperature and fuel NO_x by their organic nitrogen content.

High B.t.u.-gas is expected to follow the behavior of natural gas. Low-B.t.u. and intermediate-B.t.u. gas will generally have relatively low flame temperatures and therefore low thermal NO_x emissions. Tests with an 180 B.t.u. gas yielded NO_x levels less than a tenth the value obtained with a natural gas fired under the same conditions.[93] Ammonia, pyridine, and hydrogen cyanide and other nitrogen containing compounds are produced in the coal gasification process. The fuel NO_x emission will depend to a large extent on the gas clean-up process; it will be low when conventional low-temperature high-efficiency scrubbing systems are used but may be significant if a high temperature desulfurization step is used. The nitrogen content is more likely to be high in synthetic liquid fuels since the sulfur is more readily removed than the nitrogen by catalytic hydrogenation.

Alcohol fuels have been proposed as alternatives to Liquid Natural Gas and also as a possible product from conversion processes for coal or solid wastes. A probable composition of alcohol fuels is 90% methanol with the balance consisting of higher alcohols.[93] In studies on the combustion of methanol with 5% excess air in a hot refractory wall furnace (wall temperatures in excess of 2500°F) NO_x emission levels were found to be a factor of 4 lower than that for a distillate oil and a factor of 3 lower than that for propane,[93] reductions which could be anticipated from the differences in adiabatic flame temperatures. Blending higher alcohols in the methanol increased the emissions but even with a 50–50 blend of methanol and isopropanol the emissions were below 150 ppm (approx. 0.2 lbs/10^6 B.t.u.).[93]

10.2. *Surface Combustion*

In surface combustion very low thermal NO_x emissions are achieved by reducing the peak flame temperatures below the levels attainable in conventional combustors. The low temperatures are achieved either by very fuel lean combustion on a catalyst bed or by heat transfer from a flame layer to an adjacent cooled solid surface.[69]

Emission levels as low as 0.005 lb $NO_2/10^6$ B.t.u. have been reported for a catalytic combustor designed to simulate gas turbine operating conditions and achieving very high combustion efficiencies at an energy release rate of 3.5×10^6 B.t.u./atm h ft^3.[94] Catalytic surface combustion has been proposed for stationary gas turbines, electric utility boilers, and small space heating equipment. Emissions from non-catalytic surface combustion systems are quite low, 0.005–0.10 lb $NO_2/10^6$ B.t.u.[69] Porous plate combustors have been proposed for utility boiler and stationary gas turbine applications but several problems must be resolved before the technique is practical. These include flashback and preignition as well as the plugging of the pores by fuel impurities. The requirements for clean fuels for surface combustion may limit their applicability to large systems.

10.3. *Flue Gas Treatment*

The costs for removal of NO_x from tail-gases are generally higher than those associated with reducing NO_x formation in the combustion process. Major emphasis in the U.S. has been on combustion process modification, since it permits the construction of new equipment that can meet existing standards, and a smaller effort has been devoted to flue-gas treatment. Stringent emission standards in Japan have resulted in the active development in that country of flue gas treatment processes.[2] The processes that have proceeded farthest towards full-scale demonstration are the catalytic reduction of NO with ammonia or hydrocarbons; the oxidation of NO to NO_2 using ClO_2 followed by the reduction of NO_2 to N_2 by Na_2SO_3; the addition of NO_2 to produce an equimolal mixture of NO and NO_2 followed by absorption in $Mg(OH)_2$, a process that has the advantage of also removing SO_x.

Selective reduction of NO to N_2 using NH_3 and a noble metal catalyst is being explored on a pilot scale in the U.S. For a stack gas with a concentration of 225 ppm NO_x, 90% reduction of the NO_x was achieved, but some unreacted NH_3 (approximately 25 ppm) escaped. The selective reduction of NO by NH_3 can also be achieved by homogeneous gas phase reactions in the temperature range of 1600–2000°F.[95] In laboratory tests efficiencies of NO_x removal as high as 99% have been attained with negligible emissions of NH_3 in relatively short reaction times (Ca. 0.15 s).

Although flue-gas treatment cannot compete economically with combustion process modification, it provides, at present, an option for additional control when the need for such control is dictated by severe local environmental problems.

APPENDIX

Conversion Factors from lbs/10^6 B.t.u. to Other Units

Desired Units		Multiply lbs/10^6 B.t.u. by
grms/10^6 cal		1.8
Kgram/10^6 Kjoules		0.430
ppm at 3% O_2	gas	840*
	oil	770*
	coal	710*
ppm at 15% O_2	gas	280*
	oil	260*

* Factors will vary slightly with composition of fuel.

Acknowledgements: The material in this chapter was prepared initially for a U.S. National Academy of Sciences—National Academy of Engineering report on "Air Quality and Stationary Source Emission Control". The authors wish to acknowledge inputs to the report by the EPA, Combustion Research Laboratory and the editorial inputs of the staff officer for the Academy Reports, Dr. Raphael G. Kasper.

REFERENCES

1. MASON, H. B. and SHIMZU, A. B., Briefing document for the maximum stationary source technology (MSST) systems program for NO_x control, extracted from Aero Therm Final Report 74-123, EPA Contract No. 68-02-1318, October 1974.
2. TOHATA, H., Nitrogen oxides abatement processes in Japan, Presented at Joint U.S. Japan Symposium on Counter Measures for NO_x, Tokyo, Japan, June, 1974.
3. ZELDOVICH, Y. B., *Acta Physicochem. URSS* **21**, 577 (1946).
4. THOMPSON, D., BROWN, T. D. and BEER, J. M., The formation of oxides of nitrogen in a combustion system. *Combust. Flame* **19**, 69 (1972).
5. SAROFIM, A. F. and POHL, J. H., Kinetics of nitric oxide formation in premixed laminar flames, *Fourteenth Symposium (International) on Combustion*, pp. 739–753, The Combustion Institute, Pittsburgh, 1973.
6. LIVERSEY, J. B., ROBERTS, A. L. and WILLIAMS, A., *Combust. Sci. Technol.* **4**, 9 (1971).
7. MERRYMAN, E. L. and LEVY, A., Nitrogen oxide formation in flames: The roles of NO_2 and fuel nitrogen, paper presented at Fifteenth Symposium (International) on Combustion, Tokyo, Japan, August, 1974.
8. MARTIN, G. B. and BERKAU, E. E., An investigation of the conversion of various fuel nitrogen compounds to nitrogen oxides in oil combustion, *AICHE Symp. Ser.* **68** 126, 45 (1972).
9. MAYLAND, B. J. and HEINZE, R. C., Continuous catalytic absorption for NO_x emission control *Chem. Engng Prog.* **69**, 75 (1973).
10. POHL, J. H., Sc. D. Thesis (in preparation), Chemical Engineering Department, M.I.T., Cambridge, Mass., 1975.
11. BALL, J. S., WHISMAN, M. L. and WENGER, W. J., Nitrogen content of crude petroleums, *Ind. Engng Chem.* **143**, 2577 (1951).
12. BALL, J. S. and WENGER, W. J., How much nitrogen in crudes from your area, *Petroleum Refiner* **37** (4), 207 (1958).
13. VANDAVEER, F. F., Fuel reserves, production, gases marketed, and fuels used by utilities, in SEGELER, C. G. ed. *Gas Engineers Handbook*, p. 2–15, The Industrial Press, New York, 1965.
14. AGA *et al.*, *Reserves of Crude Oil, Natural Gas Liquids Natural Gas in the United States and Canada and United States Production Capacity as of December 1972*. AGA, Arlington, Va. 1973.
15. AVERITT, P., Coal resources of the United States. January 1, 1967, *U.S.G.S. Bull.* 1275 (1969).
16. MARK, *et al.*, ed. *Kirk-Othmer Encyclopedia of Chemical Technology*, 2nd. ed. Interscience Publishers, New York, 1964.
17. PERRY, R. H. *et al.*, eds. *Chemical Engineers Handbook*, 4th ed., McGraw-Hill, New York, 1963.
18. PARKS, B. C. and O'DONNELL, H. J., Petrography of American coals, *U.S. B.M. Bull.* **55D** (1956).
19. TURNER, D. A., ANDREWS, R. L. and SEIGMUND, C. W., Influence of combustion modifications and fuel nitrogen content on nitrogen oxides emissions from fuel oil combustion, *AICHE Symp.* Ser. **68**, 55 (1972).
20. TURNER, D. W. and SIEGMUND, C. W., Staged combustion and flue gas recycle: Potential for minimizing NO_x from fuel oil combustion, Esso Research and Engineering Company, Presented at American Flame Research Committee EPA Flame Days, Chicago, September 1972.
21. FENIMORE, C. P., Formation of nitric oxide from fuel nitrogen in ethylene flames, *Combust. Flame* **19**, 289 (1972).
22. HAZARD, H. R., Conversion of fuel nitrogen to NO_x in a compact combustor, ASME Paper No. 73-WA/GT-2, Presented at the ASME Winter Annual Meeting of the Gas Turbine Division, Detroit, Michigan, 1973.
23. MCCANN, C. R., DEMETER, J. J., ORNING, A. A. and BIENSTOCK, D., NO_x emissions at low excess-air levels in pulverized coal combustion, ASME Winter Meeting, Proceedings, November 1970.
24. PERSHING, D. W., BROWN, J. W., MARTIN, G. B. and BERKAU, E. E., Influence of design variable on the production of thermal and fuel NO_x from residual oil and coal combustion, Presented at 66th Annual AICHE Meeting, Philadelphia, Pa., November, 1973.
25. FLAGAN, R. C. and APPLETON, J. P., A stochastic model of turbulent mixing with chemical reaction: Nitric oxide formation in a plug-flow burner, *Combust. Flame* **23**, 249–267 (1974).
26. CRAWFORD, A. R., MANNY, E. H. and BARTOK, W., Field testing: Application of combustion modifications to control NO_x emissions from utility boilers, Esso Research Engineering, Report No. EPA-650/2-74-066, June, 1974.
27. BARTOK, W., CRAWFORD, A. R. and PIEGARI, G. J., Systematic field study of NO_x emission control methods for utility boilers, Esso Research and Engineering Co., Report No. GRU. 4GNOS. 71 to the Office of Air Programs, Research Triangle Park, North Carolina, December, 1971.
28. CATO, G. A., BUENING, H. J., DE VIVO, C. C., MORTON, B. G. and ROBINSON, J. M.; Field testing: Application of combustion modifications to control pollutant emissions from industrial boilers—phase I, KVB Engineering, Tustin, California, Report No. EPA-650/2-74-078a, October, 1974.
29. CATO, G. A. and ROBINSON, J. M., Application of combustion modification techniques to control pollution emissions from industrial boilers, KVB Engineering, Tustin, California, Interim Report-Phase I, under EPA Contract No. 68-02-1074 March, 1974.
30. GILLESPIE, G. R., BOYUM, A. A. and COLLINS, M. F., Catalytic purification of nitric acid tail gas: A new approach, Paper presented at 64th AICHE Annual Meeting, San Francisco, Calif., December 1971.
31. WENDT, J. O. L., STERNLING, C. V. and MATOVICH,

M. A., Reduction of sulfur trioxide and nitrogen oxides by secondary fuel injection, *Fourteenth Symposium (International) on Combustion*, Combustion Institute, Pittsburgh, 1973.
32. YAMAGISHI, K., NOZAWA, M., YOSHIE, T., TOKUMOTO, T. and KAKEGAWA, Y., A study of NO_x emission characteristics in two stage combustion, Tokyo Gas Company, paper presented at 15th Symposium (International) on Combustion, Tokyo, Japan. August, 1974.
33. LANGE, B., Jr., NO_x formation in premixed combustion: A kinetic model and experimental data, *Air Pollution and Its Control, AICHE Symp.* Ser. 126, **68**, 17 (1972).
34. BLAKESLEE, C. E. and BURBACH, H. E., Controlling NO_x emissions from steam generators, Paper presented at 65th Annual APCA Meeting, Miami Beach, Florida, June 18–22, 1972.
35. HALSTEAD, C. J., WATSON, C. D. and MUNRO, A. J. E., Nitrogen oxide control in gas fired systems using flue gas recirculation and water injection, Proceedings of the Second Conference on Natural Gas Research and Technology, Institute of Gas Technology, Chicago, Ill., June, 1972.
36. TOUSSAINT, M. and HEAP, H. P., Formation des oxydes d'azote et des particules dans une flamme d'emulsion fuel loure-eau, International Flame Research Foundation, Doc. No. G19/a/7, Ijmuiden, Holland, April, 1974.
37. WOOLRICH, P. F., Methods for estimating oxides of nitrogen emissions from combustion processes, *Am. Ind. Hyg. Ass. J.* **22**, 481 (1961).
38. ARMENTO, W. T. and SAGE, W. L., The effect of design and operation variables on NO_x formation in coal fired furnaces: Status report, Paper presented at the 66th Annual AICHE Meeting, Philadelphia, Pa., November, 1973.
39. ARMENTO, W. T., Effects of design and operating variables on NO_x from coal-fired furnaces phase I, Babcock & Wilcox Co., Alliance, Ohio, Report No. EPA-650/2-74-002a, January, 1974.
40. WASSER, J. H., HANGEBRAUCK, R. P. and SCHWARTZ, A. J., Effects of air-fuel stoichiometry on air pollution emissions from an oil-fired test furnace, *J. Air Pollution Control Ass.* **18** (5), 332 (1968).
41. WINSHIP, R. D. and BRODEUR, P. W. NO_x control in oil and gas-fired boilers, Combustion Engineering, Montreal, Canada, presented at the 1973 Power Engineering Conference and Exhibition, Hamilton, Ontario, October, 1973.
42. BARNHART, D. H. and DIEHL, E. K., Control of nitrogen oxides in boiler flue gases by two stage combustion, Paper presented at the 52nd Annual Meeting of APCA, 1959.
43. BARNHART, D. H. and DIEHL, E. K., Control of nitrogen oxides in boiler flue gases by two-stage combustion, *J. Air Pollution Control Ass.* **10**(5), 397 (1960).
44. SIEGMUND, C. W. and TURNER, D. W., NO_x emissions from industrial boilers: Potential control methods, *J. Engng Power* pp. 1–6, January (1974).
45. LOCKLIN, D. W., KRAUSE, H. H., PUTNAM, A. A., KROPP, E. L., REID, W. T. and DUFFY, M. A., Design trends and operating problems in combustion modifications of industrial boilers, Battelle Columbus Laboratories EPA Grant No. 902402, April, 1974.
46. HEAP, M. P., LOWES, T. M., WALMSLEY, R. and BARTELDS, H., Burner design principles for minimum NO_x emissions, Proceedings, Coal Combustion Seminar, Research Triangle Park, N. C., Report No. EPA 650/2-73-021, September, 1973.
47. WASSER, J. H. and BERKAU, E. E., Combustion intensity relationship to air pollution emissions from a model combustion system, *Air Pollution and its Control, AICHE Symp.* Ser. 126 **68** (1972).
48. HEMSATH, K. H., SCHULTZ, J. T. and CHOJNACKI, D. A., Investigation of NO_x emissions from industrial burners, presented at the American Flame Research Committee/EPA, American Flame Days, Chicago, September, 1972.
49. KOMIYAMA, K., Ph.D. Thesis, Mechanical Engineering Department, M.I.T., Cambridge, Mass., January, 1975.
50. LOWES, T. M., HEAP, M. P. and SMITH, R. B., Reduction of pollution by burner design, International Flame Research Foundation, Doc. No. K20/a/74, August, 1974.
51. BARRETT, R. E., MILLER, S. E. and LOCKLIN, D. W., Field investigation of emissions from combustion equipment for space heating, Battelle Columbus Laboratory Report No. EPA-R2-73-084a (API Publication 4180), June, 1973.
52. WENDT, J. O. L., STERNLING, C. V. and MATOVICH, M. A., Reduction of sulfur trioxide and nitrogen oxides by secondary fuel injection, *Fourteenth Symposium (International) on Combustion*, Combustion Institute, Pittsburgh, 1973.
53. HAMMONS, G. A. and SKOPP, A., NO_x formation and control in fluidized bed coal combustion processes, ASME Paper No. 71-WA/APC-3, Presented at ASME Winter Annual Meeting, Washington, D.C., November, 28–December 2, 1971.
54. National Electric Reliability Council, Review of overall adequacy and reliability of the North American bulk power systems, Report by Interregional Review Sub-Committee of the Technical Advisory Committee, October 1974.
55. BARTOK, W., CRAWFORD, A. R., CUNNINGHAM, A. R., HALL, H. J., MANNY, E. H. and SKOPP, A., Systems study of nitrogen oxide control methods for stationary sources, Esso Research and Engineering Co., Final Report to Division of Process Control Engineering National Air Pollution Control Administration, vol. II, GR-NOW-69, November 20, 1969, PB-192-789.
56. BAGWELL, F. A., ROSENTHAL, K. E., BREEN, B. P., BAYARD DE VOLO, N. and BELL, A. J., Oxides of nitrogen emission reduction program for oil and gas fired utility boilers, Paper presented at 32nd Annual Meeting of the American Power Conference, Chicago, Illinois, April 21–23, 1970.
57. TEIXERA, D. P. and BREEN, B. P., NO_x reductions in utility boilers, Presented at 16th Power Instrumentation Symposium, Instrument Society of America, Chicago, Illinois, May, 1973.
58. BLAKESLEE, C. E. and SELKER, S. P., Program for reduction of NO_x from tangential coal-fired boilers phase I, Combustion Engineering, Report No. EPA-650/2-73-005, August, 1973.
59. HALL, R. E., WASSER, J. H. and BERKAU, E. E., A study of air pollutant emissions from residential heating systems, Control Systems Laboratory, NERC, Research Triangle Park, N.C., Report No. EPA-650/2-003, January 1974.
60. THOMPSON, R. E., PERSHING, D. W. and BERKAV, E. E., Catalytic combustion; a pollution-free means of energy conversion, Report No. EPA 650/2-73-018, August 1973.
61. DICKERSON, R. A. and OKUDA, A. S., Design of an optimum distillate oil burner for control of pollutant emissions, Rocketdyne Division, Rockwell International, Canoga Park, California, Report No. EPA-650/2-74-047, June, 1974.
62. MARTIN, G. B., Overview of the U.S. Environmental Protection Agency's activities in NO_x control for stationary sources. U.S. EPA Research, Triangle Park, N.C., Presented at the Joint U.S.–Japan Symposium on Countermeasures for NO_x, Tokyo, Japan, June 28–29, 1974.
63. MCGOWIN, C. R., Stationary internal combustor engines in the United States, Shell Development Co., Houston, Texas, Report No. EPA R2-73-212, April, 1973.
64. JOHNSON, R. H. and SCHIEFER, R. B., Environmental compatibility of modern gas turbines, General Electric Gas Turbine Reference, Library Report No. GER-2486c, 1974.
65. ROESSLER, W. A., MORASZEW, A. and KOPA, R. D.,

Assessment of the applicability of automotive emission control technology to stationary engines, Aerospace Corp., El Segundo, California, Report No. EPA 650/2-74-051, July 1974.

66. DILMORE, J. A. and ROHRER, W., Nitric oxide formation in the combustion of fuels containing nitrogen in a gas turbine combustor, Report No. 74-GT-37 (ASME), 1974.
67. SCHAUB, F. S. and BEIGHTOL, K. V., NO_x emission reduction methods for large bore diesel and natural gas engines, Cooper-Bessemer, Mount Vernon, Ohio, ASME Paper No. 71-WA/D4P-2, presented at the Winter Annual Meeting, Washington, D.C.
68. MCGOWIN, G. R., SCHAUB, F. S. and HUBBARD, R. L., Emissions control of a stationary two-stroke spark-gas engine by modification of operating conditions. Shell Development Co., Emeryville, California, Report N. P-2136, undated.
69. ROESSLER, W. V. WEINBERG, E. K., DRAKE, J. A., WHITE, H. M. and IURA, T., Investigation of surface combustion concepts for NO_x control in utility boilers and stationary gas turbines, Aerospace Corp., El Segundo, California, Report No. EPA 650/2-73-014, August, 1973.
70. HILT, M. B. and JOHNSON, R. H., Nitric oxide abatement in heavy duty gas turbine combustions by means of aerodynamics and water injection, ASME Paper No. 72-GT-53, Presented at the Gas Turbine and Fluid Engineering Conference, San Francisco, Calif., March, 1972.
71. KLAPATCH, R. D. and KOBLISH, T. R., Nitrogen oxide control with water injection gas turbines, ASME Paper No. 71-WA/GT-9, Presented at ASME Winter Annual Meeting, Washington, D.C., November 28–December 2, 1971.
72. KOCH, H., Investigations and measurements for the reduction of gas turbine emissions, CIMAC, 1973.
73. SHAW, H., The effects of water, pressure, and equivalence ratio on nitric oxide production in gas turbines, *Trans ASME, J. Engng Power*, 240–246, July (1974).
74. SINGH, P. P., YOUNG, W. E. and AMBROSE, M. J., Formation and control of oxides of nitrogen emission from gas turbine combustion systems, Westinghouse, Pittsburgh, Pa., ASME Paper No. 72-GT-22, Presented at Gas Turbine and Fluids Engineering Conference and Products Show, San Francisco, California, March, 1972.
75. AMBROSE, M. J. and OBIDINSKI, E. S., Recent field tests for control of exhaust emissions from a 35-MW gas turbine, Report No. 72-JPG-GT-2 (ASME), 1972.
76. HUNG, W. S. Y., Accurate method of predicting the effects of humidity or injected water on NO_x emissions from industrial gas turbines, Report No. 74-WA/GT-6 (ASME), 1974.
77. LIPFERT, F. W., Correlation of gas turbine emission data, Report No. 72-GT-60 (ASME), 1972.
78. LIPFERT, F. W., SANLORENZO, E. A. and BLAKESLEE, H. W., The New York power pool gas turbine emissions test program presented at the MASS-APCA Specialty Conference on Air Quality Standards and Measurement, Kiamesha Lake, New York, October, 1974.
79. BARTZ, D. R., ARLEDGE, K. W., GABRIELSON, J. E., HAYES, L. G. and HUNTER, S. D., Control of oxides of nitrogen from stationary sources in the south coast air basin, KVB Report No. 5800-179, Tustin, California, September, 1974.
80. FORNOFF, L. L., A new molecular sieve process for NO_x removal and recovery from nitric acid plant tail gas, *Air Pollution and its Control, AICHE Symp.* Ser. 126, **68**, 111 (1972).
81. EHRLICH, S., ROBINSON, E. B., BISHOP, J. W., GORDON, J. S. and HOERL, A., Particulate emissions from a high temperature fluidized bed fly carbon burner, *AICHE Symp.* Ser. 70 **137**, 379 (1974).
82. HOY, R. H. and ROBERTS, A. G., Power generation via combined gas steam cycles and fluid-bed combustion of coal, *Gas and Oil Power* **173**, July/August (1969).
83. BISHOP, J. W., DEMING, L. F., EDGRER, R. and REINHARDT, F. W., Coal-fired packaged boilers, past, present, and future, ASME Paper No. 66-WA/FU-4, July, 1966.
84. JONKE, A. A., CARLS, E. L., JARRY, R. L., HAAS, M. and MURPHY, W. A., Reduction of atmospheric pollution by the application of fluidized bed combustion, Annual Report July 1968–June 1969 by Chemical Engineering Division, Argonne National Laboratory, Illinois, Report No. ANL/ES/-CEN-1001, 1969.
85. PEREIRA, F. J., BEER, J. M., GIBBS, B. and HEDLEY, A. B., NO_x emissions from fluidized-bed coal combustors, Paper presented at 15th Symposium on Combustion, Tokyo, Japan, August, 1974.
86. JONKE, A. A., CARLS, E. L., JARRY, R. L., ANASTASIA, L. J., HAAS, M., PAVLIK, J. R., MURPHY, W. A. and SCHOFFSTOLL, C. B., Reduction of atmospheric pollution by the application of fluidized-bed combustion, Annual Report July 1969–June 1970, by Chemical Engineering Division, Argonne National Laboratory, Illinois, Report ANL/ES-CEN-1002, 1970.
87. JONKE, A. A., VOGEL, G. J., ANASTASIA, L. J., JARRY, R. L., RAMASWAMI, D., HAAS, M., SCHOFFSTOLL, C. B., PAVLIK, J. R., VARGO, G. N. and GREEN, R., Reduction of atmospheric pollution by the application of fluidized-bed combustion, by Chemical Engineering Division, Argonne National Laboratory Annual Report July 1970–June 1971, Illinois.
88. VOGEL, G. J., SWIFT, W. M., LENC, J. F., CUNNINGHAM, P. T., WILSON, W. I., PANEK, A. F., TEATS, F. G. and JONKE, A. A., Reduction of atmospheric pollution by the application of fluidized-bed combustion and regeneration of sulfur containing additives, Argonne National Laboratory, Report No. ANL/ES-CEN 1007 and EPA 650/2-74-104, June, 1974.
89. JARRY, R. L., ANASTASIA, L. J., CARLS, E. L., JONKE, A. A. and VOGEL, J. J., Comparative emissions of pollutants during combustion of natural gas and coal in fluidized beds, Proceedings of Second International Conference on Fluidized Bed Combustion, Publication No. AP-109, Office of Air Programs, U.S. Environmental Protection Agency, Research Triangle Park, North Carolina, 1970.
90. JONKE, A. A., SWIFT, W. M. and VOGEL, G. T., Fluidized-bed combustion: Development Status, Presented at the ASME Fall Meeting, September, 1974.
91. SKOPP, A., NUTKIS, M. S., HAMMONDS, G. A. and BERTRAND, R. R., Studies of the fluidized lime-bed coal combustion desulfurization system, Esso Research and Engineering Co., Report No. PB 210-246, December, 1971.
92. DEMSKI, R. J., SNGOOGN, R. B. and CURIO, A. R., Bureau of Mines test of fluidized bed combustor of Pope, Evans and Robbins, Alexandria, Virginia, U.S. Department of Interior, Bureau of Mines, Pittsburgh Energy Research Center, March, 1973.
93. MARTIN, G. B., Environmental considerations in the use of alternate clean fuels in stationary combustion processes, presented at the EPA Symposium on the Environmental Aspects of Fuel Conversion Technology, St. Louis, Missouri, May 1974.
94. BLAZEWSKI, W. S. and BRESOWAR, G. E., Preliminary study of the catalytic combustor concept as applied to aircraft gas turbines, Air Force Aero Propulsion Laboratory, Wright-Patterson Air Force Base, Ohio, Report No. AFAPL-TR-74-32, May, 1974.
95. LYON, R. K., Verfahren zur Reduzierung von NO in Verbrennungabgasen, Germany, Patent No. 2411672, September, 1974.

Manuscript received, December 1975

Prog. Energy Combust. Sci., Vol. 2, pp. 27–60, 1976. Pergamon Press. Printed in Great Britain

THE CHARACTERIZATION AND EVALUATION OF ACCIDENTAL EXPLOSIONS

ROGER A. STREHLOW

University of Illinois, 105 Transportation Buildings, Urbana IL 61801, U.S.A.

and

WILFRED E. BAKER

Southwest Research Institute, San Antonio, Texas, 78284

Summary: This paper contains a comprehensive review of the current status of our understanding of accidental explosions. After a short historical introduction in which all explosions are characterized by type, the first section discusses the general characteristics of explosions in some detail. Here the usually defined properties of blast waves are introduced and the classical point source or ideal wave is used to discuss scaling laws and TNT or point source equivalence in some detail. Following this there is a general summary of non-ideal blast wave behavior which first discusses extant theoretical work on blast waves from non-ideal sources, i.e., sources which are extended in either space or time. Secondly, each different non-ideal source property effect is discussed in detail with examples. Thirdly, atmospheric and ground effects are discussed briefly.

In the next section the mechanisms by which blast waves produce damage are discussed in detail. In particular the new P-I (for pressure-impulse) method of evaluation is described in some detail with examples, the importance of dynamic impulse in producing tumbling and sliding is discussed, our understanding of fragment damage mechanism is presented and the classic TNT equivalence evaluation based on overpressure is described.

In the last main section of the report specific examples of accidental explosions are given by type. The types that are discussed are: simple pressure vessel failure; runaway chemical reaction or continued combustion; explosions in buildings; internal explosions; rupture followed by combustion; vapor cloud explosions; high explosives and propellants; physical explosions and nuclear reactor runaway. The length of the discussion for each case is dependent on the potential hazard and extent of our current understanding of that type of explosion.

The conclusion section summarizes the findings of the above. The main conclusions and recommendations are that 1. Accidental explosions are important and they will continue to occur. 2. Certain accidental explosions are more reproducible than others but virtually all of them are non-ideal. 3. TNT Equivalency is not a good criterion for evaluating non-ideal explosions and should be replaced, once our understanding improves. 4. Scaling laws for accidental explosions will be relatively easy to develop once our understanding of non-ideal explosions improves. 5. A considerable amount of work, both theoretical and experimental, is needed in this area.

I. INTRODUCTION

This paper is intended to provide a comprehensive review of the current state of the art relative to the characterization and evaluation of accidental explosions in the atmosphere. It was prompted in part by the recent large increase in both the frequency and destructiveness of all types of accidental explosions and in part by the lack of any comprehensive current survey of the literature in this field. It is hoped that this review will delineate, in a systematic manner, our current understanding of the various facets of explosion and damage producing processes and serve as an impetus for future research in this area.

If one examines the literature, the need for such a review becomes evident. There are only three books, Robinson (1944), Freytag (1965) and Kinney (1962), which attempt to treat the general problem. The first of these is very out of date and the second is more of a handbook of safety techniques than a description of the explosion process itself. The last of these, Kinney (1962), is the most comprehensive but is also out of date. The other texts on explosions, Glasstone (1962), Engineering Design Handbook (1972), Baker (1973) and Baker *et al.* (1973), all pertain mainly to the behavior of high explosive charges and do not really treat in detail the more general accidental explosion problem. Furthermore, the majority of the literature in this subject area is not published primarily in open journals but is buried in limited distribution reports.

In general, an explosion is said to have occurred in the atmosphere if energy is released over a sufficiently small time and in a sufficiently small volume so as to generate a pressure wave of finite amplitude traveling away from the source. This energy may have originally been stored in the system in a variety of forms; these include nuclear, chemical, electrical or pressure energy, for example. However, the release is not considered to be explosive unless it is rapid enough and concentrated enough to produce a pressure wave that one can hear. Even though many explosions damage their surroundings, it is not necessary that external damage be produced by the explosion. All that is necessary is that the explosion is capable of being heard.

There are actually many types of processes which lead to explosions in the atmosphere. Table 1 contains a comprehensive listing of all possible types of explosions including theoretical models, natural explosions, intentional explosions and accidental explosions. The list is by type of energy release and is intended to be exhaustive.

In the following sections of this review, the general nature of explosions, current theoretical models and scaling laws will be discussed. The last section will concentrate on a detailed discussion of the character-

TABLE 1. Explosion types*

Theoretical models	Natural explosions	Intentional explosions	Accidental explosions
Ideal point source	Lightning	Nuclear explosions	Pressure vessels
Ideal gas	Volcanoes	High explosives	Simple failure (no combustion)
Real gas	Meteors	Blasting	Runaway chemical reaction before failure
Self-similar (∞ source energy)		Military	Internal explosion or detonation before failure
Constant velocity piston		Pyrotechnic separators	Failure followed by immediate combustion
Accelerating piston		Guns	Gaseous explosions in enclosures (i.e., buildings, cars, etc.)
Bursting sphere		Muzzle blast	Unconfined vapor cloud explosions
Finite stroke piston		Recoilless rifle	High explosives and propellants during manufacture, transport, storage or use
Detonation driven		Exploding spark	Physical explosions
General discussions		Exploding wires	Nuclear reactor runaway
		Laser sparks	
		Continuous source†	
		Contained explosions‡	

*Each column is independent—do not read in rows.
†Hypervelocity missile and supersonic aircraft are examples.
‡Contained vessel explosions and automotive knock are examples.

istics of the accidental explosions listed in the last column of Table 1.

II. GENERAL CHARACTERISTICS OF EXPLOSIONS

A. *Wave Properties*

1. *Energy distribution*

One of the most important properties which determine the behavior of any explosion process is the energy distribution in the system and how it shifts with time as the pressure wave propagates away from the source. Initially all the energy is stored in the source in the form of potential energy. At the instant when the explosion starts, this potential energy is redistributed to produce kinetic and potential energy in different parts of the system; the system now includes all materials contained within either the lead characteristic or lead shock wave of the outwardly propagating explosion wave. The system is non-steady, both because new material is continually being processed by the lead wave front, and because the relative distribution of energy in various forms and in various parts of the system shifts with time.

In order to consider this problem in more detail in this section, we will idealize the system to some extent. We will assume (1) that the explosion is strictly spherical in an initially homogeneous external atmosphere that extends to infinity, (2) that the source of the explosion consists of both energy containing material (source material) and inert confining material, and that during the explosion process these materials do not mix to any great extent with each other or with the outside atmosphere, and (3) that shock wave formation is the only dissipative process in the surrounding atmosphere. With these assumptions, the originally stored energy is distributed among a number of distinct forms at various times and locations as the explosion process proceeds. These are:

(a) *Wave energy.* The propagating wave system contains both potential energy

$$E_p = \int_v \rho C_v(T-T_0)\,dv, \tag{1}$$

and kinetic energy

$$E_k = \int_v \tfrac{1}{2}\rho u^2\,dv, \tag{2}$$

where v is the volume of the atmosphere enclosed by the lead characteristic or lead shock wave. This volume does not include the volume occupied by the products of explosion or by the quiescent atmosphere between the products and blast wave. Furthermore, at late time when the kinetic energy of the source and confining material are zero and the wave amplitude is such that shock dissipation is negligible, the total wave energy ($E_T = E_p + E_k$) in the system must remain constant with time. This far field wave energy should therefore be a unique property of each explosion process.

(b) *Residual energy in the atmosphere (waste energy).* In most explosions a portion of the external atmosphere is treated by a shock wave of finite amplitude. This process is non-isentropic and there will be a residual temperature rise in the atmosphere after it is returned to its initial pressure. This residual energy will also reach some constant value at late time. This was first called "waste" energy by Bethe *et al.* (1947).

(c) *Kinetic and potential energy of the fragments (or confining material).* Initially the confining material will be accelerated and will also store some potential energy due to plastic flow, heat transfer, etc. Eventually all this material will decelerate to zero velocity and will store some potential energy.

(d) *Kinetic energy of source material.* In any explosion involving an extended source the source material will be set into motion by the explosion process. This source material kinetic energy will eventually go to zero as all motion stops in the near field.

(e) *Potential energy of the source.* The source originally contained all the energy of the explosion as potential energy. As the explosion process continues a portion of the energy is lost to other forms but a portion of it normally remains in the source as high temperature product gases, etc. While it is true that this stored energy eventually dissipates itself by mixing, etc., these processes are relatively slow compared to the blast wave propagation process, and for our purposes one can assume quite accurately that the residual energy stored in the products approaches a constant value at late time.

(f) *Radiation.* Radiated energy is quickly lost to the rest of the explosion system and reaches a constant value quite early in the explosion process.

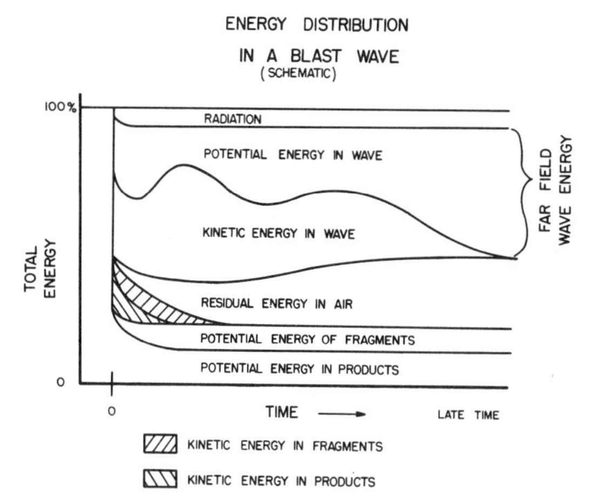

FIG. 1. Energy distribution in a blast wave as a function of time after the explosion (schematic).

Figure 1 summarizes in a schematic manner the way that energy is redistributed in a blast wave as time increases. Note that at late time, when the wave is a far field wave, the system contains potential and kinetic wave energy, residual potential energy (waste energy) in the atmosphere, potential energy in the fragments and potential energy in the products. Also, in general, some energy has been lost to the system due to radiation. However, radiation losses represent an important fraction of the total source energy only for the case of nuclear explosions. A few general statements may be made at this time about Fig. 1.

Firstly, only a fraction of the total energy which is initially available actually appears as wave energy in the far field. Secondly, the magnitude of this fraction relative to the total energy originally available must depend on the nature of the explosion process itself. This is shown for example by the fact that TNT equivalence of nuclear explosions is about 0.5–0.7 of that which one would expect on the basis of the total energy available (Lehto and Larson, 1969; Thornhill, 1960; Bethe *et al.*, 1947). More to the point, in accidental explosions the source normally releases energy relatively slowly over a sizeable volume and one would expect this effectiveness factor to be a strong function of the nature of the release process. Unfortunately, there is no extant work which yields any information on this specific problem. Brinkley (1969) and Brinkley (1970) discuss this problem using a theoretical approach but present no experimental verification of the thesis that slow release means that a larger fraction of the energy is lost to the blast wave.

2. *Usually defined properties*

As a blast wave passes through the air or interacts with and loads a structure or target, rapid variations in pressure, density, temperature and particle velocity occur. The properties of blast waves which are usually defined are related both to the properties which can be easily measured or observed and to properties which can be correlated with blast damage patterns. It is relatively easy to measure shock front arrival times and velocities and entire time histories of overpressures. Measurement of density variations and time histories of particle velocity are more difficult, and no reliable measurements of temperature variations exist.

Classically, the properties which are usually defined and measured are those of the undisturbed or side-on wave as it propagates through the air. Figure 2 shows graphically some of these properties in an ideal wave (Baker, 1973). Prior to shock front arrival, the pressure is ambient pressure p_0. At arrival time t_a, the pressure rises quite abruptly (discontinuously, in an ideal wave) to a peak value $p_s^+ + p_0$. The pressure then decays to ambient in total time $t_a + T^+$, drops to a partial vacuum of amplitude p_s^-, and eventually returns to p_0 in total time $t_a + T^+ + T^-$. The quantity p_s^+ is usually termed the peak side-on overpressure, or merely the peak overpressure. The portion of the time history above initial ambient pressure is called the positive phase, of duration T^+. That portion below p_0, of amplitude p_s^- and duration T^- is called the negative phase. Positive and negative impulses, defined by

$$I_S^+ = \int_{t_a}^{t_a+T^+} [p(t) - p_0]\,dt \tag{3}$$

and

$$I_S^- = \int_{t_a+T^+}^{t_a+T^++T^-} [p_0 - p(t)]\,dt, \tag{4}$$

respectively, are also significant blast wave parameters.

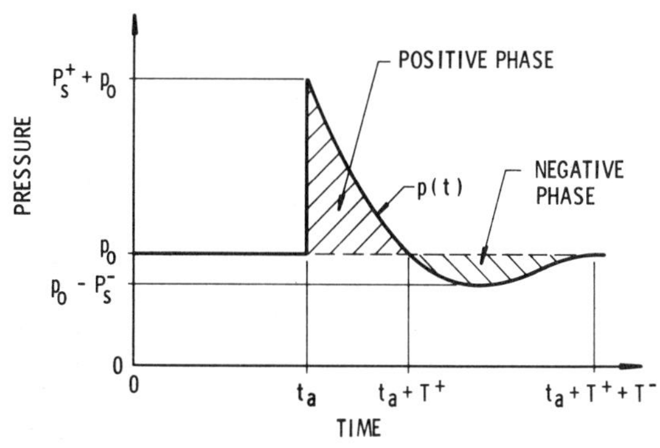

FIG. 2. Ideal blast wave structure.

In most blast studies, the negative phase of the blast wave is ignored and only blast parameters associated with the positive phase are considered or reported. (The positive superscript is usually dropped.) The ideal side-on parameters almost never represent the actual pressure loading applied to structures or targets following an explosion. So a number of other properties are defined to either more closely approximate real blast loads or to provide upper limits for such loads.

An upper limit to blast loads is obtained if one interposes an infinite, rigid wall in front of the wave, and reflects the wave normally. All flow behind the wave is stopped, and pressures are considerably greater than side-on. The peak overpressure in normally reflected waves is usually designated p_r. The integral of this pressure over the positive phase, defined similarly to eqn. (1), is the reflected impulse I_r. Durations of the positive phase of normally reflected waves are designated T_r. The parameter I_r has been measured closer to high explosive and nuclear blast sources than have most blast parameters.

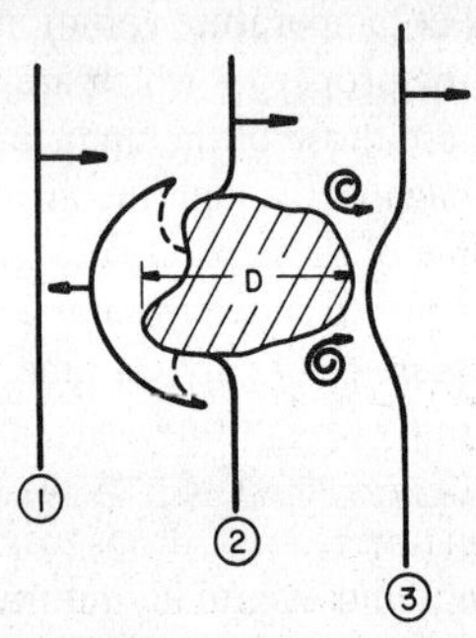

FIG. 3. Interaction of blast wave with irregular object.

A real target is subjected to a very complex loading during the process of diffraction of the shock front around the target. Figure 3 shows schematically, in three stages, the interaction of a blast wave with an irregular object. As the wave strikes the object, a portion is reflected from the front face, and the remainder diffracts around the object. In the diffraction process, the incident wave front closes in behind the object, greatly weakened locally, and a pair of trailing vortices is formed. Rarefaction waves sweep across the front face, attenuating the initial reflected blast pressure. After passage of the front, the body is immersed in a time-varying flow field. Maximum pressure on the front face during this "drag" phase of loading is the stagnation pressure.

We are interested in the net transverse pressure on the object as a function of time. This loading, somewhat idealized, is shown in Fig. 4 (details of the calculation are given by Glasstone (1962)). At time of arrival t_a, the net transverse pressure rises linearly from zero to maximum or P_r in time $(T_1 - t_a)$ (for a flat-faced object, this time is zero). Pressure then falls linearly to drag pressure in time $(T_2 - T_1)$, and then decays more slowly to zero in time $(T_3 - T_2)$. This time history of drag pressure q is a modified exponential, with a maximum given by

$$C_D Q = C_D \tfrac{1}{2} \rho_s u_s^2, \tag{5}$$

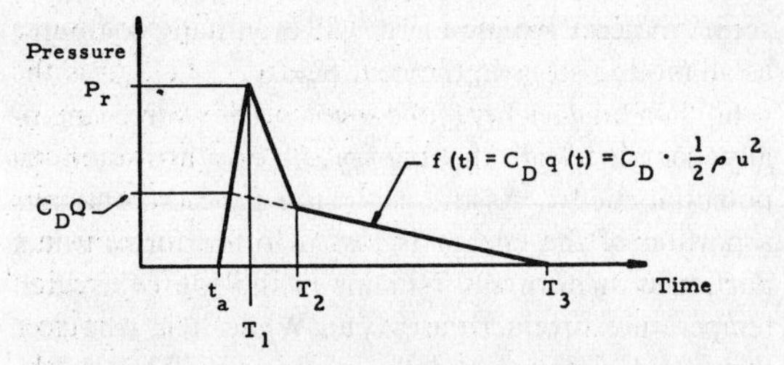

FIG. 4. Time history of net transverse pressure on object during passage of a blast wave.

where C_D is the steady-state drag coefficient for the object, Q is peak dynamic pressure, and ρ_s and u_s are peak density and particle velocity respectively for the blast wave. The characteristics of the diffraction phase of the loading can be determined if the peak side-on overpressure P_s or the shock velocity U are known, together with the shape and some characteristic dimension D of the object. The peak amplitude of the drag phase of the loading can be determined if the peak side-on overpressure P_s or the shock velocity U are known, together with the shape and some characteristic dimension D of the object. The peak amplitude of the drag phase, $C_D Q$, can also be determined explicitly from P_s or u_s.

Because of the importance of the dynamic pressure q in drag or wind effects and target tumbling, it is often reported as a blast wave property. In some instances drag impulse I_d, defined as

$$I_d = \int_{t_a}^{t_a+T} q\,\mathrm{d}t = \tfrac{1}{2} \int_{t_a}^{t_a+T} \rho u^2\,\mathrm{d}t \tag{6}$$

is also reported.

Although it is possible to define the potential or kinetic energy in blast waves, it is not customary in air blast technology to report or compute these properties. For underwater explosions, the use of "energy flux density" is more common (Cole, 1965). This quantity is given approximately by

$$E_f = \frac{1}{\rho_0 c_0} \int_{t_a}^{t_a+T} [p(t) - p_0]^2\,\mathrm{d}t, \tag{7}$$

where ρ_0 and c_0 are density and sound velocity in water ahead of the shock.

B. *The Point Source Blast Wave*

A "point source" blast wave is a blast wave which is produced by the instantaneous deposition of a fixed quantity of energy at an infinitesimal point in a uniform atmosphere. There have been many studies of the properties of point source waves, both for energy deposition in a "real air" atmosphere and for deposition in an "ideal gas" ($\gamma = 1.4$) atmosphere. Deposition in water has also been studied (Cole, 1965). Point source blast wave studies date to the second World War (Bethe *et al.*, 1944; Taylor, 1950; Brinkley and Kirkwood, 1947; Makino, 1951). They have been quite adequately summarized by Korobeinikov *et al.* (1961), Sakurai (1965), Lee *et al.* (1969) and Oppenheim *et al.* (1971) and will be briefly reviewed here. Essentially there are three regions of interest as a point source wave propagates away from its source. The first is the near field wave where pressures in the wave are so large that external pressure (or counter pressure) can be neglected. In this region the wave structure admits to a self-similar solution and analytic formulations are adequate (Bethe *et al.*, 1947; Sakurai, 1965). This region is followed at late time by an intermediate region, which is of extreme practical importance because the overpressure and impulse are sufficiently high in this region to do significant damage, but which does not yield to an analytical solution and therefore must be solved numerically (von Neumann and Goldstein, 1955; Thornhill, 1960). There have been approximate techniques developed to extend the analytical treatment from the near field. These have been summarized by Lee *et al.* (1969). The intermediate region is followed in turn by a "far field" region which yields to an analytic approximation such that if one has the overpressure time curve at one far field position one can easily construct the positive overpressure portion of the curve for large distances. In this far field region there is theoretical evidence that an "N" wave must always form and that the blast wave structure in the positive impulse phase is unaffected by the interior flow and is self-sustaining (Bethe *et al.*, 1947; Whitman, 1950). However, experimentally it is difficult to determine if such an "N" wave actually exists because atmospheric non-homogenieties tend to round the lead shock wave (Warren, 1958).

C. *Classical Experimental Work*

The classical experimental work on blast waves has mainly revolved about the use of either high explosives or nuclear weapons to produce the waves. This work is quite adequately summarized by Baker (1973). It is found in general that the intermediate and far field waves resemble quite closely those predicted using point source theory and to this extent either high explosive or nuclear explosions can be considered to be "ideal". The questions of blast wave scaling as applied to point source, high explosive and nuclear explosions will be discussed next.

1. *Scaling laws*

Scaling of the properties of blast waves from explosive sources is a common practice, and anyone who has even a rudimentary knowledge of blast technology utilizes these laws to predict the properties of blast waves from large-scale explosions based on tests on a much smaller scale. Similarly, results of tests conducted at sea level ambient atmospheric conditions are routinely used to predict the properties of blast waves from explosives detonated under high altitude conditions. It is not the purpose of this paper to review laws for scaling of blast wave properties, which are adequately summarized in Baker (1973) and Baker *et al.* (1973), but we will state the implications of the two laws most commonly used.

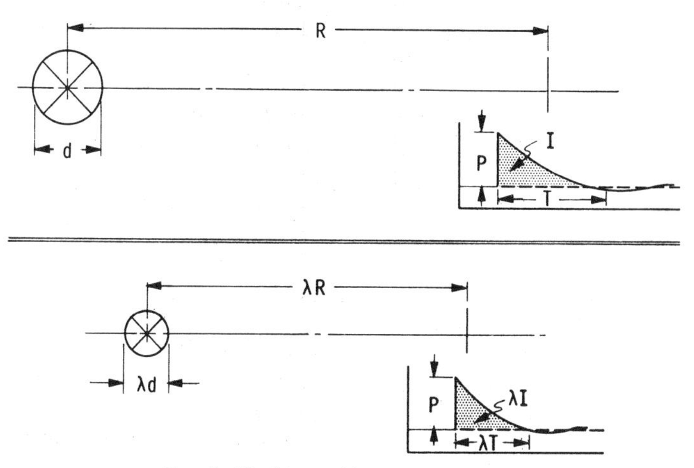

FIG. 5. Hopkinson blast wave scaling.

The most common form of blast scaling is Hopkinson or "cube-root" scaling. This law, first formulated by B. Hopkinson (1915), states that self-similar blast waves are produced at identical scaled distances when two explosive charges of similar geometry and of the same explosive, but of different sizes, are detonated in the same atmosphere.† It is customary to use as a scaled distance a dimensional parameter, $Z = R/W^{1/3}$, where R is the distance from the center of the explosive source and W is the total energy of the explosive. Figure 5 shows schematically the implications of Hopkinson blast wave scaling. An observer located at a distance R from the center of an explosive source of characteristic dimension d will be subjected to a blast wave with amplitude P, duration T, and a characteristic time history. The integral of the pressure–time history is the impulse I. Hopkinson's scaling law then states that an observer stationed at a distance λR from the center of a similar explosive source of characteristic dimension λd detonated in the same atmosphere will feel a blast wave of "similar" form with amplitude P, duration λT and impulse λI. All characteristic times are scaled by the same factor as the length scale factor λ. In Hopkinson scaling, pressures, temperatures, densities and velocities are unchanged at homologous times. Hopkinson's scaling law has been thoroughly

†In Germany, this law is attributed to Cranz (1926).

TABLE 2. Explosive properties

Explosive	Specific gravity	Density (ρ_E)	Weight specific energy (E/W)	Volume specific energy (E/V)	Radius r of 1-lb. sphere
			Imperial units		
		$lb_f\,sec^2/in^4$	in–lb_f/lb_m	in–lb_f/in^3	in
Pentolite (50/50)	1.66	1.551×10^{-4}	20.50×10^6	1.230×10^6	1.584
TNT	1.60	1.496×10^{-4}	18.13×10^6	1.048×10^6	1.604
RDX	1.65	1.542×10^{-4}	21.5×10^6	1.283×10^6	1.588
Comp B (60/40)	1.69	1.580×10^{-4}	20.8×10^6	1.271×10^6	1.575
HBX-1	1.69	1.580×10^{-4}	15.42×10^6	0.944×10^6	1.575
			S.I. units		
		kg/m^3	Joules/kg	$Joules/m^3$	m
Pentolite (50/50)		1.658×10^3	5.107×10^6	8.482×10^9	4.023×10^{-2}
TNT		1.599×10^3	4.517×10^6	7.227×10^9	4.074×10^{-2}
RDX		1.648×10^3	5.356×10^6	8.847×10^9	4.034×10^{-2}
Comp B (60/40)		1.689×10^3	5.182×10^6	8.764×10^9	4.000×10^{-2}
HBX-1		1.689×10^3	3.841×10^6	6.510×10^9	4.000×10^{-2}

verified by many experiments conducted over a large range of explosive charge energies. A much more complete discussion of this law and a demonstration of its applicability is given in Chapter 3 of Baker (1973).

The blast scaling law which is almost universally used to predict characteristics of blast waves from explosions at high altitude is that of Sachs (1944). A careful proof of Sachs' law has been given by Sperrazza (1963). Sachs' law states that dimensionless overpressure and dimensionless impulse can be expressed as unique functions of a dimensionless scaled distance, where the dimensionless parameters include quantities which define the ambient atmospheric conditions prior to the explosion. Sachs' scaled pressure is (P/P_0) (blast pressure/ambient atmospheric pressure). Sachs' scaled impulse is defined as

$$\frac{Ia_0}{(W^{1/3}p_0^{2/3})}.$$

These quantities are a function of dimensionless scaled distance, defined as

$$\frac{(Rp_0^{1/3})}{W^{1/3}}.$$

The primary experimental proof of Sachs' law is given by Dewey and Sperrazza (1950).

Hopkinson's scaling law requires that the model and prototype energy sources which drive the blast wave be of similar geometry and the same type of explosive or energy source. The law has been used in a modified form to scale the highly asymmetric blast waves generated by muzzle blasts from guns and backblasts from recoilless rifles (see Chapter 4 of Baker *et al.* (1973)). These blast sources consist of tubes of hot, high pressure gases suddenly vented to the atmosphere, and so cannot be considered as "ideal" blast sources. Important parameters in the Hopkinson law modified for weapons blast are weapon caliber c and maximum chamber pressure P_C. In contrast to the Hopkinson law. Sachs' law identifies the blast source only by its total energy W, and cannot be expected to be useful for scaling of close-in effects of non-ideal explosions.

No general laws exist for scaling of blast waves from non-ideal explosions, because not all of the physical parameters affecting such explosions are known. However, once a body of data from controlled experiments is available, or once analyses which accurately predict behavior are completed, the development of a scaling law will be straightforward.

2. *TNT or point source equivalence*

(a) *Nuclear and high explosive explosions.* The standard conversion factors for calculating equivalence of high explosive charges as given in Baker (1973) is repeated here as Table 2. With these factors and the

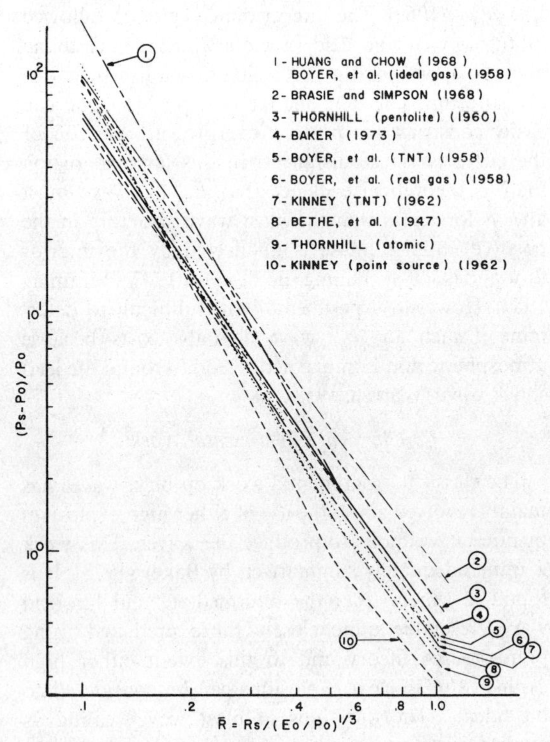

FIG. 6. Blast wave overpressure versus scaled distance taken from a number of sources (small range).

scale distance $\bar{R} = Rp_0^{1/3}/W^{1/3}$, we have plotted dimensionless overpressure, $(P_S - p_0)/p_0$, versus $\bar{R}$ on a log–log plot in Fig. 6 over a very short range, as taken from a number of published sources. It is interesting to note that the overall disagreement between these sources is approximately a factor of ± 2. This was also observed by Baker (1973) and his curve, which is based on experimental data for Pentolite (50/50), is seen to represent a good average of the other curves. Figure 7, which covers a much larger overpressure-scaled distance region, shows the overall extent of scatter. In this curve the shaded regions represent the total range covered by other curves.

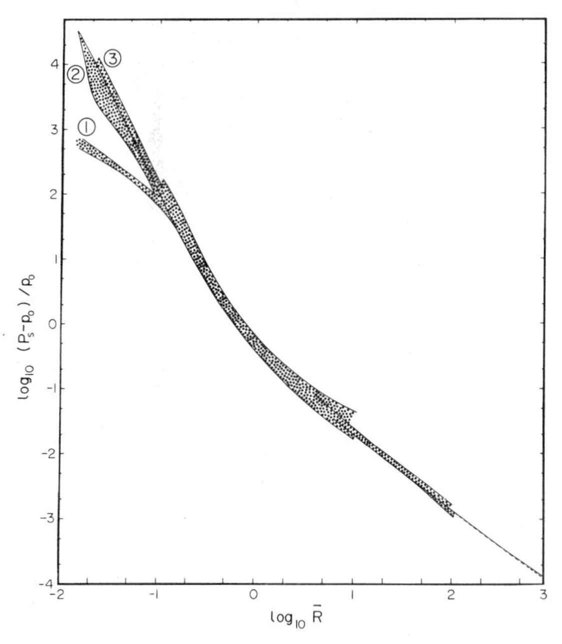

FIG. 7. Blast wave overpressure versus scaled distance for blast waves for the sources listed on Fig. 6. (1) High explosives; (2) Nuclear explosions; (3) Point source. The dotted line in the lower right is for a $1/[\bar{R}(\ln \bar{R})^{\frac{1}{2}}]$ dependence. The solid line is a $1/\bar{R}$ dependence.

There has been controversy about the far field behavior of the wave in the past. Baker (1973) opts for a $1/R$ dependence, while Bethe *et al.* (1947), Thornhill (1960) and Goodman (1960) state that the dependence should be proportional to $1/R(\ln R)^{1/2}$ and Porzel (1972) states that experimental data show an $R^{-4/3}$ dependence. For comparison we have drawn both the $1/\bar{R}(\ln \bar{R})^{1/2}$ and $1/\bar{R}$ dependences on Fig. 7 for $\bar{R} > 10^2$ as a dotted line and solid line, respectively, to show how small the differences in far field behavior really are. The question is actually moot for two reasons. Firstly, Warren (1958) has found a spread of measured overpressures in the far field of about a factor of three, this is undoubtedly due to refraction and focusing effects in the real atmosphere. He also found that the lead shock disappears in the far field and is replaced by a slower pressure rise. This is also to be expected and is due to the non-uniformity of the atmosphere. Secondly, very little damage is done in the far field and therefore it has little practical importance. Techniques for evaluating far field focusing effects due to atmospheric winds and temperature gradients will be discussed in a later section of this review.

(b) *Non-ideal explosions.* The general concept of equivalence for a non-ideal explosion is not well understood at the present time. It is true that usually the near field overpressures are much less than that of a point source explosion which produces the equivalent far field overpressure but it is not obvious exactly what the relationship between near field and far field behavior should be or how this relationship changes as the type of accidental explosion changes. It is also not obvious how one should evaluate the effectiveness for blast damage of any particular type of accidental explosion or how much effectiveness depends on type.

Three approaches to this problem have been taken to date. The first and most practical approach is an *a posteriori* approach involving real accidents. After an accident the blast damage pattern is used to determine the weight of TNT which would be required to do the observed amount of damage at that distance from the center of the explosion. If the explosion is chemical in nature, one then usually attempts to determine a per cent TNT equivalence by determining a maximum equivalent TNT weight of the fuel or chemical by calculating either the heat of reaction of the mixture or the heat of combustion of the quantity of that substance which was released. Zabatakis (1960), Brasie and Simpson (1968), and Burgess and Zabatakis (1973) have all followed this approach which is probably based on the TNT equivalence concept for high explosives, where relative damage is directly correlatable to the relative heats of explosion of different explosives measured in an inert atmosphere. The formulas are of the type

$$(W_{\text{TNT}})_{\text{calc}} = \frac{\Delta H_c W_c}{1800}, \tag{8}$$

$$(W_{\text{TNT}})_{\text{calc}} = \frac{\Delta H_c W_c}{4.198 \times 10^6}, \tag{8a}$$

and

$$\%\text{TNT} = [(W_{\text{TNT}})_{\text{Blast}}/(W_{\text{TNT}})_{\text{calc}}] \times 100,$$

where in eqn. (8) W_{TNT} = the equivalent maximum TNT weight, lbs; ΔH_c = heat of combustion of the hydrocarbon (or heat of reaction of the exothermic mixture), Btu/lb; W_c = weight of hydrocarbon or reaction mixture available as an explosive source; and 1800 = heat of explosion of TNT, Btu/lb. Equation (8a) is in SI units (energy in joules, wt in kg). In the same vein, Dow Chemical Co. (1973) in their safety and loss prevention guide advocate evaluating the relative hazard of any chemical plant operation by first calculating a ΔH of reaction or explosion for the quantity of material which is being handled and then multiplying this basic number by factors based on other known properties such as the substance's sensitivity to detonation.

There has also been a considerable amount of work in which non-ideal explosions are deliberately initiated and side-on blast pressure records obtained. The maximum TNT equivalent yield of the explosive is calculated on the basis of a formula like eqn. (8) and the per cent yield in terms of the two variables, overpressure and positive impulse, are plotted versus the scaled distance $\bar{R}$. A detailed discussion of this approach will be presented in Section IV of this report when we discuss each different type of accidental explosion in detail.

The third approach is an *a priori* approach and involves the calculation of the source energy which is available to the blast wave. Equation (8) is of this type in a sense. However, to date there has been no proof that this is the correct way to evaluate the maximum available yield for an accidental explosion. Kinney (1962) advocates the use of the work function or Helmholtz free energy, A, of the source to determine the equivalent source energy available for scaling purposes. He presents no proof, however, and for at least one case—that of an exploding frangible vessel—his formula does not yield correct far field equivalence. This will be discussed in the next section under the theory of non-ideal explosions. The only other *a priori* equivalency statements concern frangible vessels and are due to Brode (1955), Brinkley (1970), Baker (1973), and Huang and Chou (1968). They will also be discussed in the next section on non-ideal behavior.

D. *Non-ideal Behavior*

1. *Theoretical calculations or estimations*

(a) *Similarity theories.* There have been a relatively large number of attempts to find analytical solutions for the structure of the blast wave produced by different types of energy addition functions. These have all been self-similar solutions. The analytical point source solutions which were discussed above represent the only self-similar solutions which can be generated for the addition of a finite amount of energy. All other solutions which are self-similar, such as the constant velocity piston solution of Taylor (1946) as elaborated on by Kiwan (1970a) or the constant velocity flame solutions of Kuhl *et al.* (1973), Oppenheim *et al.* (1972a, b), Oppenheim (1973) and Strehlow (1975), represent the eventual addition of an infinite amount of energy if the solution is to remain self-similar. The solution of Dabora (1972) and Dabora *et al.* (1973) for a general power law piston motion has the same behavior, i.e., the solution remains self-similar only as long as one continues to add energy in the central region according to the specific power law that was chosen. While the solution of Kuhl *et al.* (1973) or its simplification by Strehlow (1975) can be used to predict maximum shock wave Mach numbers for the early stages of some deflagration explosions, they are not useful for discussing how the blast wave decays at later time because this region of the flow is no longer self-similar. To study such late behavior one must resort to numerical techniques.

(b) *Exploding vessels.* A considerable amount of effort has been expended in the calculation of the blast wave structure from exploding vessels. The numerical results of Brode (1955), Boyer *et al.* (1958), and Huang and Chou (1968) are examples of calculations in which a vessel containing high pressure quiescent gas is assumed to release its contents instantaneously at time = 0, without the interference of confining walls. The development of the blast wave is followed using a one-dimensional, time-dependent numerical technique and the resulting wave behavior, e.g., overpressure, is compared to elementary point source theory. An example of shock pressure versus scaled distance, taken from Huang and Chou (1968) is shown in Fig. 8. Typically the curves for shock pressure start at a pressure, intermediate between the initial chamber

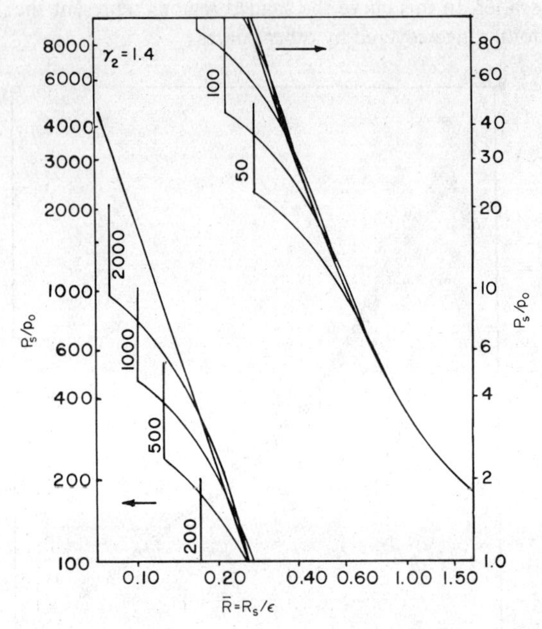

FIG. 8. Shock overpressure versus scaled distance for simple sphere bursts (after Huang and Chou (1968)). The line with no vertical section is point source. This is reproduced just as Huang and Chou presented it and the abscissa should be displaced by a factor of $\sqrt[3]{10}$ to the right as explained in the text. ε is defined in eqn. (9) in the text. R_s is the shock radius. P_s is the shock pressure. Vertical numbers by each line are initial sphere pressure ratios.

pressure and ambient pressure, which can be calculated using a standard one-dimensional flow-patching, shock tube type calculation. The pressure then drops slowly at first and eventually slightly exceeds the point source shock pressure. Then it falls more rapidly with distance than point source—eventually asymptotically approaching the point source solution. Huang and Chou (1968) found this to be true if the energy in the sphere was calculated using the formula

$$\varepsilon^3 = \frac{E_0}{p_0} = \frac{4\pi}{3(\gamma-1)} \frac{(P-p_0)}{p_0} r_0^3, \tag{9}$$

where r_0 is the sphere radius and ε is the characteristic length for point source waves. Thus the characteristic dimensionless radius for plotting is given by the formula

$$\bar{R} = R/(E_0/p_0)^{1/3} = R/\varepsilon. \tag{10}$$

The $(\gamma-1)$ term enters in eqn. (9) because Huang and Chou assumed an ideal constant gamma gas in their model. For these assumptions the $(\gamma-1)$ term essentially converts the pressure energy $(P-p_0)V_0$ held in the initial volume to the potential energy needed to raise the pressure and temperature of the stored gases to the pressure P from an initial pressure p_0 and to the burst temperature T from some low initial temperature T_0'. This means that the potential energy base for the substance held in the sphere should be the temperature T_0'. In other words

$$E_0 = nC_v(T - T_0') = (P - p_0)V_0/(\gamma - 1), \tag{11}$$

where n = the number of moles of substance in the sphere and C_v is the molar heat capacity at constant volume.

It is instructive to compare this potential energy formula to the formulas that have been suggested by Kinney (1962), Baker (1973) and Brinkley (1970). Kinney states that the work function is the available energy.

$$E_0 = A = RT \ln P \quad (\text{i.e. } p_0 = 1), \tag{12}$$

which, for an ideal gas, can be written as

$$\varepsilon^3 = \frac{4\pi}{3}\frac{P}{p_0}\left[\ln\left(\frac{P}{p_0}\right)\right] r_0^3. \tag{13}$$

Baker (1973) and Brinkley (1970) assume that the available energy is that which is released by the isentropic expansion of the gas in the sphere from the pressure P to the pressure p_0. The formula for this energy is

$$\varepsilon^3 = \frac{4\pi}{3(\gamma-1)}\left[\frac{P}{p_0} - \left(\frac{P}{p_0}\right)^{1/\gamma}\right] r_0^3. \tag{14}$$

This is the formula given in Baker (1973) and it differs from that given by Brinkley (1970) in two ways. Brinkley's formula has a factor of two in it because he assumes a surface burst. His equation also has a misplaced bracket which makes it incorrect. It should read (without the 2)

$$\varepsilon^3 = \frac{4\pi}{3(\gamma-1)}\frac{P}{p_0}\left[1 - \left(\frac{p_0}{P}\right)^{(\gamma-1/\gamma)}\right] r_0^3. \tag{15}$$

Equations (9), (13) and (14) [or (15)] have different functional form and all three cannot be correct. They are plotted in Fig. 9. As can be seen from Fig. 8, eqn. (9) agrees well with point source over a sizeable range of P/p_0. Unfortunately, Fig. 8 has an abscissa which is displaced by a factor of $\sqrt[3]{10}$ to the right in terms of the correct position for the blast wave curves. In other words, the λ of Fig. 8 is $\sqrt[3]{10}$ larger than the $\bar{R}$ of Figs. 6 and 7. This can be determined by noting that at $t = +0$ the shock is at the surface of the sphere and $x_s = x_0$. Using eqn. (9) one finds the factor described above. If the curves of Fig. 8 are displaced by this amount they uniformly asymptote the far field point source region of Fig. 7. The important point here is that one can conclude from these works that Kinney's work function is incorrect for calculating the stored energy available to the blast wave and that the potential energy formula and the isentropic expansion formula give ε/r_0 ratios which differ by a factor of 1.5 at the most. This is not sufficient to decide at present which of these two formulas is the correct one (if there is indeed a simple "correct" formula for determining the far field blast wave equivalence of an exploding sphere).

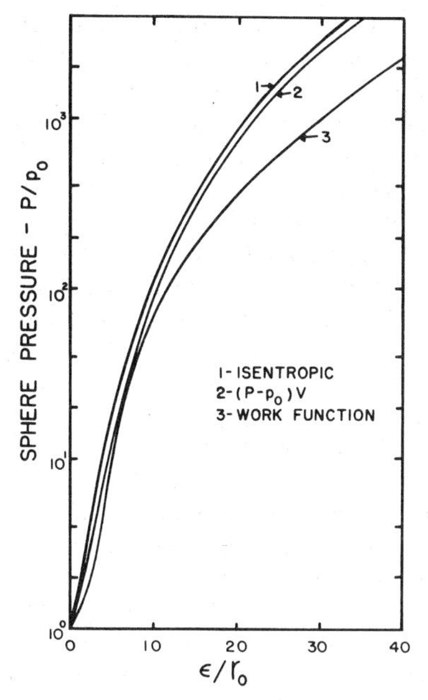

FIG. 9. A comparison of the stored energy in a sphere of high pressure gas as predicted by eqns (9), (13) and either (14) or (15) of the text. From Fig. 8, we see that eqn. (9) agrees well with point source over the range $P/p_0 = 2000$ to $P/p_0 = 50$ for room temperature air in the sphere.

In addition, two questions remain which must be resolved by further theoretical work. In the first place the velocity of sound of the gas in the sphere will determine the maximum shock pressure in the external flow, for fixed internal pressure and stored potential energy. How this change in starting shock pressure will alter the far field equivalence is not known at present. Secondly, the effect of opening time on the far field wave, as in the case of a thick walled frangible vessel, is not understood at present.

(c) *Piston (or flame) driven.* As was mentioned earlier, a constant velocity piston or flame generates a self-similar blast wave and the behavior of such a wave after the piston or the flame stop their motion can only be determined by using numerical integration techniques. Kiwan (1970b) and Guirao *et al.* (1974) have both performed such calculations. Kiwan (1970b) reports only a single calculation and unfortunately makes no comparisons with either point source or other calculations. Guirao *et al.* (1974) have performed such a comparative calculation. They calculated the rate at which a piston performed work on the surrounding atmosphere and then stopped the piston motion when a certain fixed total energy was added to the system. The resulting flow field was then used as the starting flow field for a numerical calculation. They did this for three different piston velocities and compared the resulting shock pressures to point source for the same total energy. They found that the shock pressure is always higher than that of a point source at fixed R but asymptotically approaches the point source shock pressure in the far field. More work is needed in this area to verify these early results and establish generality.

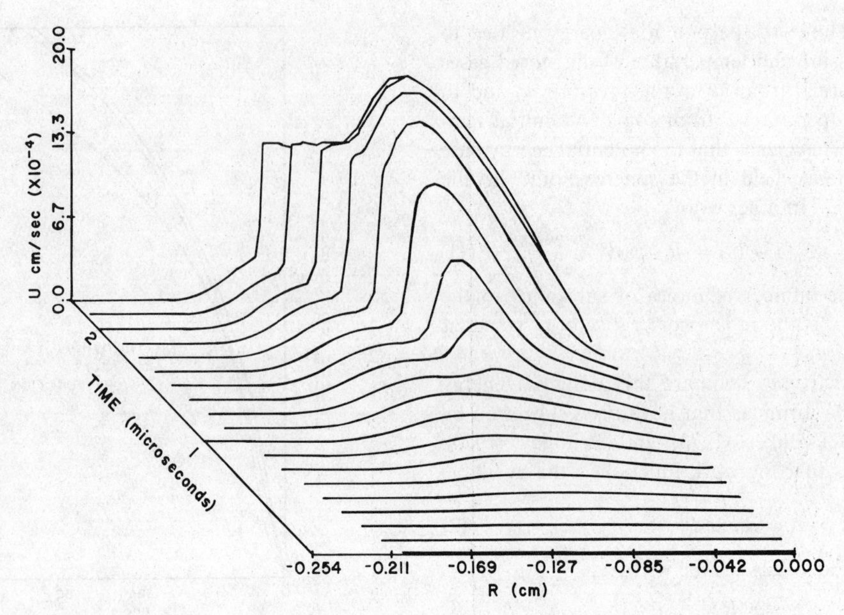

FIG. 10. Velocity distribution versus distance and time for the slow addition of energy in the central region. Spherical coordinates. Notice the imbedded shock which appears later in the flow.

(d) *More general theoretical studies.* There have been very few general studies of the behavior of non-ideal blast waves. Brinkley (1969, 1970) are the only papers which discuss the general behavior of non-ideal blast waves. In both these papers Brinkley discusses the effect of late energy addition by the source. He points out that it is well known that at later times, when the trailing portion of the blast wave contains a negative phase, further energy release by the source will not be able to reach the front and strengthen it. This contention, while interesting, has never been adequately checked.

2. *Source property effects*

(a) *"Shock up" in the near field.* There is some evidence that in combustion driven explosions which are initially unconfined the near field is shock free. Woolfolk (1971), Ablow and Woolfolk (1972), and Woolfolk and Ablow (1973) report that for the deflagrative combustion of hydrogen–nitrogen–oxygen mixtures in hemispherical balloons the near field pressure records were shock free and that the initial shock appearance occurred in the middle of a steepening compression wave. Strehlow and Adamczyk (1974) observed the same type of delayed shock formation when they calculated the blast field produced by a time dependent energy addition function. An example of the flow field associated with such an energy addition function is shown in Fig. 10. Energy addition was relatively slow for the first microsecond and then relatively rapid for the second microsecond in this figure. The weak pressure pulse produced by the slow addition of energy did not have time to coalesce into a shock wave before the shock wave produced by the later, more rapid addition of energy reached it. Zajak and Oppenheim (1971) also found this effect in a calculation of the blast wave produced by the rapid reaction of a "reactive center" placed in an inert surrounding atmosphere.

(b) *Multiple shocks.* Both Boyer *et al.* (1958) and Huang and Chou (1968) have found multiple shock waves propagating away from a bursting sphere in their calculations. A typical example is shown in Fig. 11.

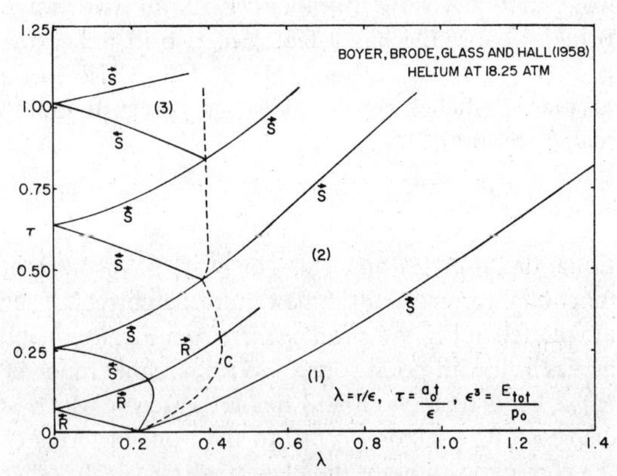

FIG. 11. Multiple shocks obtained from the burst of a sphere of helium.

These results are similar to those of Brode (1959) for TNT explosions. Brode's calculation results are shown in Fig. 12 for comparison. It appears from these results that the presence of multiple shocks is related to the finite size of the source. However, Bethe *et al.* (1947) and Whitham (1950) have shown theoretically that the far field wave in a homogeneous atmosphere should be an "N" wave and therefore should contain two shocks, even for a point source explosion. Boger and Waldman (1973) have shown that for two sequential high-explosive explosions at the same location there exists a critical delay time between the explosions below which the two lead shocks merge. For larger delays, the two shocks are found to exist as separate

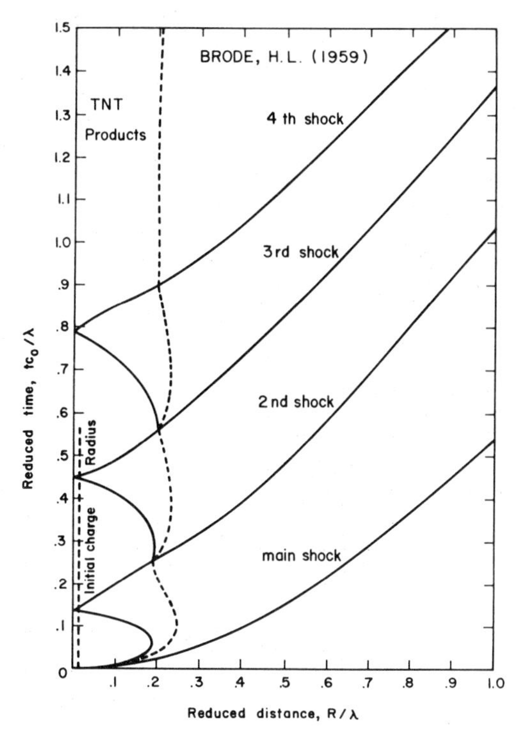

FIG. 12. Multiple shocks obtained from a TNT explosion.

shocks out to the far field region. Multiple shocks also appear when the source is non-spherical; see section (d) below. No more general statements can be made at the present time.

(c) *Variations in pressure profile and decay behavior.* It is well known that the rate of decay of the lead shock is physically related to the pressure profile immediately behind the shock and the radius of the shock. The exact and approximate mathematical relationships have been given by Brinkley and Kirkwood (1947), Bethe *et al.* (1947), and Bach and Lee (1970) to name a few sources. However, no general statements have appeared because the manner in which the profile changes shape and therefore the overall shock decay is determined by the entire flow field, not just by the profile at the shock. The problem is complex and to date only numerical solutions are available.

(d) *Non-spherical behavior.* Any explosion source which is not spherical in free air or hemispherical in contact with a reflecting plane will generate a blast wave which is, at least in its early stages, non-spherical. The wave may well have an axis of symmetry, but requires definition in at least two space coordinates and time.

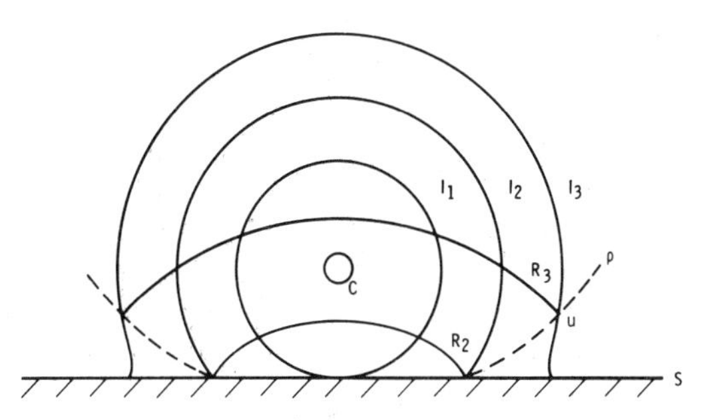

FIG. 13. Reflection of strong shock waves.

Analytically, the treatment of non-spherical waves requires more mathematical complexity, and experimentally, measurement requires many more tests than for spherical waves.

The simplest type of non-spherical behavior probably results from elevation of a spherical explosion source above a reflecting plane (usually the ground). The resulting reflection process is described in Baker (1973) and Glasstone (1962), and is illustrated schematically in Fig. 13. A structure or target on the ground feels a double shock if it is in the region of regular reflection close to the blast source, or a single strengthened shock if it is in the region of Mach reflection. Even this "simplest" case of non-spherical behavior is quite complex.

The second type of asphericity is that caused by sources which are not spherical. Most real blast sources *are* non-spherical, and can be of regular geometry such as cylindrical or block-shaped, or can be quite irregular in shape. Few analyses or experiments have been done for other than cylindrical geometry of solid explosive sources. For cylinders, the wave patterns have been shown (Wisotski and Snyder, 1965; Reisler, 1972) to be quite complex, as shown in Fig. 14. The pressure-time histories exhibit multiple shocks, as shown in Fig. 15, and decay in a quite different manner in the near field than do spherical waves.

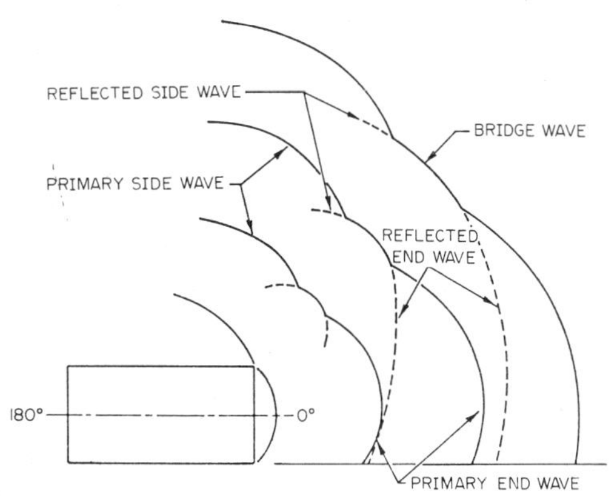

FIG. 14. Schematic of wave development for cylindrical charges (Reisler, 1972).

Another type of non-spherical behavior has been mentioned previously in the section on blast scaling. Gun muzzle blast or recoilless rifle back blast generates waves which consist of essentially single shocks, but shocks with highly directional properties. This type of asphericity is particularly pronounced behind recoilless rifles, where the shock is being driven by supersonic flow of propellant gases expanding through a nozzle (Baker *et al.*, 1971).

The above instances are only a few examples of non-spherical behavior. Let us reiterate that, close to most real blast sources, behavior is *usually* non-spherical. Fortunately, these asymmetries smooth out as the blast wave progresses, and "far enough" from most sources, the wave will become a spherical wave.

(e) *Effect of confinement or partial confinement.* The effects of confining explosion sources on blast waves

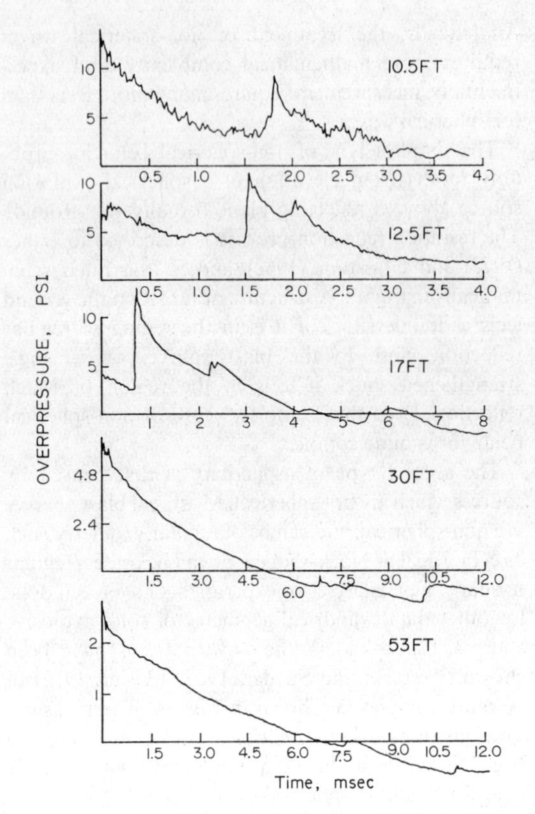

FIG. 15. Pressure time records from cylindrical charges along charge axis (Reisler, 1972).

can range from minimal to controlling, depending on the properties of the source. Nuclear weapons blasts in air are almost totally independent of the confinement provided by the weapon casing, and cased warheads or bombs filled with condensed chemical explosives produce blast waves which are relatively little affected by the confinement of the casing. On the other hand, many materials only act as explosion sources when they are confined in some manner. Some solid and liquid chemicals can act as propellants when confined in vented chambers, and as explosives when confined in unvented chambers. (Black powder is an example.) Liquid cryogenic propellants can generate blast waves when mixed and ignited (Willoughby *et al.*, 1968a–c), but the character and strength of the waves are strong functions of degree of confinement at ignition. Gaseous explosive mixtures produce blast waves which are even more strongly affected by degree of confinement, as will be evident from later discussion in this paper. Finally, the epitome of the effect of confinement is illustrated by blast waves from bursting pressure vessels —no confinement, no blast source.

The design of chambers for confinement and the testing of these designs has proceeded with two purposes in mind. In one case the confining chamber is expected to lessen blast effects in the neighborhood of the chamber or confining configuration, primarily by attenuation (Lesseigne, 1973). The simplest confining method is an overburden of earth and Nicholls *et al.* (1971) have discussed this method of confinement. Confining structures have also been designed and experimental measurements on a simple vent structure have been described by Keenan and Tancreto (1972). Discussions of the effect of internal explosions on internal pressures (Kennedy, 1946) and venting have also been presented by Sewell and Kinney (1968) and Proctor and Filler (1972). Baker and Westine (1974) and Westine and Baker (1974) have recently presented detailed discussions of how to design suppressive structures which limit the blast loading outside the structure, and Cox and Esparza (1974) present a design which is specific to a melt loading operation for high explosives. Because structures of the type discussed above are intended to strongly suppress external blast waves, the vent area ratios, usually expressed in the dimensionless form

$$\bar{A}_v = \left(\frac{A_{\text{vent}}}{V^{2/3}}\right)$$

where A_{vent} is total vent area and V is internal volume, are small, i.e., $\bar{A}_v < 0.05$. For such small venting, peak gas pressures developed within the structure are independent of vent area ratios and are entirely a function of ratio of explosive energy to volume, W/V (see Proctor and Filler (1972), and Baker and Westine (1974)).

In the other case, the problem of confinement is one of releasing the blast energy rapidly so that the confining structure (the building itself) is not damaged. Early British efforts on explosion venting are summarized in the Ministry of Labour (1965) report on explosion relief and flame arrestors. Recently in the U.S. Rune (1972) presented an elementary treatment of explosive venting of buildings which was criticized by Howard (1972). Rune's treatment is much more rudimentary than that of Proctor and Filler (1972), but it does account in an approximate way for shock-free internal pressure rises of relatively long rise times. Generally, for the very rapid venting desired to save the building, vent area ratios must be large, say $\bar{A}_v > 0.2$, and maximum internal gas pressures will be a strong function of this ratio as well as W/V. Hazards of internal blasts have recently been summarized by Mainstone (1973) and a recent review of the problem has been presented by Butlin (1975).

3. *Atmospheric and ground effects*

Ideal explosions are assumed to occur in a still, homogeneous atmosphere and to be unaffected by the presence of a ground surface. Real conditions in the atmosphere and real surface effects can modify the wave in various ways.

Variations in initial ambient temperature and pressure can affect the blast wave so that noticeably different waves would be recorded from explosions on a high mountain or mesa than from explosions near sea level, or from explosions occurring on a hot summer day versus a cold winter day. These effects are, however, quite adequately accounted for if the Sachs' scaling law described earlier is used to predict the wave properties. For very large explosions such as detonations of multi-megaton nuclear weapons, the vertical inhomogeneity of the atmosphere will cause modification of an initially spherical shock front (Lutzsky and Lehto,

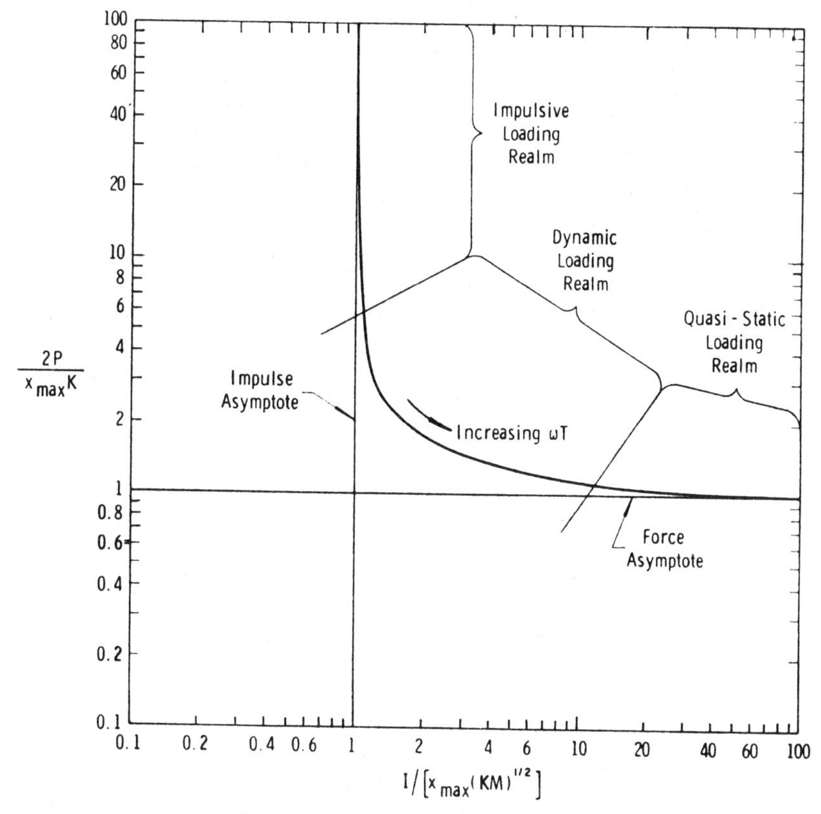

FIG. 16. Scaled P-I curve for response.

1968). Changes in relative humidity and even heavy fog or rain have been found to have insignificant effects on blast waves (Ingard, 1953).

The more significant atmospheric effects which induce non-ideal blast wave behavior are unusual weather conditions which can cause blast focusing at some distance from the source. A low-level temperature inversion can cause an initially hemispherical blast front to refract and focus on the ground in an annular region about the source (Grant *et al.*, 1967). Severe wind shear can cause focusing in the downwind direction. This effect is discussed by Baker (1973) and Reed (1973). Structural damage from accidental explosions has been correlated with these atmospheric inhomogeneities (Siskind, 1973; Siskind and Summers, 1974; Reed, 1968), and complaints of damage from explosive testing were reduced when firings were limited to days when no focusing was predicted (Perkins *et al.*, 1960). A handbook on how to perform such calculations is available (Perkins and Jackson, 1964).

Ground effects can also be important. If the ground acted as a perfectly smooth, rigid plane when explosions occurred on its surface, then it would reflect all energy at the ground plane and its only effect on the blast wave would be to double the apparent energy driving the wave. In actuality, surface bursts of energetic blast sources usually dissipate some energy in ground cratering and in ground shock, so that only partial reflection and shock strengthening occurs. A good "rule of thumb" is to multiply the effective charge energy by a factor of 1.5 to 1.8 if significant cratering occurs. For sources of low energy density such as gaseous mixtures, very little energy enters the ground, and the reflective factor of 2 is a good approximation.

A ground surface which is irregular can significantly affect the blast wave properties. Gentle upward slopes can cause enhancement, while steep upward slopes will cause formation of Mach waves and consequent strong enhancement. Downward slopes or back surfaces of crests cause expansion and weakening of shocks. These effects are usually quite localized, however, and "smooth out" quite rapidly behind the irregularities. Even deliberate obstructions such as mounded or revetted barricades produce only local effects (Wenzel and Bessey, 1969).

We have noted previously under the heading "non-spherical behavior" that blast sources located above a reflecting plane can generate Mach waves if the shocks are strong enough. The phenomenon of generation and propagation of these waves has been widely studied in blast technology (Glasstone, 1962; Baker, 1973), and will not be discussed further here, other than to note that the blast wave in a Mach stem is classical in form but differs markedly in strength from the wave from a free-air source.

III. DAMAGE MECHANISMS

A. *The P-I Relation*

1. *Simple systems*

The blast waves from accidental explosions can cause damage to structures, property or individuals by subjecting them to transient crushing pressures and transient winds which cause drag pressures. Even though the interaction of the waves with the objects they damage involves very complex phenomena, a relatively simple concept has been utilized quite effectively to correlate blast wave properties with damage to a wide variety of "targets". The concept is that damage caused by blast waves (or any transient force-time history) to a given object is primarily a function of the peak overpressure or force (P) and the applied impulse (I). Therefore, for any object, curves of constant damage level can be plotted on a P-I diagram, or empirical or analytical equations developed to describe a P-I relation. An example is shown in Fig. 16.

To illustrate this concept, we will first consider a simple system, characterized by a mass (inertia) and a linear spring (resisting force). The development of the P-I diagram for this system is given in Baker *et al.*

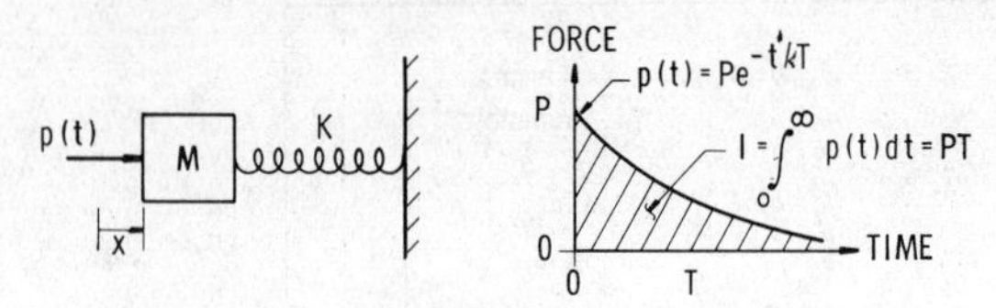

FIG. 17. Schematic of spring-mass system under time-varying force.

(1973), and will be paraphrased here. Figure 17 shows schematically the linear spring-mass system to which is applied a specific time-varying force $p(t)$. The equation of motion can be obtained by considering a free-body diagram for the mass. Summing the forces in the x-direction gives

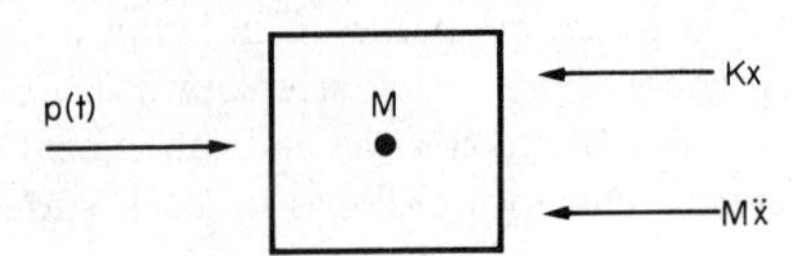

FIG. 18. Summation of forces.

Inertial force / Spring force / External force

$$M\ddot{x}+Kx = p(t) = P\,e^{-t/T}. \tag{16†}$$

The initial conditions describing the dynamic state of the mass at zero time must be written. For the mass initially at rest, these are:

$$x(0) = 0, \tag{17}$$

$$\dot{x}(0) = 0. \tag{18}$$

Referring to any standard text on mechanical vibrations, we can find that the solution to eqns (16) through (18) is

$$x(t) = \frac{1}{\omega M}\int_0^t p(\tau)\sin\omega(t-\tau)\,d\tau, \tag{19}$$

where

$$\omega = \left(\frac{K}{M}\right)^{1/2} \tag{20}$$

†Dot denotes differentiation with respect to time, i.e., $\dot{x} = dx/dt$, $\ddot{x} = d^2x/dt^2$.

is the "natural" frequency of the system. Or, for the specific form of $p(t)$ which we have assumed,

$$x(t) = \frac{PT^2}{M(1+\omega^2T^2)}\left(\frac{\sin\omega t}{\omega T} - \cos\omega t + e^{-t/T}\right). \tag{21}$$

By simple manipulation and use of eqn. (20), we can render eqn. (21) dimensionless, as follows:

$$\frac{x(t)}{(P/K)} = \frac{(\omega T)^2}{[1+(\omega T)^2]}\left[\frac{\sin\omega t}{\omega T} - \cos\omega t + e^{-(\omega t)/(\omega T)}\right]. \tag{22}$$

The left-hand side is a ratio of transient displacement to static deflection of the spring under unit load, ωt is dimensionless time, and ωT is a ratio of a characteristic response time ω^{-1} to a characteristic loading time T.

Response is characterized by the maximum displacement x_{max} of the single-degree-of-freedom system. We can operate on eqn. (22) to determine the *time for maximum displacement*. This time is obtained by differentiating eqn. (22) with respect to $\omega t = \bar{t}$ for specific values of ωT, setting it equal to zero, and solving by trial-and-error. The resulting transcendental equation for $\bar{t}_{max}$ is

$$\frac{\cos\bar{t}_{max}}{\omega T} + \sin\bar{t}_{max} - \frac{\exp[-\bar{t}_{max}/\omega T]}{\omega T} = 0. \tag{23}$$

The maximum displacement is then obtained by substitution of $\bar{t}_{max}$ in eqn. (22).

TABLE 3. Maximum response of single-degree-of-freedom system to exponential force

ωT	$\bar{t}_{max}$	$\bar{x}_{max}$
0.01	1.580 ≈ π/2	0.01
0.1	1.670	0.1
1.0	2.283	0.754
10.0	2.969	1.728
100	3.122	1.969
∞	π	2

The process just described yields the results shown in Table 3 and Fig. 19. An excellent empirical fit can be made to the curve of $\bar{x}_{max}$ as a function of ωT, as can be seen in Fig. 19, over the entire range by the formula

$$\bar{x}_{max} = \left[2-\exp\left(-\frac{\omega^2T^2}{100}\right)\right]\tanh\omega T. \tag{24}$$

The asymptotic values for x_{max} at large and small ωT give, in dimensional form,

$$x_{max} = \frac{PT}{(KM)^{1/2}}\left(\frac{K}{M}\right)^{1/2} \quad \text{for} \quad T < 0.2, \tag{25}$$

$$x_{max} = \frac{2P}{K}\left(\frac{K}{M}\right)^{1/2} \quad \text{for} \quad T > 100. \tag{26}$$

In eqn. (26) the product PT is exactly the *integral under the force-time curve*, which we call the impulse, I. It can be rewritten

$$x_{max} = \frac{I}{(KM)^{1/2}}\left(\frac{K}{M}\right)^{1/2} \quad T < 0.2. \tag{27}$$

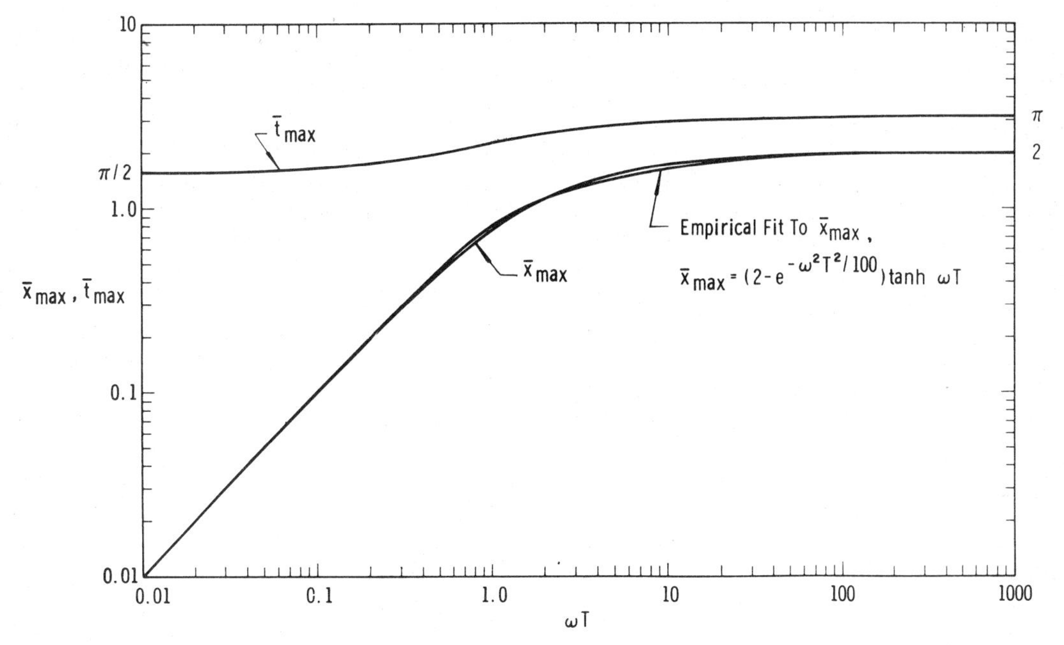

FIG. 19. Maximum response to force pulse. The characteristics of the response are as follows: (1) For small ωT, $\bar{x}_{max} = T$. (2) For large ωT, $\bar{x}_{max} = 2$. (3) For intermediate ωT, $\bar{x}_{max}$ is a more complex function of ωT. (4) Scaled time for maximum response changes relatively slowly from $\bar{t}_{max} \approx \pi/2$ to $\bar{t}_{max} \approx \pi$ as ωT increases.

For rapidly-decaying force (ωT small), the response is *proportional to impulse* I, while, for slowly decaying force, the response is *proportional to peak force* P. In this latter case, the response is just twice that for static application of the force P, i.e., we have a dynamic load factor of two. Possible dimensionless forms for peak force and impulse are

$$\bar{P} = \frac{2P}{x_{max}K}, \tag{28}$$

$$\bar{I} = \frac{I}{x_{max}(KM)^{1/2}}. \tag{29}$$

If the empirical fit of eqn. (24) is used, these become

$$\bar{P} = \frac{2}{[2-\exp(-\omega^2T^2/100)]\tanh \omega T}, \tag{30}$$

$$\bar{I} = \frac{\omega T}{[2-\exp(-\omega^2T^2/100)]\tanh \omega T}. \tag{31}$$

From these last two equations, one can finally generate a scaled response curve, or P-I curve by varying ωT. This has already been shown in Fig. 17. The curve represents the *combinations* of scaled force and scaled impulse which cause the *same* scaled response $\bar{x}_{max}$ of the system. It is then an *isoresponse* curve. A similar curve for a system undergoing a given level of damage would be an *isodamage* curve. We can divide the curve into the three indicated regions. In the impulsive loading realm, impulse alone correlates with response. In the quasi-static loading realm, *peak force alone* correlates with response. In the intermediate dynamic loading realm, both the impulse and the force must be known, i.e., we must know the entire time history of the loading. For the simple system just described, a good fit to the P-I relation is

$$(\bar{P}-1)(\bar{I}-1) = C. \tag{32}$$

This equation described a rectangular hyperbola in the $\bar{P}$-$\bar{I}$ plane, with asymptotes (1, 1).

Arguments similar to those just presented can be given for the simplest kind of a dynamic permanently deforming system, i.e., a rigid-plastic system with inertia. This is done by Baker *et al.* (1973), who show that a scaled $\bar{P}$, $\bar{I}$ curve applies for this system also. The scaled forces and impulses are defined differently for plastically-deforming systems, but the same concept holds.

2. *Complex systems*

The P-I curve for describing a given level of damage to a system has also been shown experimentally and analytically to apply for a wide variety of blast-loaded systems. For high explosive blast sources, given combinations of P and I are unique functions of standoff R and charge energy W (see Section II). Westine (1972) has shown that damage to a number of complex targets such as trucks, houses, and aircraft can be presented on an R-W plane, and that such a presentation is equivalent to presentation in the P-I plane. Sewell and Kinney (1968) have also presented a method which is a modified form of the P-I concept.

In a number of instances, the behavior of complex systems under blast loading is too complex to be described by a single hyperbolic P-I diagram. As the combinations of P and I change for such systems, the mechanisms of damage also change. Two examples for widely different "systems" follow.

Thin cylindrical shells subjected to external blast loading from the side will be damaged by plastic buckling. Depending on the duration of loading, two basically different types of buckling failure occur. In one, the shell exhibits longitudinal wrinkles or lobes; in the other, the shell collapses by creasing in the middle

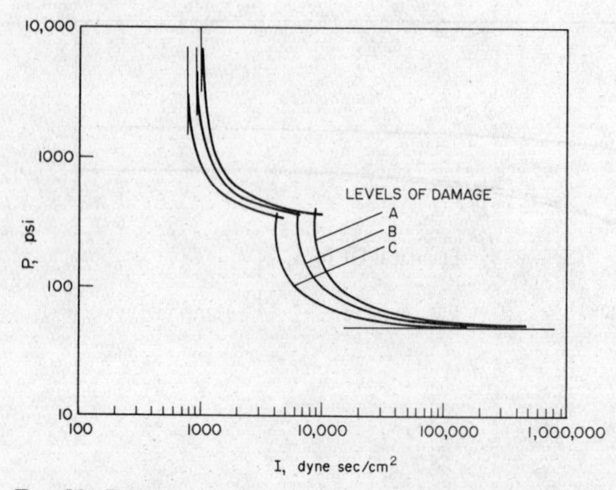

FIG. 20. Pressure-impulse relations for blast-loaded cylindrical shells (Lindberg *et al.*, 1965).

of its length. Figure 20 shows a P-I diagram for this dual behavior. This figure, from Lindberg *et al.* (1965) also shows different curves for different levels of damage.

The second example of dual damage mechanisms is the threshold response of humans to blast waves. For relatively short-duration waves, the governing criterion is threshold of eardrum damage. For long-duration waves, a standing individual is knocked down. The resulting P-I diagram, from Custard *et al.* (1970) is shown in Fig. 21. This figure also indicates that the impulse asymptote for eardrum damage is very low, because the ear has a very short characteristic response time, i.e., responds to quite high frequencies.

A number of analysis methods have been developed to compute response of and damage to a variety of complex structures. These represent a higher degree of sophistication than the P-I concept, but also usually require a large expenditure of manhours and computer time to yield answers. Typical of these methods are Norris *et al.* (1959), Baker *et al.* (1969), Leigh (1974), and Crocker and Hudson (1969).

B. *Dynamic Impulse*

The dynamic impulse in blast waves has been defined earlier, and a typical curve given for net transverse pressure applied to an object immersed in a blast wave. For certain systems or objects, the initial diffraction phase of blast loading is unimportant, and the time-history of drag force controls. An object resting on the ground can be accelerated by this loading, and slide or overturn. (An example has already been given in discussing the P-I diagram for humans.) If the object is massive, it will respond slowly to the drag forces, and drag impulse, multiplied by a drag coefficient dependent upon the shape of the object, will determine incipient overturning or sliding. Response of light bodies will depend on the entire history of drag force. An example of a "target" in this latter category is a camper-pickup, which has a lot of side area and a high center of gravity. Custard *et al.* (1970) show a P-I diagram for such a vehicle, with overturning being the critical mode of damage (Fig. 22).

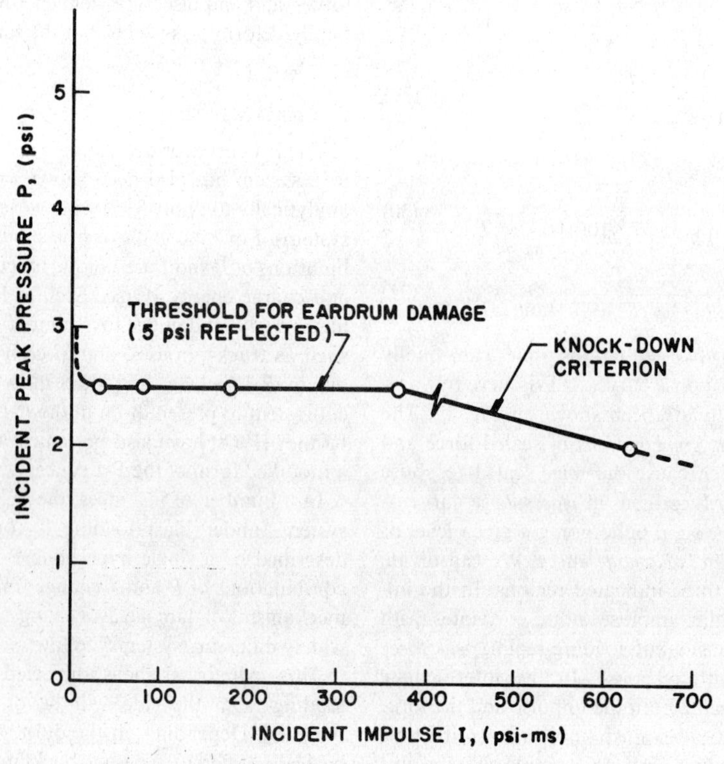

FIG. 21. Acceptable incident peak pressure-impulse relationship for a 168-pound man exposed to explosive blast (Custard *et al.*, 1970).

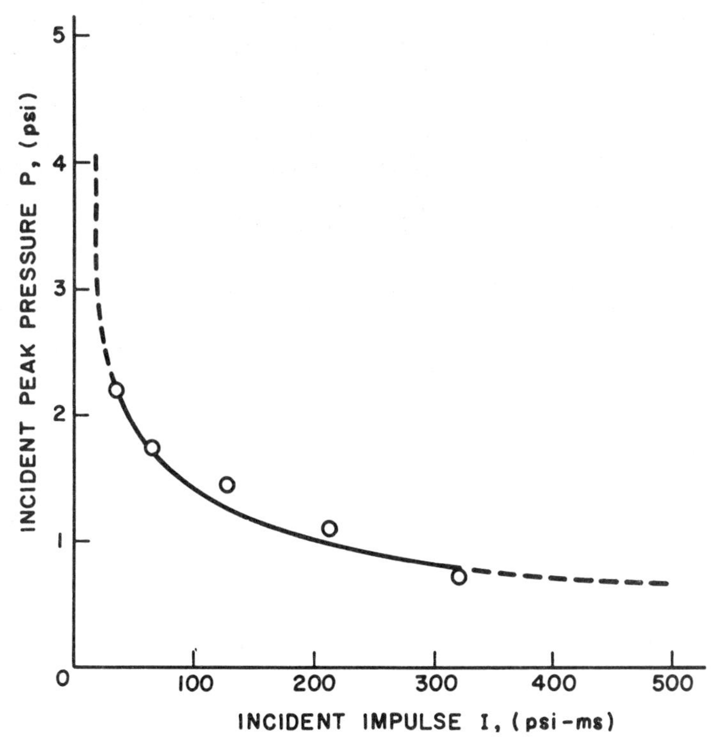

FIG. 22. Acceptable incident peak pressure-impulse relationship for constant damage (80% of overturning impulse) to the camper-pickup (Custard *et al.*, 1970).

C. *Fragments, Primary and Secondary*

An important factor in damage from accidental explosions can be the fragmentation of the container or fracture and acceleration to high velocity of nearby objects or parts of structures. Fragments from containers are usually called *primary fragments*, and those resulting from fracture and acceleration of nearby objects are called *secondary fragments*.

The state of knowledge and our ability to apply it to damage predictions for accidental explosions is much less satisfactory for fragmentation effects than for blast effects. Some of the reasons for this are:

Fragmentation is inherently statistical in nature,
Primary fragmentation is very dependent on details of the explosion process,
Effects of fragments on important "targets" such as humans are highly classified.

Some unclassified studies have been conducted by Feinstein (1972) for fragments from bursting piles of bombs; by Baker *et al.* (1974a, b), fragments from bursting liquid propellant vessels; by Pittman (1972a, b), and Taylor and Price (1971) for fragments from bursting high pressure tanks; and by Siewert (1972) for the fragments from ductile tank cars which contain volatile chemicals and fail by pressure burst in fires. Rather than attempt to completely cover this difficult field in this paper, we merely mention the above related papers and note that much further study seems to be required before adequate predictions can be made of fragmentation effects.

D. *TNT Equivalence Evaluation Based on Blast Damage*

Even though the relatively sophisticated P-I and related techniques that have been developed recently have great generality and give good correlation for blast damage there is still a tendency to evaluate blast damage from accidental explosions by using simpler techniques. One example of a very thorough evaluation from specific indicator evidence is the recent open publication of a paper on the yields from the Hiroshima and Nagasaki bombs by Penny *et al.* (1970). Because the blasts from these bombs were of long duration, the structures whose damages were used to predict yield all fall within the pressure asymptote of their P-I diagrams.

It is usual, however, to relate blast damage patterns directly to overpressure or to scaled distance based on overpressure, thus neglecting positive impulse. Brasie and Simpson (1968) present a typical graph, reproduced here as Fig. 23, listing overpressure effects. Notice in particular the large range for glass breakage from about 0.1 to 0.006 psi side-on overpressure. Reed (1968) in an evaluation of glass breakage from a munitions explosion near San Antonio, Texas, plots his data on a breakage probability versus log overpressure scale and shows, as one would expect, that large plate glass panes, thin glass panes and stressed glass panes are most vulnerable to breakage. Usually if one is using glass breakage, one determines the distance to the location of 50% breakage and uses 0.1 psi as the pressure level. This yields a scaled range

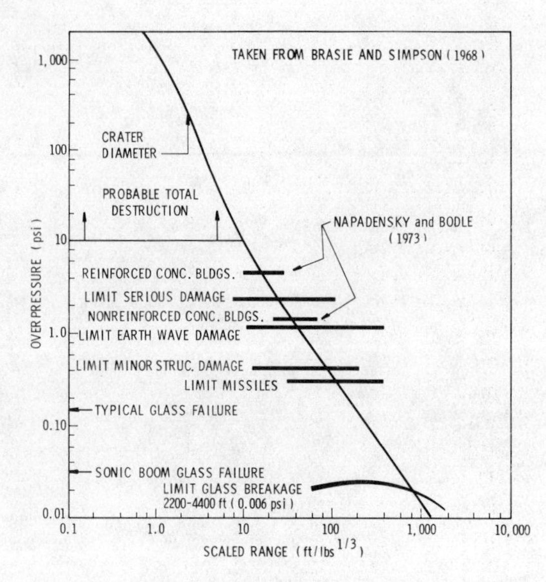

FIG. 23. Overpressure scaled distance plot showing typical levels for blast damage.

of about 100–200 ft/(lb TNT)$^{1/3}$ so that TNT equivalent weight = (ft/200)^3 where ft is the distance in feet from the explosion to the location of 50% glass breakage. The same approach is used in estimating distances for other types of breakage. Errors using this method will be greatest when the actual explosion yield is small, and when the accidental explosion is markedly non-ideal.

E. *Non-Ideal Effects*

From our previous discussion of P-I relations, it seems apparent that predictions of blast damage from accidental explosions are possible, if the blast wave characteristics are known as a function of distance from the explosion. But, non-ideal effects discussed earlier in this paper can render prediction of these characteristics somewhat uncertain. Similarly, the relatively small amount of information on fragmentation effects indicates that these effects can only be accurately predicted for relatively ideal explosions such as explosions of cased munitions. If one uses TNT equivalency concepts, prediction or correlation of damage effects with real non-ideal explosions may introduce large and unknown errors, given the present state of knowledge.

IV. SPECIFIC EXAMPLES OF ACCIDENTAL EXPLOSIONS

In this section we wish to discuss either experimental or theoretical results or simply observations that have been made about each of the specific accidental explosion types listed in Table 1 on page 28. The discussion will range from rather precise to quite vague because of our current understanding of the mechanisms by which these various sources produce non-ideal blast waves. Furthermore, the relative importance or potential hazard of each different type will determine the length of each discussion.

Before we turn to individual explosion types we will mention two compilations of case histories of accidents which involved explosions. Doyle (1969) has reviewed eighty-three incidents involving explosions in chemical plants. He found that approximately 50% could be traced to combustion reactions—primarily due to leakage of combustibles from a vessel into a building, with a few due to ignition of the combustible material in the vessel itself. He also found that approximately 40% were due to a runaway chemical reaction in a reactor and that 10% could be labeled as metal failure explosions under otherwise normal operating conditions. Ordin (1974) has compiled information on over 200 accidents involving hydrogen. However, he does not concern himself directly with the occurrence or absence of an explosion or its type; he is more interested in cause. Most of the cases he discusses involve a release during handling.

We now turn our attention to specific cases.

A. *Simple Pressure Vessel Failure*

1. *Frangible vessels*

The blast wave produced by the rupture of a frangible spherical vessel is by far the most reproducible of all possible accidental explosions. Also it has been studied in more detail than the others. The theoretical work of Huang and Chou (1968) and of Boyer *et al.* (1958) has already been discussed in some detail in section II.D.1.b., including the problem of how to define the source energy for such a vessel explosion. The arguments of that section show that it is probably best to use the total stored pressure energy, $(P-P_0)V_0/(\gamma-1)$, if one wants to compare far field behavior to point source wave behavior.

At the present time theoretical work still remains to be done. In particular there has been no systematic study of the effect of the velocity of sound and heat capacity ratio of the gas continued in the vessel on the near field shock produced when it bursts. Since the internal velocity of sound dictates the maximum shock velocity at the time of burst through contact surface balance requirements, and since the velocity of sound can be varied in a manner which is independent of the energy contained in the vessel, it must represent an important additional variable in the determination of near field effects.

Recently there has been some experimental work by Pittman (1972a, b) to complement that of Boyer *et al.* (1958). In this work he considers not only the blast wave but also fragmentation patterns from a number of different vessel explosions. He makes comparisons of blast wave overpressure and positive impulse to point source values, but he uses the isentropic relationship given by eqns (14) or (15) of this text for this comparison. Conversion to the recommended energy relation would move his experimental data points closer to the theoretical point source curves. Baker *et al.* (1974a, b) have presented calculations of fragment acceleration behavior and review other work in the field on this subject.

There have been a number of accidental explosions which undoubtedly could be represented as simple frangible vessel bursts. A few examples will be discussed. Stephens and Livingston (1973) report on a frangible rupture disk burst due to an exothermic H_2+Cl_2 reaction. Evidence was that there was no contribution to the blast from subsequent reaction. It is true of course that the rupture of a relief disk is strikingly non-ideal because of the explosion's directionality. In this respect they are somewhat like the muzzle blast from a gun or the back blast from a recoilless rifle. Munday (1973) is currently working on this problem and there is a publication by the Ministry of Labour (1965) which discusses the design of flame arrestors and explosion relief devices.

Another example is the explosion that occurred on *Apollo* 13, endangering the lives of the astronauts (Anon, 1970). In this case an oxygen storage vessel for a fuel cell burst because of overpressure due to an internal fire of electrical insulation and blast and fragment damage to neighboring equipment was extensive. Fortunately, the ground crew and astronauts were able to respond successfully to the crisis and a safe return to earth was effected. This incident dramatically points out the need for safety and hazard evaluation based on the best available information as well as the need to improve our understanding of the near field, non-ideal behavior of blast waves.

Two other examples are the explosion of a liquid oxygen truck (National Transportation Safety Board, 1971), and the explosion of a filter containing chlorine and organics (Statesir, 1973). In both cases the evidence led to the conclusion that a simple pressure vessel burst was involved.

2. *Ductile vessels*

There are few examples of ductile failure where subsequent combustion of the products is not involved. Freese (1973) reports one such example of a thin-walled vessel with ductile failure and no subsequent combustion. Ductile failure followed by combustion of the products will be discussed later, in section IV.E.

B. *Runaway Chemical Reaction or Continued Combustion*

A runaway chemical reaction or continued combustion explosion is in some sense similar to the bursting vessel explosion. However in this case there is the possibility that heat addition due to continued reaction or to flame propagation after the vessel bursts may alter the properties of the non-ideal blast wave that is produced by the burst. Andersen and Louie (1975) have performed a very simplified one-dimensional (i.e., planar) blast wave calculation for the case where the total amount of energy Q is kept constant but is added in different ways. Firstly, they assumed some fraction of Q trapped as pressure energy in a pressure vessel which was assumed to burst at time $t = 0$. Then they added the remainder of the energy at a constant rate, homogeneously, to all elements of fluid originally contained in the vessel. This represents the continuing chemical reaction or flame propagation. They studied the planar blast wave which is produced as a function of both the fraction of the energy which is added instantaneously and the time required to add the remainder of the energy. Interestingly enough, they found stronger blast waves in the near field when about half of the energy was added over a relatively short period of time after the initial burst. These results are quite intriguing and point to the need for a systematic study of the effects of adding energy over a finite time period in spherical geometry.

Three examples of accidental explosions involving runaway chemical reactions in a pressure vessel are described by Angiullo (1975), Dartnell and Ventrone (1971), and Vincent (1971). Nickerson (1975) discusses a case which involved afterburning in a dryer explosion. In this case the dryer ductwork released the explosive mixture at relatively low pressure and the dryer was not damaged significantly. However, rapid afterburning produced significant blast damage to the building and to a neighboring building.

C. *Explosions in Buildings*

Explosions in buildings are of three main types. In the first type there is a spill of some combustible material and a slow deflagration wave or "flash back" fire which causes a relatively slow buildup of pressure in the building. In the second type, a piece of equipment explodes, thus producing a blast wave inside the building which damages the structure and/or is relieved by venting. In the third case a leak occurs but the combustible mixture that forms detonates. Severity of damage increases from case 1 to 3. In case 1 or 2 explosive relief or vent design can save the building, as was discussed in Section II.D. Case 3 will be discussed extensively in the following section. For case 3, relief or venting is, in general, not very useful.

D. *Internal Explosions*

These can be very dangerous. In this case the contents of the pressure vessel, reactor, distillation column, building, car or whatever detonate. It is important to realize that these explosions are uniquely different than those discussed in sections A, B and C above. In those cases the degree of confinement or "bursting pressure" of the vessel or building, etc. determined the nature of the blast wave which is generated and the damage patterns. However in the case of detonative combustion or reaction, the blast wave behavior and the damage patterns are primarily determined by the behavior of the detonation and are only *modified* by the confinement.

It appears that very little useful research can be done on these explosions. The major question here is the sensitivity of the exothermic substance or mixture to transition to detonation under confined conditions. Once the transition occurs damage levels are high and have usually been found to correlate well with detonation overpressures. There is the possibility that the

FIG. 24. Damage from acetylene–air explosion in a car. Leaking tank in trunk. Car parked in sun about 1 h. Ignition source unknown. From Baker (1974).

FIG. 25. Methane air detonation in 3 elevator shafts, central portion, left side of building. New York, NY. Shepard (1975).

P-I technique discussed in section III.A. may be more generally useful for this type of explosion. It appears, though, that whichever technique is used, point source approximations are probably adequate up to overpressure levels which yield light structural damage. Relative to heavy structural damage, the vessel or building, etc., that could not "contain" the explosion is usually extensively damaged and causes major structural damage to nearby equipment, vessels and/or buildings both by fragment and blast. Normally if the explosive material is gaseous (e.g., has low density) cratering does not occur. However, if it has a high density because it is solid or liquid, cratering does occur.

There is one very interesting report by Burgess *et al.* (1968) which covers this subject very well. It describes the results of numerous experiments and presents guidelines for evaluating such explosions using the point source-overpressure-scaled distance technique discussed in section III.D. Also the nature of the process of acceleration to detonation in pipes and overpressures connected with these detonations is discussed by Craven and Grieg (1968). Howard (1975) discusses the testing of flame arrestors in pipes to stop a propagating detonation in a hydrogen–air mixture. Explosion protection for processing vessels have been discussed by Charney (1967, 1969), and by Peterson and Cutler (1973).

Examples of case histories in the literature of incidents which involved detonations are numerous: Smith (1959), oil in a high pressure air line; Jarvis (1971a, b), and Freeman and McCready (1971a, b), distillation tower containing vinyl acetylene; Zabatakis (1960), air in dephlegmator; Brasie and Simpson (1968), buildings (3 incidents discussed); Baker (1974), acetylene in a car (see Fig. 24); Shepard (1975), methane in an elevator shaft (27 story building, 3 shafts involved, see Fig. 25); National Transportation Safety Board (1972c), dynamite in a truck; and Wilse (1974) and Halverson (1975), "empty" super tankers during cleaning or partially full tankers during off loading, to name a few.

FIG. 26. Fireball from the rupture of one tank car originally containing 33,000 gallons (120 m^3) of LPG. Crescent City, Illinois, June 21, 1970. Notice water tower on left and train on the right side for scale. The size of the fireball agrees well with the prediction of High (1968). Champaign Fire Department (1972).

E. *Rupture Followed by Combustion*

This very special type of explosion occurs primarily when a tank of liquefied fuel, under pressure, is heated by an external fire following an accident, until it vents and torches. For an explosion to occur the subsequent heating of the venting tank must be sufficiently intense to cause the internal pressure to rise above the tank's bursting pressure, even with venting. This type of explosion produces three distinct damage producing effects. These are (1) a blast wave due to internal pressure relief, (2) a fireball due to subsequent massive burning of the contents of the tank in the air, and (3) large fragments scattered for large distances due to the ductile nature of the tank's rupture and the rocketing of pieces by reaction forces.

The blast from such explosions is usually minor because vessel bursting pressures are in the 200–400 psi range and only a portion of the vessel contains high pressure gas. Estimates of the blast can be made using simple pressure burst formulas if one knows the fraction of the vessel's contents that are in the gas phase and the burst pressure. One can assume that the energy is equal to the pressure burst of a vessel equal to the size of the vapor space plus some contribution from flash evaporation of at least a portion of the liquid phase in the vessel. Flash evaporation is rapid, except as modified by the inertia of the liquid, and contributes to the blast wave. There is no extant work on this aspect of the explosion process and it is doubtful if any will be performed because the blast produced by these explosions is the least damaging of the three effects.

Fireball damage can be severe, particularly if the release of material is large. High (1968) has documented the size and duration of fireballs from a large variety of explosions and finds that they can be predicted quite well with the equations

$$D = 3.86\, W^{0.320},$$

for size and

$$T = 0.299\, W^{0.320},$$

for duration. Here D is diameter in meters, W is weight of combustible in kg and T is duration of fireball in seconds. The exponent should properly be 1/3, not 0.320; the 0.320 value was obtained from least squares fit to the data. Figure 26 is an example of such a fireball taken by the Champaign Fire Department (1972). The size of this ball agrees well with the correlation of High (1968).

Large fireballs radiate energy at levels which are sufficient to cause severe flash burns to exposed skin and ignite cellulosic materials over a large area. Also, depending on the circumstances, the fireball may entrain firebrands which can ignite multiple fires at a later time.

Fragments from this type of incident can travel large distances. Baker *et al.* (1974a, b) have discussed fragmentation patterns for explosions of liquid propellant vessels, and Siewert (1972) has collected fragment distribution data for eighty-four tank car explosions. He finds that the data for terminal position correlate well on a cumulative probability versus logarithmic radius plot and recommends a safe evacuation radius of 2000 feet (610 m) for all cases where a tank car containing a liquid combustible is being heated by an external fire. Only 5% of the fragments travel beyond this distance.

Recent incidents involving this type of explosion include: Crescent City, Illinois (National Transportation Safety Board, 1972b); New Jersey Turnpike Exit 8 (National Transportation Safety Board, 1973c); Houston, Texas (National Transportation Safety Board, 1972d); and Oneonta, New York (National Transportation Safety Board, 1974). There has recently been some research in this area relative to developing techniques to protect tank cars from fire by applying an insulating coating (Phillips, 1975).

F. *Vapor Cloud Explosions*

Unconfined vapor cloud explosions have been occurring for as long as man has handled large quantities of combustible liquids with high vapor pressure. The usual sequence of events is (1) a massive release of a combustible fuel, (2) a reasonable delay to ignition, of the order of 30 s to 30 min, and (3) ignition of the cloud to detonation. Strehlow (1973b) reviewed the state of the art relative to our understanding of these explosions 2 years ago and showed that their frequency and magnitude has increased markedly in the past 10 years. Most of his references will not be repeated here. In addition, Coevert *et al.* (1974) have also discussed their behavior in general terms.

There is currently a great deal of interest in these explosions, primarily because of their frequency of occurrence and the damage produced by recent incidents in which damaging blast waves were produced. While this list is not exhaustive it does represent typical vapor cloud explosion behavior. Table 5 lists five recent incidents involving ignition without explosion. In these cases there were also massive releases and delays to ignition without the production of damaging blast waves. All that occurred for these later cases was a flash of fire back to the leak site followed by torching of the leak. It is interesting to note that one of the incidents listed in Table 5 (Anon, 1972) was actually a controlled experiment to determine flame velocity. Raj and Emmons (1975) have recently presented a theory to calculate the flame thickness during flash back of such flames.

TABLE 4. A few recent vapor cloud explosions which produced blast damage

Location and date	Fuel and quantity	Delay time to ignition	Loss Dollars and fatalities	TNT yield based on overpressure	Evidence for detonation	References
Berlin, N.Y. July 26, 1962	LPG 1,500 kg	minutes	$200,000 10	unknown	Dwelling exploded	Walls, 1963
Lake Charles, LA August 6, 1967	Butane Butylene 9,000 kg	unknown	$35 M 7	9,000–11,000 kg (10%)	Not reported	Goforth, 1969
Pernis, The Netherlands Jan. 20, 1968	H.C. slops	> 13 min.	$46 M 2	18,000 kg (—)	Fire before severe explosion	MSAPH, 1968* Fontein, 1970
Franklin Co., MO Dec. 9, 1970	Propane 30,000 kg	13 min.	$1.5 M 0	45,000 kg (10%)	Pump house destroyed by internal explosion	Burgess and Zabatakis, 1973 NTSB, 1972a†
East St. Louis, IL Dec. 22, 1972	Propylene 65,000 kg	> 5 min.	$7.6 M 0	1,000–2,500 kg (0.3%)	Box car destroyed by internal explosion	Strehlow, 1973a NTSB, 1973a
Flixbourough, England June 1, 1974	Hot Cyclohexane 50,000 kg	> 1 min.	> $100 M 28	18,000–27,000 kg (5%)	Fire before severe explosion	Kletz, 1975 Kinnersly, 1975 Slater, 1974
Decatur, IL July 19, 1974	Propane 65,000 kg	> 5 min.	$15 M 7	5,000–10,000 kg (2%)	Fire before explosion. Box car destroyed by internal explosion	Benner, 1975

*MSAPH = Ministry of Social Affairs and Public Health.
†NTSB = National Transportation Safety Board.

The fact that vapor cloud ignition can lead to two very different types of behaviors relative to blast wave production leads one to the conclusion that detonative combustion must always occur before a destructive blast wave is produced. Brown (1973) argues this position rather convincingly and evidence from actual incidents invariably shows that a detonation or detonations had occurred locally in the cloud when severe blast damage was found.

Research which is currently being performed relative to the behavior of these explosions is really of two major

TABLE 5. Release, delay to ignition, no explosion

Location and date	Fuel and amount	Delay to ignition	Fatalities		
Lynchburg, Virginia March 19, 1972	Propane, about 1500 kg	3–4 min	1	Flashback fire to torch. Driver killed inside fireball.	NTSB, 1973b*
Griffith, Indiana September 13, 1974	Propane > 100,000 kg	7 h	0	0.45 m vertical pipe at 10^6 Pa connected to a 36,000 m^3 storage cavern blew continuously until ignition. Workmen heard minor "pops" and then plume torched.	Schneidman and Strobel, 1975 Adderton, 1974 Shepard, 1975
Black Bayou Junction, Mississippi September 11, 1969	Vinyl cloride, monomer ?	6 h	0	Tank leaked after derailment—no explosion at ignition. Subsequent explosion of tank cars bathed in fire.	Kolger, 1971
Austin, Texas February 23, 1973	Natural gas liq. (Primarily C_3H_8, C_4H_{10})	15 min	8	Pipe burst—auto engine started fire—800 m × 66 m burn area—no explosion.	NTSB, 1973d*
France September 1972	LNG ?	?	0	Experiment to measure flame speed—27 m dike with ignition source 50 m away—no explosion.	Anon, 1972

*NTSB = National Transportation Safety Board.

types. Firstly, there is considerable effort in the area of assessing the behavior of a deflagrative explosion of the cloud. Here there are two major thrusts to the work. These are (1) overpressures from normal flame propagation are being evaluated, and (2) mechanisms for acceleration to detonation for various degrees of "confinement" are being investigated. Measurement of overpressures and flame behaviors for centrally ignited spherical or hemispherical clouds have been made by Woolfolk (1971), Ablow and Woolfolk (1972), and Woolfolk and Ablow (1973) using weather balloons of 15 and 90 ft^3 (0.42 and 2.5 m^3) capacity filled with hydrogen–oxygen–nitrogen mixtures. These mixtures all had relatively large burning velocities compared to ordinary hydrocarbon fuels. For deflagrative (weak) ignition they observed no shocks in the near field and far field overpressures which approached or even exceeded those produced by detonative combustion of the mixture. Unfortunately these experiments were on such a small scale that it is difficult to extrapolate to meaningful cloud sizes. Also, their weather balloons had rather thick walls and quite certainly interfered with the late flow and combustion processes.

Lind (1975) has performed experiments with a number of hydrocarbons in 5 and 10 m hemispheres of 0.202″ (5×10^{-5} m) thick polyethylene (volumes of 261 and 2094 m^3 respectively) ignited centrally at the ground level. He observed a number of interesting phenomena. Firstly, the flame propagated very rapidly ahead of the main flame ball along the concrete pad. This is probably due to a boundary layer-flame interaction similar to that which is observed in tubes. Secondly, the vertical propagation rate was always somewhat higher than the bulk horizontal rate and was accelerating. This can be attributed to a buoyant rise of the hot combustion products. Thirdly, he observed a very rough flame surface with both large and small roughness and he measured burning velocities (space velocities) which are three to five times the space velocity one would calculate from the normal burning velocity or would observe in a laboratory scale experiment. The mechanism which leads to these enhanced burning velocities is not understood at the present time. There is no turbulence ahead of the flame in the bulk gases. There can only be acoustic level disturbances ahead of the flame to possibly trigger flame accelerations. However, there is no theory available for this phenomenon at the present time. Fourthly, pressure levels measured in the flame ball or near it agreed quite well with those calculated using the theory of Kuhl *et al.* (1973) while the ball was burning. Finally, they did not observe transition to detonation under any circumstances and they did put obstacles and a number of different shapes to simulate enclosures, etc., in the flame region.

Wagner (1975) has also reported experiments on a relatively small scale using a 1 ft^3 (0.028 m^3) box containing thin transparent walls and central ignition. He finds that when he completely surrounds the ignition source with a spherical coarse mesh screen with a relatively small blockage factor, the flame accelerates instantaneously to a very high velocity as it passes through the screen. He sees accelerated flame velocities as high as twelve times the normal space velocity. Unfortunately, his experiment is rather small scale. However, he did observe that the flame velocity started to decrease after the flame passed entirely through the gas which was rendered turbulent by the screen. Again, acceleration to detonation was not observed.

Calculations of the self-similar pressure and flow

fields associated with constant velocity flame propagation have been made by Kuhl *et al.* (1974). These require numerical integration of differential equations for solution, and Strehlow (1975) and Guirao (1975) have both obtained approximate solutions which do not require numerical integration. Strehlow's solution is very simple and yields agreement within about 20% while Guirao's solution is more accurate but yields a more complex analytical relationship. Williams (1974) and Ablow and Woolfolk (1972) have also presented very crude treatments of the blast wave behavior. In addition to these, some calculations for blast behavior after the flame stops burning have been performed by Kiwan (1970b, 1971) and Guirao *et al.* (1974). Williams (1974) and Sichel and Hu (1974) have also looked in a relatively crude way at blast wave generation from non-spherical clouds both for deflagrative and detonative combustion and Strehlow *et al.* (1973) has suggested a way to use shock-free experimental pressure-time curves to estimate the rate of energy release during the deflagrative explosion of a cloud.

The second major area that has been looked at extensively is the area of cloud dispersion. Burgess *et al.* (1975) used the usual atmospheric dispersion equations and determined that the maximum fraction of the fuel that would be in the combustible range at any one time from either a continuous or massive spill would be about 10%. This agrees quite satisfactorily with the data shown in Table 4, where the maximum yield based on TNT equivalent weight is about 10%. In one of these cases (Franklin County, MO), it is known that the entire cloud detonated as a unit.

In addition to the research mentioned above there have been three important papers evaluating the hazard of unconfined vapor cloud explosions and recommending safe distances. Doyle (1970) presents charts giving overpressure levels as a function of distance based on a 2% yield from the massive release. Napadensky and Bodle (1973) specify that petrochemical plants should not be located any closer than 3/4 miles apart based on safety conditions and possible survivability. Finally, Iotti *et al.* (1972) discuss safe siting of nuclear plants based on estimates of yield from an unconfined vapor cloud explosion.

The final recent development in the area of vapor cloud explosions has to do with the release and vaporization of liquefied natural gas (LNG) particularly when it is spilled on water. The concern is that the contemplated large shipments of LNG into ports located near large population concentrations could lead to a truly massive spill followed by a catastrophic explosion. After all, ships with three containers of 50,000 m^3 volume each and single containers of 300,000 m^3 on land near water are either in operation or nearing completion. This development has led to "popular" papers concerning this new "danger" by Crouch and Hillyer (1972), Fay and MacKenzie (1972) and Fay (1973). The boiloff rate of LNG on ground has been treated by Burgess and Zabatakis (1962), while spills on water have been measured and dispersion calculations for different size spills have been made by Burgess *et al.* (1970a, b) and Opschoor (1975). Furthermore, it has been found that methane is extremely difficult to detonate. Foster (1974) and Kogarko *et al.* (1965) both found that at least 1 kg of C4 or TNT were required to produce a sustained detonation in a stoichiometric methane–air mixture. The import of this work is that the danger of an accidental unconfined methane–air detonation may be slight. However, the final answer to this complex problem has not been reached as yet and more work on flame acceleration processes in large size clouds plus independent verification that transition to detonation is difficult or impossible is necessary before the danger of handling LNG can be fully assessed.

G. *High Explosives and Propellants*

The blast waves produced by the accidental explosion of high explosives, black powder, high explosive intermediates or liquid propellants which are accidentally mixed are in general quite unreproducible and difficult to model adequately. This is reflected in the extensive discussion of liquid propellant explosions by Baker *et al.* (1974a, b), concerning the results of Willoughby *et al.* (1968a–c) (i.e., the Project PYRO tests), the modeling work of Farber and Deese (1968), Farber *et al.* (1968), and Farber (1969) and the work of Fletcher (1968a, b). A portion of the conclusions by Baker *et al.* (1974a) are reproduced here directly from their report because they are rather concise and it would be difficult to paraphrase them adequately.

Liquid propellant explosions differ from TNT explosions in a number of ways, so that the concept of "TNT equivalence" quoted in pounds of TNT is far from exact. Some of the differences are described below.

(1) The specific energies of liquid propellants, in stoichiometric mixtures, are significantly greater than for TNT (specific energy is energy per unit mass). In fact, *all* energy ratios are greater than 1, and can range as high as 5.3.
(2) Although the *potential* explosive yield is very high for liquid propellants, the *actual* yield is much lower, because propellant and oxidizer are never intimately mixed in the proper proportions before ignition.
(3) Confinement of propellant and oxidizer, and subsequent effect on explosive yield, are very different for liquid propellants and TNT. Degree of confinement can seriously affect explosive yield of liquid propellants, but has only a secondary effect on detonation of TNT or any other solid explosive.
(4) The geometry of the liquid propellant mixture at time of ignition can be quite different than that of the spherical or hemispherical geometry of TNT usually used for generation of controlled blast waves. The liquid propellant mixture can, for example, be a shallow pool of large lateral extent at time of ignition.
(5) The blast waves from liquid propellant explosions show different characteristics as a function of

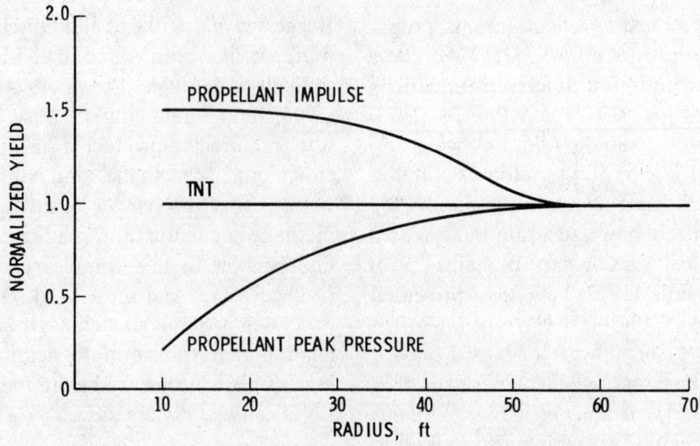

FIG. 27. Normalized pressure and impulse yields from explosion of N_2O_4/Aerozine 50. Fletcher (1968b).

distance from the explosion than do waves from TNT explosions. This is undoubtedly simply a manifestation of some of the differences discussed previously, but it does change the "TNT equivalence" of a liquid-propellant explosion. Fletcher discusses these differences (we show his curves as Figs. 27 and 28). These differences are very evident in the results of the many blast experiments reported in Project PYRO. They have caused the coinage of the phrase "terminal yield", meaning the yield based on blast data taken at great enough distance from the explosion for the blast waves to be similar to those produced by TNT explosions. At closer distances, two different yields are usually reported; an overpressure yield based on equivalence on side-on peak overpressures, and an impulse yield based on equivalence of side-on positive impulses.

There exist at present at least three methods for estimating yield from liquid propellant explosions, which do not necessarily give the same predictions. One method is based on Project PYRO results and the other two are the "Seven Chart Approach" and the "Mathematical Model" of Farber and Deese (1968).

In addition, Baker *et al.* (1974) observed that for liquid propellant explosions:

(1) The yield is very dependent on the mode of mixing of fuel and oxidizer, i.e., on the type of accident which is simulated. Maximum yields are experienced when intimate mixing is accomplished before ignition. For all cases the yield was found to range over the very large range of from 0.01% to 3.5% based on propellant weight.
(2) Blast yield per unit mass of propellant decreases as total propellant mass increases.
(3) The character of the blast wave as a function of distance differs between propellant explosions and TNT explosions, as noted before. There is some evidence that these differences are greatest for low percentage yield explosions.
(4) On many of the LH_2/LO_2 tests (regardless of investigators), spontaneous ignition occurred very early in the mixing process, resulting in very low percentage yields.
(5) Yield is very dependent on time of ignition, even ignoring the possibility of spontaneous ignition.
(6) Yield is quite dependent on the particular fuel and oxidizer being mixed.
(7) Variability in yields for supposedly identical tests was great, compared to variability in blast measurements of conventional explosives.

In an earlier report Bracco (1966) presents a calculation for the near field overpressure from the detonation of a mixture of liquid oxygen and liquid hydrogen normalized to far field behavior. He developed a rule of thumb which states that far field equivalence should obtain at a radius which is ten times the charge radius and calculates near field behavior normalized to this far field behavior. Willoughby, in a response which is attached to Bracco (1966), is critical of this simple approach, primarily of the fact that Bracco assumed that the entire mixture detonates as a unit. He points to some of the (then) current PYRO results to show that some of Bracco's estimations may underestimate blast levels in the near field.

Recently, Sutherland (1974) has presented a simplified technique for estimating near field overpressures from liquid propellant explosions and Farber (1974)

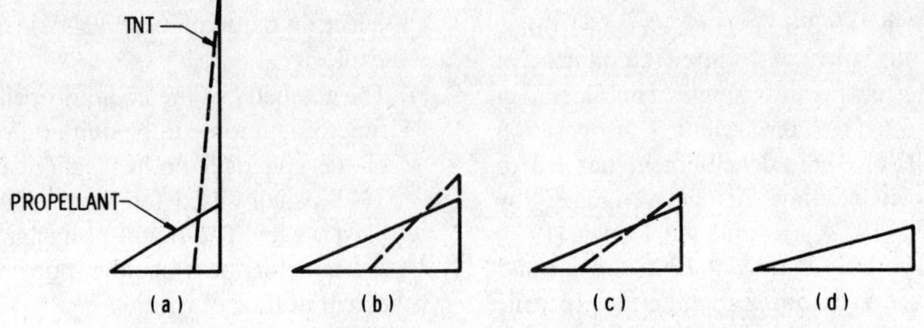

FIG. 28. Representative shock impulses showing coalescence of shock waves from dissimilar sources (Stages (a) through (d)), Fletcher (1968b).

has summarized his earlier work which involves techniques he developed for estimating yield. Also Mastromonico (1974) has presented some new data on the carbon monoxide–nitrous oxide system. In short he found that he could not detonate gaseous CO–N_2O mixtures in 33 and 60 m^3 thin walled containers (balloons or tents of thin mylar) but that with proper explosive squib initiation liquid–slush mixtures of CO–N_2O detonated with up to a 60% TNT yield based on propellant weight and overpressure.

There have also been studies of the yield of explosions involving propellants and explosives in configurations which represent manufacturing, transport and storage. Napadensky has been particularly active in this area. In her studies a charge of material is initiated at a site instrumented with blast pressure gauges and the per cent TNT yield based on overpressure and positive impulse are determined using the weight of propellant or explosive only and not the calculated total energy contained therein. In this respect their technique differs from that used with liquid propellants where the energy is normalized. Per cent yield is calculated at each gauge station and is then plotted against scaled TNT distance. Napadensky *et al.* (1973) contains a good summary of the technique and results. Two figures from that report are reproduced here as Figs. 29 and 30.

Figures 29 and 30 show three interesting general effects. Firstly, the per cent positive impulse curves are quite flat for all cases. This is to be expected because for relatively low overpressures the positive impulse represents wave energy to a good first approximation and this is a conserved quantity except for shock dissipation. Thus the ratio between TNT and non-ideal explosion impulse should be a constant to a good first approximation.

Secondly, per cent yield based on overpressure uniformly increases as one travels away from the source.

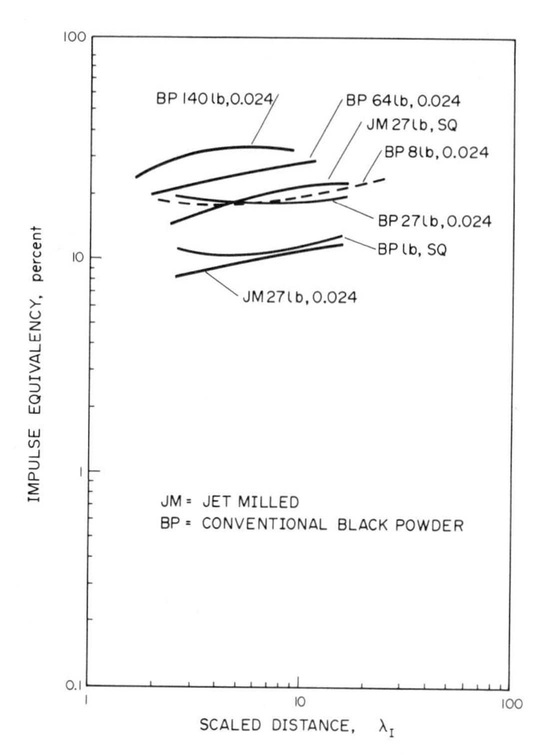

FIG. 30. Effect of black powder charge weight on TNT impulse equivalency; confined tests, Squib (SQ), or 0.024 lb booster. Napadensky *et al.* (1973).

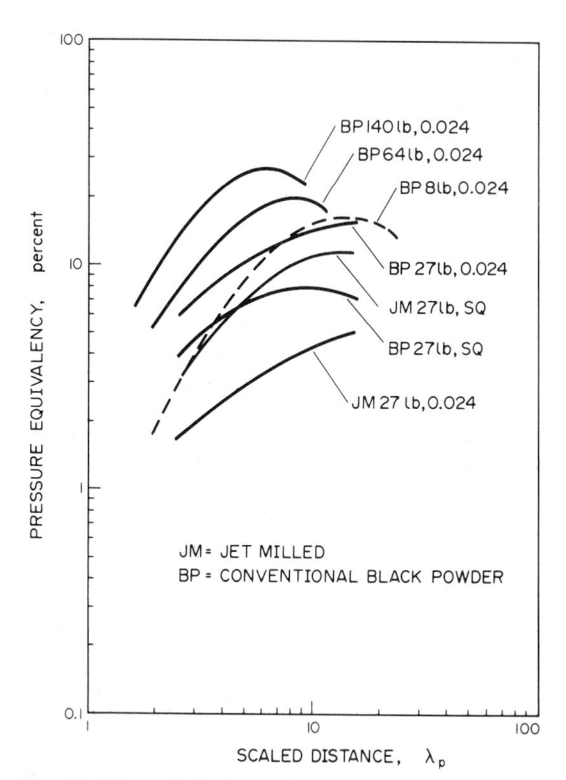

FIG. 29. Effect of black powder charge weight on TNT pressure equivalency; confined tests, Squib (SQ), or 0.024 lb booster. Napadensky *et al.* (1973).

This is also to be expected because near field overpressure curves can never be as high as TNT curves because of the slower energy release rate. However because these low pressure curves contain a relatively high impulse they decay more slowly than the TNT curves. Bursting sphere data, Fig. 11 and Fletcher's curve in Fig. 27, show the same effects, only the method of data presentation was different. Finally, the results of Napadensky *et al.* (1973) do not show the same far field equivalent TNT yield for overpressure and positive impulse as Fletcher's curve, Fig. 27, implies. Analysis of Project PYRO data also shows this type of non-equivalence in the far field for liquid propellant explosions. Fugelso *et al.* (1974) present a computational aid for estimating the damage effects due to the accidental explosion of stored munitions.

In summary it appears that the highly non-ideal and very irreproducible results which are obtained from liquid propellant and accidental high explosive explosions must complicate the construction of an adequate theoretical model for these explosion processes which fits all the facts. It appears that more work is needed in this area.

H. *Physical Explosions*

One class of accidental explosion does not involve chemical reaction or the release of the stored energy of a compressed gas. Instead, the explosion occurs upon flash boiling of a cold high vapor pressure liquid when it contacts a high temperature material, usually another liquid. These "physical explosions" or "vapor formation explosions" have occurred in the past when molten or very hot solid metals have been violently mixed with water, or vice versa. They have resulted in

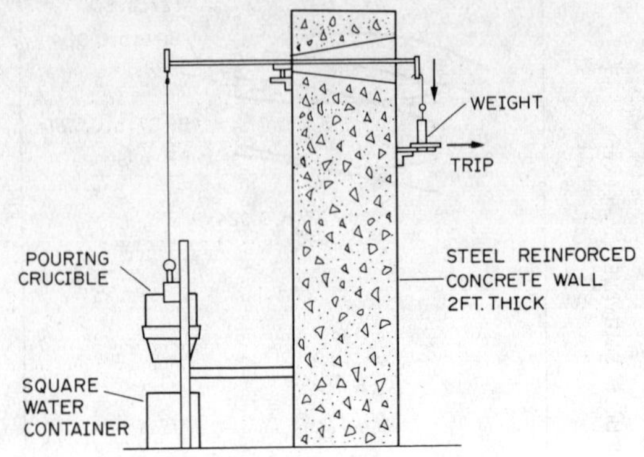

FIG. 31. Experimental apparatus for molten aluminum explosions in water. Witte *et al.* (1970).

a number of serious accidents over the years, primarily in foundries and other industries employing molten metals. Some serious accidents are summarized by Witte *et al.* (1970). These authors also describe the nature of this class of explosion, and summarize experiments designated to simulate accidental vapor formation explosions. An apparatus and some test results reported by Witte *et al.*, for molten aluminum being poured into water are shown in Figs. 31 and 32. Other accidental explosions of this type are discussed by Flory *et al.* (1969). Nelson (1973) discusses the theory of these physical explosions.

Today, large quantities of the very cold cryogenic liquid LNG (Liquefied Natural Gas) are being shipped in specially built and insulated tanker vessels and the projected increase in numbers is large (Hale, 1972). Because these vessels navigate in crowded harbors as

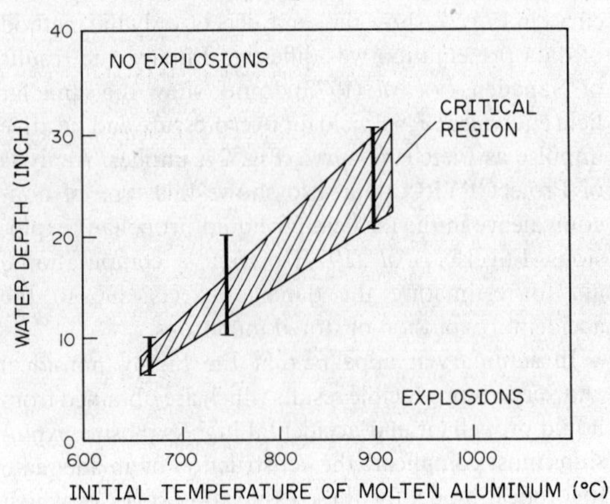

FIG. 32. Results for molten aluminum dropped into water. Witte *et al.* (1970).

well as the open sea, there is a very real possibility of collision or storm damage, and rapid release and mixing of LNG with water. Several years ago, there was serious concern that violent physical explosions could occur during such mixing. The first observations of such explosions were made at the U.S. Bureau of Mines by Burgess *et al.* (1970a, b). Since then the problem has been investigated by a number of investigators including Nakanishi and Reid (1971) and Enger and Hartman (1971 and 1972). Katz and Sliepcevich (1971) and Enger (1972) both discuss the mechanism and come to the conclusion that the phenomenon is caused by the occurrence of sufficient superheat of the colder fluid at the liquid–liquid interface to cause homogeneous nucleation and "explosive" formation of vapor bubbles. Enger has measured shock pressures in the liquid as high as 1.37 MPa near the "explosions". He reports that LNG must contain less than 40% CH_4 before an explosion will occur. Furthermore, he finds that if the mole ratio of propane to ethane in the LNG is greater than 1.3 "explosions" will not occur. The total energy released by these explosions is rather small, of the order of only 20,000 J/m^2, and normally in a large spill situation there will be many small explosions rather than one large explosion. The general conclusion is that these explosions do not produce dangerous blast waves in air.

I. *Nuclear Reactor Runaway*

At first glance, runaway reactions in nuclear reactors would appear to represent very serious explosion hazards. The total amount of energy which could potentially be released is enormous, and the energy source is confined to a relatively small volume. But, reactors are designed so that the magnitudes and rates of real maximum possible energy releases are many orders of magnitude less than for nuclear weapons. Furthermore, they employ many redundant safety features, including massive containment structures designed to withstand strong internal blast and missile impact. The chances of venting to the atmosphere of an explosion resulting from reactor runaway are therefore very remote.

It is quite possible that accidents can occur to nuclear plants which will, however, cause internal explosions. Reactor runaway is essentially an uncontrolled power excursion which increases exponentially until some physical process causes disruption of the reactor core. This disruption can be explosive, as has been predicted analytically (Stratton *et al.*, 1958; Corben, 1958) and observed in deliberate reactor runaway tests (Dietrich, 1954). Secondary explosions can occur, either chemical explosions such as sodium–air (Humphreys, 1958), or metal–water reactions as the metallic reactor core melts (Janssen *et al.*, 1958; McCarthy *et al.*, 1958; Owens, 1959; Bendler *et al.*, 1958). All experimental and analytical evidence to date indicate that explosion hazards in nuclear reactors, although real, can be contained and the effects confined to the containment structure (Baker, 1958).

V. CONCLUSIONS AND RECOMMENDATIONS FOR FUTURE WORK

The authors draw a number of conclusions from the survey reported here, and also make some recommendations for further work.

We conclude that:

1. Many damaging accidental explosions have occurred and will occur in industry, transportation and other fields. These explosions are almost always "non-ideal", i.e., they are significantly different than point source or chemical explosive (TNT) detonations primarily because of their low energy density and the slow addition of energy.
2. Different types of accidental explosions lead to different types of blast waves. Furthermore, certain accidental explosions, like the simple pressure vessel burst for example, are more reproducible than others and therefore much more amenable to analysis.
3. Because the comparison between ideal and accidental explosions is inexact, the concept of "TNT equivalence", which is widely used in safety studies, is also very inexact and may be quite misleading. But, this concept will undoubtedly be used to estimate "yields" of accidental explosions until better measures are available.
4. There is quite a lot still to be learned about the formation and transmission of blast waves from non-ideal explosions.
5. Scaling laws for non-ideal explosions are not now known exactly but, they can be easily developed once the physics of such explosions are well-known. They will likely be variants on Sachs' Law.
6. If blast wave characteristics can be defined for accidental explosions, correlation with damaging effects on buildings, vehicles, humans, etc. can be made based on existing methods and data in the literature.
7. Fragmentation patterns from accidental explosions, and the damaging effects of these fragments, are both quite difficult to predict.

Some recommendations for further work seem in order. Some of these studies are already in progress, but others are not. The former are indicated by an asterisk.

1. *Analytical study of the physics of non-ideal explosions, and comparisons with test data.
2. *Development of scaling laws for non-ideal explosions.
3. Establishment of a method or methods for estimating blast energies of accidental explosions to replace "TNT equivalency*.
4. Careful review of fragmentation effects from accidental explosions, and better definition of these effects.

Acknowledgement—This work was supported by NASA Grant NSG 3008, Technical Monitor, Dr. Robert Siewert, Aerospace Safety and Data Institute, NASA.

REFERENCES

Ablow, C. M. and Woolfolk, R. W. (1972) Blast effects from non-ideal explosions, SRI Final Report, Contract No. 0017-71-C-4421, Stanford Research Institute, Menlo Park, CA (December 1972).

Adderton, D. V. (1974) Gas inferno rages, *The News Gazette*, Champaign, Illinois, Sunday, September 15, 1974.

Anderson, W. H. and Louie, N. A. (1975) Effect of energy release rate on the blast produced by fuel–air explosions, SH-TR-75-01, Shock Hydrodynamics Division, Whittaker Corporation, North Hollywood, CA (January 1975).

Angiullo, F. J. (1975) Explosion of a chloronitrotoluene distillation column, Paper No. 90a, Presented at the AICHE Symposium on Loss Prevention in the Chemical Industry, Houston, TX, March 18–20, 1975; to be published in the *Loss Prevention Journal*, Vol. 8.

Anon. (1972) Essais d'ependage de gaz naturel liquefie sur de sol, Report on the experiments conducted by Gas de France, September, 1972.

Anon. (1970) Spacecraft incident investigation, Panel I, Vols. I, II & III, June, July, and September (1970), NASA TMX-66922, 66921 and 66934.

Bach, G. G. and Lee, J. H. S. (1970) An analytic solution for blast waves, *AIAA Jl* **8**, 271–275.

Baker, W. E. (1958) Scale model tests for evaluating outer containment structures for nuclear reactors, *Proceeding of the Second International Conference on the Peaceful Uses of Atomic Energy*, United Nations, Geneva, Vol. II, pp. 79–84.

Baker, W. E. (1973) *Explosions in Air*, University of Texas Press, Austin, TX.

Baker, W. E. (1974) An accidental acetylene–air explosion, letter report dated September 1974.

Baker, W. E., Parr, V. B., Bessey, R. L. and Cox, P. A. (1974a) Assembly and analysis of fragmentation data for liquid propellant vessels, Minutes of the 15th Explosives Safety Seminar, Department of Defense Explosives Safety Board, Washington, D.C., Vol. II, pp. 1171–1203, September 1973.

Baker, W. E., Parr, V. B., Bessey, R. L. and Cox, P. A. (1974b) *Assembly and Analysis of Fragmentation Data for Liquid Propellant Vessels*, NASA Contractors Report 134538, NASA Lewis Research Center, Cleveland, OH (January 1974).

Baker, W. E., Silverman, S., Cox, P. A., Jr. and Young, D. (1969) Methods of computing structural response of helicopters to weapons' muzzle and breech blast, *Shock and Vibration Bull.* **40**, 2, 227–241.

Baker, W. E., Westine, P. S. and Dodge, F. T. (1973) *Similarity Methods in Engineering Dynamics: Theory and Practice of Scale Modeling*, Spartan Books, Rochelle Park, NJ.

Baker, W. E. and Westine, P. S. (1974) Methods of predicting loading and blast field outside suppressive structures, Minutes of 16th Annual Explosion Safety Seminar, Department of Defense Safety Board.

Baker, W. E., Westine, P. S. and Bessey, R. L. (1971) Blast fields about rockets and recoilless rifles, Final Technical Report Contract No. DAAD05-70-C-0170, Southwest Research Institute, San Antonio, TX (May 1971).

Bendler, A. J., Roros, J. K. and Wagner, N. H. (1958) Fast transient heating and explosion of metals under stagnant liquids, AECU-3623, Contract AT(30-3)-187, Task II, Columbia University, Department of Chemical Engineering (February 1958).

Benner, L. (1975) Private communication to R. A. Strehlow, National Transportation Safety Board, Washington, D.C.

Bethe, H. A., Fuchs, K., Hirschfelder, H. O., Magee, J. L., Peierls, R. E. and von Neumann, J. (1947) Blast wave, LASL 2000, Los Alamos Scientific Laboratory (August 1947), distributed March 27, 1958.

Bethe, H. A., Fuchs, K., von Neumann, J., Peierls, R. and Penney, W. G. (1944) Shock hydrodynamics and blast waves, AECD 2860 (October 1944).

Boger, R. C. and Waldman, G. D. (1973) Blast wave interactions from multiple explosions, Paper #XII, *Proceed-*

ings of the Conference on Mechanisms of Explosion and Blast Waves, J. Alstor, Editor, Sponsored by The Joint Technical Coordinating Group for Air Launched Non-Nuclear Ordnance Working Party for Explosives (November 1973).

BOYER, D. W., BRODE, H. L., GLASS, I. I. and HALL, J. G. (1958) *Blast From a Pressurized Sphere*, UTIA Report No. 48, Institute of Aerophysics, University of Toronto.

BRACCO, F. V. (1966) Air blast parameters close to a liquid propellant explosion, ICRPG Hazards Working Group Publication 113 (June 1966); 2nd Meeting Bulletin CPIA, Silver Spring, MD (Includes response by Willoughby, A. B.).

BRASIE, W. C. and SIMPSON, D. W. (1968) Guidelines for estimating damage from chemical explosions, Preprint 21A, Paper presented at the Symposium on Loss Prevention in the Process Industries, 63rd National Meeting AICHE, St. Louis, MO, February 1968.

BRINKLEY, S. R. (1969) Determination of explosion yields, *AICHE Loss Prevention* **3**, 79–82.

BRINKLEY, S. R. (1970) Shock waves in air generated by deflagration explosions, Paper presented at Disaster Hazards Meeting of CSSCI, Houston, TX, April 1970.

BRINKLEY, S. R. and KIRKWOOD, J. G. (1947) Theory of the propagation of shock waves, *Phys. Rev.* **71**, 606.

BRODE, H. L. (1955) Numerical solutions of spherical blast waves, *J. appl. Phys.* **26**, 766–775.

BRODE, H. L. (1959) Blast wave from a spherical charge, *Physics Fluids* **2**, 217.

BROWN, J. A. (1973) A study of the growing danger of detonation in unconfined gas cloud explosions, John Brown Associates, Inc., Berkeley Heights, NJ (December 1973).

BURGESS, D. S., MURPHY, J. N., HANNA, N. E. and VAN DOLAH, R. W. (1968) Large scale studies of gas detonations, Report of Investigations 7196, U.S. Department of the Interior, Bureau of Mines, Washington, D.C. (November 1968).

BURGESS, D. S., MURPHY, J. N. and ZABETAKIS, M. G. (1970a) Hazards of LNG spillage in marine transportation, SRC Report #S-4105, Final Report February 1970 MIPR No. Z-70099-9-92317. Project 714152 U.S. Coast Guard, Washington, D.C.

BURGESS, D. S., MURPHY, J. N. and ZABETAKIS, M. G. (1970b) Hazards Associated with spillage of liquefied natural gas on water, U.S. Bureau of Mines, RI 7448, November 1970, 27 p.

BURGESS, D. S., MURPHY, J. N., ZABETAKIS, M. G. and PERLEE, H. E. (1975) Volume of flammable mixtures resulting from the atmospheric dispersion of a leak or spill, *Fifteenth International Symposium on Combustion*, The Combustion Institute, Pittsburgh, PA, paper #29 (in press).

BURGESS, D. S. and ZABETAKIS, M. G. (1962) Fire and explosion hazards associated with liquefied natural gas, U.S. Department of the Interior (1962) RI 6099, 34 p.

BURGESS, D. S. and ZABETAKIS, M. G. (1973) Detonation of flammable cloud following a propane pipeline break, the December 9, 1970, explosion in Port Hudson, MO, Report of Investigations 7752, U.S. Department of the Interior, Bureau of Mines, Washington, D.C.

BUTLIN, R. N. (1975) A review of information on experiments concerning the venting of gas explosions in buildings, Fire Research Note No. 1026, February 1975.

Champaign Fire Department (1972) Private communication to R. A. Strehlow.

CHARNEY, M. (1967) Explosive venting vs. explosion venting, Presented at AICHE Petrochemical and Refining Exposition, Houston, TX, February 2, 1967.

CHARNEY, M. (1969) Flame inhibition of vapor air mixture, Presented to AICHE Petrochemical and Refining Exposition, New Orleans, LA, March 18, 1969.

COEVERT, K., GROOTHUIZEN, TH. M., PASMAN, H. J. and TRENSE, R. W. (1974) Explosions of unconfined vapour clouds, Presented at the Loss Prevention Symposium, The Hague, Netherlands (May 1974).

COLE, R. H. (1965) *Underwater Explosions*, Dover, New York.

CORBEN, H. C. (1958) Power bursts in nuclear reactors, RWC 22-127, Contract AT (04-3)-165 with U.S. AEC, The Ramo-Wooldridge Corporation, Los Angeles, CA (September 1958).

COX, P. A. and ESPANZA, E. D. (1974) Design of a suppressive structure for a melt loading operation, Minutes of 16th Annual Explosion Safety Seminar, Department of Defense Safety Board.

CRANZ, C. (1926) *Lehrbuch der Ballistik*, Springer-Verlag, Berlin.

CRAVEN, A. D. and GRIEG, T. R. (1968) The development of detonation over-pressures in pipelines, *I. Chem. E. Series* No. 25, Institution of Chemical Engineers, London.

CROCKER, M. J. and HUDSON, R. R. (1969) Structural response to sonic booms, *J. Sound Vibration* **9**, 454–468.

CROUCH, W. W. and HILLYER, J. C. (1972) What happens when LNG spills? *Chemtech*, 210–215.

CUSTARD, G. H. and THAYER, J. R. (1970) Evaluation of explosive storage safety criteria, Falcon Research and Development Co., Contract DAHC 04-69-C-0095, March 1970; also target response to explosive blast (September 1970).

DABORA, E. K. (1972) Variable energy blast waves, *AIAA Jl* **10**, 1385.

DABORA, E. K., DIRECTOR, M. N. and STOY, R. L. (1973) Analytical and experimental studies of variable energy blast waves, Paper #III, *Proceedings of the Conference on Mechanisms of Explosion and Blast Waves*, J. Alstor, editor, Sponsored by The Joint Technical Coordinating Group for Air Launched Non-Nuclear Ordnance Working Party for Explosives (November 1973).

DARTNELL, R. C. and VENTRONE, T. A. (1971) Explosion of a para-nitro-meta cresol unit, *Loss Prevention* **5**, 53–56.

DEWEY, J. M. and SPERRAZZA, J. (1950) The effect of atmospheric pressure and temperature on air shock, BRL Report 721, Aberdeen Proving Ground, MD.

DIETRICH, J. R. (1954) Experimental investigation of the self-limitation of power during reactivity transients in a subcooled, water-moderated reactor (Borax-I experiments), AECD-36y8, Argonne National Laboratories, Lemont, IL.

Dow Chemical Company (1973) *Fire and Explosion, Dow's Safety and Loss Prevention Guide. Hazard Classification and Protection*, prepared by the Ed. of Chem. Eng. Progress, American Institute of Chemical Engineering, NY.

DOYLE, W. H. (1969) Industrial explosions and insurance, *Loss Prevention* **3**, 11–17.

DOYLE, W. H. (1970) Estimating losses, Paper presented at CSSCI Meeting in Houston, TX, April 1970.

ENGER, T. (1972) Explosive boiling of liquefied gases on water, Presented at the National Research Council Conference on LNG Importation and Terminal Safety, Boston, MA, June 13, 1972. National Academy of Sciences, Washington, D.C.

ENGER, T. and HARTMAN, D. E. (1971) LNG spillage on water. I. Exploratory research on rapid phase transformations, Shell Pipe Line Corporation, Research and Development Laboratory, Technical Progress Report No. 1-71 (February 1971).

ENGER, T. and HARTMAN, D. E. (1972) LNG spillage on water. II. Final report on rapid phase transformations, Shell Pipe Line Corporation, Research and Development Laboratory, Technical Progress Report No. 1-72 (February 1972).

Engineering Design Handbook (*1972*) *Principles of Explosive Behavior*, AMCP 706-180, Headquarters, U.S. Army Material Command, Washington, D.C. (April 1972).

FARBER, E. A. (1969) Characteristics of liquid rocket propellant explosion phenomena, in *Engineering Progress at the University of Florida*, Vol. 23, #11, Technical paper series, November 1969.

FARBER, E. A. (1974) Explosive yield limiting self-ignition phenomena in LO_2/LH_2 and LO_2/RP-1 mixtures, Minutes of the 15th Explosives Safety Seminar, San Francisco, CA, 18–20 September 1973, Department of Defense Explosives Safety Board, 1287–1304.

FARBER, E. A. and DEESE, J. H. (1968) A systematic approach for the analytical analysis and prediction of the yield from liquid propellant explosions, *Ann. N. Y. Acad. Sci.* **152**, I, 654–665.

FARBER, E. A., KLEMENT, F. W. and BONZON, C. F. (1968) Prediction of explosive yield and other characteristics of liquid propellant rocket explosions, Final Report Contract NAS 10-1255, October 31, 1968, University of Florida Engineering Experiment Station.

FAY, J. A. (1973) Unusual fire hazard of LNG tanker spills, *Combust. Sci. and Tech.* **7**, 47–49.

FAY, J. A. and MACKENZIE, J. J. (1972) Cold cargo, *Environment* **14**, 9, 21–29.

FEINSTEIN, D. I. (1973) Fragmentation hazard evaluations and experimental verification, Minutes of 14th Explosives Safety Seminar, New Orleans, 8–10 November 1972, Department of Defense Explosives Safety Board, 1099–1116.

FLETCHER, R. F. (1968a) Characteristics of liquid propellant explosions, *Ann. N.Y. Acad. Sci.* **152**, I, 432–440.

FLETCHER, R. F. (1968b) Liquid-propellent explosions, *J. Spacecraft Rockets* **5**, 10, 1227–1229.

FLORY, K., PAOLI, R. and MESLER, R. (1969) Molten metal–water explosions, *Chem. Engng Prog.* **65**, 12, 50–54.

FONTEIN, R. J. (1970) Disastrous fire at the Shell Oil Refinery, Rotterdam, *Inst. Fire Engr. Q.* (Ref. 3), 408–417.

FOSTER, J. (1974) Private Communication to R. A. Strehlow, Eglin AFB, Florida.

FREEMAN, R. H. and MCCREADY, M. P. (1971a) Butadiene explosion at Texas City—2, *Chemical Engng Prog.* **67**, 45–50.

FREEMAN, R. H. and MCCREADY, M. P. (1971b) Butadiene explosion at Texas City—2, *Loss Prevention* **5**, 61–66.

FREESE, R. W. (1973) Solvent recovery from waste chemical sludge—An explosion case history, *Loss Prevention* **7**, 108–112.

FREYTAG, H. H., (ed.) (1965) *Handbuch der Raumexplosionen*, Verlag Bhemie, GMBH Weinheim/Bergstr.

FUGELSO, L. E., WEINER, L. M. and SCHIFFMAN, T. H. (1974) A computation aid for estimating blast damage from accidental detonation of stored munitions, Minutes of 14th Explosives Safety Seminar, New Orleans, 8–10 November 1973, Department of Defense Explosives Safety Board, pp. 1139–1166.

GLASSTONE, S., (ed.) (1962) *The Effects of Nuclear Weapons*, U.S. AEC.

GOFORTH, C. P. (1969) Functions of a loss control program, *AICHE Loss Prevention* **4**, 1.

GOODMAN, H. S. (1960) Compiled free air blast data on bare spherical pentolite, BRL Report 1092, Aberdeen Proving Ground, MD.

GRANT, R. L., MURPHY, J. N. and BOWSER, M. L. (1967) Effect of weather on sound transmission from explosive shots, Bureau of Mines Report of Investigation 6921, 13 pp.

GUIRAO, C. M. (1975) Private communication to R. A. Strehlow, McGill University, Montreal, Canada.

GUIRAO, C. M., LEE, J. H. and BACH, G. G. (1974) *The Propagation of Non-Ideal Blast Waves*, AFOSR Interim Scientific Report, AFOSR-TR-74-0187, Department of Mechanical Engineering, McGill University, Montreal (January 1974).

HALE, D. (1972) LNG continuous spectacular growth, *Pipeline and Gas J.* 41–46 and following articles.

HALVERSTON, LCDR F. H. (1975) A review of some recent accidents in the marine transportation mode, Paper No. 52e, Presented at the AICHE Symposium on Loss Prevention in the Chemical Industry, Houston, TX, March 18–20, 1975; to be published in the *Loss Prevention J.* Vol. 8.

HIGH, R. W. (1968) The Saturn fireball, *Ann. N.Y. Acad. Sci.* **152**, I, 441–451.

HOPKINSON, B. (1915) British Ordnance Board Minutes 13565.

HOWARD, W. B. (1972) Interpretation of a building explosion accident, *Loss Prevention* **6**, 68–73.

HOWARD, W. B. (1975) Tests of orifices and flame arresters to prevent flashback of hydrogen flames, Paper No. 37c, Presented at the AICHE Symposium on Loss Prevention in the Chemical Industry, Houston, TX, March 18–20, 1975; to be published in the *Loss Prevention J.* Vol. 8.

HUANG, S. L. and CHOU, P. C. (1968) *Calculations of Expanding Shock Waves and Late-State Equivalence*, Final Report, Contract No. DA-18-001-AMC-876 (X), Report 125-12, Drexel Institute of Technology, Philadelphia, PA (April 1968).

HUMPHREYS, J. R., Jr. (1958) Sodium–air reactions as they pertain to reactor safety and containment, *Proceedings of Second International Conference on the Peaceful Uses of Atomic Energy*, United Nations, Geneva, Vol. II, pp. 177–185.

INGARD, U. (1953) A review of the influence of meterological conditions on sound propagation, *J. Acoust. Soc. Am.* **25**, 405–411.

IOTTI, R. C., KROTIUK, W. J. and DEBOISBLANC, D. R. (1974) Hazards to nuclear plants from on (or near) site gaseous explosions, Ebasco Services, Inc., 2 Reactor Street, New York, NY.

JANSSEN, E., COOK, W. H. and HIKIDO, K. (1958) Metal–water reactions: I. A method for analyzing a nuclear excursion in a water cooled and moderated reactor, GEAP-3073, General Electric Atomic Power Equipment Department, San Jose, CA (October 1958).

JARVIS, H. C. (1971a) Butadiene explosions at Texas City—I, *Chem. Engng Prog.* **67**, 6, 41–44.

JARVIS, H. C. (1971b) Butadiene explosions at Texas City—1, *Loss Prevention* **5**, 57–60.

KATZ, D. L. and SLIEPCEVICH, C. M. (1971) LNG/water explosions: cause and effect, *Hydrocarbon Processing* **50**, No. 11, 240–244 (November 1971).

KEENEN, W. A. and TANCRETO, J. E. (1973) Effects of venting and frangibility of blast environment from explosions in cubicles, Minutes of 14th Explosives Safety Seminar, New Orleans, 8–10 November 1972, Department of Defense Explosives Safety Board, 125–162.

KENNEDY, W. D. (1946) Explosions and explosives in air, in *Effects of Impact and Explosion*, M. T. WHITE (ed.), Summary Technical Report of Div. 2, NDRC, Vol. I, Washington, D.C. AD 221 586.

KINNERSLY, P. (1975) What really happened at Flixborough? *New Scient.* **65**(938), 520–522.

KINNEY, G. F. (1962) *Explosive Shocks in Air*, Macmillan, New York, NY.

KINNEY, G. F. (1968) Engineering elements of explosions, NWC TP4654, Naval Weapons Center, China Lake, CA (November 1968). AD 844 917.

KIWAN, A. R. (1970a) Self-similar flow outside an expanding sphere, BRL Report 1495, Ballistic Research Laboratories, Aberdeen Proving Ground, MD (September 1970).

KIWAN, A. R. (1970b) Gas flow during and after the deflagration of a spherical cloud of fuel–air mixture BRL-R1511, Ballistic Research Laboratories, Aberdeen Proving Ground, MD, November 1970.

KIWAN, A. R. (1971) FAE flow calculations using AFAMF code (U), BRL Report 1547, Ballistics Research Laboratories, Aberdeen Proving Ground, MD (September 1971).

KLETZ, T. A. (1975) Lessons to be learned from Flixborough,

Paper No. 67f, Presented at the AICHE Symposium on Loss Prevention in the Chemical Industry, Houston, TX, March 18–20, 1975; to be published in the *Loss Prevention J.* Vol. 8.

KOGARKO, ADUSHKIN and LYAMIN (1965) Investigation of spherical detonation of gas mixtures, *Combust. Explosion and Shock Waves* **1**, 2, 22–34.

KOGLER (1971) Vinyl tank car incident, *Loss Prevention* **5**, 26–28.

KOROBEINIKOV, V. P., MIL'NIKOVA, N. S. and RYAZANOV, Ye. V. (1962) *The Theory of Point Explosion*, Fizmatgiz, Moscow, 1961; English translation, U.S. Department of Commerce, JPRS: 14, 334, CSO: 69-61-N, Washington, D.C.

KUHL, A. L., KAMEL, M. M. and OPPENHEIM, A. K. (1973) Flame generated self-similar blast waves, *14th Symposium (International) on Combustion*, pp. 1201–1214, The Combustion Institute.

LASSEIGNE, A. H. (1973) Static and blast pressure investigation for the chemical agent munition demilitarization system: sub-scale, Report EA-FR-4C04, Edgewood Arsenal Resident Office, Bay Saint Louis, MS (November 1973).

LEE, J. H., KNYSTAUTAS, R. and BACH, G. G. (1969) Theory of explosions, Department of Mechanical Engineering, McGill University, AFOSR Scientific Report 69-3090 TR.

LEHTO, D. L. and LARSON, R. A. (1969) Long range propagation of spherical shock waves from explosions in air, NOLTR 69–88, Naval Ordnance Laboratory, White Oak, MD.

LEIGH, B. R. (1974) Lifetime concept of plaster panels subjected to sonic boom, UTIAS Technical Note 91, Institute for Space Studies, University of Toronto.

LINE, D. (1975) Unconfirmed vapor cloud explosions studies, Paper No. 67e, Presented at the AICHE Symposium on Loss Prevention in the Chemical Industry, Houston, TX, March 18–20, 1975; to be published in *Loss Prevention J.* Vol. 8.

LINDBERG, H. E., ANDERSON, D. L., FIRTH, R. D. and PARKER, L. V. (1965) *Response of Reentry Vehicle-type Shells to Blast Loads*, Final Report, SRI Project FGD-5228, P.O. 24-14517 under contract AF 04(694)-655, Stanford Research Institute, Menlo Park, CA (September 1965).

LUTZSKY, M. and LEHTO, D. (1968) Shock propagation in spherically symmetric exponential atmospheres, *Phys. Fluids* **11**, 1466.

MAINSTONE, R. J. (1973) The hazard of internal blast in buildings, Building Research Establishment Current Paper CP11/73, April 1973.

MAKINO, R. (1951) The Kirkwood–Brinkley theory of propagation of spherical shock waves and its comparison with experiment, BRL Report #750 (April 1951).

MASTROMONICO, C. R. (1974) Blast hazards of CO/N_2O mixtures, Minutes of the 15th Explosives Safety Seminar, San Francisco, CA, 18–20 September 1973, Department of Defense Explosives Safety Board, 1305–1357.

MCCARTHY, W. J., JR., NICHOLSON, R. B., O'KRENT, D. and JANKUS, V. Z. (1958) Studies of nuclear accidents in fast power reactors, Paper P/2165, Second United Nations International Conference on the Peaceful Uses of Atomic Energy (June 1958).

Ministry of Labour (1965) Guide to the use of flame arresters and explosion reliefs, Safety, Health and Welfare Series No. 34, Her Majesty's Stationery Office, London.

Ministry of Social Affairs and Public Health (1968) Report concerning an inquiry into the cause of the explosion on 20th January 1968, at the premises of Shell Nederland Raffinaderij N.V. in Pernis, State Publishing House, The Hague.

MUNDAY, G. (1973) Imperial College, London, Private communication to R. A. Strehlow.

NAKANISHI, E. and REID, R. C. (1971) Liquid natural gas–water reactions, *Chem. Engng Prog.* **67**, 12, 36–41.

NAPADENSKY, H. S. and BODLE, W. W. (1973) Safe siting of petrochemical plants, Presented at the National Symposium on Occupational Safety and Health, Carnegie Institution, Washington, D.C., June 4–6, 1973.

NAPADENSKY, H. S., SWATOSH, J. J., JR. and MORITA, D. R. (1973) TNT equivalency studies, Minutes of 14th Explosives Safety Seminar, New Orleans, 8–10 November 1972, Department of Defense Explosives Safety Board, pp. 289–312.

National Transportation Safety Board (1971) Highway accident report, liquefied oxygen tank explosion followed by fires in Brooklyn, New York, May 30, 1970, NTSB-HAR-71-6, Washington, D.C.

National Transportation Safety Board (1972a) Pipeline accident report, Phillips Pipe Line Company propane gas explosion, Franklin County, Missouri, December 9, 1970, Report NTSB-PAR-72-1, Washington, D.C.

National Transportation Safety Board (1972b) Railroad accident report, derailment of Toledo, Peoria and Western Railroad Company's train No. 20 with resultant fire and tank car ruptures, Crescent City, Illinois, June 21, 1970, NTSB-RAR-72-2, Washington, D.C.

National Transportation Safety Board (1972c) Highway accident report, automobile-truck collision followed by fire and explosion of dynamite cargo on U.S. Highway 78, near Waco, Georgia on June 4, 1971, NTSB-HAR-72-5, Washington, D.C.

National Transportation Safety Board (1972d) Railroad accident report, derailment of Missouri Pacific Railroad Company's train 94 at Houston, Texas, October 19, 1971, NTSB-RAR-72-6, Washington, D.C.

National Transportation Safety Board (1973a) Railroad accident report, hazardous materials railroad accident in the Alton and Southern Gateway Yard in East St. Louis, Illinois, January 22, 1972, NTSB-RAR-73-1, Washington, D.C.

National Transportation Safety Board (1973b) Highway accident report, propane tractor-semitrailer overturn and fire, U.S. Route 501, Lynchburg, Virginia, March 9, 1972, NTSB-HAR-73-3, Washington, D.C.

National Transportation Safety Board (1973c) Highway accident report, multiple-vehicle collision followed by propylene cargo-tank explosion, New Jersey Turnpike, Exit 8, September 21, 1972, NTSB-HAR-73-4, Washington, D.C.

National Transportation Safety Board (1973d) Pipeline accident report, Phillips Pipeline Co., natural gas liquids fire, Austin, Texas, February 22, 1973, NTSB-PAR-73-4.

National Transportation Safety Board (1974), Railroad accident report, derailment and subsequent burning of Delaware and Hudson Railway freight train at Oneonta, New York, February 12, 1974, NTSB-RAR-74-4, Washington, D.C.

NELSON, W. (1973) A new theory to explain physical explosions, *Combustion*, 31–36.

NICHOLLS, R. W., JOHNSON, C. F. and DUVALL, W. I. (1971) Blast vibrations and their effects on structures, Bureau of Mines Bulletin 656, 105 pp.

NICKERSON, J. I. (1975) Dryer explosion, Paper No. 90b, Presented at the AICHE Symposium on Loss Prevention in the Chemical Industry, Houston, TX, March 18–20, 1975; to be published in *Loss Prevention J.* Vol. 8.

NORRIS, C. H., HANSEN, R. J., HOLLEY, M. J., BIGGS, J. M., MAMYET, S. and MINAMI, J. V. (1959) *Structural Design for Dynamic Loads*, McGraw-Hill, New York.

OPPENHEIM, A. K. (1973) Elementary blast wave theory and computations, Paper #I, *Proceedings of the Conference on Mechanisms of Explosion and Blast Waves*, J. ALSTOR, ed., Sponsored by the Joint Technical Coordinating Group for Air Launched Non-Nuclear Ordnance Working Party for Explosives.

OPPENHEIM, A. K., LUNDSTROM, E. A., KUHL, A. L. and KAMEL, M. M. (1971) A systematic exposition of the conservation equations for blast waves, *J. Appl. Mech.* 783–794.

OPPENHEIM, A. K., KUHL, A. L. and KAMEL, M. M. (1972a) On self-similar blast waves headed by the Chapman Jouguet detonation, *J. Fluid Mech.* **55**, 2, 257–270.

OPPENHEIM, A. K., KUHL, A. L., LUNDSTROM, E. A. and KAMEL, M. M. (1972b) A parametric study of self-similar blast waves, *J. Fluid Mech.* **52**, 4, 657–682.

OPSCHOOR, IR. G. (1975) Investigations into the spreading and evaporation of LNG spilled on water, Subproject 4, Report 2, Central Technical Institute T.N.O., Rijswijk (January 1975).

ORDIN, P. M. (1974) Review of hydrogen accidents and incidents in NASA operations, NASA Technical Memorandum X-71565.

OWENS, J. I. (1959) Metal–water reactions: II. An evaluation of severe nuclear excursions in light water reactors, GEAP-3178, General Electric Atomic Power Equipment, San Jose, CA (June 1959).

PENNY, W. G., SAMUELS, D. E. J. and SCORGIE, G. C. (1970) The nuclear explosive yields at Hiroshima and Nagasaki, *Phil. Trans. R. Soc.* A **266**, 357.

PERKINS, B. J. and JACKSON, W. F. (1964) Handbook for prediction of air blast focusing, BRL Report 1240, Ballistics Research Laboratories, Aberdeen Proving Ground, MD, 100 pp.

PERKINS, B., JR., LORRAIN, P. H. and TOWNSEND, W. H. (1960) Forecasting the focus of air blasts due to meteorological conditions in the lower atmosphere, BRL Report 1118, Aberdeen Proving Ground, MD.

PETERSON, P. and CUTLER, H. R. (1973) Explosion protection for centrifuges, *Chem. Engng Prog.* **69**, (4), 42–44.

PHILLIPS, E. A. (1975) RPI-AAR tank car safety research and test project—status report, Paper No. 52c, Presented at the AICHE Symposium on Loss Prevention in the Chemical Industry, Houston, TX, March 18–20, 1975; to be published in *Loss Prevention J.* Vol. 8.

PITTMAN, J. F. (1972a) Blast and fragment hazards from bursting high pressure tanks, NOLTR 72-102.

PITTMAN, J. F. (1972b) Pressures, fragments, and damage from bursting pressure tanks, Minutes of 14th Explosives Safety Seminar, New Orleans, 8–10 November 1972, Department of Defense Explosives Safety Board, 1117–1138.

PORZEL, F. B. (1972) *Introduction to a Unified Theory of Explosions* (UTE), NOLTR 72-209, Naval Ordinance Laboratory, White Oak, Silver Spring, MD (14 September 1972). AD 758 000.

PROCTOR, J. R. and FILLER, W. S. (1973) A computerized technique for blast loads from confined explosions, Minutes of 14th Explosives Safety Seminar, New Orleans, 8–10 November 1972, Department of Defense Explosives Safety Board, 99–124.

RAJ, P. P. K. and EMMONS, H. W. (1975) On the burning of a large flammable vapor cloud, Paper presented at the Joint Meeting of the Western and Central States Sections of the Combustion Institute, San Antonio, TX, April 21–22, 1975.

REED, J. W. (1968) Evaluation of window pane damage intensity in San Antonio resulting from Medina Facility explosion on November 13, 1963, *Ann. N.Y. Acad. Sci.* **152**, I, 565–584.

REED, J. W. (1973) Distant blast predictions for explosions, Minutes of the 15th Explosives Safety Seminar, Department of Defense Explosives Safety Board, Washington, D.C., Vol. II, pp. 1403–1424.

REISLER, R. C. (1973) Explosive yield criteria, Minutes of 14th Explosives Safety Seminar, New Orleans, 8–10 November 1972, Department of Defense Explosives Safety Board, 271–288.

ROBINSON, C. S. (1944) *Explosions, Their Anatomy and Destructiveness.* McGraw-Hill, New York.

RUNES, E. (1972) Explosion venting, *Loss Prevention* **6**, 63–67.

SACHS, R. G. (1944) The dependence of blast on ambient pressure and temperature, BRL Report 466, Aberdeen Proving Ground, MD.

SAKURAI, A. (1965) Blast wave theory, in *Basic Developments in Fluid Mechanics*, Vol. I, pp. 309–375. MORRIS HOLT (ed.), Academic Press, New York.

SCHNEIDMAN, D. and STROBEL, L. (1974) Indiana gas leak capped; 1700 Return, *Chicago Daily Tribune*, Sunday, September 15, 1974.

SEWELL, R. G. S. and KINNEY, G. F. (1968) Response of structures to blast: a new criterion, *Ann. N.Y. Acad. Sci.* **152**, I, 532–547.

SEWELL, R. G. S. and KINNEY, G. F. (1974) Internal explosions in vented and unvented chambers, Minutes of 14th Explosives Safety Seminar, New Orleans, 8–10 November 1973, Department of Defense Explosives Safety Board, pp. 87–98.

SHEPARD, H. M. (1975) Private communication to R. A. Strehlow, National Transportation Safety Board, Washington, D.C.

SICHEL, M. and HU, C. (1973) The impulse generated by blast waves propagating through combustible mixtures, Paper VIII, *Proceedings of the Conference on Mechanisms of Explosions and Blast Waves*, J. ALSTOR (ed.), Sponsored by the Joint Technical Coordinating Group for Air-Launched Non-Nuclear Ordnance Working Party for Explosives.

SIEWERT, R. D. (1972) Evacuation areas for transportation accidents involving propellant tank pressure bursts, NASA Technical Memorandum X-68277.

SISKIND, D. E. (1973) Ground and air vibrations from blasting, Subsection 11.8 *SME and Mining Engineering Handbook*, VI, pp. 11–99 to 11–112. A. B. CUMMINGS and I. A. GIVEN (eds.), Soc. of Min. Eng. of the Am. Inst. of Min. Metallur. and Pet. Eng. Inc., New York.

SISKIND, D. E. and SUMMERS, C. R. (1974) Blast noise standards and instrumentation, Bureau of Mines Environmental Research Program, Technical Progress Report 78, U.S. Department of the Interior (May 1974).

SLATER, D. H. (1974) Private communication to R. A. Strehlow, Cremer & Warner, Consulting Engineers, London, England.

SMITH, T. L. (1959) Explosion in wind tunnel air line, BRL Memorandum Report 1235, Ballistic Research Laboratories, Aberdeen Proving Ground, MD.

SPERRAZZA, J. (1963) Modeling of air blast, in *Use of Modeling and Scaling and Vibration*, 65–78, W. E. BAKER (ed.), ASME, New York (November, 1963).

STATESIR, W. A. (1973) Explosive reactivity of organics and chlorine, *Loss Prevention* **7**, 114–116.

STEPHENS, T. J. R. and LIVINGSTON, C. B. (1973) Explosion in a chlorine distillate receiver, *Loss Prevention* **7**, 104–107.

STRATTON, W. R., COLVIN, T. H. and LAZARUS, R. B. (1958) Analysis of prompt excursions in simple systems and idealized fast reactors, Paper P/431, Second United Nations International Conference on the Peaceful Uses of Atomic Energy.

STREHLOW, R. A. (1973a) Equivalent explosive yield of the explosion in the Alton and Southern Gateway, East St. Louis, Illinois, January 22, 1972, AAE TR 73-3, Department of Aeronautical and Astronautical Engineering, University of Illinois, Urbana, IL (June 1973).

STREHLOW, R. A. (1973b) Unconfined vapor-cloud explosions—an overview, *14th Symposium (International) on Combustion*, pp. 1189–1200. The Combustion Institute, Pittsburgh.

STREHLOW, R. A. (1975) Blast waves generated by constant velocity flames—a simplified approach, *Combust. Flame* **24** (in press).

STREHLOW, R. A. and ADAMCZYK, A. A. (1974) On the nature of non-ideal blast waves, Technical Report AAE 74-2, UILU-ENG-740502, University of Illinois, Urbana, IL.

STREHLOW, R. A., SAVAGE, L. D. and VANCE, G. M. (1973) On the measurement of energy release rates in vapor cloud explosions, *Combust. Sci. Tech.* **6**, 307–312.

SUTHERLAND, L. C. (1974) A simplified method for estimating the approximate TNT equivalent from liquid propellant explosions, Minutes of the 15th Explosives Safety Seminar, Department of Defense Explosives Safety Board, Washington, D.C. Vol. II, pp. 1272–1277.

TAYLOR, D. B. and PRICE, C. F. (1971) Velocities of fragment from bursting gas reservoirs, *ASME Trans., J. Engng Ind.* **93B**, 981–985.

TAYLOR, G. I. (1946) The air wave surrounding an expanding sphere, *Proc. R. Soc.* A **186**, 273–292.

TAYLOR, G. I. (1950) The formation of a blast wave by a very intense explosion, *Proc. R. Soc.* A **201**, 159.

THORNHILL, C. K. (1960) Explosions in air, ARDE Memo (B) 57/60, Armament Research and Development Establishment, England (1960).

VINCENT, G. C. (1971), Rupture of a nitroaniline reactor, *Loss Prevention* **5**, 46–52.

VON NEUMANN, J. and GOLDSTINE, H. (1955) Blast wave calculation, *Commun. Pure appl. Math.* **8**, 327–353; reprinted in *John von Neumann Collected Works*, Vol. VI, pp. 386–412, A. H. TAUB (ed.), Pergamon Press, New York.

WAGNER, H. G. (1975) Private communication to R. A. Strehlow, University of Gottingen, West Germany.

WALLS, W. L. (1963) LP-gas tank truck accident and fire, Berlin, N.Y., *Nat. Fire Protect. Ass. Q.* **57**, 3–8.

WARREN, A. (1958) Blast pressures from distant explosions, ARDE Memo 18/58 AD 305 732.

WENZEL, A. B. and BESSEY, R. L. (1969) Barricaded and unbarricaded blast measurements, Contract No. DAHC04-69-C-0028, Subcontract 1-OU-431, Southwest Research Institute, San Antonio, TX.

WESTINE, P. S. (1972) R-W plane analysis for vulnerability of targets to air blast, *Shock Vibration Bull.* **42**, 5, 173–183.

WESTINE, P. S. and BAKER, W. E. (1974) Energy solutions for predicting deformations in blast loaded structures, Minutes of 16th Annual Explosion Safety Seminar, Department of Defense Safety Board.

WHITHAM, G. B. (1950) The propagation of spherical blast, *Proc. R. Soc.* A **203**, 571–581.

WILLIAMS, F. A. (1974) Qualitative theory of non-ideal explosions, in Phase I. Final Report entitled, *Explosion Hazards Associated with Spills of Large Quantities of Hazardous Materials*, by D. C. LIND, Naval Weapons Center, China Lake, CA (August 5, 1974), U.S. Coast Guard, Washington, D.C.

WILLOUGHBY, A. B., WILTON, C. and MANSFIELD, J. (1968a) Liquid propellant explosive hazards. Final report—December 1968. Vol. I—technical documentary report, AFRPL-TR-68-92, URS-652-35, URS Research Co., Burlingame, CA.

WILLOUGHBY, A. B., WILTON, C. and MANSFIELD, J. (1968b) Liquid propellant explosion hazards, final report—December 1968. Vol. II—test data, AFRPL-TR-68-92, URS 652-35, URS Research Co. Burlingame, CA.

WILLOUGHBY, A. B., WILTON, C. and MANSFIELD, J. (1968c) Liquid propellant explosion hazards, final report—December 1968. Vol. III—prediction methods, AFRPL-TR-68-92, URS-652-35, URS Research Co., Burlingame, CA.

WILSE, T. (1974) Fire and explosions onboard ships, *Veritas* No. 80, pp. 12–16.

WISOTSKI, J. and SNYDER, W. H. (1965) Characteristics of blast waves obtained from cylindrical high explosive charges, University of Denver, Denver Research Institute.

WITTE, L. C., COX, J. E. and BOUVIER, J. E. (1970) The vapor explosion, *J. Metals* **22**, 2, 39–44.

WOOLFOLK, R. W. (1971) Correlation of rate of explosion with blast effects for non-ideal explosions, SRI Final Report for Contract No. 0017-69-C-4432, January 25, 1971, Stanford Research Institute, Menlo Park, CA.

WOOLFOLK, R. W. and ABLOW, C. M. (1973) Blast waves from non-ideal explosions, Paper #IV, *Proceedings of the Conference on Mechanisms of Explosion and Blast Waves*, J. Alstor, Editor, Sponsored by The Joint Technical Coordinating Group for Air Launched Non-Nuclear Ordnance Working Party for Explosives.

ZABATAKIS, M. G. (1960) Explosion of dephlegmator at Cities Service Oil Company refinery, Ponca City, Oklahoma, 1959, Bureau of Mines Report of Investigation 5645, 16 pp.

ZAJAK, L. J. and OPPENHEIM, A. K. (1971) The dynamics of an explosive reaction center, *AIAA Jl* **9**, 545–553.

Manuscript received, 12 September 1975

Prog. Energy Combust. Sci., Vol. 2, pp. 61–82, 1976. Pergamon Press. Printed in Great Britain

FLUIDIZED BED COMBUSTION OF COAL FOR POWER GENERATION

D. Anson
C.E.G.B., Marchwood Engineering Laboratories, currently seconded to Electric Power Research Institute, Palo Alto, California

INTRODUCTION AND BACKGROUND

The principle of the fluidized bed is simple. It relies on the fact that if a fluid is forced upwards through a bed of relatively dense particles, a point will be reached at which the drag forces levitate the particles. Expansion of the bed occurs permitting an increase in the flow rate of fluid. At this point the particles can move freely, the bed behaves like a heavy fluid, and the total drag force equals the total particulate weight. Further increases in fluid velocity are compensated initially by increasing particle spacing due to bed expansion, and then by the formation of "bubbles" of fluid moving vertically in the relatively dense fluidized solid or "emulsion" phase. These bubbles account for both considerable secondary bed expansion and greatly enhanced mixing, giving the bed the appearance of a bath of boiling liquid. At very high gas velocities the fluid-like motion becomes very violent and exhibits slugging and splashing behaviour akin to that observed in true two phase fluid flow.

For over 30 years, fluidized bed reactors have been well known to the chemical engineering industry as highly effective devices for contacting gases with solids. The most extensive applications are in the petrochemical industry, and the classic example is the fluidized catalyst cracking unit, in which hydrocarbon vapours pass through a fluidized bed of catalyst beads. The hydrocarbons crack endothermically on contact with the beads, with the deposition of carbon. The coated beads are directed through a side stream in which the carbon is burned off, and the beads, thus regenerated and reheated, are recycled to the cracker unit. This process utilizes very well all the advantages of the fluidized bed, namely good solid/gas contacting, temperature uniformity, solids transport mobility, and heat exchange. Surface combustion occurs only in the regeneration stage.

The treatment of coal in a fluidized bed goes back further than the catalytic cracker, Winkler having described it in his patent for a gasifier in 1921.[1] The Winkler gasifier was used extensively as a source of synthesis gas from coal, successfully handling a wide variety of coal types.

There were also several developments in Europe of fluidized combustion systems which exploited the inherently good gas/solid contacting to deal with relatively unreactive low grade fuels, especially those with a very high ash content (up to 90%). These were reported in BCURA Information Circular 301 and also in a later review of the subject (see Skinner[2]). There seems to have been little attempt to sophisticate these systems, which were handling such cheap fuels that efficiency was of secondary importance, and for which there was perhaps only limited application. Godel's Ignifluid combustion system is probably the best known of them, and was patented in 1950.[3]

During the 1950s several major US and European manufacturing corporations were active in examination of fluidized bed combustion, and a number of patents were filed. Again, these are described in Skinner's review[2] but some will be referred to further in later sections, as it is apparent that, in some cases at least, systematic experimental work must have been done to permit optimization of the various processes, and serious consideration was given to their application to power generation. Much of this work ran parallel to the work carried out in Britain, and may well have anticipated some of the results of that work.

The British National Coal Board (NCB) adopted fluidized bed reactors in the 1950s, to process fine bituminous coal by partial oxidation with air. The method gave close control of the reaction temperature, and permitted the removal of volatile tars released at relatively low temperatures (300–600°C), without extensive oxidation of the solid combustible material, which was subsequently briquetted to yield smokeless fuels. The agitation of the particles prevented caking, the good solids mixing ensured a uniform product and rapid reaction, and the overall process could be controlled by the oxygen availability in the fluidizing gas and by the solids throughput rate.

Elliott, at that time working for the British Central Electricity Generating Board (CEGB), conceived the idea of using the process to fractionate coal into an ash free volatile part which would be used as a gas turbine fuel, and a carbonaceous part, or char, which could be used as a boiler fuel. Essentially, this simply involved raising the process temperature to increase volatiles yield. The problems facing the practical application of the technique were concerned with the behaviour of the volatiles in the turbine feed pipes and combustor, and with the small amount of mineral matter that would inevitably carry over. There was scope to optimize the fluidized bed conditions to provide a good yield of volatiles while leaving a reactive char that would be an acceptable boiler fuel.

Subsequent experimental work at CEGBs Marchwood Engineering Laboratories provided the bases for a variety of designs of power plants which incorporated both gas turbine plant and steam plant. Indeed a scheme was considered for the installation of a gas turbine, to operate on coal distillation products, but the idea was abandoned because of a lack of economic incentive, after a careful engineering evaluation.

The experimental work at Marchwood turned to the fluidized combustion both of the char produced by coal devolatilization and of the raw coal. As has been noted, fluidized bed combustion systems had already been developed, patented as early as 1950, and applied successfully to industrial boilers. In general, there was no heat exchange in the fluidized bed. Godel's "Ignifluid" combustor was really a two stage device in which high ash coal was gasified in the fluidized bed where the ash sintered into large agglomerates and was removed. The heat transfer process was conventional, essentially convective.

The fluidized combustion beds used in the CEGB work differed from Godel's arrangement in two essential respects. Firstly, they were composed essentially of inert particles of refractory material, with only minor quantities (below 1%) of combustible, and secondly they contained immersed heat transfer surface for bed cooling and steam raising. The low combustible levels permitted complete combustion without the escape of combustible gases above the bed, and the arrangement of cooling surface exploited the excellent heat transfer properties of the fluidized material. This combination appeared to offer the prospect of an efficient compact boiler without risk of high temperatures that could lead to ash fusion or bonded tube deposits, or to dangerous excursions in tube temperature during fault or transient conditions.

These potential advantages were checked in brief laboratory tests, and CEGB and NCB jointly sponsored design studies of water tube boilers of 120 MW(e) to 660 MW(e) size for electric power generation applications. In considering units of this size a fundamental design problem was encountered, and this problem still exists. To keep the pressure drop through the fluidized bed to an acceptable level, the bed depth must be restricted to about 1 m or less. The heat generation per unit base area is proportional to the air flow rate or superficial air velocity, which is restricted to 1–4 $m s^{-1}$ to avoid excessive carry over of bed solids including unburnt coal; the early studies were based on fine bed material and hence to velocities at the lower end of this range. The net result of these limitations is that, for large boilers, the bed plan area becomes very large, and there are difficulties in arranging heat exchange surface and coal feed points without incorporating a great deal of mechanical complication. The majority of schemes put forward for these larger boilers incorporated "stacked" beds to reduce plan area, and all were based on subdivision of the total bed area into manageable modules.[4] The effect of this on economics was that boilers of moderate size, say 50–200 MW(e), appeared more attractive than the largest units, although all were judged to be competitive with equivalent pulverized coal-fired boilers.

So far as the British market was concerned, there was very limited incentive to develop the idea further. The CEGB regarded its generating capacity to be overdependent on coal, especially as in the 1960s oil prices were comparable to coal prices, boilers designed for oil firing were cheaper in first cost and maintenance, and were better suited to load following or two shift operation. Moreover, the power industry's research and development capability was too heavily involved in pursuing a vigorous nuclear programme and in improving the reliability of new plant to be capable of undertaking a new line of development. During the late 1960s, therefore, the CEGB withdrew from active work in this field, and after 1964 most experimentation in Britain was done by NCB at the Leatherhead (BCURA) site or at Stoke Orchard (Coal Research Laboratories). The work at Leatherhead included operation of an industrial boiler of 3600 kg/h steaming capacity[5] which was commissioned during 1969. Although technically successful this design could not compete commercially with oil or gas fired units. The most important aspect of the continued NCB work was the accumulation of a great deal of systematic data on coal burn up, dust carry over, and heat transfer for a range of bed operating conditions which included variations in air velocity, coal size, temperature and excess air. The work at Leatherhead included studies of fluidized combustion systems operating at pressures up to 6 atm. The NCB work up to about 1970 is summarized in the review by Skinner[2] already referred to, and which includes a useful bibliography of the earlier work.

During this period it was realized that the excellent gas/solids contact which could be achieved in the fluidized bed provided a means for absorbing gaseous sulphur compounds in a suitable solid receptor. In the case of coal, it was observed that naturally occurring basic minerals, such as limestone, were sulphated. Systematic studies of SO_2 absorption by added limestone were reported by Jonke *et al.*[6] of the Argonne National Laboratory (USA) and by Skopp[7] of Esso Research and Engineering Company at the First International Conference on Fluidized Bed Combustion. This has since proved to be the main incentive for further study and development of fluidized combustion for power generation in conventional steam plant.

The BCURA pressurized work, mentioned above, is aimed at demonstrating the feasibility of running direct-fired gas turbines on the off gas from a fluidized bed combustor. It is claimed that particulates carried over from the fluidized bed combustor are far less abrasive than those from a pulverized fuel-fired unit, which are often glassy in nature following fusion in the flame. Negligibly low erosion has been reported[8] and deposits of potentially corrosive salts are also minimized since at the low combustion temperature (800°C) the volatilization of alkali metals (e.g. as sulphates or chlorides) is slight. The pressurized fluidized bed overcomes many of the design problems of the atmospheric bed; the required bed cross section is reduced by the inverse of the pressure, the bed depth may be increased without an unacceptable pressure loss, so that the overall physical proportions improve and carbon burn up increases, and the volume of gas to be cleaned of dust is reduced. There are obvious mechanical complications affecting fuel feed, dust removal, etc., but the possibility of including gas turbine machinery in fluidized bed fired electric generators opens up a number of new options.

Summarizing, by the late 1960s the basic elements of fluidized bed combustion of coal had been explored, and it had been established that sulphur oxide emissions could be controlled by reaction with suitable alkaline bed material or additives, while the low combustion temperatures reduce the formation of nitrogen oxides from atmospheric nitrogen. Application to steam boilers, gas turbines, and combined cycles appeared feasible.

US CLEAN AIR LEGISLATION

In 1967 the US Government, in the Air Quality Act, reinforced existing legislation by introducing the element of Federal responsibility, and followed this up in 1970 by amendments which, amongst other things, set national air quality standards. This stimulated public awareness, and in some cases led to stringent state or local regulations which required sweeping changes in the quality of primary fuels or their usage. While most of the more populous states on the eastern seaboard were able to find ample supplies of low sulphur oil to meet these regulations, it was foreseen that those inland areas which were dependent on the cheaper high sulphur Appalachian, Kentucky and Illinois coals would find oil expensive, and would need to find alternative ways of meeting the regulations. The possibilities of fluidized bed combustion were obviously relevant, and research and development received active support from the National Air Pollution Control Administration (NAPCA), later the Environmental Protection Agency (EPA) and other government bodies such as the Office of Coal Research (OCR). Both British and American organizations were involved, and fluidized bed combustion entered a new phase of development. The work was given further point by the realization that oil and natural gas resources were limited, a fact that has been driven forcefully home by Middle East oil export restrictions and rapid price rises in recent years.

The same agencies promoted research in other areas, aimed at the conversion of coal to more versatile or convenient fuels which were compatible with the clean air legislation. The main programmes were in gasification or liquefaction, both of which require quite extensive chemical processing, hydrogenation, and considerable consumption of primary fuel within the conversion process. Coal utilization efficiency was seen to be anything from 60–80% with available processes, although long term developments could produce improvements. Against these, fluidized bed combustion did not involve any conversion penalty when compared with other coal fired plant, although if limestone were used as a sulphur neutralizing additive it would introduce a significant cost element. In terms of time scale, it was more immediately feasible, although not without risk. It is pertinent at this point to reassess how the state of the art has developed in the last 5 years as a result of the increased research impetus. The main topics of direct interest to power generation applications are combustion, emissions, heat transfer, materials and thermodynamic cycles.

COMBUSTION

In its simplest form, the solid fuel fluidized bed combustor may be regarded as a plug flow gas/solid contactor in which the solid reactant is continuously replaced. Because particles in the bed are in more or less random movement, the fuel, which is normally present in concentrations around 1%, may be regarded as uniformly distributed in an inert matrix. The probability of an elementary volume of oxygen in the fluidizing air contacting a fuel particle before it leaves the bed will then depend largely on the number of fuel particles in its path as it passes vertically (on average) through the bed, i.e. on the numerical coal particle population per unit bed cross section area. The extent of reaction on any given contact will depend on the usual rate determining factors.

Normal diffusion processes are likely to be modified under fluidization conditions, but in any case the reaction of the smaller particles ($< 300\,\mu m$) will probably be chemically controlled. Usual bed temperatures (750–950°C), and the range of particle sizes, are such that both diffusion and chemical rate control will be significant. At the lower end of this temperature range the effect of bed temperature will be especially important, as chemical control will be predominant.

Two important parameters in determining the efficiency of oxygen conversion are thus the number of fuel particles per unit bed base area, and the bed temperature. Air velocity will determine the overall rate of combustion, but particle size and fuel mass concentration *per se* should not exert major influences. However, since the number of fuel particles is proportional to d^{-3} for a given fuel mass concentration, the mean particle size d exerts a powerful influence on the population density. The range of coal particle sizes that can remain effectively dispersed in a given bed is to some extent limited by considerations of settling of large particles and elutriation of fines, but inasmuch as the bulk of the bed material, which determines the fluidization behaviour, is inert, coal particle size is not critical. The equilibrium bed carbon content can be controlled via the mean particle size and the size distribution in the feed coal.

This model is complicated by the fact that, with an imperfect fuel feed system, the fuel particles will not be uniformly dispersed. With coarse coal feed the individual particles burn relatively slowly and any initial maldistribution will be less critical than with fine coal sizes.

Combustion data are frequently reported in terms of the feed stoichiometry. In the fluidized bed the oxygen consumption depends equally on the preexisting fuel particle population. Thus a sudden drop in the air rate, with no change in fuel rate, would not be expected to result in an abrupt change in flue gas composition, but there will be a change in the combustion rate of the fuel with a progressive build up in fuel particle population in the bed. Eventually this will lead to all available oxygen being consumed before it reaches the top of the bed, with the likelihood of combustible gases being formed in the upper part, e.g. by reaction of CO_2 with excess carbon. Conversely, a sudden increase in air rate will lead to a progressive fall in the bed carbon content and the appearance of increasing oxygen levels in the off gas. The time for new equilibria to establish will depend largely on the size of coal feed, since particle size determines the mean life of the fuel under the prevailing combustion conditions.

The broad features of fluidized bed combustion behaviour described above are well illustrated by Gibbs *et al.*[9] who included observations of in bed gas composition in their work.

Where the range of coal particle sizes is large, or is markedly different from the bed material, vertical size segregation may occur. This does not in itself invalidate the simple model described above, although fine fuel elutriated and carried over with the reacted gases will reduce the population of fuel particles in the bed. Since this fine material would provide a very large number of reaction sites if it remained within the bed, the oxygen levels in the off gas would be higher than might be expected from the feed rate, and combustion would continue above the bed.

Much of the early work on fluidized bed combustion was concerned with fine coal, with maximum particle sizes from about 0.5 to 2 mm and fluidization velocities below 0.5 m s^{-1}. In such cases the concentration of coal in the bed could be very low (~0.1%), and losses by incomplete reaction of the mixture or carbon carry over could be very small. These conditions contrast with the commercial trend to velocities around 4 m s^{-1} and coal feed size 0–15 mm. Here much of the fine coal may escape incompletely reacted and the bed carbon content can be 1–5%.

Loss by carry over of small unburnt particles is commonly the major source of inefficiency, and the behaviour of fine coal (i.e. such that its free fall velocity is less than the nominal air velocity) needs to be distinguished from that of other material in the bed. This material is not in random motion in the bed, but tends to be transported with the rising air stream, being concentrated above the feed points and hence in a deficiency of air relative to the bulk material. Significant concentrations of combustible gases, especially CO, may appear in the effluent gases via the same route, leading to secondary combustion as they find the excess air in other zones.

For the case where the air supply rate and bed temperature are constant and the bed fuel concentration is low, some oxygen will pass through the bed unreacted; fine coal, which is capable of being elutriated, will, if fed to the bottom of the bed, burn in its

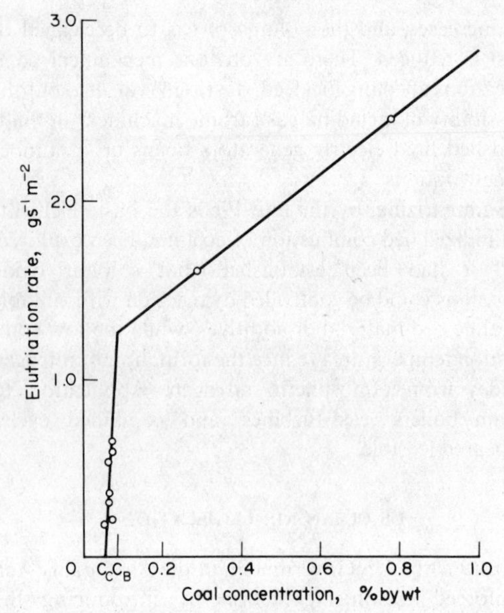

FIG. 1. Variation in coal elutriation rate with coal concentration in a 0.6 m deep bed with fluidizing velocity of 0.6 m s^{-1} and a temperature of 800°C (from ref. 10).

passage to the bed surface and only ash may be elutriated. If the coal concentration in the bed is now increased, the oxygen availability will decrease at all levels, and fine coal may then reach the top of the bed unreacted, to be elutriated. Further increase in coal feed rate results in more fine coal reaching the surface, with correspondingly greater elutriation loss. Unlike the case of the pulverized coal combustor, the losses do not normally derive from the coarser coal, which, in the absence of severe splashing, remains in the bed until burnt unless it is reduced in density and size to the point at which it can be elutriated. The situation is well illustrated by Fig. 1 from a paper by Highley and Merrick.[10]

Combustion loss by elutriation of fines is likely to depend on the feed arrangement at any given temperature and velocity; it will decrease with increase in temperature, and increase with increasing velocity due to reduced residence times of the fine particles and the increased range of sizes that can be elutriated. With the formation of bubbles at higher gas velocities a considerable proportion of the available oxygen can also bypass the emulsion phase. Vertical mixing is invariably much more rapid than lateral mixing and, as already noted, fuel and air deficiencies may exist side by side above the bed to the detriment of combustion performance.

Figure 2, from Skinner,[2] shows the variation in flue gas composition as a function of fuel concentration in the bed. At low fuel concentrations significant oxygen levels exist, but fall rapidly as the fuel concentration increases, while CO concentrations increase. This leads to an optimum condition for operation illustrated by Fig. 3 from a paper by Bailey *et al.*[11] Figure 4, from Highley and Merrick, shows the dependence of combustion rate on coal feed rate up to the point where virtually all oxygen is reacted within the bed, after which the air supply rate is limiting.

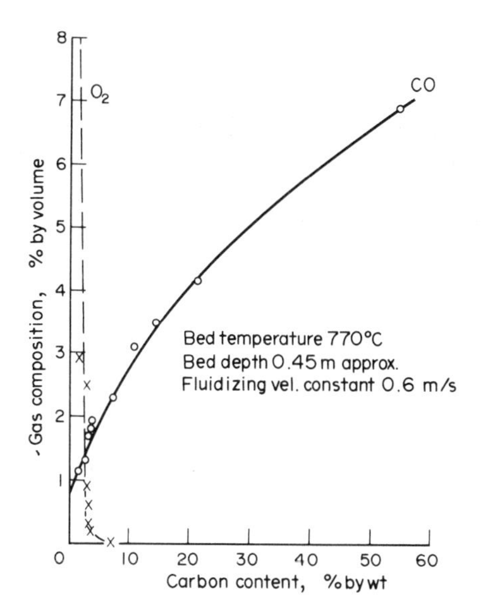

FIG. 2. Gas composition/bed carbon content for combustion of char in CEGB fluidized combustor. Bed temperature 770°C; bed depth 0.45 m; fluidizing vel. 0.6 m s^{-1} (from ref. 2).

In the absence of heat abstraction, the bed temperature will depend on the balance of heat produced by combustion and that lost to the flue gases or taken up in endothermic (gasifying) reactions. In practical coal combustion systems the bed temperature must be kept below the ash sintering temperature, and of course flue gas losses as sensible heat or gas calorific value (combustibles) must be minimized. Control of bed temperature is then achieved by heat transfer to a load (i.e. to water or steam in a boiler). The higher the temperature the more rapid is the air/coal reaction and the smaller the particle population required to take up the available oxygen. The effect of bed temperature is marked below ~750°C for bituminous coal combustion systems, although combustion can be sustained at much lower temperatures given adequate coal concentrations. The upper temperature limit depends mainly on ash constitution and bed material, which together determine

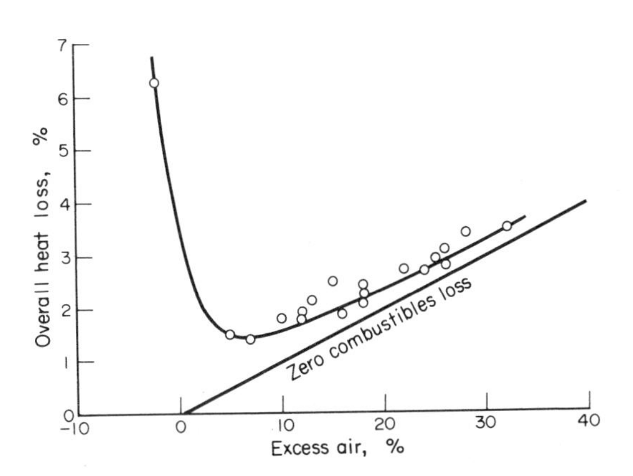

FIG. 3. Variation in overall heat loss in CO, unburnt carbon and excess air with % excess air. 800°C bed fluidized at 0.6 m s^{-1} with fines recycle (from ref. 11).

sintering behaviour, and is typically in the range 900–950°C.

Temperature effects are important during start up and load changing. Start up normally involves the use of a relatively expensive auxiliary fuel (oil or gas) and it is economically important to minimize the start up fuel consumption. A high coal concentration is desirable to offset the low specific reaction rate and to

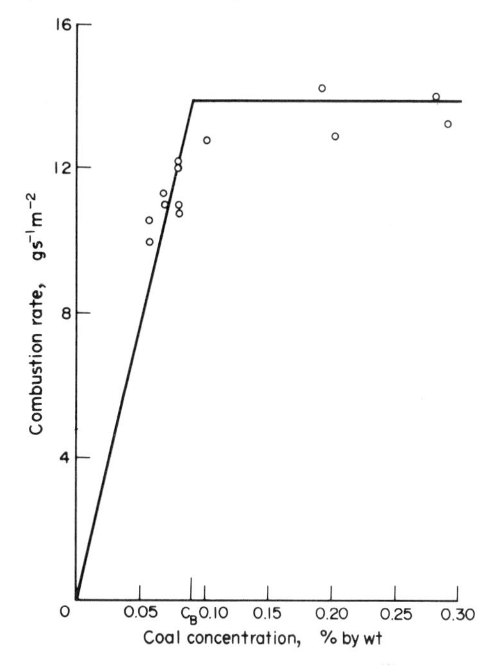

FIG. 4. Variation in coal combustion rate with coal concentration in a 0.6 m deep bed with a fluidizing velocity of 0.6 m s^{-1} and a temperature of 800°C (from ref. 10).

increase the total reaction rate. As combustion is initiated, the bed becomes self-heating and, with rapidly increasing temperature, it can pass through a stage when the coal concentration is too great for efficient combustion at the prevailing temperature and allowable air rate. To a lesser extent the same limitations apply during load changes. The effects can be minimized by the elimination of coarse coal, since this reduces the total coal inventory of the bed for a given coal particle population.

The transient behaviour of fluidized beds has been studied in relation to boiler designs,[4] which will be referred to later. However, it appears that the reaction rates and efficiencies during the start up period are still in some doubt, although practical start up problems apparently have been readily overcome on experimental units. In such cases excursions from optimum operating conditions are of minor importance, but proper full scale evaluations of the problems will be necessary for commercial plant.

Economically, it is desirable to work with high superficial gas velocities to keep down the size of plant, with relatively low bed depths to reduce system pressure drop, and with few coal feed points in the interests of simplicity. It is, therefore, understandable that these practical parameters have figured prominently in reported work on fluidized bed combustion as such. The earlier NCB/BCURA work in the combustion area is summarized in Skinner's review,[2] while Highley and Merrick[10] have analyzed the effect of coal feed conditions. All this work indicates that at superficial fluidization velocities above 1 m s^{-1} and bed depths around 0.5 m fine carbon is elutriated in quantities large enough to produce unacceptably high losses, the effect being aggravated with increased feed point spacing. In a practical world, while an upper coal size can be stipulated, there will be no way of eliminating dust, and in fact this may be formed in the bed by attrition or thermal decrepitation of coal, so that there will always be a range of sizes of combustible material which is small enough to be elutriated at these higher velocities before it is completely burnt. The effect is illustrated by work reported by Coates and Rice[12] of the US Bureau of Mines, and has been analysed by Merrick and Highley.[13]

More recent work on combustion efficiency has accepted this limitation and attempts to burn effectively in one pass have been virtually abandoned. In some respects this is unfortunate, in that it has meant that there has been relatively little refinement of data or of firing techniques, but the need to refire unburnt carbon has been addressed. Both Ehrlich[14] in the US and Hoy[15] in England reported work on this subject at the Second International Conference on Fluidized Beds in 1970. Ehrlich produced a model for the particular system with which he was concerned. This showed that burn out was most affected by bed temperature and excess air.

Further work by Coates and Rice[16] supported the need for carbon recycle, and again illustrated the importance of excess air and temperature, although their published data did not provide a sufficient basis for systematic analysis.

Although the loss in combustion efficiency in carbon elutriated from a fluidized bed combustor can be recouped by recirculation or in a burn up bed, it is important not to overlook the importance of uniform combustion in the bed. Large variations have been observed in both gas temperature and unburnt fuel gases above fluidized beds, and these clearly relate to the observations of Highley and Merrick[10] already referred to. From this observation it may be inferred that conditions within the bed will also vary widely in terms of gas composition, to the detriment of immersed surfaces, which may suffer corrosive attack. The unburnt gases will eventually find oxygen above the fluidized bed, but turbulent mixing in this zone will in general be relatively slow and as cooling proceeds, the probability of effective combustion will fall. In any practical application of fluidized bed combustion to steam generators, these effects will need careful assessment in considering fuel distributor design and bed geometry, including immersed heat transfer surface.

In pressurized fluidized bed combustors combustion efficiency has been observed to be less of a problem. To some extent this is probably due to the enhanced oxygen partial pressure, but most reported experience has been for systems with low superficial velocities and particle sizes, high excess air, and relatively deep beds. Hoy and Roberts, for instance,[8] reported efficiencies of 98–99% without carbon recycle when operating at superficial velocities of about 0.6 m s^{-1} but Hoy and Locke foresaw the need for recycle at higher velocities.[17]

EMISSIONS

Since the recent US interest in fluidized bed combustion arose from the possibility of removing sulphur from the combustion products, it is almost axiomatic that the SO_2/CaO reactions, and practical ways of achieving them, have been a major subject for research. At the same time the emission of other important pollutants has received study; essentially these are oxides of nitrogen, hydrocarbons, carbon monoxide, and particulates. With the exception of particulate emissions the prospects of operating a fluidized bed coal combustor with low emission levels are good. Hydrocarbons and carbon monoxide are a direct consequence of combustion inefficiency, and can be reduced by allowing adequate excess air and ensuring proper mixing.[18] Particulate emissions have received attention mainly because of their high carbon content and because in any application they would need to be removed by cyclones, electrostatic precipitators or filters. For gas turbine applications this involves hot cleaning. Although it has been claimed that the low combustion temperatures produce a non abrasive ash[8] the detailed chemical nature of elutriated material and associated condensed or adsorbed species will also be important in deciding the corrosive properties and any possible environmental hazards arising from them.

The usual arrangement for sulphur absorption by limestone is to use stone crushed to the desired size (depending on fluidization velocity) as the bed material; thus, unlike the combustion reactions, the limestone reactions are not regulated by the limestone population so much as by the available reactive surface. Products of reaction are not removed as gas, as is reacted coal, and if not removed by natural elutriation the solid product, mainly $CaSO_4$, must be removed from the bed mechanically. The equilibrium composition of the bed material will therefore depend on the sulphur loss rate from the coal, the reactivity of the bed material, and its rate of replacement. The composition of the bed material and its reactivity are interrelated. As the limestone becomes sulphated the unreacted material within a given particle becomes less accessible to SO_2 and the reaction rate falls.

The most important parameter in deciding the reactivity and capacity of limestones appears to be pore size. Borgwardt and Harvey[19] reported on a variety of natural rocks and correlated both initial reaction rate of SO_2 uptake with the BET surface area of the calcined material. The decline in reactivity in lime is due to progressive pore blockage by reacted material which has a specific volume some 170% greater than that of calcium oxide. Hartman and Coughlin[20] were able to develop a mathematical model on this basis, and showed good agreement with experimental data. They concluded that pore blockage could be expected to restrict limestone/sulphate conversion to about 50%.

The behaviour of limestone as a sorbent for SO_2 was extensively reported in the Second International Conference on Fluidized Bed Combustion. Workers at Argonne National Laboratory[21] employed direct experimentation in a 150 mm diameter bed with substantial microscopic and chemical examination of particles. They studied both very fine limestone (25 μm) which would have elutriated rapidly from the bed (which was largely alumina), and coarser limestone (500–600 μm) as the bed material itself. In both cases they found a strong dependence of SO_2 absorption on the Ca/S ratio in the feed, with typical values around 80% at 2.5 Ca/S. The correspondence between the two sets of results was regarded as fortuitous, since conditions for contacting gas and limestone would have been quite different. It is not impossible, however, that the fine limestone coated the alumina particles temporarily, the alumina then providing a gas contacting medium similar to coarse stone.

The fundamental studies carried out at Argonne showed that, as suggested above, the reaction rate of limestone decreased with time of exposure (i.e. with SO_2 take up), (see Fig. 5) and gave an insight into particle penetration and usage by SO_2 via microprobe analyses.

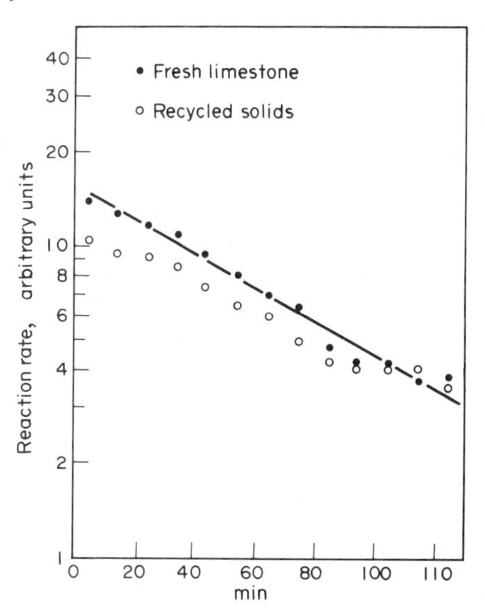

FIG. 5. Reaction rates at an SO_2 partial pressure of a P440 for fresh limestone and recycled solids (from ref. 21).

Similar results on absorption efficiency have been reported by other workers—for example Glenn and Robison[18] at the same Conference, reported 80% reduction of SO_2 levels with 2.5 Ca/S ratio with one limestone, but found that the numerical values varied with different materials, suggesting that physical properties such as pore size are critical in deciding stone reactivity. Figure 6 illustrates this finding, which was confirmed elsewhere for a variety of British and US coals.[22] The same workers studied the effects of particle size and bed depth and found a "worst"

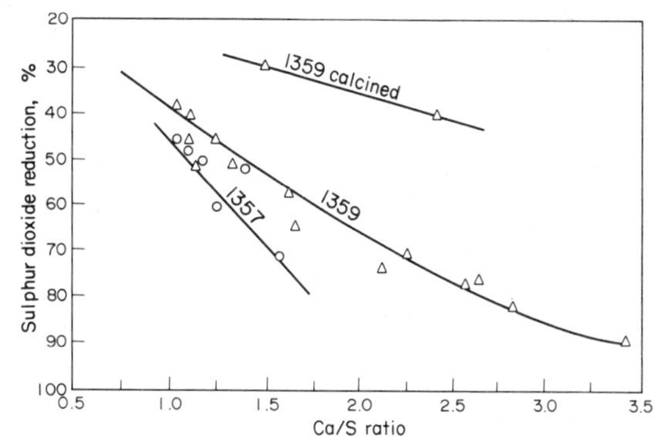

FIG. 6. SO_2 reduction—effect of limestone to sulphur ratio (from ref. 18).

particle size for each particular bed condition. This seems to be explained by the fact that finer particles reacted more effectively before being elutriated, while large bed depth was beneficial in providing longer solids contact time. Only very fine limestone (44 μm) gave good absorption efficiency when used as an entrained additive, and it was concluded that the bed material should itself be limestone to provide a high contact area. As already noted, most subsequent work has been on this basis.

The use of limestone as bed material means that, irrespective of the limitations imposed by progressive pore blockage, very high stone utilization is unlikely. To achieve a steady bed level and reactivity, fresh stone must be continually added and bed material removed. Because of the inherently good mixing in the fluidized bed the material being removed tends to be a time average sample, and may include anything between fresh material and fully reacted material. Fortunately, because the initial rate of limestone reaction is high, and declines exponentially, the probability is small of stone being withdrawn from the bed before it is significantly reacted, and relatively uniform conversion is achieved even though the ideal countercurrent reactor condition cannot normally be approached. It may be significant that better stone utilization was reported in experiments in which limestone movement was downward.[16] Higher air velocities and more rapid mixing were deleterious in this case.

A further complication arises from the fact that the reactions between lime and SO_2 are strongly dependent on the availability of oxygen, and on the temperature. Glenn and Robison[18] reported that when the flue gas oxygen fell to 1% and the bed temperature was raised above 1000°C, SO_2 was released rapidly, giving some 8% SO_2 in the flue gas; they pointed out that this suggested a means of limestone regeneration.

These observations agree with earlier work by Moss on gasification and combustion of oil in fluidized beds[23] and by Jarry *et al.* at Argonne.[24] Moss pointed out that for SO_2 uptake at temperatures above 850°C, the presence of oxygen was essential to ensure the formation of calcium sulphate rather than the sulphite, and demonstrated (Fig. 7) that an optimum temperature for the reaction lay between 850°C and 900°C. He suggested that this could relate to an optimum con-

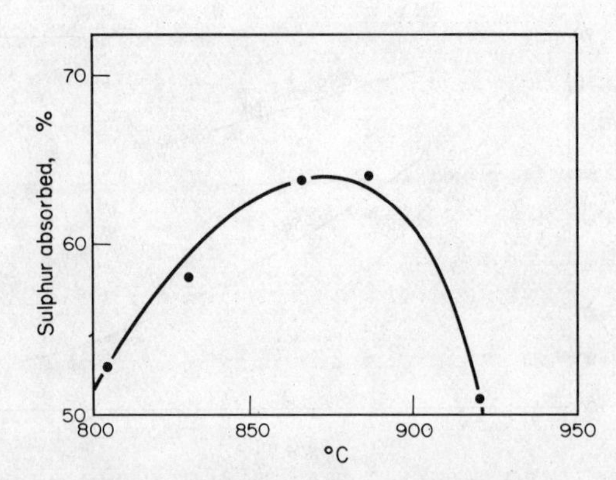

FIG. 7. Optimum temperature for sulphur oxide absorption by lime (31% of bed reacted) (from ref. 23).

dition for the formation of SO_3. He also recognized that (in his gasifier) local conditions in the fluidized bed could be either oxidizing or reducing, and that stone circulating between such zones could undergo alternating sulphation and regeneration. The same is true in a combustor. Apart from the physical effect of the high mixing rate, therefore, the degree of conversion of lime to the desired sulphate form will depend on the availability of oxygen at different positions in the bed, and the relative rates of the sulphation/regeneration reactions under local bed conditions. Stone conversions in the order of 40%, as generally reported, are not surprising.

Sulphur sorption at elevated pressures was reported by Wright[25] in 1973, together with further low pressure data. Pressure has an important effect in that limestone does not calcine at bed temperatures that are near optimum for SO_2 reaction. Dolomite, in which the magnesium carbonate does calcine, provides a more porous structure for access to the calcium carbonate, and Wright found it to be a markedly superior sorbent. Dolomite was also superior in low pressure tests at temperatures below 800°C. High gas velocities produced a drop in sulphur uptake which was explained by reduced residence time but could equally have been due to by-passing in bubbles and to the increased elutriation of the finer stone. Recycle of this stone produced a considerable improvement in SO_2 retention, and the performance of fine stone was strongly dependent on the gas velocity, which decided whether or not the stone was elutriated before being reacted. Kinetic studies by Westinghouse confirmed the effectiveness of dolomite in removing SO_2 at pressures of 10 atm.[26]

Later work on sulphur retention in pressurized systems has been reported by Vogel *et al.*,[27] Roberts *et al.*[28] and Hoke and Bertrand.[29] A significant feature of the first two papers is that dolomite at around 800°C apparently achieved sulphur retention efficiencies of around 85% for Ca/S ratios near unity when the bed superficial velocity was low ($\sim 0.6\,m\,s^{-1}$). This suggests that under such conditions, in relatively deep beds, the particle movement was more ordered than it would have been under low pressure high velocity systems, and indeed this would be expected. Also, in such conditions, with good carbon burn up, the chances of sulphate being reduced before leaving the bed would be relatively low—one would expect uniformly high oxygen availability near the top, and very little carbon. Of considerable significance was the fact, observed by Roberts *et al.*[28] that sulphur retention efficiency was uniformly good over the range 800–950°C. If generally confirmed, this increases the scope for achieving good combustion efficiency and for high thermodynamic efficiencies in gas turbine applications.

While the retention of sulphur in limestone offers an apparently simple solution to the sulphur emission problem, the quantities of stone required by a "once through" system are considerable, typically 10–20% of the coal burnt, and this would incur a significant cost in handling and waste disposal, apart from the cost of the material itself. There has therefore been a good deal of effort expended in looking for limestones with superior sulphur capacity, and in studying the regeneration of "spent" limestone (calcium sulphate) to recover sulphur and provide lime for recycling.

In thermogravimetric studies,[30] it has been found that the sorbent calcination conditions have a major effect on the sorbent capacity. Slow calcination (under high CO_2 partial pressures) apparently gives time for the oxide lattice structure to rearrange to provide relatively more large pores for SO_2 penetration. Slowly calcined stones were observed to have more than double the sulphur capacity of stones calcined rapidly in nitrogen. The same study concluded that the initial sulphation rate is determined by mass transfer of SO_2 to the particle surface. As surface is progressively sulphated, in-pore diffusion takes over as the rate limiting factor, and finally diffusion through the sulphate layer becomes limiting. The effect of system pressure is most marked in the second phase, when there is a direct effect of SO_2 partial pressure on reaction rate. The thermogravimetric work offered a consistent explanation of the observations of earlier workers, and confirmed the temperature for maximum absorption at about 850°C, with reasonable absorption to at least 900°C. The decline in rate at higher temperatures was attributed to reactions involving CaS as well as $CaSO_4$. This work suggests ways of improving sorbent performance, which could be of considerable economic value.

One way in which regeneration may be achieved is by heating the sulphate to around 1000°C in contact with a reducing gas such as hydrogen or carbon monoxide.

$$CaSO_4 + CO \rightarrow SO_2 + CaO + CO_2.$$

Other reaction routes are also possible.

Desorption of SO_2 was observed in natural gas combustion experiments by Jarry *et al.*[24] when excess air was low and temperatures were ~ 950°C, and the regeneration of $CaSO_4$ in coal-fired experiments has already been noted.[18] A good deal of time was devoted to controlled regeneration experiments during 1968–1972 and the results were reported at the 3rd International Symposium on Fluidized Bed Combustion and elsewhere.

The reaction route chosen by Vogel *et al.*[31] involved two stage regeneration; at lower temperatures (~850°C) $CaSO_4$ can be reduced to CaS which can be reacted with CO_2/H_2O at about 550°C to yield $CaCO_3$ and H_2S. A major problem appeared to be progressive deterioration in regeneration efficiency with successive cycles. Somewhat similar difficulties were found by O'Neill *et al.*, although CaS produced by sulphidation of limestone by H_2S could be converted to $CaCO_3$.[26] Hoke *et al.*[32] demonstrated that the direct reduction of $CaSO_4$ by partial combustion products of propane was feasible at 1000–1150°C in line with the reaction given above.

Moss[33] has put forward a good practical explanation of both the sulphation and direct regeneration processes; both his own work and that of workers at Pope Evans and Robbins in the US had suggested that regeneration was more effective in a carbon rich bed with combustion reactions proceeding than in a sulphate bed fluidized by reducing gas. The carbon may be important in ensuring adequate CO levels in contact with the sulphate; it is apparent from Moss' data that $CaSO_4$ is a powerful oxidizing agent. He has observed off gas compositions of 8–9% SO_2 and 25% CO_2, compared with only about 2% SO_2 reported by Hoke and Bertrand[29] for a gas reduction system, and the solid carbon reactor would appear to be a good basis for a commercial sulphur recovery unit.

Questions which are not yet properly resolved are, whether the majority of limestones will stand recycling without rapid decrepitation, and under what conditions this can be achieved.

A practical fluidized bed combustor utilizing limestone for sulphur removal, and having a stone regenerator, could possibly be built on the lines indicated in Fig. 8. A number of practical difficulties become evident, and as yet there does not appear to have been any serious attempt to deal with these on an engineering scale. Most obvious are the recovery of usable stone from the stone and ash carried over or removed from the bed during normal operation and the circulation of sulphate and regenerated material from bed to bed at temperatures of 800–1000°C.

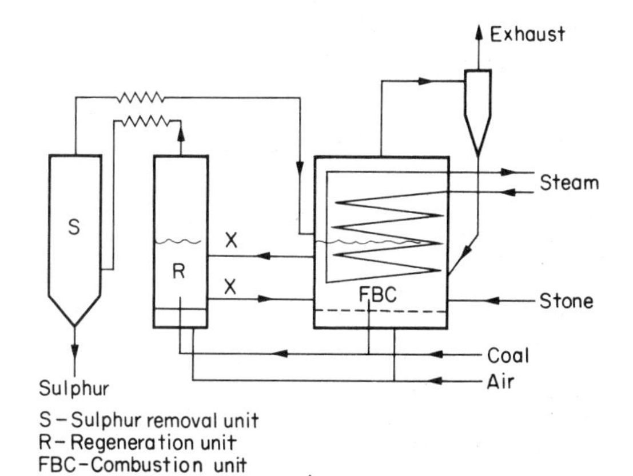

FIG. 8. Limestone fluidized bed with stone regeneration. S: sulphur removal unit; R: regeneration unit; FBC: combustion unit; X: bed solids exchange lines.

The emission of oxides of nitrogen from combustion sources is of considerable concern in regions where photochemically produced "smogs" may occur—two well known locations being Los Angeles and Tokyo. The phenomenon has led to restrictive legislation in the US and this trend may spread. It is known that atmospheric nitrogen can be oxidized in flames at high temperature, but that at temperatures below ~1500°C the conversion is small. The operating temperatures of fluidized beds are low enough for this mechanism to be unimportant. Nitrogen compounds are also present in solid fuels and oil, and these may form oxides of nitrogen during combustion at much lower temperatures, so that emissions from fluidized beds will depend on the fuel composition. These facts were established experimentally by Jarry *et al.*,[24] who observed also that NO levels in the off gas could be reduced some 40% when calcium sulphate was present in a coal burning fluidized bed. They suggested that $CaSO_4$ (but not $CaCO_3$ or CaO) reacted in some way with the NO. It is also possible that nitrogen compounds released from the coal reacted under these conditions without forming NO. Attempts to control NO formation catalytically were not successful.

Glenn and Robison[18] found a slight dependence of NO levels on excess air, which is in line with experience on other forms of combustor. Hoke *et al.*[32] found a slight effect of temperature and oxygen levels, both of which increased NO production, and went on to explore interactions between NO, CO and SO_2 with a variety of bed conditions. They found the NO+CO reaction to be dependent on the availability of H_2O in a packed column of $CaSO_4$, and to be strongly catalysed by lime and dolomite. NO and SO_2 were found to react in contact with partially sulphated limestone, and it was suggested that NO acted as an oxidant for $CaSO_3$. It was pointed out that, in a two stage combustion process, NO could be reduced by CO in the primary stage. Work on pressurized systems by the National Coal Board[25] showed that NO emissions were substantially reduced as compared with atmospheric operation but reasons for this were not suggested. This has been confirmed by Vogel *et al.*[27] who also reported a positive relationship between NO levels and the Ca/S mole ratio in the bed. This ratio defines the degree of sulphation of the bed material, and again suggests that $NO/SO_2/CaSO_4$ reactions occurred.

While NO levels in fluidized bed systems are lower than those found in normal pulverized coal systems, and lower again in pressurized systems, the explanation of this behaviour is not entirely clear. The reactions concerned may involve nitrogen compounds other than the oxides, simple reactions between NO and SO_2, or more complex routes in which sulphates and sulphites are intermediaries. Work at Sheffield University[9] suggests yet another possibility—namely that oxides of nitrogen are produced within the bed in relatively high concentrations, but are reduced in reactions above the bed.

Although the Sheffield experiments were confined to

small scale apparatus (0.3 m × 0.3 m), the published results show convincingly that NO_x can be reduced 40–50% in freeboard reactions. Much of the variation in other reported NO_x data seems to have been attributable to variations in bed temperature, since between 700 and 850°C an approximately linear four fold increase, from 200–800 ppm NO_x, was observed, with relatively slight sensitivity to excess air, for the conditions of the Sheffield reactor. This work was supported by further work in a 7.5 cm dia bed to explore mechanisms. It was concluded that NO can be reduced by both heterogeneous reactions with char and homogeneous reactions with CO, H_2, hydrocarbons and other nitrogenous compounds. It should be noted that in all these experiments the bed material was inert silica clinker, so that reactions involving the bed material were discounted, and the conditions were not directly comparable with the other work discussed above.

Without commenting on the possible mechanisms, Roberts *et al.*[34] published data which suggested a continuous correlation between NO and O_2 in the exhaust gases, from which negligible amounts of NO would be expected with less than 1% O_2. This is not inconsistent with the Sheffield theory, and indeed if most of the inherent nitrogen in the coal is initially oxidized, some such explanation must apply. However, the Sheffield results show more temperature dependence than oxygen dependence.

While it may be impossible to avoid formation of NO_x from fuel constituents, the feasibility of reducing them in subsequent gas phase reactions warrants much more study in relation to fluidized bed systems. There is little doubt that in America NO_x emission regulations will present an increasing obstacle to direct coal combustion, and other countries may be expected to follow.

On particulates, it is apparent that in all cases employing fluidized beds operating above $1\,m\,s^{-1}$ gas velocity, in which bubble formation will be experienced, particles carried over from the bed are considerably larger than would be expected from simple consideration of terminal velocities. Splashing by bursting bubbles is apparently the cause, and the only practical correctives appear to be baffles or the provision of long disengagement heights. Suitable arrangements of heat transfer surface may act as baffles in this context.

It has been confirmed by Vogel *et al.*[27] that the release of volatile inorganic impurities from coal is less in fluidized bed systems than in conventional combustors, as had been expected. From the environmental point of view, therefore, particulate burdens, for gases subjected to standard cleaning procedures such as filtration or electrostatic precipitation, should be less of a hazard than those from pulverized coal units. Elements of particular concern in this respect include lead, mercury, arsenic and fluorine; of these, lead and arsenic were strongly retained in the fluidized bed.

HEAT TRANSFER

Surfaces immersed in a bed of fluidized solid material will exchange heat with the bed material by both convection and radiation; since the bed will constitute something approaching a black body, radiative exchange will be very efficient and should be calculable from the usual Stefan Boltzman equations. At most temperatures of practical interest a more important mechanism is that of convection from the fluidized bed material to the immersed surfaces, and it is the dependence of this mechanism on particle properties, gas properties, and fluidization behaviour that has stimulated most investigation. Convection, in relation to fluidized beds, is not a very definitive term, and really covers a variety of closely interacting non radiant mechanisms involving both convection and conduction.

When a gas is passed through a bed of fine solids which are initially at a different temperature, it is to be expected that, because of the large surface available for heat transfer, and the higher thermal capacity of the solid, the gas temperature will rapidly approach that of the solid. Thus, in the absence of solids movement (packed bed) the system is characterized, after a given time, by a zone of solids near the gas entry in which the temperature is near that of the entering gas, a relatively narrow zone in which most of the heat is being transferred, and a third zone in which the solids temperature is virtually unchanged but the gas is nearly at its leaving temperature, which will approach the temperature of the solids. If, on the other hand, the solids are in a state of continuous movement due to fluidization, particles actively exchanging heat with the gas will be continuously dispersed in the bulk solids. Under these conditions the gas, because of its low specific heat, will be effective only as a "short range" heat transfer medium as it moves from particle to particle, and will not be a bulk heat carrier except insofar as it conveys heated solid particles.

In practice the important heat transfer stage is between the bed (solids and gas) and heat transfer surface immersed in it. Direct conduction between solids in contact, convection by gas moving alternately over bed particles and adjacent heat transfer surface, and conduction through the gas boundary layer between surfaces will all play a part. These mechanisms will be strongly dependent on the particle size (which determines the number of points of contact, the surface for particle/gas heat transfer, and the gas transport distance), and on the particle movement (which determines the rate at which heat is transported from the bulk of the bed to the vicinity of the surface). To a lesser extent the properties of the particles themselves (as heat carriers), and of the gases (as conducting films), will be important, but in coal combustion systems the only likely variation of any importance is that due to pressurization, which will directly affect gas density.

Strictly, the effects of radiative and convective heat transfer should be additive, but where bulk material properties are the basis of calculation there will be interference. Thus, especially with coarse bed material, particles exchange heat by radiation as they approach a surface, and the effective particle temperature will then be lower than the bulk bed temperature. For large temperature differences between bed material and immersed surfaces this accounts for a lower apparent emissivity and convective heat transfer coefficient than would be found for small differences.

Botterill[35] has recently reviewed the very extensive literature on heat transfer in fluidized beds and his book gives a very full background. Unfortunately the importance of particle movement is such that it is impractical to produce a rigorous generalized equation describing heat transfer to surfaces immersed in gas fluidized beds. Correlations such as that put forward by Elliott *et al.* (Fig. 9)[36] for the effect of particle size must therefore only be accepted as guides. In this

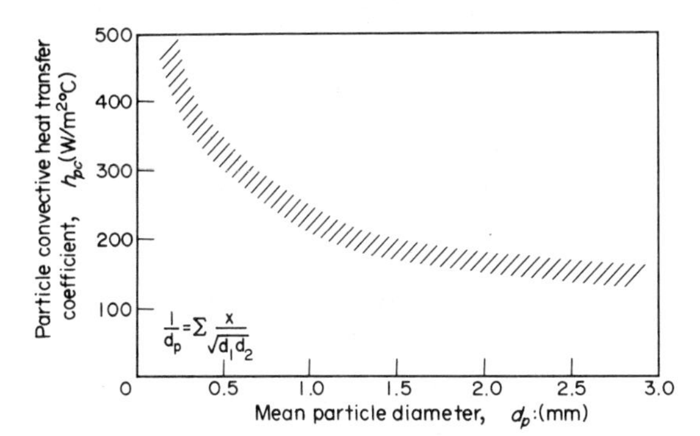

FIG. 9. Approximate correlation between particle convective component of heat transfer coefficient and the mean particle size (from ref. 36).

correlation the heat transfer coefficient is proportional to about $d_p^{-0.5}$ where d_p is the mean particle size defined by

$$\left[\frac{1}{d_p} = \Sigma \frac{x}{\sqrt{(d_1 d_2)}}\right]$$

in which d_1 and d_2 define the size range contained in fraction x. Other workers have proposed values for the exponent of d_p between -0.25 and -1, and this variability may be ascribed to differences in the bed design, differential temperature, and particle size distribution. Even in the same apparatus there may exist significant point to point differences in particle movement due to wall effects, the immersed surface, and the geometry of the bed.

Thus Hoy,[15] reporting on some of the earlier BCURA work, mentioned the fact that higher heat transfer coefficients were found with higher tube temperature (i.e. with lower temperature differentials), and gave values of convective heat transfer coefficient in the range 140–700 $Wm^{-2}K^{-1}$ for mean particle sizes 2.5 to 0.1 mm, and tube temperature 100°C. For operation under 5 bar pressure he quoted values around 300 $Wm^{-2}K^{-1}$ for 0.5 mm particles showing that the effect of pressure was small.[8]

Skinner[2] has reproduced results of some of the earlier work in which small particle sizes were used, and this suggests that there is little further increase in heat transfer coefficients for particle size reductions below 0.05 mm. Where a very wide range of particle sizes exists non uniform fluidization conditions may exist and the behaviour may become very sensitive to gas velocity.

It is commonly observed that the heat transfer coefficient increases with fluidizing gas velocity, up to a point where excessive void formation reduces particle to surface contact significantly. With wide ranges of particle size both settlement of large particles at low velocities, and elutriation of fines at high velocity, may produce additional effects. These possibilities were discussed by McLaren and Williams.[37] Their work, with fluidization velocities between 0.3 and 1.2 $m s^{-1}$ demonstrates, in general, an increase of the order of 10% in heat transfer coefficient over that range. The coefficient was also studied in relation to tube arrangement and the variation was correlated using as parameter the narrowest gap between (horizontal) tubes. For gaps increasing from 50–300 mm the coefficient increased about 20%. These results were obtained in a bed at approximately room temperature and may not be quantitatively applicable to a combustion system, especially one having larger particle sizes, with significant radiant heat transfer.

Although the published data illustrate trends in heat transfer coefficients, and order of magnitude, there are so many apparent anomalies that some full scale experimentation may be necessary to support any specific design. It is also clear that control of particle size in the plant will need to be reasonably good, but conversely, deliberate variation of particle size may offer an acceptable way of correcting heat transfer behaviour so long as it is not inconsistent with combustion and sulphur retention.

The velocity/heat transfer characteristic of the fluidized bed may raise operating problems in plant such as steam boilers because at a given temperature of bed the characteristic is near linear over a restricted range only, and is not very steep. Control of heat transfer may therefore require techniques such as variable immersion of heat transfer tubes using the natural increase in bed depth with increasing gas velocity or variation of the solids inventory of the bed. There is, therefore, interest in heat transfer in the "splash zone" where tubes are subject to intermittent contact with the bed material. NCB work[37] indicated that such tubes would receive about 70% of the heat transferred to fully immersed tubes. Elliott and Virr investigated heat transfer from very shallow beds to finned tubes. These systems are virtually all splash zone. The reported overall heat transfer coefficients[38] were slightly higher than those for fully immersed tubular surfaces. This result is surprising, and its relevance to conditions in the splash zone of a normal deep bed is doubtful.

MATERIALS

The application of fluidized bed combustion to power generation requires assurances about the compatibility of materials. Most high temperature corrosion problems in conventional fossil-fired boilers arise from the action of salt melts.[39,40] Reactive elements such as the alkali metals may be volatilized during combustion and condensed onto heat transfer surfaces. The addition of reducing species such as partly burnt coal modifies the ensuing reactions, so that sulphur penetration (sulphidation) may occur as well as, or alternatively to, accelerated oxidation. The lower combustion temperatures of fluidized bed combustion will reduce the volatilization of salts, and the surface temperatures will also tend to be free from some of the major non uniformities which may be found in such components as radiant superheaters. Ash fusion should not occur, and relatively few coal impurities will be subject to thermal breakdown.

Despite these advantages there is some fear that at higher velocities the fluidized bed solids may erode protective oxides, and that chemically reducing conditions may be found in parts of the bed adjacent to fuel feed points. Erosion is very strongly velocity dependent, $\propto$ vel^3 at least, and despite the high solid loadings in and above fluidized beds, velocities are generally less than half those in conventional systems.[41] Further, the ash particles are soft and friable since they have not been heated to temperatures at which fusion would occur to form hard dense glasses. Corrosion is therefore considered to offer more cause for concern than erosion in most applications.

The NCB have carried out fairly extensive materials testing,[42] the results of which are on the whole encouraging. A range of possible boiler tube materials was included in both immersed and freeboard test samples. Fluidizing velocities were up to 2.4 m s^{-1}, bed particle sizes up to 3 mm. These tests indicated that protective oxides were formed and were generally effective up to the normal maximum operating temperatures of the materials. High chromium and austenitic materials performed exceptionally well. However, attention must be drawn to certain observations, which require further investigation: (1) $2\frac{1}{4}$Cr steel, which is commonly used for boiler pressure parts suffered an unacceptable metal loss at temperatures approaching its normal maximum service temperature (590°C). (2) Sulphide penetration was observed in a number of specimens, but was not generally serious. Again, $2\frac{1}{4}$Cr steel appeared most susceptible, especially at 600°C. (3) Ash deposits built up to equilibrium thickness of about 1.5 mm on the undersides of tubes in the vicinity of coal feeds, and there was evidence of enrichment in sulphur and sodium in these deposits. (4) Tubes above the bed suffered less attack under oxidizing conditions than immersed tubes, but about the same in some cases under reducing gas conditions. (5) In one case with a high chlorine coal where the occurrence of fluidization problems may have complicated the issue, markedly higher corrosion rates were observed. This may be no more than confirmation that this coal was potentially corrosive. (6) Limestone addition to the bed appeared to enhance sulphidation rates.

Fluidizing velocity, operating pressure, coal type, and combustion conditions did not have any major effect. These results are encouraging, but much more long term experience is required to give an assurance that tube gas side corrosion will not occur, and to permit the choice of the best materials.

Gas turbine applications involve much higher velocities and metal temperatures above 800°C. The erosion problem is linked with the efficacy of hot gas filtration, but evidence from NCB work, although restricted to velocities between Mach 0.6 and 1.0 and mostly at temperatures below 800°C, suggests that moderate filtration efficiencies may be adequate. No erosion has been observed when gases were cleaned by two stage centrifugation,[34] neither was oxidation a problem. However, at temperatures above 900°C sulphidation of blade materials occurred. Even with sulphur retention in the bed sulphur penetration into blade materials may be the limiting factor, and more information, over long running hours, is required. The mechanism of sulphur transport and attack in this instance is not clear, although $CaSO_4$ would inevitably be present in the gas stream. It may be significant that the addition of limestone to the bed in the low temperature atmospheric pressure tests appeared slightly to increase the corrosion rate.

THERMODYNAMIC CYCLES

Power generation prime movers may be broadly classified as being either of internal combustion or external combustion type. So far as fluidized bed combustion is concerned, internal combustion engines are gas turbines, external combustion engines may be vapour turbines or gas turbines with heat exchangers. A wide range of combinations of both types is possible, and to some extent the virtues of each one depend on local conditions such as the availability of cooling water, the load characteristics of the system, the market for waste heat (e.g. as low pressure steam or hot water) etc.

In the present context, the important characteristics of the steam turbine (Rankine) cycle are that, because of the high pressures involved, the maximum economic steam temperatures are limited, currently below 600°C, possibly rising to 650°C, and all heat has to be transferred from the combustion gas to water or steam which is at high pressure (up to 300 bar) or used for air preheat. Fluidized bed combustion temperatures in the range 800–900°C afford a good temperature differential for direct heat transfer and permit efficient combustion with optimum conditions for sulphur retention if this is desired. At 800°C the combustion gases still contain about one third of the useful heat of combustion and this must be taken up in conventional convection surface, i.e. above the fluidized bed.

The immersed surface would operate at uniformly high heat transfer loadings up to 60 kW m^{-2}. This is only a tenth of the maximum normal radiant flux found in conventional oil-fired boilers, but it would apply around the whole tube circumference. The average heat flux per unit length of fully immersed tube would be about $\pi/10$, or say three tenths, of the maximum found in the oil-fired boiler case, and near the maximum found in pulverized coal boilers. The fluidized bed combustor/heater therefore achieves efficient and economical use of surface without involving dangerously high local heat fluxes. The low corrosion potential noted earlier also holds out the possibility of achieving higher steam temperatures provided that steam side problems due to oxide exfoliation, or turbine materials problems, do not present limitations. On reheat cycles, which are usual in large steam plant, it may be possible to make use of the fluidized bed's adaptability to modular construction, to introduce more than one reheat stage without being confronted by serious layout and control problems. In summary, the fluidized bed boiler applied to Rankine cycle plant could lead to minor advances in cycle efficiency by 5–10% to over 40% overall, or to lower boiler costs, without major changes in plant design apart from the boiler itself.[43] At the same time the combustor could use low grade (high ash, high sulphur) coals and still meet requirements for low emissions of pollutants. It is the latter feature that constitutes the main spur to development of the fluidized bed atmospheric pressure combustor in America.

The direct fired (internal combustion) gas turbine is a cheaper machine than the steam turbine since in its simplest form it avoids heat transfer surface. For power generation its use has in the past been limited by fuel costs since turbine efficiency and blade life are seriously affected by fuel contaminants such as sodium, vanadium and other ash forming substances. In the presence of sulphur oxides a range of fusible salts may be formed, leading to fouling and corrosion at the economic operating temperature level. Currently, therefore, gas turbines are operating on what are normally high cost fuels such as gas, kerosene, or selected high grade crude oils, and attempts to burn solid fuels have been unsuccessful. The fluidized bed may be able to alter this situation because, as has been noted, the volatilization of potential corrodants is slight at the relatively low combustion temperatures, and the unfused ash tends to be so friable that blade erosion is likely to be acceptably slow. The limitations are that combustion temperature must be limited not only from the above considerations, but, where emission control is important, to permit sulphur oxide capture. Turbine inlet temperatures in the range 850–900°C appear feasible; these compare with those in current use in industrial turbines, and are above those that can be tolerated with residual fuel oil. However, they are well below the temperatures currently being considered (up to 1300°C) for advanced turbines burning clean fuels.

To apply the solid fuel fluidized bed combustor to the direct fired gas turbine requires that the combustor operate at a pressure of 10–20 atm. This introduces a need to feed solid fuel at these pressures, and a need to remove solid particulates from the hot gas stream to a level that the turbine can tolerate. BCURA, in their work, which represents most of that reported on the subject, have used lock hoppers and cyclones respectively for these purposes, but there are also plans to use such devices as pebble bed filters for hot gas clean up. At 850–900°C there is obviously a materials problem, and these components will inevitably add to the plant cost and reduce operating efficiency. When these factors are taken into account the incentive to develop the fluidized bed direct-fired gas turbine is relatively small; the cost (including development cost) is unattractive for peaking units, while the efficiency is generally too low for base load operation.

Combined cycles offer a means of achieving high thermal efficiency at capital costs between those of simple gas turbines and Rankine cycle plant. The basis of the high efficiency is that the gas turbine part of the cycle utilizes high grade heat (i.e. high temperature gas) while the steam turbine is capable of using the low grade heat rejected by the gas turbine. Since only the steam generator involves heat transfer, total plant costs are lower than for 100% steam plant. The boiler gas side may operate at atmospheric or elevated pressure. Three alternative arrangements are shown in Figs. 10 a–c.

These illustrations do not cover all of the possibilities—for instance the hybrid gas turbine in Fig. 10d can be used with exhaust heat recovery in a steam turbine as in Fig. 10a.

The exhaust heat recovery cycle can be built onto an existing boiler to increase the total plant output and efficiency. Since there is no heat transfer surface in the fluidized bed, the bed temperature must be controlled by operation with a large excess of air, which can subsequently be used for combustion of more fuel in the boiler. In this mode the cycle would use about twice as much fuel in the boiler as in the turbine, so that for a new plant the cost would not be greatly below that for a Rankine cycle plant, and the efficiency gain would also be small, of the order of 5–10%, depending on the steam conditions. At the other extreme the exhaust from the turbine can be used to generate steam without supplementary firing. This minimizes the role of the steam generator and so reduces the capital cost and requirements for cooling water. Heat is rejected largely via the excess air in the combustion products. Keairns *et al.*[44] estimated that the efficiency for this type of plant would be 1–5% lower than for a fully integrated plant of the type shown in Fig. 10b, in which the boiler is pressurized on the gas side.

The exhaust heat recovery cycle with supplementary firing has been applied successfully in oil or gas fired units where it is attractive because of the wide load range that can be covered by varying the secondary (boiler) fuel input. Potentially, oil- or gas-fired gas turbines can be added to existing boilers at low cost and without major demands on space. In fluidized bed fired arrangements this is not easily done, and a low cost flexible unit for load following applications seems unlikely to take this form.

The fully integrated concept illustrated at Fig. 10b is not only capable of a lower heat rate (heat consumption for unit output), but the inclusion of most of the boiler heat transfer surface in the pressurized part of the gas circuit leads to a more compact design. Bed temperature is controlled by this surface so that high excess air levels are not needed. Most heat rejection from the cycle is to cooling water in the steam turbine condenser. Where cooling water is not easily

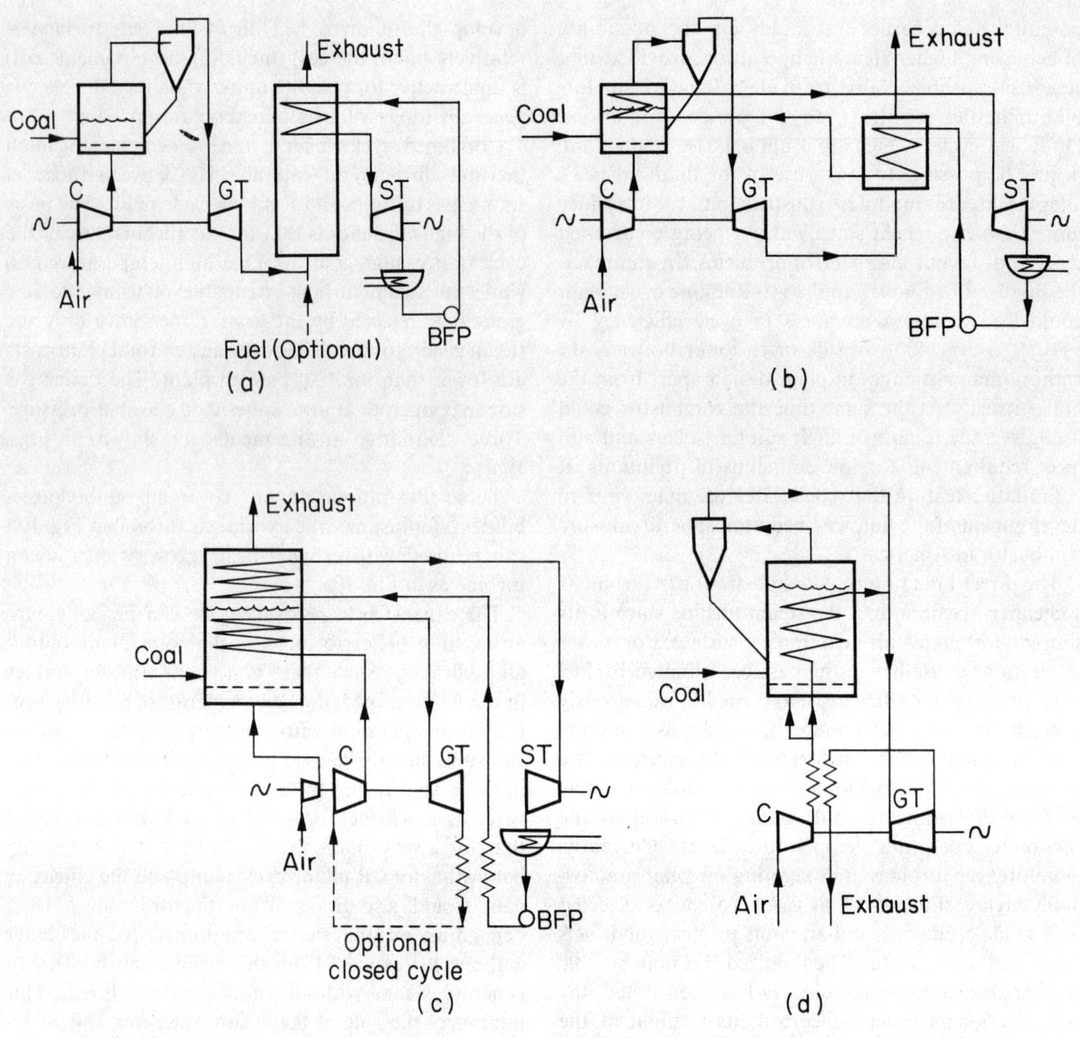

FIG. 10a–d. a. Exhaust heat recovery cycle. b. Integrated direct-fired combined cycle. c. Integrated indirect-fired combined cycle. d. Hybrid direct/indirect-fired combined cycle.
C: Compressor; GT: Gas turbine; ST: Steam turbine; BFP: Feed pump.

available, this is a disadvantage, but the good overall heat economy and power contribution from the gas turbine mean that cooling water requirements are still less than for a steam turbine plant.

A good analysis of this type of cycle was provided by Keairns *et al.*[44] for a range of operating conditions. According to this analysis the efficiency increased as excess air was increased, i.e. with increasing gas turbine capacity (Fig. 11). The combustor shell and heat transfer surface costs rose steeply with excess air up to about 40% but did not change much thereafter. The choice of conditions for overall optimum economy would depend on the relative importance, in a particular situation, of such factors as fuel costs, the cost of providing cooling water and emission control. It was pointed out that heat rate could be improved by about 1% for an increase in turbine inlet temperature of 100°C, but with the need to maximize sulphur capture a design temperature of 870°C (1600°F) was selected. Higher temperatures could be achieved by permitting combustion in the freeboard (i.e. above the fluidized bed) or by deliberate secondary combustion, for which purpose it was suggested that part of the coal could be gasified and injected above the bed. It is doubtful whether complications of this sort are warranted; although better heat rates can be achieved with very high temperature cycles, these depend on availability of clean fuels which will inevitably be expensive. The technical factors limiting development of the coal-fired combined cycles in this form are essentially the same as those limiting the coal-fired gas turbine.

To avoid interaction between "dirty" combustion gases and the turbine blades, indirect cycles of the type illustrated in Fig. 10c have been proposed. These may be either open cycle, with exhaust to atmosphere, or closed cycle, with the exhaust being recompressed as indicated by the dotted line. In such designs potential turbine corrosion problems are transferred to the heat transfer surface, where conditions are somewhat less onerous in terms of stress and mass transfer. The advantage is questionable, however, inasmuch as turbine blade cooling is an established technique, and in any case the relatively small amount of material used for blading permits the use of exotic alloys without

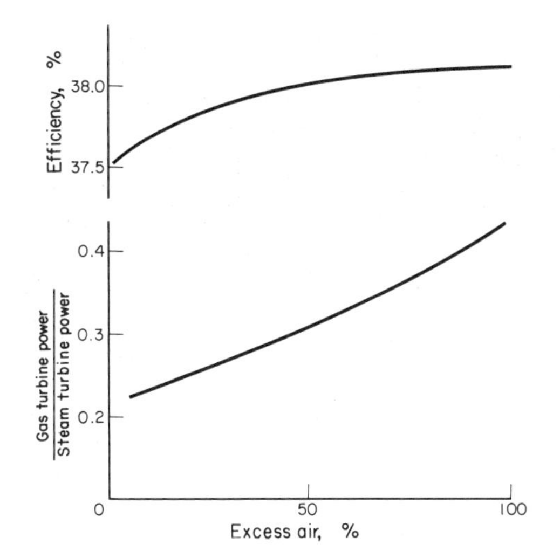

FIG. 11. Combined cycle plant. Effect of excess air on power ratio and efficiency.

severe cost penalties. Neither approach is applicable to heat exchangers.

Thermodynamically, the closed cycle indirect-fired unit is the only form showing an advantage, because it permits the use of a working fluid such as argon or helium with better thermodynamic properties than air. Such (monatomic) gases have a ratio of 1:1.6 between their specific heats at constant volume and constant pressure respectively, and this determines the amount of work that can be extracted for a given pressure ratio, and the cycle efficiency. The value for diatomic gases including air, is 1:1.4. For a given turbine inlet temperature the monatomic gas turbine is consequently more efficient than an air turbine, so that it is able to achieve good thermodynamic performance within the temperature limitations of the coal burning fluidized bed combustor. At turbine inlet temperatures around 800°C a helium turbine has a performance similar to that of an air turbine with inlet temperature of 1100°C.

As with open cycle turbines, the overall efficiency is improved by using the exhaust heat either in a recuperator or in a secondary cycle. The closed cycle machine inevitably requires heat transfer surface for heat rejection, so that these refinements follow naturally and with relatively little incremental complication. To reduce plant size the whole turbine cycle would normally be pressurized, and an important concern would be the containment of expensive working fluids such as helium.

A form of combined cycle that has received a good deal of attention in the US is the dual vapour cycle.[45] It is worthy of special mention in relation to the fluidized bed coal combustor because it has the capability of achieving very high efficiency without very high temperatures or pressures. Essentially this is because most of the heat input to the cycle is taken up as latent heat of evaporation at constant temperature, which can be chosen to match the constant temperature of the fluidized bed itself. The principle is illustrated in Fig. 12 which refers to a potassium/steam combination. Since potassium vapour pressure is only about 2 bar at 800°C the containment stresses are much less than for high pressure steam, and after doing work the potassium is still at a temperature high enough to pass on its latent heat to feed an efficient steam cycle. Efficiencies in the range 45–50% are possible in this way without compromising the ability of the fluidized bed primary heater to avoid ash fusion and to retain sulphur.

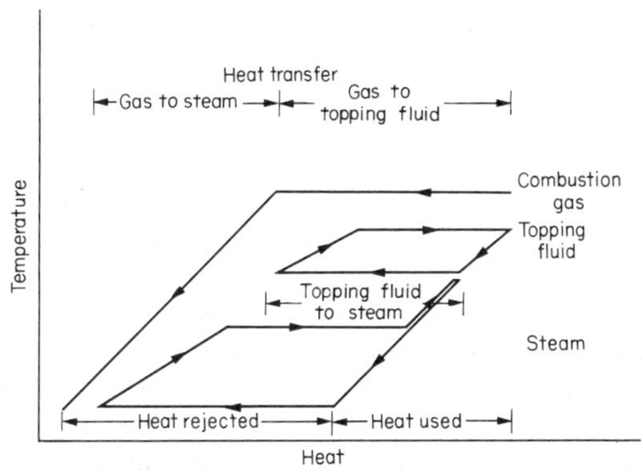

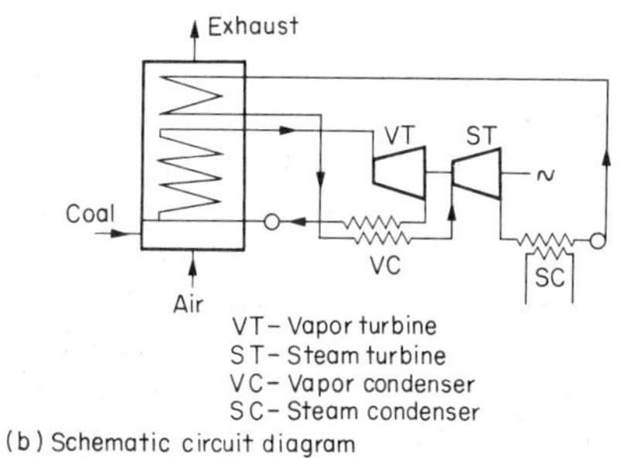

FIG. 12. Vapour topping cycle. (a) Heat flow diagram; (b) Schematic circuit diagram. VT: Vapour turbine; ST: Steam turbine; VC: Vapour condenser; SC: Steam condenser.

DEVELOPMENT PROBLEMS

The application of atmospheric fluidized bed combustion to steam generators is mainly a question of engineering development. Small combustors are fairly easily made and operated, and there is a good deal of experience from the many experiments quoted in this note and elsewhere. However, the successful operation of a large unit requires not only optimization of the basic combustion and heat transfer processes, materials selection, and effluent control, on all of which data exists, but also the availability of reliable means of distributing fuel and controlling the boiler/turbine combination. Control relates not only to coal feed rate, but also to the coal size and hence fluidizing velocity. Coarse coal, say 5–15 mm, will burn relatively slowly; it does not require precise feeding, as it will be distributed by the continual bed movement, and the response to a change in feed rate will be slow. To ensure an adequate coal particle population in the bed the total mass of coal in place at any one time must be relatively large. Such systems will therefore be relatively stable when operating, but present problems when rapid load changes are required. Conversely, systems fed with fine coal will require precise feeding arrangements as the coal will react rapidly near the point of injection, and they will respond rapidly to load changes. Volatile release from coal is much more rapid than carbon burn up, so that beds burning high volatile coals will be more sensitive to coal distribution than others burning non volatile fuels.

Start up and shut down are particularly important phases in a control scheme, and may seriously affect the operating economics in a power scheme. The fluidized bed has considerable thermal inertia, and current start up schemes rely on initiating combustion in part of the bed using oil or other auxiliary fuel. On shut down the bed may be "banked" only in the absence of cooling surface, or at temperatures near the normal operating level of that surface.

A good deal of careful analysis and development is needed to achieve anything approaching optimum design, but there would appear to be no significant unknowns.

Where limestone is used as a bed material for sulphur capture it is again clear that, although workable designs may be drawn up, the stone utilization efficiency is likely to be low. This may be unimportant where the amount of sulphur to be removed is small, but in other cases there is a need for a clearer understanding of the process and/or development of a practical scheme for the regeneration of spent material, from both economic and environmental points of view.

With these provisos, the atmospheric pressure fluidized bed combustor can be accepted as a practical (though not necessarily economic) proposition for steam raising. The position with regard to gas turbines is less certain, but it is clear that, in addition to the specific points discussed below, the problems of feed, control, and additive regeneration will need to be addressed in all applications.

In all applications of the coal-fired fluidized bed combustor, and especially where sulphur removal is important, the upper limit of temperature in the combustor is low relative to that possible with other devices. Where open cycle gas turbines are involved (simple, recuperative, or combined cycles) this limits the maximum cycle temperature to a level which is near that attainable with current turbine technology. However, current development programmes of nearly all the major turbine manufacturers will probably make possible turbine entry temperatures in the range 1000–1300°C when burning clean fuels. Thermodynamically, fluidized bed units will not be competitive. On the other hand, the only naturally occurring clean fuel is gas, and this is becoming scarce. The production of clean fuels from coal or oil is both expensive and to some extent inefficient, so that the overall advantage to be gained by their use is less significant and possibly negative.

Of possibly greater significance are the problems inherent in the use of solid fuel in a gas turbine. Feeding solids to a pressurized system is relatively difficult, while the combustor itself is bulky. Altogether, the fuel system is considerably more costly than that for a liquid or gas. The clean up of the combustion gas ahead of the turbine poses a further problem. As indicated earlier, the solid particulates appear to be much less abrasive than pulverized coal ash and acceptable clean up may be met in a number of ways. However this is done, it will again require large filters capable of operation at turbine inlet temperatures ($> 800°C$), or multi stage cyclones with significant pressure drops. The success of the direct coal-fired gas turbines rests primarily on ability to develop and prove such systems economically.

Inasmuch as fluidized bed coal-fired turbines are certain to be more costly than those designed for liquid or gaseous fuels, they are unlikely to qualify for peaking duties, or moderately low load factor operation. Furthermore, the large size and thermal inertia of the fluidized bed combustor and of the particulate removal system make them unsuitable for such applications. Combinations of gas turbine and steam cycle are more suitable for operation at high load factor because of their high efficiency, and in terms of capital cost they would compete favourably with steam turbine plant. Their technical limitations are similar to those of the gas turbine, with the addition of a somewhat more difficult control system.

The paper by Keairns *et al.*[44] gives a very good review of the various problems and costs at that time (1973) and concludes that, compared with conventional steam turbine plant, a pressurized fluidized bed combined cycle plant could expect to produce overall cost savings of 5–10% for power generation. The design approach on which this was based assumed the use of two stages of dust removal by cyclones and a conventional lock hopper coal feed system. It was acknowledged that these were capable of refinement, but the study did not anticipate any major technical obstacles. At the same time we have noted the incidence of sulphidation of high temperature materials, and the possibility that this may be aggravated by the presence of calcium sulphate from the fluidized bed. This could pose a development barrier, denying the achievement of overall efficiencies in the order of 45%, which Keairns *et al.* regarded as a possible development goal.

Schemes utilizing indirect gas heating avoid the turbine blade erosion problem. The heat transfer surface must operate near the bed temperature, and is still exposed to a possible sulphidation problem. In general, there will be a significant pressure differential between the working fluid and the combustion gases, so that creep resistance of heat exchanger materials will be important. Solutions to this tube materials problem are likely to be as difficult as, and perhaps more costly than, the turbine blade problem in direct-fired turbines. The "hybrid" gas turbine, in which about two thirds of the air is indirectly heated, has the virtues that the pressurized combustor not only reduces the dimensions of the fluidized bed but also obviates the pressure differential and hence the need for creep resistance, and this broadens the choice of available materials. Such machines do not eliminate possible blade erosion/corrosion problems, although they do reduce the amount of gas to be cleaned, as the combustion gases do not include a large excess of air. They may represent a good compromise, or they may encounter both blade and tube corrosion. Testing is needed to prove the point.

Because of the low specific heat and density of air compared with those of water, and the smaller temperature differentials, overall heat transfer rates to air cooled tubes are much lower than those applying to water cooled tubes. In pressurized combustion indirectly-fired gas turbines this leads to some difficulty in accommodating the required surface within the fluidized bed without adopting tube spacings which are so small as to inhibit particle circulation within the bed. Bed depths of the order of 3 m have been proposed to overcome this restriction, are acceptable from a pressure drop point of view, and may be expected to give good combustion efficiency and sulphur absorption. However, it may be impossible to maintain uniformity of bed temperature in such designs and this can only be established experimentally on a large scale.

The accommodation of heat transfer surface in other forms of indirect-fired gas turbines is not likely to be so difficult, in that the bed volume is inherently large. The problem here will probably be more concerned with support of the tubes over large bed base areas, a problem which is more severe than in the case of a boiler because of the higher operating temperature. In the helium closed cycle machine the problems are reduced by the excellent thermal conductivity of helium, but its low density would call for high operating pressures (~ 50 bar) to achieve economical designs.

From the fluidized bed point of view the dual vapour cycles, using fluids such as potassium, are not much different from steam cycles, except that the higher cycle temperatures reduce the available temperature differentials and hence heat transfer rates. The main problem otherwise appears to be the risk of liquid carry over from the vapour generator, since superheat is not practicable. Surprisingly, potassium does not pose severe materials problems and containment is amenable to conventional approaches at the relatively low pressures involved. On the other hand a large machine, in the range of even tens of megawatts output, has yet to be demonstrated.

The development status of the main thermodynamic options are summarized below.

Steam cycles with unpressurized combustors—
- Can be built from available knowledge.
- Need further development of coal feed, control systems, sulphur capture where applicable.
- Permit fuel efficiencies slightly higher than pulverized coal systems (over 40%). Sulphur absorbents add a cost penalty.

Direct-fired gas turbines and combined cycles—
- Have uncertainties on turbine blade life.
- Need development of particulate removal devices.
- Require continuous coal feed system.
- Permit combined cycle efficiencies 40–43%.

Indirect-fired gas turbines and combined cycles—
- Have uncertainties on mechanical design of heat transfer surface.
- Need materials to operate near bed temperature with good corrosion and creep properties.
- Permit efficiencies 40–43% with closed systems.

Dual vapour cycles—
- Have uncertainties on heat transfer surface.
- Require development of metal vapour turbines.
- Need materials to resist corrosion at bed temperatures (stresses low).
- Need to eliminate liquid metal carry over to turbines.
- Need development of containments for metal vapour turbines.
- Offer cycle efficiencies above 45%.

THE CURRENT STATUS OF FLUIDIZED BED POWER SYSTEMS

With the exception of Ignifluid designs, there are no fluidized bed coal-fired power generation units in service. As noted in the introduction, the Ignifluid design was more concerned with pregasification and ash removal than with the steam generation process, and has not had a major effect on large power generation systems. The most active interest in fluidized bed combustion for large power generating units is in the US where declining gas and oil reserves are enforcing a greater use of the vast indigenous coal supplies, while emission controls require the removal of sulphur from the fuel or from the combustion gases. The first pilot scale plant, of 30 MW capacity, is expected to be fired in 1976, at the Monongahela Power Co's Rivesville plant in West Virginia. This unit incorporates most of the present technology applicable to atmospheric pressure fluidized bed combustors for steam generation, and is regarded as a prototype for a 200 MW demonstration unit. Both of these have been described in a number of presentations [4,46,47,48] and are shown in Figs. 13 and 14. Foster Wheeler Inc. and Pope Evans and Robbins have formed the Fluidized Bed Combustion Co. to develop such boilers.

The 30 MW unit is a single drum boiler, to serve an existing steam turbine, and is laid out at a single level, the fluidized bed base covering an area roughly 12 m × 3.5 m (38 ft × 12 ft). This area is subdivided into four, providing three cells of roughly equal size and one rather smaller module. The containing and division walls are water cooled and form the bulk of the steam generation surface as in pulverized fuel boilers. The depth of the fuel bed during normal operation (the "expanded" depth) will be about 1–1.2 m and within this there will be immersed steam superheating surface in two of the modules, and evaporative surface in a third one. The fourth (small) module has no immersed surface. All four modules have economizer surfaces above the fluidized bed level.

This unit is based on the use of superficial gas velocities around 4 m s^{-1} using crushed limestone as the bed material, and will burn coal crushed to about 6 mm maximum size. Under these conditions the bed mixing will be very rapid, and it is envisaged that

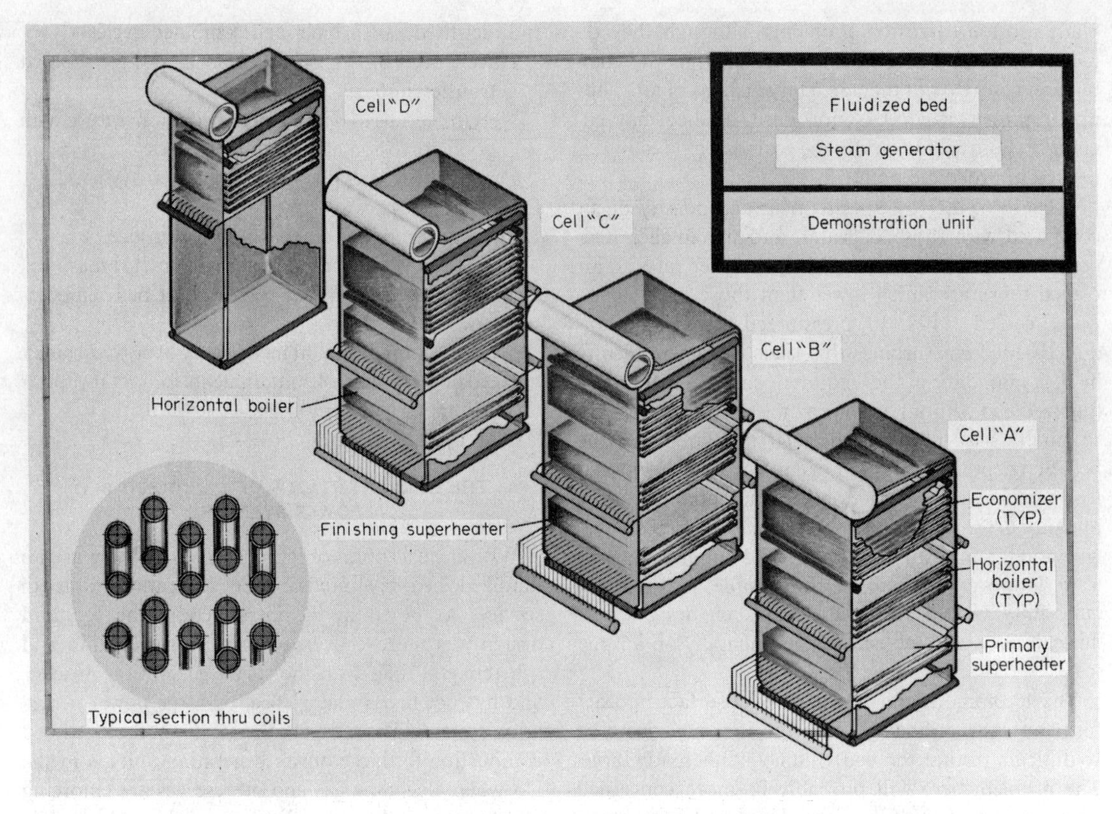

FIG. 13. 30 MW demonstration unit (from ref. 48).

simple gravity coal feed, with about one feed point per $1.5\,m^2$ will suffice. It seems likely that the base and feed design will be the subject of further development, depending on experience.

The small module in this design is a carbon burn up cell, and is fed with solids carried over from the three larger cells and separated by cyclones. The absence of cooling surface within the bed allows this bed to operate at a higher temperature, and an excess air rate of around 50% will be used to provide conditions under which residual carbon from the primary cells will burn efficiently.

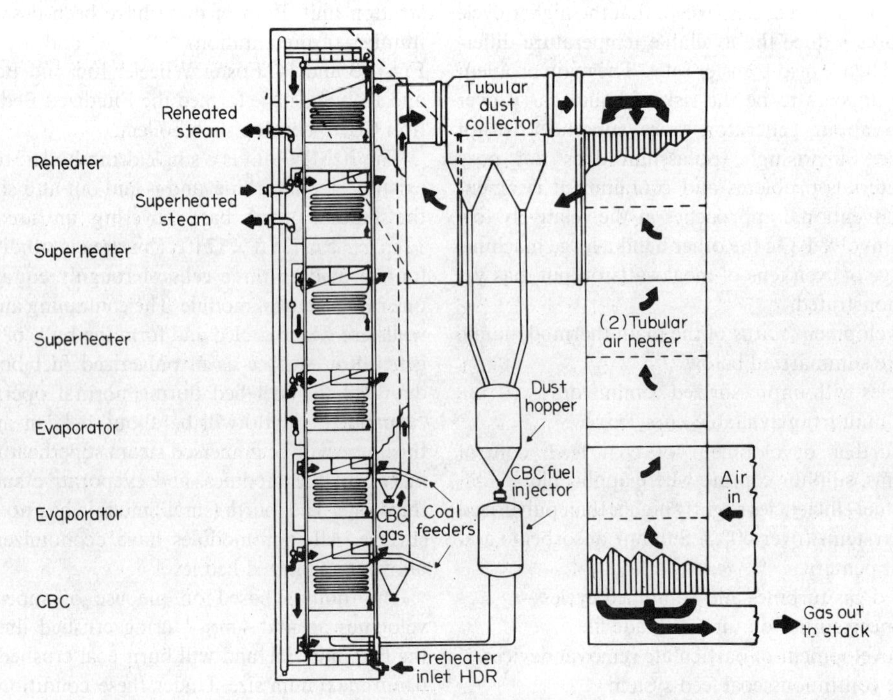

FIG. 14. Atmospheric pressure fluidized steam generator module (from ref. 47).

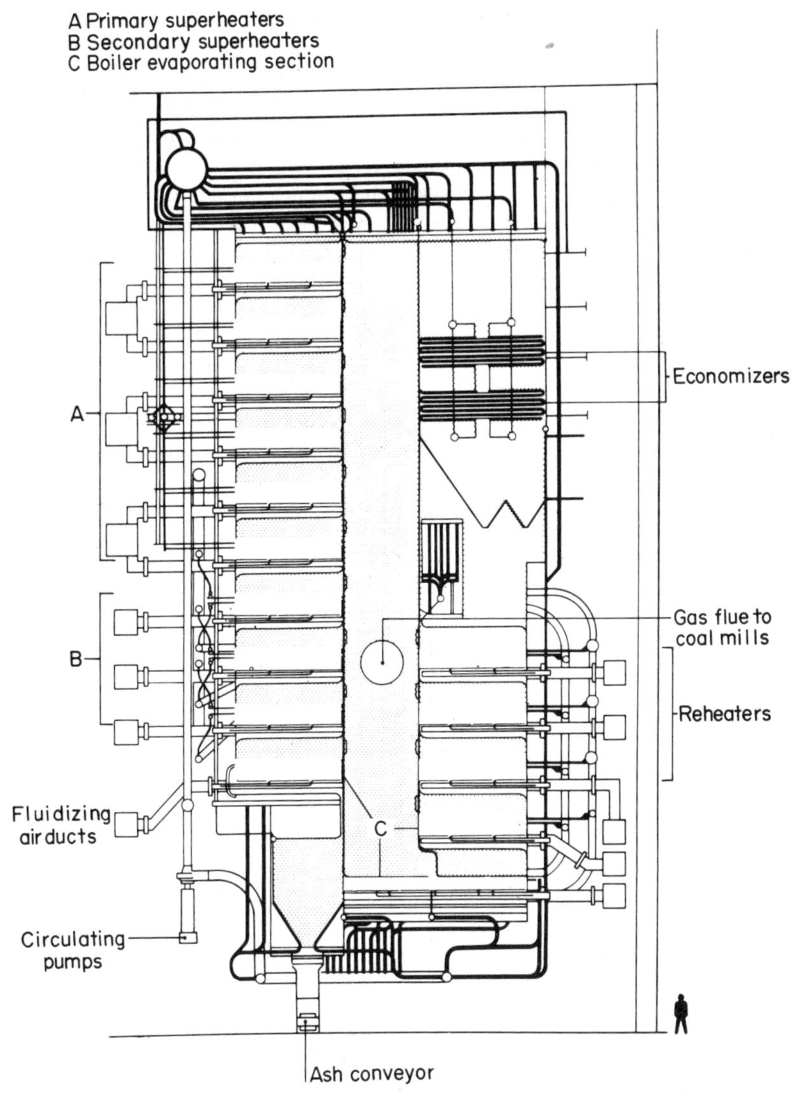

FIG. 15. Arrangement of 660 MW(e) atmospheric fluidized combustion boiler (from ref. 49).

Ignition of this unit will be achieved by oil burners above the surface of the carbon burn up cell. This cell will be gently fluidized during the start up period to ensure mixing. Coal will be admitted when ignition is assured. This may be at any temperature above ~400°C, but coal injected at low temperatures will burn slowly, so that the build up of carbon in the bed must be watched carefully. Ignition of the other cells is by transfer of material from the first, via communicating ports in the division walls. The cells will thus be lit in cascade from the burn up cell.

The American electric utilities will be watching this experimental unit with considerable interest, especially from the operational and control point of view, and to see how well the emissions of sulphur and NO_x are controlled. Compared with existing pulverized fuel boilers in the same boiler house its compactness is already impressive.

The conceptual design of a 200 MW unit shown in Fig. 14 is based on a vertically stacked arrangement of modules, and is laid out for a once through water/steam circuit. This is well suited to the modular or cellular concept of the fluidized bed boiler, allowing extensive use of horizontal tubes, relatively complex shapes, and high water side pressure drops. In other respects the 200 MW unit uses the same design ideas as the 30 MW unit; extending the idea still further, larger boilers of up to 800 MW size are envisaged as multiples of the 200 MW design. Current thinking in the US is that a 200 MW boiler could be operational by 1980. While this is a possibility, it is unlikely that any organization will accept the financial risk without further experience and assurances of technical success.

In Britain, conceptual designs for fluidized bed boilers have been published by Combustion Systems Ltd., a consortium of National Coal Board, British Petroleum, and the National Research Development Corporation.[49] Their brochure shows both single storey and stacked arrangements. The latter, conceived by Babcock and Wilcox Ltd., is illustrated in some detail (Fig. 15) and shows fifteen fluidized beds in a 660 MW unit. Of these three are evaporators, six primary superheaters, three secondary superheaters, and three reheaters. The enclosure comprises evaporative surface, and the economizers are conventional convective surface. The water circuit is of the assisted circulation single drum type. Although details are not quoted, the design appears to use lower fluidizing velocities than the American design and has no specially designated burn up cell. The overall size would seem to be substantially less than a corresponding pulverized coal unit.

In support of this concept, Babcock and Wilcox are turbine inlet temperature (~750°C) this experience is not particularly meaningful so far as material life is concerned, especially as improved particulate removal systems are still under development. One point of interest to power generation applications is that the large volume of the combustor and clean up train necessitated the addition of a flywheel to the generator shaft to reduce overspeed on loss of load. Multi shaft turbines may largely avoid this problem, and it is also unlikely that any commercial design would use such large capacity vessels.

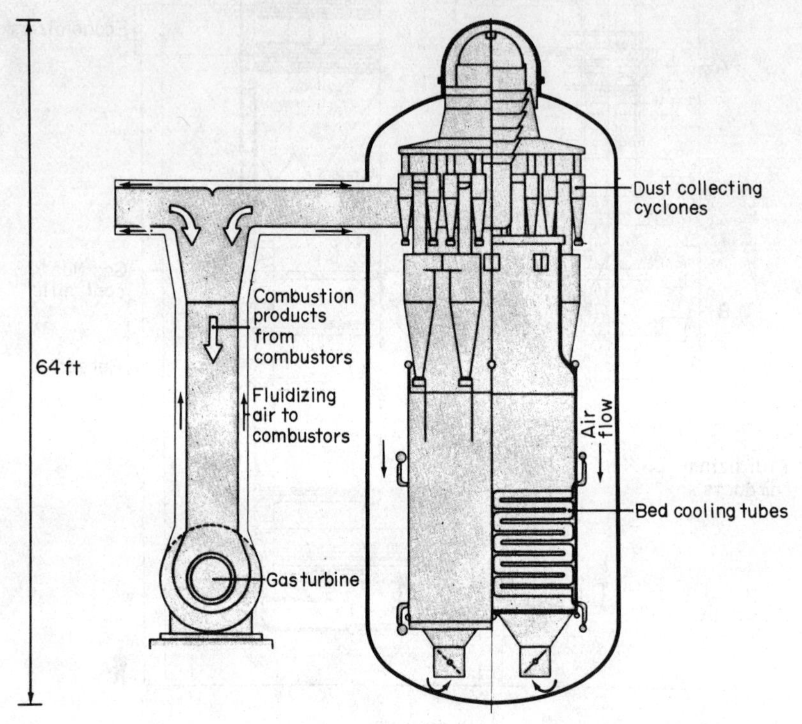

FIG. 16. Steam raising pressurized fluidized combustor and gas turbine plant. Only one of the two combustors in this plant has been shown. Capacity is 200 MW(e) with 160 MW(e) being generated from steam turbines and 40 MW(e) by way of the gas turbine (from ref. 49).

operating, at their Renfrew works, a commercial scale fluidized bed boiler which was built by conversion of a stoker-fired unit. The base size is approximately 3 m × 3 m (10 ft × 10 ft), with an evaporative rating equivalent to about 5 MW. This again suggests a moderate fluidizing velocity compared with that of the Rivesville plant.

An experimental direct-fired coal burning fluidized bed gas turbine has been operated by the Combustion Power Company, Menlo Park, California, for the Office of Coal Research.[50] This unit, designated the CPU 400, was originally used in garbage burning experiments. The turbine is a modified Ruston TA-1500. The fluidized bed combustor has operated at temperatures between 750°C and 1000°C, demonstrating 80–90% SO_2 suppression with calcium additive rates (as dolomite) as low as one and a half times the sulphur feed rate, and power production in the order of 900 kW. This unit uses three stages of cyclones for particulate removal ahead of the turbine. In view of the low

There have been a number of studies of combined cycle systems embodying pressurized fluidized bed combustors. Conceptual designs, prepared by Foster Wheeler, of possible arrangements have been published.[4] They embrace both vertically and laterally stacked arrangements. For pressure containment the beds are housed in cylindrical vessels together with the cyclones for dust recycling. The vertically stacked arrangement was chosen by Keairns *et al.*[44] to illustrate the combined cycle concept. A slightly different concept, with a group of (apparently) four deep fluidized beds arranged side by side in a pressurized containment (Fig. 16), appears in the brochure produced by Combustion Systems Ltd.[49] The hybrid indirect/direct-fired gas turbine system has been studied by Babcock and Wilcox and associates, and the CSL brochure refers to both this and gas/steam cycles.

Common factors in all these concepts are the use of multiple beds, contained in a single pressure shell, and internal cyclone dust separators. The designs do not address the problems of improved fuel feeding under pressure or improved dust separation. There are as yet no large scale demonstrations of integrated plant of this kind, but they are being seriously considered in the US and to some extent in Europe.

The immediate future for the fluidized bed boiler for large scale power generation will depend on the demand. Both Europe and the US which represent the largest market areas at present, may in the longer term expect to rely on nuclear power for base load generation, and on coal for the balance of their needs, although oil will certainly have a major stake for the next decade. In such a scheme low capital cost and flexibility will be very important for the load following duty. The other factor is emission control, and here a great deal depends on the coal type.

The fluidized bed offers potential advantages over the pulverized coal boiler with post combustion emission control, in economic terms, for both steam and combined cycle systems. The real advantages need demonstration before widespread use can be expected. At least in the US the more significant factor is the ability to control sulphur emission, and it is in the US where we may expect to see major advances in fluidized bed technology in the near future, provided that prototypes of significant scale enter service before flue gas scrubbing techniques are perfected.

For the longer term, the US is studying many alternatives which include coal based gaseous and liquid fuels. Bearing in mind the importance of fuel distribution in such a large country, and the availability of cheap coals remote from the load centres, power generation strategies which include a coal conversion stage (to a more readily transported form) may ultimately have the economic edge over direct coal usage for many regions. In areas where transport is a secondary cost item, the direct coal burning fluidized bed steam or combined cycle plants have a good chance of adoption for base load and intermediate load duties.

In Europe, the absence so far of general excessive restrictions on emissions, and the likelihood that nuclear development will take up the base load more rapidly mean that the same pressures for rapid adoption of fluidized beds do not exist, neither is research funding available at anything approaching American spending levels. Here the longer term attraction of the fluidized bed may well be its ability to utilize lower grade coals. Where district heating or process heat can be associated with power generation, moderate power generation heat rates would be no great penalty so long as overall energy utilization were good. Schemes of this sort are not uncommon and are likely to increase.

Acknowledgements: Figures 1, 2, 3 and 4 are reproduced by permission of the National Coal Board; Fig. 5 by permission of Argonne National Laboratory; Fig. 6 by permission of Pope Evans & Robbins Inc.; Fig. 7 by permission of Esso Petroleum Co.; Figs. 13 and 14 by permission of the Foster Wheeler Energy Corporation; and Figs. 15 and 16 by permission of Combustion Systems Ltd. Thanks are due to Dr. George Hill and Dr. Mike Maaghoul for their comments and suggestions on the first draft.

REFERENCES

1. WINKLER, US patent No.: 1687118 (1921).
2. SKINNER, D. G., *The Fluidized Combustion of Coal.* National Coal Board, London (1970).
3. GODEL, A. A., *Revue gen. Therm.* **5**(52), 349–359 (1966).
4. ARCHER, D. H. *et al.*, E.P.A. Report PB 213-152, Vol. III, National Technical Information Service, US Dept. of Commerce (November 1971).
5. BARKER, M. H., ROBERTS A. G. and WRIGHT, S. J., *Steam Plant for the 1970's*, Institute of Mechanical Engineers, London (1969).
6. JONKE, A. A., JARRY, R. J. and CARLS, E. L., *Proceedings of the 1st International Conference on Fluidized Bed Combustion*, NAPCA (1968).
7. SKOPP, A., *Proceedings of the 1st International Conference on Fluidized Bed Combustion*, NAPCA (1968).
8. HOY, H. R. and ROBERTS, A. G., *AICh. E. Symp. Ser.* **68**, 126 (1972).
9. GIBBS, B. N., PEREIRA, F. J. and BEÉR, J. M., *Institute of Fuel Fluidized Combustion Conference*, London (1975).
10. HIGHLEY, J. and MERRICK, D., *AICh. E. Symp. Ser.* **67**, 116, 219 (1971).
11. BAILEY, R., COOKE, M. J. and WILLIAMS, D. F., *7th International Conference on Coal Science*, Prague, C.R.E. Tech. Memo 361 (1968).
12. COATES, N. H. and RICE, R. L., *Proceedings of the 2nd International Conference on Fluidized Bed Combustion*, US E.P.A. (1970).
13. MERRICK, D. and HIGHLEY, J. Control of particulate emissions from gaseous fluidized beds. *AICh. E. Symp. No. 2*, N.Y. (1972).
14. EHRLICH, S. *Proceedings of the 2nd International Conference of Fluidized Bed Combustion.* US E.P.A. (1970).
15. HOY, H. R., *Proceedings of the 2nd International Conference on Fluidized Bed Combustion.* US E.P.A. (1970).
16. RICE, R. L. and COATES, N. H., *Proceedings of the 3rd International Conference on Fluidized Bed Combustion.* US E.P.A. (1973).
17. HOY, H. R. and LOCKE, H. B., Paper to Swedish Institute of Engineers, Stockholm (1972).
18. GLENN, R. D. and ROBISON, E. B., *Proceedings of the 2nd International Conference on Fluidized Bed Combustion.* US E.P.A. (1970).
19. BORGWARDT, R. H. and HARVEY, R. D., *Environ. Sci. Tech.* **6**, 4, 350–360 (April 1972).
20. HARTMAN, M. and COUGHLIN, R. W., *Ind. Eng. Chem. Proc. Des. Dev.* **13**, 3, 248–253 (1974).
21. ANASTASIA, E. L., CARLS, E. L., JARRY, R. J., JONKE, A. A. and VOGEL, G. J., *Proceedings of the 2nd International Conference on Fluidized Bed Combustion.* US E.P.A. (1970).
22. DAVIDSON, D. C. and SMALE, A. W., *Proceedings of the 2nd International Conference on Fluidized Bed Combustion.* US E.P.A. (1970).
23. MOSS, G. *Proceedings of the 2nd International Conference on Fluidized Bed Combustion.* US E.P.A. (1970).
24. JARRY, R. J., ANASTASIA, L. J., CARLS, E. L., JONKE, A. A. and VOGEL, G. J., *Proceedings of the 2nd International Conference on Fluidized Bed Combustion.* US E.P.A. (1970).
25. WRIGHT, S. J., *Proceedings of the 3rd International Conference on Fluidized Bed Combustion.* US E.P.A. (1973).
26. O'NEILL, E. P., KEAIRNS, D. C. and KITTLE, W. F. *Proceedings of the 3rd International Conference on Fluidized Bed Combustion.* US E.P.A. (1973).
27. VOGEL, G. J., SWIFT, W. M., MONTAGNA, J. C., LENC, J. F. and JONKE, A. A., *Inst. F. Symp. Series No. 1*, Proc. Vol. 1 (1975).
28. ROBERTS, A. G., STANTON, J. E., WILKINS, D. M., BEACHAM, B. and HOY, H. R., *Institute Fuel Fluidized Combustion Conference.* London (1975).
29. HOKE, R. C. and BERTRAND, R. R., *Institute Fuel Fluidized Combustion Conference.* London (1975).

30. O'NEILL, E. P., KEAIRNS, D. C. and KITTLE, W. F., *North American Thermal Analysis Society.* Trent University, Toronto (1975).
31. VOGEL, G. J., CARLS, E. L., ACKERMAN, J., HAAS, RIHA, J. and JONKE, A. A., *Proceedings of the 3rd International Conference on Fluidized Bed Combustion.* US E.P.A. (1973).
32. HOKE, R. C., SHAW, H. and SKOPP, A., *Proceedings of the 3rd International Conference on Fluidized Bed Combustion.* US E.P.A. (1973).
33. MOSS, G., *Institute of Fuel Fluidized Combustion Conference.* London (1975). *Inst. F. Symp. Series No. 1*, Proc. Vol. 1 (1975).
34. ROBERTS, A. G., LUNN, H. G., HOY, H. R. and LOCKE, H. B., *Coal Processing Tech.* **2**, 34–39, 1975.
35. BOTTERILL, J. S. M., *Fluid Bed Heat Transfer.* Academic Press, London (1975).
36. ELLIOTT, D. E., HEALEY, E. M. and ROBERTS, A. G., *Paper No. 17, Joint Conference on Heat Exchangers*, Francais des Comb. et de l'Energie and Inst., Paris (1971).
37. MCLAREN, J. and WILLIAMS, D. F., *J. Inst. Fuel* **42**, 303 (August 1969).
38. ELLIOTT, D. E. and VIRR, M. J., *Proceedings of the 3rd International Conference on Fluidized Bed Combustion.* US E.P.A. (1973).
39. JACKSON, P. J., *J. Inst. Fuel* **40**, 335 (1967).
40. HALSTEAD, W. D. and RAASK, E., *J. Inst. Fuel* **42**, 344 (1969).
41. RAASK, E. and WEAR, **13**, 301 (1969).
42. COOKE, M. J. and ROGERS, E. A., *Institute Fuel Fluidized Combustion Conference.* London (1975). *Inst. F. Symp. Series No. 1*, Proc. Vol. 1 (1975).
43. ELLIOTT, D. E. and HEALEY, E. M., *Proceedings of the 2nd International Conference on Fluidized Bed Combustion.* US E.P.A. (1970).
44. KEAIRNS, D. L., YANG, W. C., HAMM, J. R. and ARCHER, D. H., *Proceedings of the 3rd International Conference on Fluidized Bed Combustion.* US E.P.A. (1973).
45. FRAAS, A. P., *World Energy Conference.* Detroit, Paper 4, 1–12 (1974).
46. EHRLICH, S., *Institute Fuel Fluidized Combustion Conference.* London (1975). *Inst. F. Symp. Series No. 1*, Proc. Vol. 1 (1975).
47. Foster Wheeler Corp., house journal, *Heat Engineering.* Foster Wheeler, New Jersey (July-September 1972).
48. GAMBLE, R. L. and WARSHANY, F. R., *ASME/IEEE Joint Power Generation Conference.* Portland (1975).
49. Combustion Systems Ltd., London, Brochure—*Fluidized Combustion* (1975).
50. US Dept. of Interior O.C.R. R & D Report No. 94 (September 1974).

Manuscript received, 9 March 1976.

Prog. Energy Combust. Sci., Vol. 2, pp. 83–95, 1976. Pergamon Press. Printed in Great Britain

COMBUSTION AND HEAT TRANSFER IN LARGE BOILER FURNACES

A. M. GODRIDGE* and A. W. READ†
Central Electricity Generating Board, England

CONTENTS

1. INTRODUCTION

A dramatic feature in the development of power station boiler design in Great Britain is the tenfold increase in unit size within 20 years. This is illustrated in Table 1 where it can be seen that in addition to the increase in unit size, there has also been a move to higher steam temperatures and pressures.

TABLE 1. Details of standardized unit coal-fired boiler designs adopted by the CEGB since 1950

Unit size, MW (electrical output)	Boiler capacity (kg/s)	Steam pressure (Bar)	Final steam temperature (°C)	Commissioned
30	38	41	454	1950–55
60	69	62	482	1952–60
120	108	103	538	1958–62
500	435	166	568	1965–73
660	560	166	568	1973 onwards

* Marchwood Engineering Laboratories.
† North Eastern Region, Scientific Services Department.

Although some boilers were converted to oil-firing in the early 1960's the C.E.G.B. did not operate power stations specifically designed to burn oil until about 1970. The CEGB have two oil-fired stations with 500 MW (electrical output) units which together with the coal fired plant make nearly fifty 500 MW units in service. The 500 MW units were the subject of extensive boiler trials which had the following aims:

(1) to provide the basic data required for improving future designs;
(2) to validate the use of models for predicting the performance of boilers; and
(3) to determine any design deficiencies as soon as possible and allow modifications to be made to subsequent units of the same design.

This paper outlines the combustion and gas side heat transfer behaviour in the furnace chamber of the larger units and refers to operating procedures and design changes. Furnace modelling is discussed and reference is made to some of the work that has gone on elsewhere.

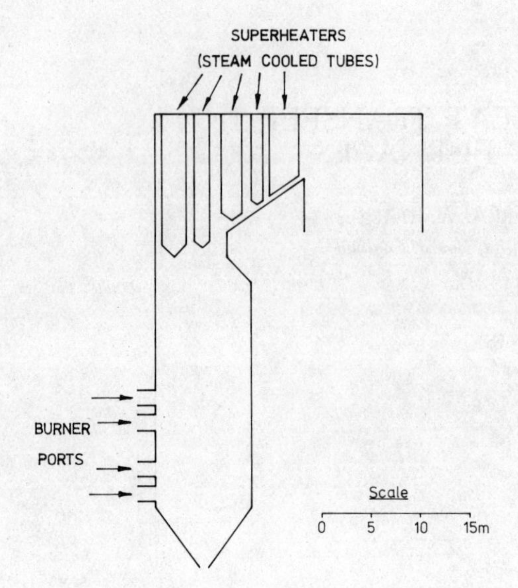

FIG. 1. Cross-section of a Front-Wall Fired 500 MW Coal Burning Boiler at Ferrybridge (Babcock & Wilcox Ltd.).

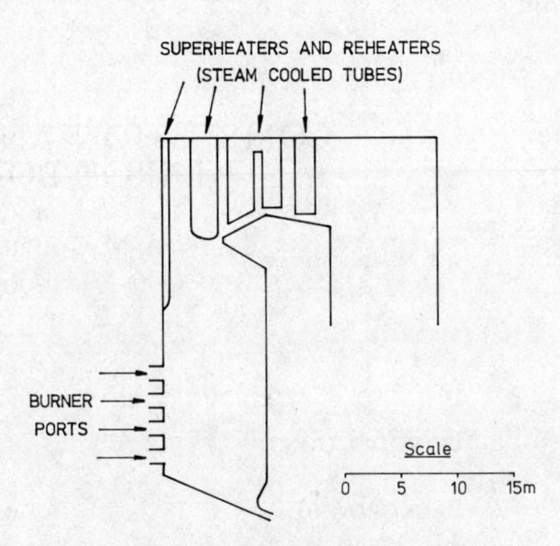

FIG. 3. Cross-section of Front-Wall Fired 500 MW Oil Burning Boiler at Fawley (Clarke Chapman Ltd.).

2. BACKGROUND

The first point to note in considering modern water-tube boiler furnace chambers is that they are very large, typically 35 m high with a 300 m^2 cross-section. Into this large chamber fuel and air, again in large quantities (i.e. 220 tonnes/h of coal and about ten times as much air for a 20% ash coal), are fired through twenty to fifty burners, depending on the particular design and operation of the plant. Figures 1 to 4 show the box-like arrangement of the furnace chamber. At the bottom is the sloping floor where the ash collects and is removed in coal-fired plant, and at the top the outlet restriction, or nose, around which flow the gases leaving the furnace. These figures illustrate the general furnace design and are considered in more detail later.

2.1. *Plant Design*

In a drum-type water-tube boiler, steam is generated in the vertical tubes arranged to form the walls of the furnace chamber. This steam and accompanying water then passes to the boiler drum where the steam is separated and the remaining water, together with the incoming feedwater, is returned to the furnace water-tubes through pipes outside the furnace (called down-comers). An outline drawing of the main components and the flow paths in a water-tube boiler is shown in Fig. 5. Here it can be seen that steam leaving the boiler drum enters the superheater where it receives further heat before passing to the turbines. Steam which has partly expanded through the turbine may be returned to the plant and superheated again in the

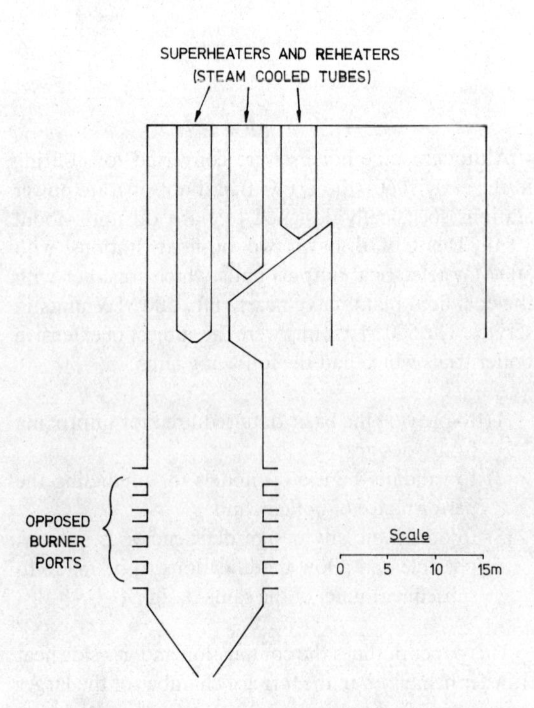

FIG. 2. Cross-section of an Opposed-Fired 660 MW Coal Burning Boiler at Drax (Babcock & Wilcox Ltd.).

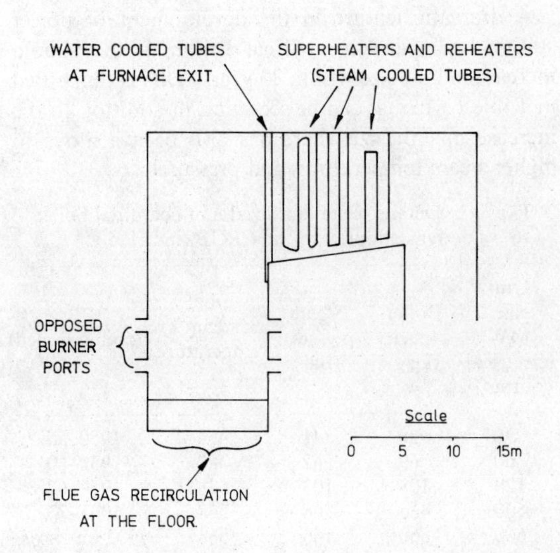

FIG. 4. Cross-section of an Opposed-Fired 660 MW Oil Burning Boiler at the Isle of Grain (Babcock & Wilcox Ltd.).

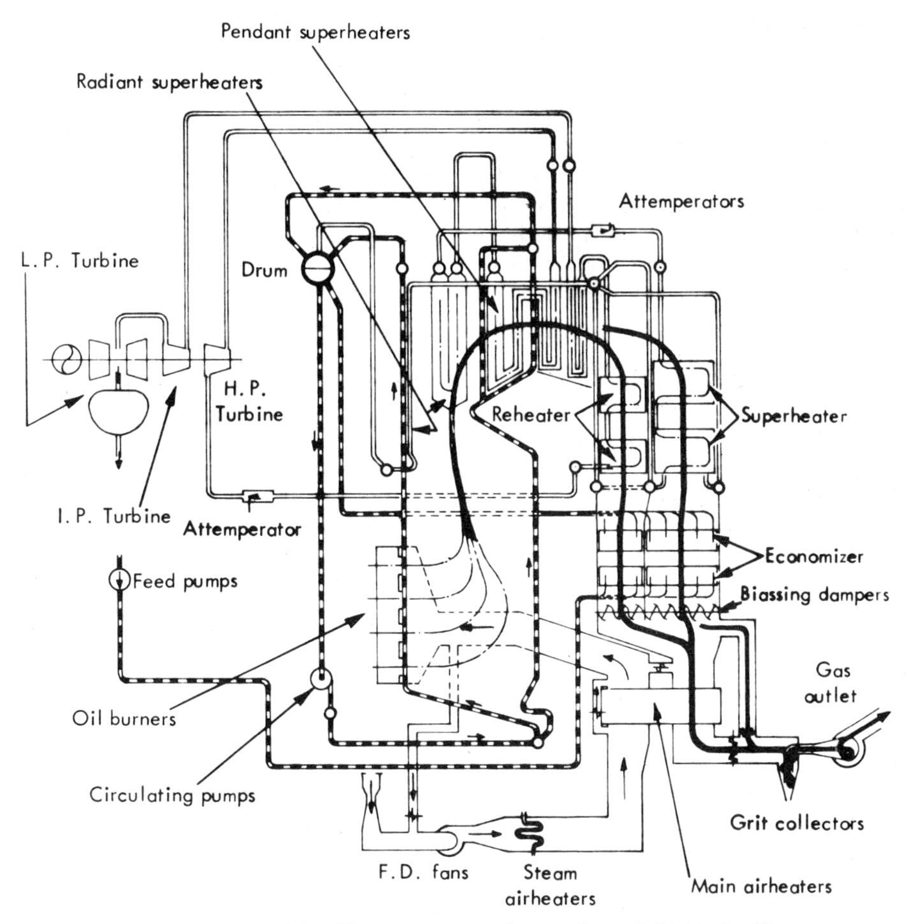

FIG. 5. Steam and Gas Flow arrangement in a modern oil-fired boiler.[58]

reheater. Superheaters generally receive heat by convection from gases that have left the furnace chamber but they may also be located within or adjacent to the furnace chamber in order to receive extra radiant heat from the flames (see Fig. 1–4).

Furnace size and shape needs to be such as to provide an adequate residence time within the furnace in order to achieve virtually complete combustion of the fuel. At the same time the temperatures of the gases leaving the furnace must be low enough to avoid high temperature deposits in the superheater zone, but have sufficient heat content to enable a suitable design of superheater, reheater, economizer and air heater to be derived. The flue gas leaving the air heater must have a sufficiently low temperature to give a minimum boiler heat loss, yet not so low as to result in corrosion at the back end of the boiler. The tubes that completely cover the interior surfaces of the furnace can vary in outside diameter from about 40 mm in an assisted circulation boiler of the type shown in Fig. 5, to about 80 mm in a natural circulation system (see Table 2). In the past, the usual form of wall construction has involved the tubes loosely touching each other in a so called "tangent tube" arrangement. On the outside of the furnace the tubes are cased in with steel sheets which are designed to be air-tight or, on earlier units, faced with panels of hard insulating material, air sealing being helped by the use of additional refractory.[1] Also, in some earlier furnaces tubes were finned (see the second column in Table 2) with two external longitudinal fins diametrically opposite to each other and each of about 25 mm in width. Currently fins are being used in a different manner in the "membrane wall". In this arrangement panels of tubes are fabricated at the manufacturer's works by welding to give a continuous longitudinal strip about 12 mm wide between tubes. Thus a solid panel is built up of a size limited only by transport considerations. The panels are welded together on site to form the complete furnace.

As furnace size increases, the surface to volume factor decreases. Thus a situation has been arrived at where furnaces may need to include water-cooled division walls in order to ensure sufficient heating surfaces without excessively large furnace volumes. A further complication to an increase in furnace width for large volume furnaces is the limitation imposed by the maximum available length of retractable sootblower. For example, the radiant superheater (platen) zone can only be about 30 m wide if it is to be adequately covered by sootblowers from each side.[1]

It is possible in an oil-fired boiler to operate with a furnace having a volume less than that required for a p.f. furnace, with the same boiler output (compare Figs. 2 and 4). There is a variety of reasons for this, including shorter burning times for oil droplets compared with coal particles, and better radiant heat transfer because of higher flame temperatures with less excess air and less ash coverage on the tubes. Where both coal and oil are to be burnt as alternative fuels, however, it is necessary to size the furnace for the coal-fired conditions. The adoption of a smaller furnace suitable for oil firing would result in a highly rated furnace when burning coal with the consequent risk of furnace slagging and, because of the shorter residence times, unburnt coal passing out of the furnace chamber.

2.2. *Previous Experimental Studies*

Over the years there has been a number of investigations of furnace chamber performance on pulverized-coal-fired water-tube boilers and these have been listed by Godridge.[2] The only well-documented extensive and systematic studies on front-wall and corner-fired furnaces have been those sponsored by the American Society of Mechanical Engineers, at Tidd and Paddy's Run and by the Central Electricity Generating Board at Brunswick Wharf Generating Station. Details of these plants are given in Table 2 and a comparison with modern plant (e.g. Ferrybridge and Fawley, see columns 4 and 5 of Table 2) shows that conditions, particularly with regard to output, steam pressure and the number of burners, are very different.

Reported systematic tests on coal burning plant are limited, but no trials on oil-fired plant have been reported in detail. The position in Great Britain is not too difficult to understand for it is only in the last 10 years that oil has been used to any great extent in large power station boilers. The early problems arising from the conversion of coal-fired plant to oil-firing have been discussed by Dadswell and Thompson[3] and Meyler and Lang,[4] but the number of measurements of combustion and heat transfer properties of oil flames in these operating plants have been small.

2.3. *Modern Large Plant*

Details of the CEGB's most modern coal- and oil-fired plant have been given by Lamb[5] and four of these units may be briefly considered here as their operating and design features are of major interest. Because the latest CEGB coal burning boilers are opposed-fired (Drax Power Station) and the next two oil fired boilers will be opposed-fired (Grain Power Station) and front-wall-fired (Littlebrook "D" Power Station) little mention is made in the text of corner-fired boilers.

Ferrybridge (Coal-fired)

A typical cross-section of a front-fired 500 MW unit is shown in Fig. 1. This is the Babcock & Wilcox design installed at Ferrybridge. These are natural circulation boilers with no division walls. The furnace width is about 30 m and this is tapered in at the sides local to the superheat platens to approximately 25 m. The depth of this furnace is 10 m and the height from the hopper throat to the nose is about 35 m. It was on these boilers that the membrane type of furnace casing was first used by the CEGB. There are forty-eight front wall burners each with a capacity of about 1 kg/s of coal. Eight vertical spindle mills grind the coal to a fine powder with a typical mean particle size of 0.04 mm or less.

TABLE 2. Plant trials—comparative boiler characteristics

		Brunswick Wharf 1960	Tidd U.S.A. 1948	Paddy's Run U.S.A. 1950	Ferrybridge "C" 1970	Fawley 1970
Output	kg/s	40	51	72	430	430
Steam pressure	Bar	63	96	64	165	165
Steam temperature	°C	496	496	482	568	538
Furnace cross-section	m	7 × 8.5	7 × 7	8 × 10	10 × 30	10 × 20
Furnace height	m	17	22	30	35*	27*
Furnace projected area	m²	570	710	990	3100	2000
Furnace tubes	mm	82 + 25 mm fins	76 Tangent	76 Tangent	63.50D, 49.51D Membrane 12 mm	41 Tangent
Firing		Coal, tangential	Coal, tangential	Coal, front wall	Coal, front wall	Oil, front wall
Number burners		12	8	8	48	32
Height above ash pit	m	9	8	8	7.5	3
Range of heat fluxes	kW/m²	150–270	160–270	140–260	250–350	300–550
Fuel analyses (as received)						
Heat content	kj/g	23.0–25.0	24.0–26.0	24.0–25.0	21.0–26.0	43
Moisture	%	9.7–12.4	5.3–10.8	7.8–11.6	8.0–12.0	—
Ash	%	13.5–20.0	10.7–13.5	11.4–15.5	16.0–24.0	0.025
V.M.	%	28.8–31.6	33.0–37.2	32.5–36.0	26.0–30.0	—
References		CEGB (55)	Reid *et al.* (56)	Corey and Cohen (57)		

* to the furnace nose level

Drax (Coal-fired)

660 MW units have been constructed by Babcock & Wilcox at Drax and a cross section of these is shown in Fig. 2. They are basically a similar design to Ferrybridge, the main difference being that opposed firing is used from the front and rear walls instead of the front wall only. The furnace width is about 30 m, the depth is about 13 m and the height from the hopper throat to the nose is 40 m. The furnace walls are of membrane construction. Natural circulation is used for furnace wall cooling and the radiant superheater is in the form of platens. There are sixty burners divided equally between front and rear walls each with a capacity of just over 1 kg/s of pulverized coal, produced by ten vertical spindle mills.

Fawley (Oil-fired)

These 500 MW boilers designed and manufactured by Clarke Chapman/John Thompson are designed to burn residual oil. A cross-section is shown in Fig. 3 and it should be noted that the furnace is only 20 m wide compared with approximately 30 m for a front-wall coal-fired unit; the furnace height to the nose is 27 m and the depth 10 m. Thirty-two oil burners of the pressure atomizing type are located in the front wall each with a capacity of approximately 1 kg/s.

Isle of Grain (Oil-fired)

The Generating Board have placed an order with Babcock & Wilcox for the supply of five 660 MW oil-fired boilers for the Isle of Grain and a cross section of this unit is shown in Fig. 4. The arrangement is considerably different from the previous 500 MW design. The divided furnace is only 25 m wide, 12 m deep and 22 m high to the centre of the gas outlet and is completely devoid of any superheating surface. All the superheat and reheater tubes are in pendent form in the relatively long horizontal section outside the furnace. As with Drax, opposed-firing is used with a total of twenty-four burners in the front and rear walls, each with a capacity of 2 kg/s. Flue gas can be recirculated through the ash pit.

3. COMBUSTION

3.1. Burners

How[6] has reviewed the papers on coal-fired water-tube boilers published in the *Journal of the Institute of Fuel* since 1950. In the period 1950–75 only one paper was published on p.f. burner development.[7] This highlights the fact that the basic principles have not changed substantially over the last 25 years.

Pulverized coal is transported to the furnace by preheated primary air which supplies approximately 25% of the total combustion air requirements. The remaining 75% of the air is also preheated and enters the burner through swirl vanes. The flame is stabilized by the internal and external gas recirculation patterns set up by the swirling and expanding jet. On large boilers an oil burner, firing co-axially, is used to light each coal burner.

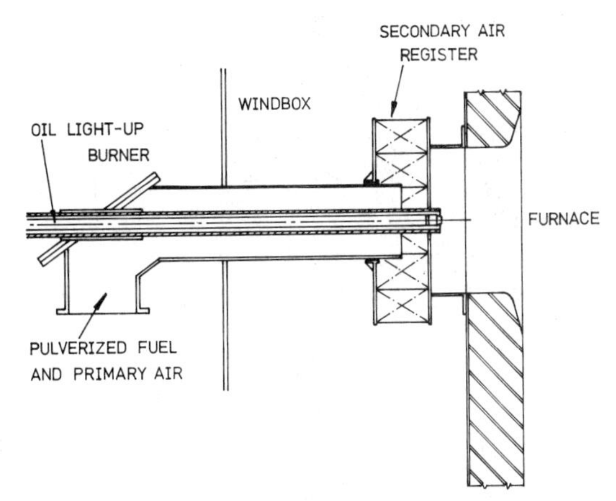

FIG. 6. P.F. Burner.

Although the basic principles have remained unchanged, there have been many developments to produce simpler designs with improved service life and more reliable performance. In a modern burner, the p.f. is fired as an annular jet around the light-up oil burner carrier tube, and the swirl vanes are preset at the optimum angle. The secondary air is controlled by an external damper. A typical design for a boiler with a common windbox is shown in Fig. 6.

When low volatile fuel (VM $< 15\%$) has to be burnt, it is usual to use a downshot fired boiler, with stabilization by radiation and convection from the tail of the flame. The only other readily available paper referring

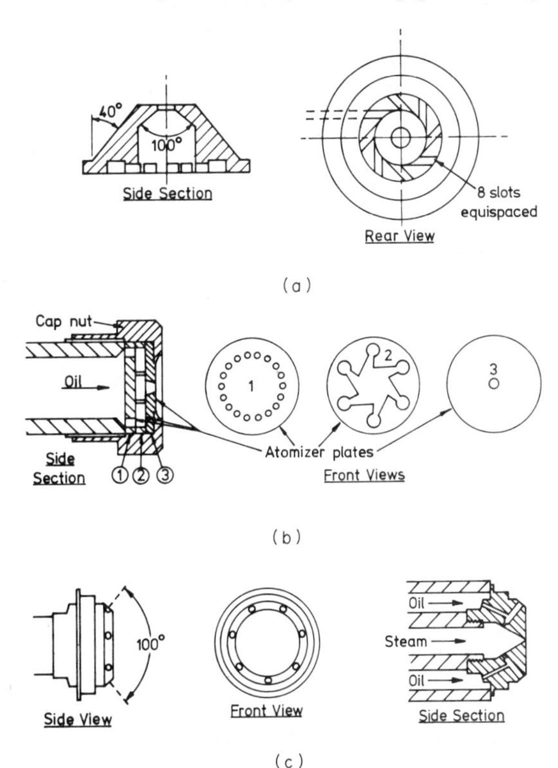

FIG. 7. Oil Atomizer Types used in CEGB Stations. (a) Fawley type, (b) Atomizer type for converted CEGB stations, (c) Two fluid atomizer type.

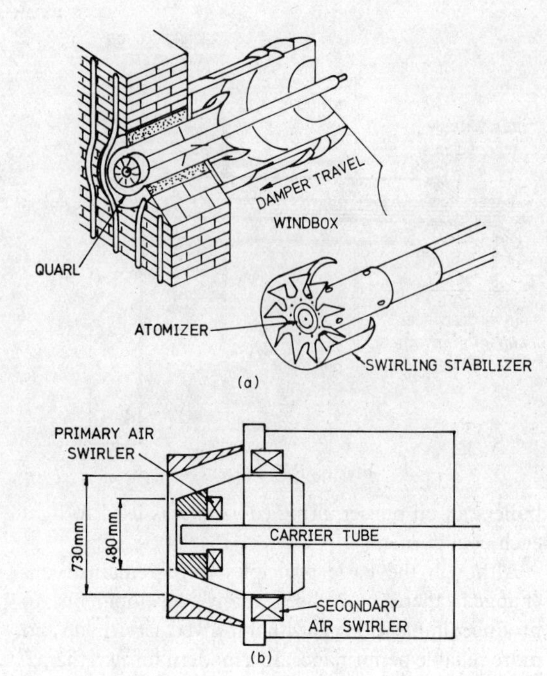

FIG. 8. CEGB Standard and Fawley Type Oil Burners. (a) Burner Register with stabilizer—(CEGB Standard Oil Burner),[10] (b) Burner with swirl air register.[11]

to coal burners for water-tube boilers is by Juniper,[8] who has reported work on flames in brown coal-fired boilers. This work is concerned with the design of burners for acceptable ignition and combustion stability with high moisture coals.

Oil-fired power stations normally use residual fuel oil (R.F.O.) and it is necessary to heat the oil and to break it into fine sprays by means of atomizers. This atomization is done either by high pressure ejection from an orifice or by entrainment in a blast of air or steam (see Fig. 7). Oil burner systems for power stations have been discussed in detail by Brooks *et al.*[9] and the development of a standard pressure-jet oil burner, also for use in power stations, described by Anson and Tindall[10] (Fig. 8a). These papers consider burners having oil throughputs in the range 0.1–0.4 kg/s and the latter authors suggest that their results can be extrapolated to 0.75 kg/s burners. Present practice has advanced beyond these reported tests and burners firing 1.0 kg/s of residual fuel oil (Fig. 8b) are in use on modern plant.[11] Equipment firing up to about 2.0 kg/s has been tested and will shortly be in use within the CEGB.

3.2. Combustion Efficiency

Combustion efficiency in a large water-tube boiler is typically over 98%. The relationship between boiler losses and the overall air/fuel ratio is well known. On the one hand, a fuel rich situation leads to losses of unburnt fuel and on the other, the use of excessive air results, amongst other things, in losses due to heating too much gas. If a boiler test shows that at the correct air/fuel ratio the efficiency is low, then other factors need to be considered. For example, pulverized fuel quality (or, in the case of oil, atomization quality), burner design, fuel and air distribution to the burners and air leakage. With regard to the last two, major burner to burner variations in p.f. flow (occasionally ±30%) can be overcome with the use of riffles in the p.f. transport lines,[12] and by close examination with modern techniques of faulty casings, or ash hoppers, for air leaks.[13]

One very satisfactory outcome of the CEGB's work on boiler plant is the recognition of CO monitoring as a direct indicator of satisfactory combustion conditions. Carbon monoxide measurements, unlike those for oxygen, are only slightly affected by air in-leakage so that sampling can be done anywhere and in particular as far downstream as the ID fan where mixing is good, giving representative samples. Following earlier work,[14] the system is now standard for oil-fired plant and is also in use on many large coal-fired boilers.

The operation of oil-fired plant is influenced by the sulphur trioxide in the combustion products which can, in addition to corroding the low temperature metallic parts, lead to the emission of acid smuts, which are carbonaceous material containing excess acid. The reaction of sulphur dioxide to give sulphur trioxide is oxygen-dependent and thus a function of excess air.

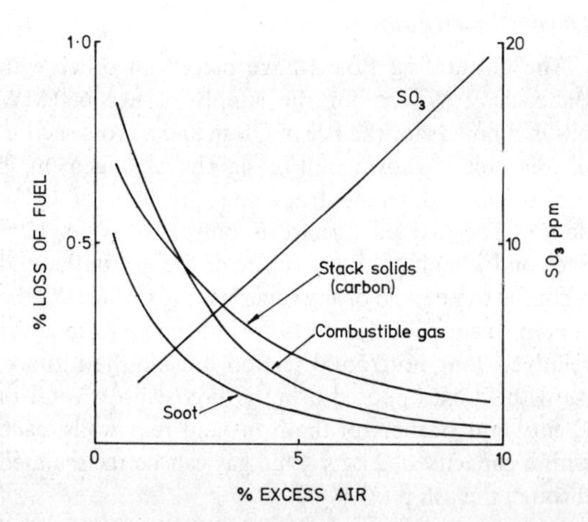

FIG. 9. Acid Production (SO_3) added to the curves of Chemical Heat Loss (illustrating that acid production may restrict the use of excess air).

Figure 9 shows that acid production (as SO_3) may restrict the use of excess air and tends to bias operation towards the very steeply rising part of the heat loss curve. This figure also differentiates between the forms of combustion loss which arise from the presence of burnable gases (CO, H_2 and hydrocarbons), carbon in the form of "stack solids", or coarse, cokey material, and very fine carbon or soot which is responsible for dark smoke.[15]

Figure 9 cannot be taken as applying quantitatively to any specific plant: the scales give a realistic order of magnitude for the losses concerned, as a function of the excess air supplied, but the exact figures will depend on the plant and its operating condition.

3.3. *Combustion Control*

The primary task of a combustion control system is to provide enough heat to satisfy at all times the steam demand from the boiler, whilst minimizing the fuel costs per unit of steam produced. Combustion control of a large boiler by measurement and control of the input fuel and air flows has not, so far, been possible. Coal flows to individual burners, which can vary substantially, cannot be measured on-line, and the quality of control of air flow to individual burners is unlikely to be better than $\pm 5\%$ and could be substantially worse (this is caused by windbox geometry, unequal burner diameters, temperature variations in the windbox and buoyancy forces). Therefore, combustion control has to be achieved by an indirect method. Fortunately, large multi-burner furnaces even out many of the fuel and air maldistributions by cross mixing within the furnace chamber and this reduces the control problem.

A consideration of the flue gas constituents shows that only carbon monoxide and oxygen are directly related to efficiency and can be measured accurately and reliably. The other quantity which can be readily measured is the total air input to the boiler. A boiler load against air input relationship can be derived, assuming a typical fuel and average boiler condition, and this relationship is then used to set the input air figure. At the optimum operating point the carbon monoxide concentration is very low, but the concentration rises very rapidly just below the optimum (Fig. 10). The required fuel flow can be achieved by linking the fuel flow control loop, operating on the mill feeders, to the steam flow indicator. The load line is derived to produce an air-rich situation which can be trimmed back to the optimum by the oxygen and carbon monoxide in flue gas measurements.

Oxygen levels are seriously affected by in-leakage of air to the boiler system. At 1% oxygen, which is a typical figure for an oil-fired boiler, a 1% air in-leakage produces 20% error in the recorded oxygen level and the effect is greater at lower oxygen levels. Because of this effect, oxygen sampling must be carried out as close to the furnace exit as possible, which poses the further complication of obtaining representative samples from very large ducts. In spite of these problems the retention of oxygen analysis is well worthwhile provided a reliable mean value can be obtained. Comparison of the two figures for carbon monoxide and oxygen will indicate at once whether the boiler is operating normally or whether casing leaks or other faults are developing.

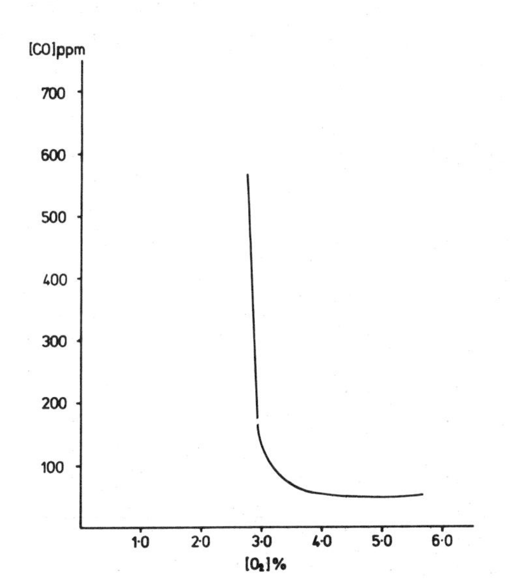

FIG. 10. Oxygen–carbon monoxide relationship in a large p.f. boiler.

Most of the other parameters have rather greater limitations, although a number have been used (see Table 3). The opacimeter (smoke meter) is widely used on oil-fired plant, being simple and directly relevant to the clean air restrictions. This is, however, a qualitative, rather than quantitative, instrument. Also, it is possible, especially in well-mixed combustion systems, to have unburnt gases in significant amounts without soot so that the smoke meter is not a reliable indicator of efficiency.

3.4. *Furnace Fouling*

Deposits which can corrode boiler tubes and form slag layers can occur, particularly in coal fired systems. The larger furnaces run hotter than the previous smaller units because the surface to volume factor is less and this has increased the tendency to slagging. In addition, fouling and slagging can be promoted by low oxygen concentrations or reducing atmospheres. Many of the coal-fired 500 MW units have been short of air at full load and this has led to fouling. Steam sootblowers have traditionally been used to control fouling, but compressed air sootblowers are used at Drax. In other countries, including Australia, Germany and U.S.A., water lances have been developed to remove deposits without causing tube failures due to thermal shock.[16]

Fouling and slagging are often quite localized effects and can occur in boilers where the overall balance of fuel and air is satisfactory. There are problems in investigating local effects because the size of the combustion chambers makes it difficult to reach the area of interest and obtain representative samples.

The gas temperature at the furnace exit of the large boilers can be above the fusion point of the ash which is typically 1100°–1200°C for British coals (see Section 4.3). In spite of this, slagging of pendent tubes has not been a major problem due to the relatively wide spacing of the elements in a modern boiler, and the provision of adequate sootblowers.

3.5. *Multi-Fuel Firing*

Recent events, particularly the 1973 oil crisis, have emphasized the dangers of relying exclusively on one fuel. An organization such as the CEGB, which operates a large and fully integrated system, can respond to the

TABLE 3. Performance of representative instruments for combustion control (Oil-fired plant, Reference 14)

Parameter	Concentration in dry gases (%)	Detector	Limit of discrimination	Response time (s)	Effect of 1% air in-leakage (% of amount present)	Notes
CO_2	15.75–14.25	Infra-red absorption	0.2	1	−1	Affected by other gases, especially H_2 at 1% O_2 level
		Thermal conductivity	0.3	2	−1	
O_2	0–2.2	Paramagnetic	0.05	5	+20	
CO	0–0.1	Infra-red absorption	0.001	1		
H_2	0–0.05	Mass spectrometer	0.001	1	−1	Affected by hydrocarbons and water
SO_3	0–0.01	MEL monitor	0.0001	60	−1	Not directly related to efficiency
CO_2	As above	Gas chromatograph	0.03	300	−1	
CO	As above	Gas chromatograph	0.05	300	−1	
O_2	As above	Gas chromatograph	0.03	300	−1	
H_2	As above	Gas chromatograph	0.01	300	−1	
Smoke	0–100% opacity	Opacimeter	5%	1	slight	Semi-quantitative only
	0–9 Shell Smoke Number	Recording filter unit	5%	1	slight	Semi-quantitative only
Solids	0.02–0.20 g/Nm^3	Automatic sampling				Not fully tested

General notes: Fuel analysis taken as C 86%, H_2 11.56%, S 1.9%. Effect of air leakage assumes no reaction with flue gases.

fluctuating price and availability of fuels by shifting the generating load from one type of plant to another. However, even in a large system there are many occasions when all the plant is required to meet the load. Smaller systems, such as those supplying steam or electricity to an industrial complex, usually have insufficient capacity to move load from boiler to boiler to match a changing fuel situation. Therefore, the idea of firing a boiler with a variety of fuels, multi-fuel firing, has become increasingly attractive.

Coal-fired plant has been converted to fire either coal or natural gas.[17] In a typical installation, the gas is fired through the same burner as the coal, either using a central gas injector, or a series of gas injectors placed around the coal tube. Experience has shown that the heat-transfer performance of the boiler can be maintained when firing gas and some earlier worries about the non-luminous nature of the gas flame have proved largely unfounded. The burning of crude oil in power stations has been extensively used in Japan[18] and some instances have been reported from America.[19]

However, there is a developing world wide shortage of natural gas, and crude oil is increasing in price and therefore both these fuels are unlikely to form a major part of any multi-fuel firing policy. Attention has, therefore, turned to the mixed firing of RFO and coal. It was pointed out in Section 2.1 that the combustion chamber of an oil/coal boiler must be of a sufficiently large size to burn the coal. Some large boilers have been constructed which have separate coal and oil burners. However, there are a number of problems with this approach, such as maintaining burner availability and achieving low excess air operation when firing 100% oil.[20] An alternative solution is to use dual fuel burners which have a 100% capability on each fuel. A number of burner designs can be proposed, but development work is required to ensure a satisfactory combustion performance on both fuels.

One advantage of oil/coal plant is its generally superior emission characteristics when used in the dual firing mode. The pulverized coal ash acts as an additive to the products of oil combustion, reducing their acidity[21] and tendency to form acid smuts. At the same time the absorption of SO_3 on the ash reduces the resistivity of the particles so that they are more easily captured in the electrostatic precipitators.

4. HEAT TRANSFER

4.1. *Radiant Heat Transfer*

In the furnace chambers of water-tube boilers the heat transfer process is almost wholly radiant, with less than 5% of the heat transferred by convection.[22] Unlike the radiation emitted from a solid the radiation from a flame emanates from within the volume of the flame. The radiant energy dE, emitted from an elemental volume of gas dV at a constant temperature T_g and of constant absorption coefficient k is[23]

$$dE_g = 4k\,dV\sigma T_g^4, \tag{1}$$

where σ is the Stefan–Boltzmann constant, and where the radiation is assumed to be distributed uniformly over a total solid angle of 4π. However, for practical volumes of gas only a fraction of the energy originating within the gas volume escapes from the bounding surfaces, the remaining energy being self-absorbed. For a cube of gas of side B the energy radiating through the six bounding surfaces is, from eqn. (1)

$$E_g = 4kB^3\phi\sigma T_g^4 \tag{2}$$

where ϕ is the fraction of energy originating within the cube that escapes, ϕ being dependent upon k and B only.[24]

Beér and Howarth[25] have discussed absorption and scattering of radiation by particles in flue gases and have made reference to the parameter X where $X = \pi d/\lambda$ (d is the particle diameter and λ the wavelength of the incident light). Three characteristic ranges of X are considered as follows:

(a) $X \gg 1$ which is the case for coal, chars and fly ash. Here scattering is commensurable with absorption and is anisotropic so that measurements of attenuation coefficients in p.f. flames using narrow angled instruments are open to error. Moreover, it is very difficult to make line of sight measurements in large furnace chambers so that information on absorption coefficients has to be taken from published measurements on single small flames,[26,27] or calculated for char and ash mixtures at known burn-out fractions.

(b) $X \simeq 1$ which will apply to the smaller oil cokes, coal residues (including cenospheres) and some fly ash. In these cases scattering is less important and can be considered isotropic so a simple correction can be made to emissivity at a given wavelength, e_λ.[25]

(c) $X \ll 1$ occurs for the very small particles, such as soot, which gives the flame its luminosity. Scattering is negligible and the total flame emissivity e_T can be readily measured and represented by an equation of the following form:

$$e_T = 1-(\exp -kCL)(1-e_l),$$

where e_l is the non-luminous gas emissivity, C the soot particle concentration and L the flame thickness. There have been many measurements of e_T on single flames[28] and some measurements on plant which will be referred to later.

4.2. Furnace Models

Much of the earlier work on furnace chamber performance in both coal- and oil-fired furnaces was concerned only with predicting the overall furnace heat absorption. This work has been reviewed by Godridge[2] and found to be mainly empirical. This empirical work was followed by semi-theoretical analyses applied to specific types of furnace, for example, the well-stirred furnace[29] and the plug-flow furnace.[30] A very useful addition to these models and one which has been used to predict the performance of water tube furnaces is the zone method described by Hottel and Cohen.[24] This work makes use of eqn. (2) from the previous section and considers a furnace to consist of a number of discrete cubes of gas, each with a particular temperature and absorption coefficient. The radiation from each cube to the walls can then be calculated, provided their relative geometrical orientation is known. The temperatures are obtained from individual heat balances so that for this and other models, which are essentially mathematical, the following inputs are required: (1) flow distribution, (2) heat release distribution, (3) flame absorption coefficients and (4) furnace wall optical properties.

There are only a limited number of applications of Hottel and Cohen's method readily available in the literature. Of these Field *et al.*[27] have illustrated the method for a coal-fired water-tube boiler. Yardley and Patrick[31] and Hottel and Sarofim[32] deal theoretically with gas-fired furnaces. Field's work on the coal-fired water-tube boiler used only five big gas zones and assumed the input data. Also the other workers assumed certain gas flow patterns, including recirculation, and determined temperature from a system of heat balances. Thornycroft and Thring[33] did similar work on a petroleum heater which was checked experimentally and showed only a fair degree of agreement. They calculated a value for the flame emissivity and assumed that it did not vary throughout the plant. Johnson and Beér[34] have since shown how to calculate the emissivities of gas–soot mixtures for zone calculations taking account of temperature variations.

It was said earlier that input data are required for flow distribution, heat release distribution, flame absorption coefficients and furnace wall optical properties. Within the CEGB the present arrangement is that the flow patterns are obtained from isothermal models of the furnaces although there is work in hand to provide these data by mathematical means. The heat release data and the flame absorption coefficients are obtained from experiments with single full-sized burners of the type that are installed in the actual plant and wall properties have been measured on operational plant. The earlier CEGB models used the calculation procedure outlined by Hottel and Cohen[24] with the exception that cube temperatures were calculated from a gas cooling curve. These predictions were reasonable although they over-estimated the maximum heat fluxes at the walls and failed to pick out the high heat fluxes at the furnace floor, in front wall oil-fired units. In the latest methods the radiant heat transfer step may be tackled by two separate methods. These are based on an adaption of non-equilibrium diffusion theory and the Monte-Carlo technique, both methods being incorporated into a single computer program. Arscott *et al.*[35] have compared their predictions with measurements from full scale oil-fired plant and shown very good agreement (see Fig. 11).

The method of calculation for steady turbulent flow developed by Spalding and others[36] has been extended by Gibson and Morgan[37] to include solid radiating particles that react at a finite rate with the gas in a system with cooled isothermal walls. Whilst the comparisons of predicted two-dimensional values with measured flame characteristics is reasonable for swirling flames the results only apply to a 660 kg/h coal-fired rig. Similar comparisons have been made between the Spalding model and single flames in the IFRF furnace[38] which give realistic but not precise agreement. Work is in hand to compute three-dimensional furnace flows[39] and Zuber and Konecny[40] have reported work relevant to furnace chambers but using a very coarse grid.

4.3. Instrumentation and Experimental Work

Single Flames

The performance of new burners and the characteristics of flames have been investigated systematically

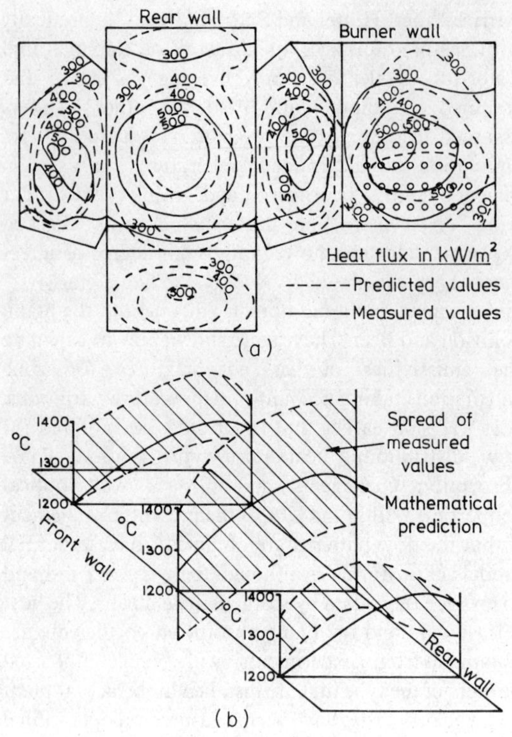

FIG. 11. Heat Fluxes and Gas Temperatures in the Fawley Furnace.

in test facilities by, for example, the burner manufacturers, the IFRF and the CEGB. In practice, burners operating in water-tube boilers typically fire in groups, and work has been done to compare measurements of flame temperatures, emissivities, heat fluxes and carbon concentrations measured on large (i.e. up to 1 kg/s throughput) single-oil flames and the same flames firing in groups on plant.[41] It was observed in the CEGB test facility that the heat fluxes close to the burner are less than in the boiler by a factor of two to three, even though the flame emissivities are similar and the maximum temperatures only up to 10% less. This decrease in heat fluxes occurs because in the test facility the flame is surrounded by rig surfaces and sees more surface than flame, whilst in a front-wall-fired boiler the burners, with the exception of the wing burners, face a sheet of flame. In a corner-fired furnace the burners see more water cooled surface than in the front-wall-fired case and the heat fluxes are that much less, but still greater than in the test facility. Thus apart from providing data on flame stability and combustion efficiency (referred to earlier) the main use of single flame work in the field of heat transfer has been to provide heat release data and flame absorption coefficients for furnace models.

Furnace Measurements

In full scale plant it is relatively easy to calculate the total heat absorbed in the furnace chamber (i.e. by subtracting the heat leaving the furnace on the gas side from the total heat input or by waterside measurement and calculations) but it is more difficult to determine local heat absorption rates. At the same time it is more important than ever to determine these local fluxes at the walls because of their effect on the waterside conditions within the tubes. This has led to the extensive use of heat flux meters attached directly to the furnace tubes. These have been described in the literature[42] but it only became apparent during and shortly after the boiler trials that all instruments of this type are liable to a variety of errors. These occur because the meters run at a higher temperature than the tubes to which they are attached and are involved in a different ash deposition pattern, both of which lead to surface emissivity changes. In general any meter can be calibrated to read correctly under clean conditions, but in dirty, corrosive conditions large and often unpredictable errors are possible.[43] In practice the overall performance of the heat flux meters was checked by comparing the integrated measured fluxes with the calculated total heat absorption, using the methods outlined previously. The results usually agreed to within 5–10% in units where between 190 and 250 heat flux meters were installed. In the highly rated operational oil-fired plant the maximum heat fluxes on the wall are about 550 kW/m² (heat flux contours are given in Fig. 11). In coal-fired plant the equivalent fluxes are about 320 kW/m² which normally increase by about 10% after sootblowing.

Access for probing the furnace chambers of large boilers is very restricted. Because of this, only a limited number of determinations of gas temperature, composition and velocity have been made and almost without exception these were done at the furnace exit. In future plant experiments it is intended that more flame data be collected within the furnace chamber. The probes consist of a venturi pneumatic temperature measuring instrument and five-hole pitot tubes combined into one water cooled probe. These differ from the usual instruments described by Chedaille and Braud[44] only in that they are very long, being up to 10 m in length. This leads to considerable problems in handling, providing adequate cooling and keeping the gas passages clean, but these can be overcome. A set of temperature results for the outlet planes of oil-fired and coal-fired boilers is shown in Figs. 11 and 12. The

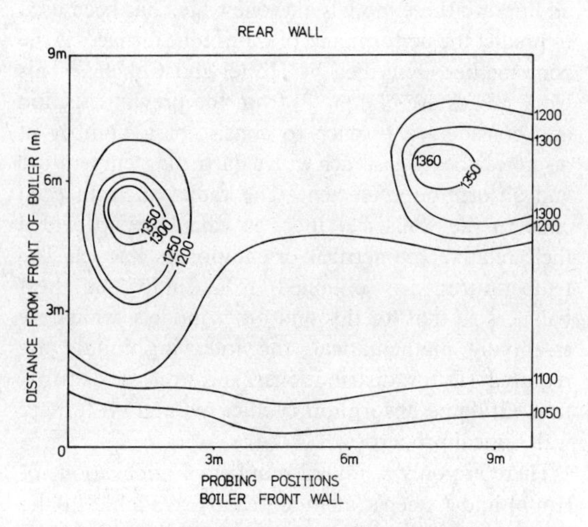

FIG. 12. Typical Gas Temperature Contour at the furnace exit plane of a large p.f. boiler.

maximum values are some 250°C higher than would normally be observed in smaller units[45] but they are in line with present practice, since a greater percentage of the heat in the gases is going to the superheater and reheater sections.[5]

Where it is necessary to know the emissivity and temperature of the furnace walls or ash deposits in order to calculate heat fluxes these may be determined using a Land hemispherical pyrometer. The theory and use of the instrument has been given by Drury *et al.*[46] On much of the CEGB operational plant the arrangement of the furnace tubes in relation to the ports is ideal for making measurements on the tube surfaces. This is because the tubes that would normally pass over the ports stand proud to one side and can easily be reached by the surface pyrometer although in these circumstances the probe may need to be placed in a water cooled jacket. Emissivity values for coal and oil ashes in operational plant have been reported by Godridge and Morgan.[47]

A method of determining flame emissivities that does not require a hot background, but only a cold non-reflecting one, has been described by Daws and Thring.[48] It utilizes the fact that most radiation is in the infra-red (say 1–5 μm) and comes from incandescent carbon particles whose variation of absorption coefficient with wavelength is known. The comparison of readings with a total radiation pyrometer and an optical pyrometer operating at 0.65 μm, together with the variation of spectral absorption coefficient, are used to compute the mean radiant flame temperature and mean emissivity. This is equivalent to a two-colour measurement, the theory of which has been discussed by McAdams.[23] For emissivity the method has an accuracy of ±10% and for temperature about ±5% in oil-fired water-tube boilers.[49] The method has not been used in coal-fired boilers where account needs to be taken of multiple scattering[25] by the many ash particles which make it difficult to see into the furnace for any distance.

5. DISCUSSION

With increasing fuel inputs to the larger boilers and the extra burners needed it has been necessary to locate burners in two opposing walls instead of one. This may not be an ideal solution and the problem of accommodation can be eased by increasing the fuel throughput per burner. In oil-fired plant this is happening (e.g. Fawley has thirty-two burners each of 1 kg/s throughout whilst Grain has twenty-four burners each of just under 2 kg/s oil flow). There are additional advantages in reducing the number of burners in that it simplifies the fuel and air layouts and reduces the number of burner components to be maintained.

With larger fuel throughputs at each burner more attention has to be paid to mill or atomizer performance and to careful matching of the air flow in order to maintain efficient combustion. Present oil-fired plants (but not coal-fired units) are built without electrostatic precipitators and future units without even mechanical grit arrestors. This means that stringent standards for oil-burners will have to be achieved continuously. Present CEGB standards demand less than 0.12 g/N m^3 of solids, a Shell smoke number of 5 or less and a CO level of less than 100 ppm at, in future plant, an excess O_2 level of 0.5%. Recent studies[50] have led to an improvement in the droplet size distribution from twin-fluid atomizers whilst at the same time reducing steam or air consumption.[51] Work on air flow distribution has been mainly concerned with the turndown problem. Here the difficulty is to provide enough momentum to the air to achieve good mixing for complete combustion. This can be solved by using a dual air flow burner[52] which passes the air through a central core at high velocity during turndown but which has a second supply of air through an outer annulus, controlled by a linearized flow sleeve damper, as the load is increased.

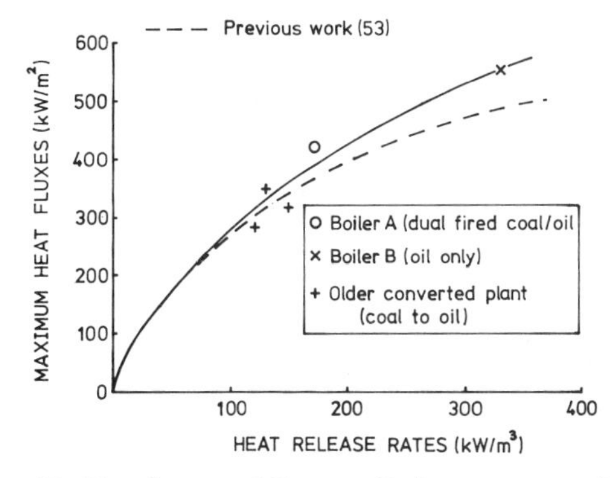

FIG. 13. Heat fluxes and Furnace Ratings—a comparison between older plant and present (A and B) 500 MW units.

Examination of heat fluxes and heat release rates shows that maximum local heat fluxes in the larger boilers are higher than might have been expected from previous experience with CEGB plant and foreign experience.[53] For oil-fired plant this is well illustrated in Fig. 13 where as expected, heat fluxes increase as the furnace rating increases but the latest maximum heat fluxes are well above the curve representing previous work. It is likely that larger absolute furnace sizes giving a reduced surface to volume factor and generally higher air preheat temperatures contribute to higher mean flame temperatures and hence to increased heat fluxes. These higher heat fluxes can damage inadequately-cooled tubes so that in future plant furnace floors are to be put further away from the lowest row of burners (other remedies would be to cover the floor with refractory material or use rifled tubes) and superheater tubes will not be subjected to the very hot furnace conditions (see the design for Grain Power Station—Fig. 4). It is also present practice not to have tube butt welds in the highest heat flux zones by using continuous runs of pipe through these zones and to limit tube distortions (e.g. port holes) in these areas. Both these measures are taken because at high heat fluxes and rapid rates of evaporation boiler water solutes may concentrate and deposit where tube geometries change. This can lead to rapid internal pitting (on load corrosion) and tube failure. Apart from damaging tubes, high heat fluxes, if they occur close to off-load burners, can also readily damage the burner components (cooling the off-load burners by admitting large amounts of air is not acceptable in oil-fired plant because high excess air in the furnace leads to SO_3 formation). Methods for cooling off-load burners and materials able to withstand high temperatures are being studied.

The opposed-fired combustion systems should give good mixing in the tail of the flames and hence improved carbon burnout. However, Droin and Perthuis[54] have shown, using an oil-fired rig, that firing burners in direct opposition can lead to unstable conditions in the furnace with one burner dominating the other for a while and then the position being reversed. This could lead to difficulties with combustion control.

The swirl-type coal burners used in p.f. boilers have a high inherent stability and there have been no combustion instabilities in the Drax boilers. These boilers have shown good combustion performance. One interesting effect is the larger than normal amount of ash falling into the furnace ash hopper. The reasons for this change are not yet established but it has beneficial effects on the gas borne solids emission from the furnace.

Flue gas recirculation through the ash pit is also incorporated in future oil-fired plant in order to assist operation at part load. This is necessary because at low loads gas residence times in the chamber are longer so that too much heat may be absorbed in the furnace chamber leaving insufficient for superheating. Recirculated gas decreases the plug flow residence times and increases the mass of gas flowing through the convective passes of the superheaters. The boiler makers and the CEGB are collaborating in a full-scale plant study of recirculation.

6. CONCLUSIONS

The early troubles with the large water-tube boilers have been identified and largely overcome. Troublesome items of plant have been altered or omitted from new designs. The most recently designed coal-fired boiler has shown good heat transfer and combustion performance.

Furnace performance predictions remain important and are worthy of further development, particularly on oil-fired plant where high heat fluxes can occur. Experiments on the large boilers have been valuable, but more will be learnt from operating plant given more reliable instrumentation. However, detailed research investigations are best carried out on fully commissioned plant where reliable operation can be assured. The CEGB and the manufacturers are at present involved in such an exercise at Richborough Power Station.

Acknowledgements—The authors are grateful to many colleagues for useful discussions, and the Central Electricity Generating Board for permission to publish this paper.

REFERENCES

1. ANON, *Modern Power Station Practice*, 2nd edn., Pergamon Press, 1971.
2. GODRIDGE, A. M., *J. Inst. Fuel* **40**, 300 (1967).
3. DADSWELL, K. E. and THOMPSON, F. R., *Conference on Major Developments in Fuel Firing*, by Institute of Fuel, 1959.
4. MEYLER, J. J. and LANG, J. F., *J. Inst. Fuel* **36**, 361 (1963).
5. LAMB, L., *J. Inst. Fuel* **45**, 443 (1972).
6. HOW, M. E., *Energy Wld* **13**, 7 (1975).
7. MCKENZIE, E. C., *J. Inst. Fuel* **44**, 468 (1971).
8. JUNIPER, L. A., *Comb. Inst., 1st European Symp., Sheffield*, Academic Press, New York, 1973.
9. BROOKS, W. J. D., HOLMES, L. and LEASON, D. B., *J. Inst. Fuel* **38**, 218 (1965).
10. ANSON, D. and TINDALL, D., *J. Inst. Fuel* **40**, 551 (1967).
11. CHEETHAM, H. (in press) (1976).
12. LUCAS, D. H., *J. Inst. Fuel* **36**, 203 (1963).
13. READ, A. W. and UMPLEBY, R. A., *J. Inst. Fuel* **34**, 249 (1971).
14. ANSON, D., CLARKE, W. H. N., CUNNINGHAM, A. T. S. and TODD, P., *J. Inst. Fuel* **44**, 191 (1971).
15. GODRIDGE, A. M. and HORSLEY, M. E., *J. Inst. Fuel* **44**, 599 (1971).
16. BIEBER, K. H., *Mitt. V.G.B.* **50**, 83 (1970).
17. SNELL, P. A. and CRESSWELL, P. J., *J. Inst. Fuel* **43**, 248 (1970).
18. BURNETT, D. J., Babcock & Wilcox Canada Ltd., Bulletin 70–75. Also Institute of Combustion and Fuel Technology Meeting, Toronto, December 1970.
19. ANON, *Electrical Wld Week*, p. 2 December 1970 and p. 6 February 1971.
20. DIXON, R., Joint CEGB/Institute of Fuel/Institute of Petroleum Conference, Harrogate November 1975 (To be published).
21. PROUX, J., *A.I.M. Symposium on Power Stations*, Liege, 1974.
22. THRING, M. W., *Verfahrenstechnik*, 544 (1965).
23. MCADAMS, W. H., *Heat Transmission*, 3rd ed. McGraw-Hill, New York, 1954.
24. HOTTEL, H. C. and COHEN, E. G., *A.I.Ch.E. Jl* **4**, 3 (1958).
25. BEER, J. M. and HOWARTH, C. R., *12th Symp. Combustion*, Combustion Institute, Pittsburgh, 1968.
26. BEÉR, J. M., *J. Inst. Fuel* **37**, 286 (1964).
27. FIELD, M. A., GILL, D. W., MORGAN, B. B. and HAWKSLEY, P. G. W., Comb. Pulverized Coal BCURA, Leatherhead, *Inst. Fuel* (1967).
28. BEÉR, J. M., *J. Inst. Fuel* **35**, 3 (1962).
29. HOTTEL, H., *J. Inst. Fuel* **34**, 220 (1961).
30. THRING, M. W., *The Science of Flames and Furnaces*, 2nd ed., Chapman & Hall (1962).
31. YARDLEY, B. E. and PATRICK, E. A. K., *1st Inst. Fuel Symp.* (1963).
32. HOTTEL, H. and SAROFIM, A. F., *Radiative Transfer*, McGraw-Hill, New York, 1967.
33. THORNYCROFT, W. T. and THRING, M. W., *Trans. Inst. chem. Engrs.* **38**, 63 (1960).
34. JOHNSON, T. R. and BEÉR, J. M., *J. Inst. Fuel* **46**, 301 (1973).
35. ARSCOTT, J. A., GIBB, J. and JENNER, R., *Combustion Institute, 1st European Symp. Sheffield*, Academic Press, New York, 1973.
36. GOSMAN, A. D., PUN, W. M., RUNCHAL, A. K., SPALDING, D. B. and WOLFSHTEIN, M. W., *Heat and Mass Transfer in Recirculating Flows*, Academic Press, New York, 1969.
37. GIBSON, M. M. and MORGAN, B. B., *J. Inst. Fuel* **43**, 517 (1970).
38. BARTELDS, H., LOWES, T. M., MICHELFELDER, S. and PAI, B. R., *Combustion Institute, 1st European Symp., Sheffield*, Academic Press, New York, 1973.
39. PATANKER, S. V. and SPALDING, D. B., *J. Inst. Fuel* **46**, 279 (1973).
40. ZUBER, I. and KONECNY, V., J. *Inst. Fuel* **46**, 285 (1973).

41. GODRIDGE, A. M., *Combustion Institute, 1st European Symp., Sheffield*, Academic Press, New York, 1973.
42. NORTHOVER, E. W. and HITCHCOCK, J. A., *J. Scient. Instrum.* **44**, 371 (1963).
43. MORGAN, E. S., *J. Inst. Fuel* **47**, 113 (1974).
44. CHEDAILLE, J. and BRAUD, Y., *Industrial Flames*, Vol. 1, Edward Arnold, 1972.
45. DOLEZAL, R., *Large Boiler Furnaces*, Ed. J. M. BEÉR. Elsevier Publ. Corp., 1967.
46. DRURY, M. D., PERRY, K. P. and LAND, T., *J. Iron Steel Inst.* **169**, 245 (1951).
47. GODRIDGE, A. M. and MORGAN, E. S., *J. Inst. Fuel* **44**, 207 (1971).
48. DAWS, L. F. and THRING, M. W., *J. Inst. Fuel* **25**, 528 (1952).
49. ANSON, D., GODRIDGE, A. M. and HAMMOND, E. E., *J. Inst. Fuel* **47**, 83 (1974).
50. MULLINGER, P. J. and CHIGIER, N. A., *J. Inst. Fuel* **47**, 251 (1974).
51. PYE, J. W. and WAY, R. E., CEGB Tech. Disc. Bull., No. 193.
52. ANSON, D., HOW, M. E., STANDEVEN, P. R. C. and TURTLE, J., CEGB Tech. Disc. Bull. No. 169.
53. LEDINEGG, M., *Mitt. V.G.B.* **97**, 251 (1965).
54. DROIN, R. and PERTHUIS, E., I.F.R.F. 3rd Members Conference, 7 (1974).
55. C.E.G.B., Flame Radiation Committee, Brunswick Wharf Report (1960).
56. REID, W. T., COHEN, P., and COREY, R. C., *Trans. Am. Soc. mech. Engrs.* **70**, 569 (1948).
57. COREY, R. C. and COHEN, P., *Trans. Am. Soc. mech. Engrs.* **72**, 925 (1950).
58. ANON, Boiler House Rev. Supplement, May 1963.

Manuscript received, 19 February 1976

Prog. Energy Combust. Sci., Vol. 2, pp. 97–114, 1976. Pergamon Press. Printed in Great Britain

THE ATOMIZATION AND BURNING OF LIQUID FUEL SPRAYS

NORMAN A. CHIGIER
Department of Chemical Engineering and Fuel Technology, University of Sheffield

1. INTRODUCTION

Liquid fuel requires to be broken up into small droplets in order that it can effectively burn in combustion chambers. Atomization of the liquid fuel is most commonly carried out either by injecting the fuel through small orifices at high pressure or by mixing the fuel with high pressure air or gas. The most effective atomization is achieved when thin liquid sheets are formed which subsequently become unstable and then break up to form ligaments and large drops, which then break down further into small droplets. Because of the importance of the atomization process for completion of burning, the twin-fluid and pressure jet atomizers are described in detail. Despite the fact that sprays are so extensively used in industry, they have only been examined in detail in the last few years. The burning of single droplets has been studied extensively and it has been assumed that this information is directly relevant to the burning of liquid fuel sprays. This paper examines the most recent experimental information, which shows fundamental differences between liquid spray combustion and single droplet combustion. On the basis of this evidence, physical models are developed and an idealized spray flame is postulated.

The physical processes involved in air blast atomization are examined in detail and empirical formulae are provided for the determination of mean droplet size on the basis of atomizer geometry and flow variables. High-speed photographic techniques have provided much of the detailed information of spray characteristics, both under nonburning and burning conditions in an extensive series of experiments on twin-fluid and pressure jet spray burning carried out at the University of Sheffield, which served as a basis for postulating physical models of spray combustion. Since single droplet burning does occur when droplets leave the main spray, some of the most recent information on the effects of combustion on flow around vaporizing droplets is examined. Also included in this article are studies of fuel droplets vaporizing in gas flames and the special problems associated with diesel spray combustion.

Onuma and Ogasawara[1] made a direct comparison between the structures of a spray combustion flame and a turbulent gas diffusion flame in a vertical cylindrical furnace. For the spray measurements, an air atomizing nozzle with kerosene fuel was used with a secondary air supply. Spatial distributions of droplet concentration and size were measured by inserting a probe containing a magnesium oxide coated slide, covered by a shutter. Measurements were also made of temperature, velocity and gas concentrations, by using a thermocouple, pitot tube and sampling tubes with gas phase chromatography. Further measurements were made in a turbulent gas diffusion flame, using the same apparatus and replacing the liquid kerosene fuel by propane gas. This set of experiments serves as a test of the validity of the hypothesis regarding the structure of spray flames put forward by Chigier and Roett[2] on the basis of their measurements made in an unconfined air/kerosene spray flame.

Onuma and Ogasawara[1] found that the spray flame could be sub-divided into three main regions: an initial region consisting of a two-phase mixture in which most of the droplets evaporate and where soot is formed,

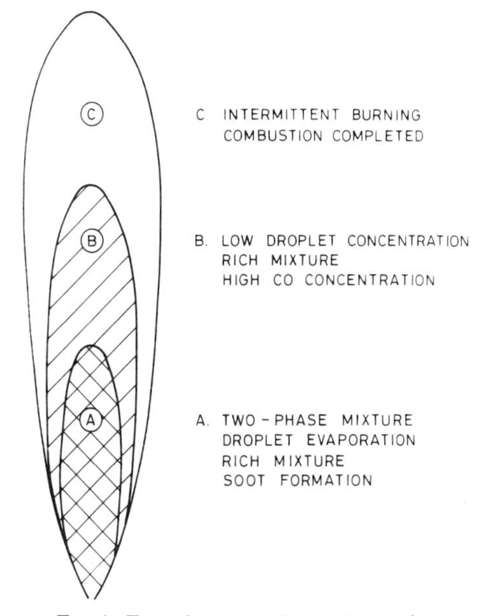

FIG. 1. Zones in a spray flame (Onuma[1]).

due to the very rich mixture ratios; in the intermediate region concentration and size of droplets is very low but concentrations of combustible gas are high, with particularly high concentrations of CO; in the final region intermittent burning takes place as combustion is completed. These three regions are shown schematically in Fig. 1. The direct comparisons made by Onuma and Ogasawara[1] between temperature, velocity, O_2, CO, CO_2 and HC profiles in diffusion flames and spray flames show that they are very similar and the differences in the measured profiles were due to differences in emissivity and the initial conditions between the two flames. It was concluded that most of the droplets do not burn individually but that the vapour-cloud from the evaporated droplets burned like a diffusion flame in the turbulent state. It was demonstrated that reduction in droplet size could be calculated on the basis of single droplets vaporizing in a hot gas environment.

Mellor[3] has made measurements in spray flames in a simulated gas turbine combustor and has concluded that droplet burning does not take place. The same basic results were obtained when running propane and, subsequently, kerosene through the same atomizer. We thus have, to date, a series of separate experiments in which measurements have been made in burning spray flames: the twin-fluid, air-assist kerosene unconfined spray flame of Roett;[2] the hollow cone pressure jet unconfined spray flame in the wake of a stabilizer disc of McCreath;[4] the confined spray flame of Onuma and Ogasawara;[1] and, finally, the spray flame in an experimental gas turbine combustor of Mellor.[3] In each of these experiments it has been concluded that there is no evidence of flames surrounding individual droplets; there is no evidence of burning within the dense spray region where temperatures are low and the mixture ratio is rich; and the structure of the flame is similar to that of a gas diffusion flame. There is no justification for investigators[5] to use results obtained from single droplet burning tests in order to make predictions in spray flames. In the single droplet tests with envelope flames surrounding the drop, many investigators[6] have shown good agreement between predictions and measurements where the assumption is made that heat is transferred by conduction from the flame to the drop surface and vapour diffuses by molecular diffusion from the drop surface to the flame front. Application of the results from these single drop experiments to spray flames will only be valid when evidence is shown of envelope flames around individual droplets. From the above tests which have been carried out on spray flames, droplets have been shown to be vaporizing in a low oxygen, rich fuel gas environment at temperatures substantially lower than the flame temperature. Heat transfer from the flame front to the drop surface takes place over distances of hundreds or tens of drop diameters by both radiation and turbulent convection. Mass transfer from the drop surface is restricted by the high concentrations of fuel in the gas surrounding the droplet and concentration gradients of fuel vapour can be expected to be much smaller than in the case of single drop burning. Turbulence levels in the surrounding gas also need to be taken into account when predicting rates of vaporization. Most practical spray flames are, thus, mixing controlled and the rate of burning is mainly dependent upon the rates of turbulent diffusion of fuel vapour and air to the flame front. In the initial regions of the flame, where mixtures are rich, vaporization of droplets will play no significant role in the rate of combustion.

It is possible that, in the outer regions and towards the end of the spray, a small proportion of droplets are in relative isolation in a predominantly air surrounding, resulting in the formation of a flame around the individual drops. Further, in the burning of heavy fuel oils, droplets having a low volatility persist for long periods and distances in a combustor with the subsequent formation of envelope flames. More evidence is clearly required of the burning mechanisms and structure of spray flames but the balance of evidence reported in the literature supports the theory that individual droplet burning is not significant and that most spray flames burn similarly to gas diffusion flames.

2. HIGH-SPEED PHOTOGRAPHY OF DROPLETS IN SPRAYS

High-speed photography of droplets in sprays provides a direct measure of droplets in flight and obviates the introduction of probes into the spray. When a high-powered light source from a spark is focused on a small area within the spray, shadow photographs of the droplets can be obtained, provided that the spark duration is short, so as to "freeze" the droplet in flight. The intensity of the light source requires to be sufficiently high to penetrate through the spray, in order to give a sharp image on the photographic plate. When two sparks are fired in rapid succession a double-image is obtained of a single droplet on the photographic plate, from which the velocity of the droplet can be determined by measuring the distance travelled by the droplet and dividing this distance by the time interval between the two sparks. The direction of movement of the droplet can also be directly determined from the photographs as an angle of flight with respect to the central axis of the spray. The method provides instantaneous measurements for individual droplets and from a series of such measurements time-average and space-average quantities can be determined as well as standard deviations. Measurement techniques based on this principle have been used by York and Stubbs,[7] DeCorso,[8] Briffa and Dombrowski,[9] Finlay and Welsh[10] and Mellor,[11,12] Roett,[2] McCreath[4,14] and Chigier.[11–18] The technique has been successfully applied to measurements in spray flames,[14] except for the regions of high radiation. The method has also been developed for the determination of size distribution from large industrial atomizers for utility boilers.

The optical and photographic system, as used for measuring a liquid spray in a uniform stream wind tunnel,[11] is shown in Fig. 2. The optical system consists of two sparks and lenses, a diffusing screen and a camera. The camera is focused onto a plane inside the spray with a depth of field of approximately 1 mm. The spark units have capacitors charged to 10 kV which, on discharging to the earth electrode, ionize the air in the spark gap. The light emitted from the sparks is mainly in the blue and near ultra-violet regions of the spectrum with some small traces of green and red light. For nonburning measurements, monochromatic plates with $\frac{1}{2}$ s exposure times could be used in normal daylight conditions. For measurements in flames, orthochromatic film is used and a filter is introduced between the spray and the camera so as to reduce radiation from the flame and to match the spark light with the most sensitive wavelength range of the photographic plate. The depth of field must be kept small

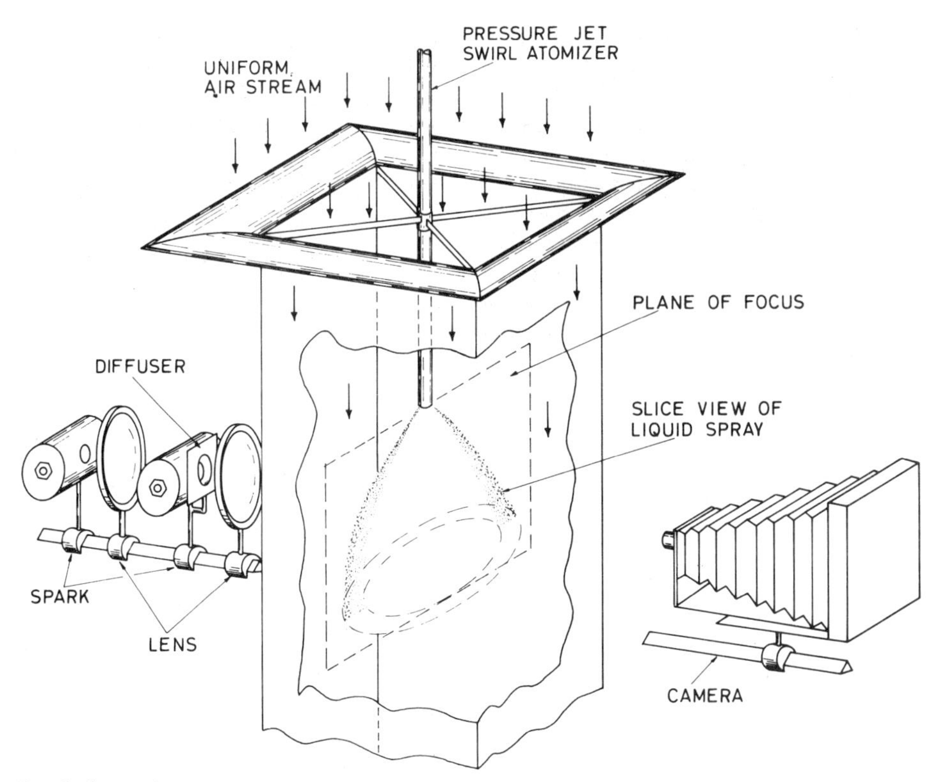

FIG. 2. System for measuring velocity, size and trajectory of droplets in a spray using double-image high-speed photography.[11]

and the relative aperture is required to be large in order to reduce "fogging" of the photographic plate by flame radiation. In order to achieve these conditions, the objective distance of the camera lens requires to be small, resulting in radiation from the flame to the lens. A high-pressure curtain of compressed air is, therefore, introduced to the front of the lens, which also serves to prevent wetting by stray droplets.

An electronic system controls the double-flash light source, providing a variable inter-spark time interval so that double-image photographs of the spray can be obtained. An oscilloscope camera records the photo cell output, providing a direct measure of the actual time interval between each set of double sparks. The control system generates a trigger pulse for the oscilloscope, fires each of the two sparks in sequence so that time intervals can be controlled within the range from 10 to 4000 μs.

The photographic negatives of any area of the spray can be analysed directly by projection on to a translucent screen, so as to give an overall magnification of 100. Alternatively, the photographs can be analysed automatically by an electronic fast-scanning Image Analyser (Quantimet) which provides a size distribution on the basis of an automatic pattern recognition analysis.

3. AIR BLAST ATOMIZATION

The term twin-fluid atomization is used for systems in which a high velocity gas stream is used to atomize fuel in a relatively low velocity liquid fuel stream. Atomization is most effective when very thin liquid sheets are sandwiched between, at either higher or lower velocities than the liquid stream, gas streams with high rates of shear and velocity gradients across the liquid–gas interfaces. The atomizing fluid is generally high pressure air, in gas turbine systems, and high pressure steam in land-based and marine systems where steam is readily available from a boiler. The selection of steam or air is based upon its availability and cost, since, from the point of view of atomization, there appears to be no significant difference in using steam or air. From the point of view of combustion efficiency, air is preferable to steam. When mass flow rates of steam are less than 10% of that of the fuel, steam atomization has been found to cause no significant deterioration in combustion efficiency. In addition to the general overall reduction in size of droplets that can be achieved by twin-fluid atomization, as compared to pressure jet atomization, production of large droplets can be avoided over a wide range of fuel flow rates. In practice, combustion systems are required to operate effectively at idle conditions as well as at maximum power and difficulties have been experienced in achieving adequate atomization at both the very low and very high levels of fuel flow rate. The ratio of the maximum to minimum fuel flow rates is referred to as the turn-down ratio. For marine applications, turn-down ratios of 20:1 are required and, in gas turbines, ratios of 50:1 are needed. In the past, atomization systems were designed to cope with "cruise" or "normal" operating conditions and it was accepted that during start-up and idle, as well as during rapid acceleration, atomization would be poor and high rates of emissions of smoke and unburned hydrocarbons would occur for periods which are short, relative to the total period of combustion. In order to satisfy the strict limitation of total emission of pollutants, it is necessary to obtain good atomization over a wide range of operating conditions and the high rates of emission obtained during start-up and idle require to be reduced. The most effective practical means of achieving this is by the use of twin-fluid atomization. The specific terms "air blast" and "steam blast" are used in the literature but the more general term is "twin-fluid atomization".

In gas turbine combustion chambers the air blast atomizer is replacing the pressure jet atomizer, which has been the most commonly used atomizer. An additional advantage of air blast atomizers over pressure atomizers is that small droplets are air-borne in the high velocity air stream so that their distribution throughout the combustion zone is dictated by the air flow pattern, which remains fairly constant under all operating conditions. In consequence, the temperature profile at the combustion chamber exhaust tends to remain constant, thereby extending the life of the turbine blades. Mixing of fuel and air is greatly improved, resulting in blue flames of low luminosity compared to the yellow flames of high luminosity, which can be found with pressure jet atomization. Reduction in radiative heat transfer results in cooler flames and the reduction in quantity and dimensions of soot particles leads to overall reductions in exhaust smoke. The air blast atomizer has an important advantage over the gas turbine vaporizing tube atomizer, which is normally immersed in flame and very susceptible to overheating. The air blast atomizer is continuously cooled by high velocity air flowing over it at compressor outlet temperatures.

The physical process of air blast atomization is composed of the following steps[19]:

(1) formation of thin liquid sheets on a plate or along the inner walls of an internal-mix atomizer or free sheets unattached to walls;
(2) disintegration of the liquid sheet by aerodynamic forces to form ligaments, large drops and droplets;
(3) break-up of ligaments and large drops into droplets;
(4) acceleration of droplets by high-speed gas stream and/or deceleration of droplets by low velocity and recirculation flows;
(5) formation of two-phase, liquid-gas spray, followed by spreading of spray jet and entrainment of gas from surroundings;
(6) evaporation of droplets as a result of temperature and vapour pressure differentials between droplet surface and surroundings; and
(7) agglomeration of droplets by collision can occur but, except under conditions of rapid deceleration in the regions of a spray close to the nozzle, this mechanism is not considered to be significant.

Nukiyama and Tanasawa[20] studied air blast atomization for application in piston engines. Fuel was injected through a fine orifice into a venturi of a carburettor, through which air was flowing, Fig. 3. They examined the effect of variation in fluid flow properties by using mixtures of water, alcohol and glycerine and they also varied the relative velocity of air and liquid. They found that droplet size depended on the surface tension, density and absolute viscosity of the liquid; on the ratio of liquid to air flow rate; and on the initial relative velocity of air and liquid. They derived the following empirical equation:

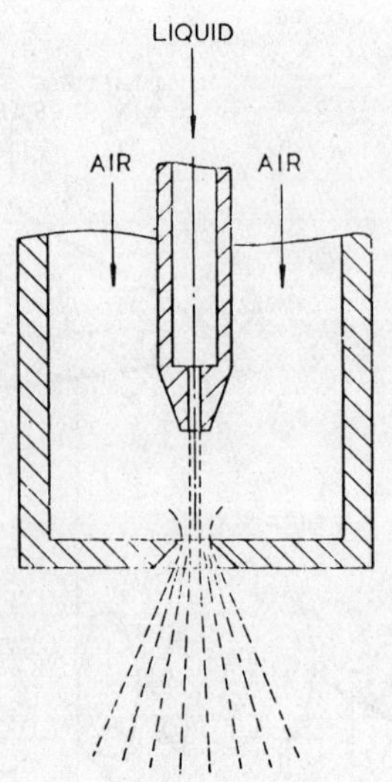

FIG. 3. Air blast atomizer (Nukiyama and Tanasawa[20]).

$$d_0 = \frac{0.585\sqrt{\sigma}}{\Delta u\sqrt{\rho_l}} + 1.415\times 10^6\left(\frac{\mu_l}{\sqrt{\sigma\rho_l}}\right)^{0.45}\left(\frac{Q_l}{Q_a}\right)^{1.5}, \quad (1)$$

where: d_0 = Sauter mean diameter of droplet, μm; Δu = relative velocity of air and liquid, m/s; σ = surface tension of liquid, N/m; ρ_l = density of liquid, kg/m^3; μ_l = viscosity of liquid, N s/m^2; Q_l = liquid flow rate, m^3/s, and Q_a = air flow rate, m^3/s.

For kerosene with specific gravity of 0.78 at 15°C, eqn. (1) reduces to

$$d_0 = \frac{10.57}{\Delta u} + 7\left(\frac{Q_l}{Q_a}\right)^{1.5}. \quad (2)$$

Equation (2) is not dimensionally correct but Lewis *et al.*[21] have shown that this simple equation can be used for air blast atomizers when the liquid density is between 700 and 1200 kg/m^3, the surface tension between 0.019 and 0.073 N/m, viscosity between 3×10^{-4} and 5×10^{-2} N s/m^2 and the air velocity is subsonic. In the derivation of eqn. (1) changes in air properties have not been taken into account so that its application is limited to conditions close to standard temperature and pressure. Nukiyama and Tanasawa[20] concluded, on the basis of their results, that the effect of change in nozzle shape can be incorporated in changes in the discharge coefficient.

Wigg[22] measured size distributions by using the freezing wax method, in which molten wax is atomized and the solid particles formed in the air are then collected and measured. On the basis of his own measurements, and those of a number of other investigators, he concluded that eqn. (1) gives too high a droplet size at some velocities and over-estimates the effect of a change in the ratio of liquid to air flow rates. On the basis of the energy required to effect atomization, Wigg developed an equation that was dimensionally correct and could be used for velocities greater than 100 m/s. The Wigg equation is

$$d_m = 0.004 v^{0.5} W^{0.1}\left(1+\frac{W}{A}\right)^{0.5} h^{0.1}\sigma^{0.2}\rho_a^{-0.3}\,\Delta u^{-1}, \quad (3)$$

where d_m = mean median diameter, μm; v = kinematic viscosity, m²/s; W = liquid mass flow rate, kg/s; A = air mass flow rate, kg/s; and, h = height of air annulus, mm.

This equation shows that the greatest changes in droplet diameter can be obtained by changing the relative velocity between liquid and air. In general, for a fixed nozzle, variation in the liquid/air mass flow rate ratio also affects the relative velocity. Amongst the physical properties kinematic viscosity has the exponent 0.5, while surface tension has the exponent 0.2. This is explained by the shear forces predominating over the surface tension forces, which is generally true for the high velocities used in twin-fluid atomization. Changes in dimensions of the atomizer do not necessarily change the thickness of the liquid film and, thus, atomizer dimensions have an exponent which is only 0.1.

Mullinger and Chigier[17] made a comprehensive study of the effect of changing geometrical variables in atomizer design as well as the variables which affect the distribution and the thickness of the fuel film in the mixing chamber of twin-fluid atomizers. They found that Wigg's equation for the determination of mean drop size agreed closely with their experimental results. They carried out tests with air/fuel mass ratios as low as 0.005, which is a factor of 10 smaller than that tested by Wigg.[22] They thus showed that eqn. (3) is still applicable at much lower values of atomizing air/fuel mass ratio than those for which it was originally developed. Furthermore, this agreement with eqn. (3) for a wide variety of different designs of twin-fluid atomizer, both internal and external mixing, confirmed that atomizer geometry has relatively little effect upon droplet size, except for the way in which it affects atomizing fluid velocity and density.

The type of atomizer tested by Wigg,[22] and Mullinger and Chigier,[17] is commonly used in boilers for both land and marine applications. Rizkalla and Lefebvre[23,24] undertook a detailed experimental investigation of air blast atomization, using a specially designed form of air blast atomizer which is representative of modern gas turbine practice. In this atomizer (Fig. 4), the liquid flows through six tangential ports into a weir, from which it spills over the prefilming surface before being discharged at the atomizing edge. One airstream flows through a central circular duct and is deflected radially outwards by a pintle before striking the inner surface of the liquid sheet, while another airstream flows through an annular passage surrounding the main body of the

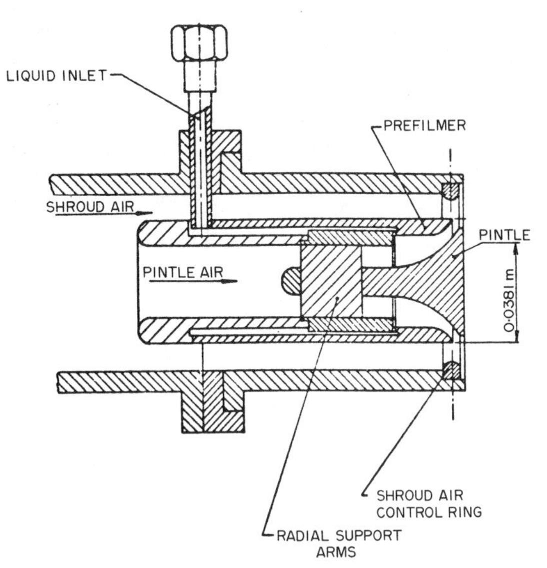

FIG. 4. Air blast atomizer for gas turbines (Rizkalla and Lefebvre[23]).

atomizer. This passage has its minimum flow area in the plane of the atomizing lip in order to impart a high velocity to the air where it meets the outer surface of the liquid sheet. In the first phase of their program, Rizkalla and Lefebvre[23] examined the effects of changes in liquid properties, namely viscosity, surface tension and density on mean droplet size. These studies were carried out using air supplied from a fan at atmospheric pressure. In the second phase of their work they examined the effect of changing air flow properties by varying the air temperature using a kerosene-fired preheater located upstream of the atomizer. Changes in air density were thus obtained by varying the air temperature between 23 and 151°C. They measured mean droplet size by using a light scattering technique based on the forward scattering of a parallel beam of monochromatic light which had been passed through the spray. Rizkalla and Lefebvre[24] drew certain general conclusions concerning the main factors governing air blast atomization. For liquids of low viscosity, average droplet diameters are inversely proportional to both air velocity and air density and these are the dominant factors. From results obtained over a wide test range they concluded that liquid viscosity had an effect which is quite separate and independent from that of air velocity. This suggested a form of equation for droplet diameter expressed as the sum of two terms, the first term being dominated by air velocity and density and the second term by liquid viscosity. Dimensional analysis was used to derive the following equation with the various indices in the expression being deduced from the experimental data.

$$\mathrm{SMD} = 7.33\times 10^3 \frac{(\sigma_l\rho_l D)^{0.5}}{V_a\rho_a}(1+W_l/W_a) + 785\left(\frac{\eta_l^2}{\sigma_l\rho_a}\right)^{0.425} D^{0.575}(1+W_l/W_a)^2, \quad (4)$$

where: SMD = Sauter mean diameter = $\Sigma nd^3/\Sigma nd^2$, μm; n = number of droplets; W = mass flow rate, kg/s;

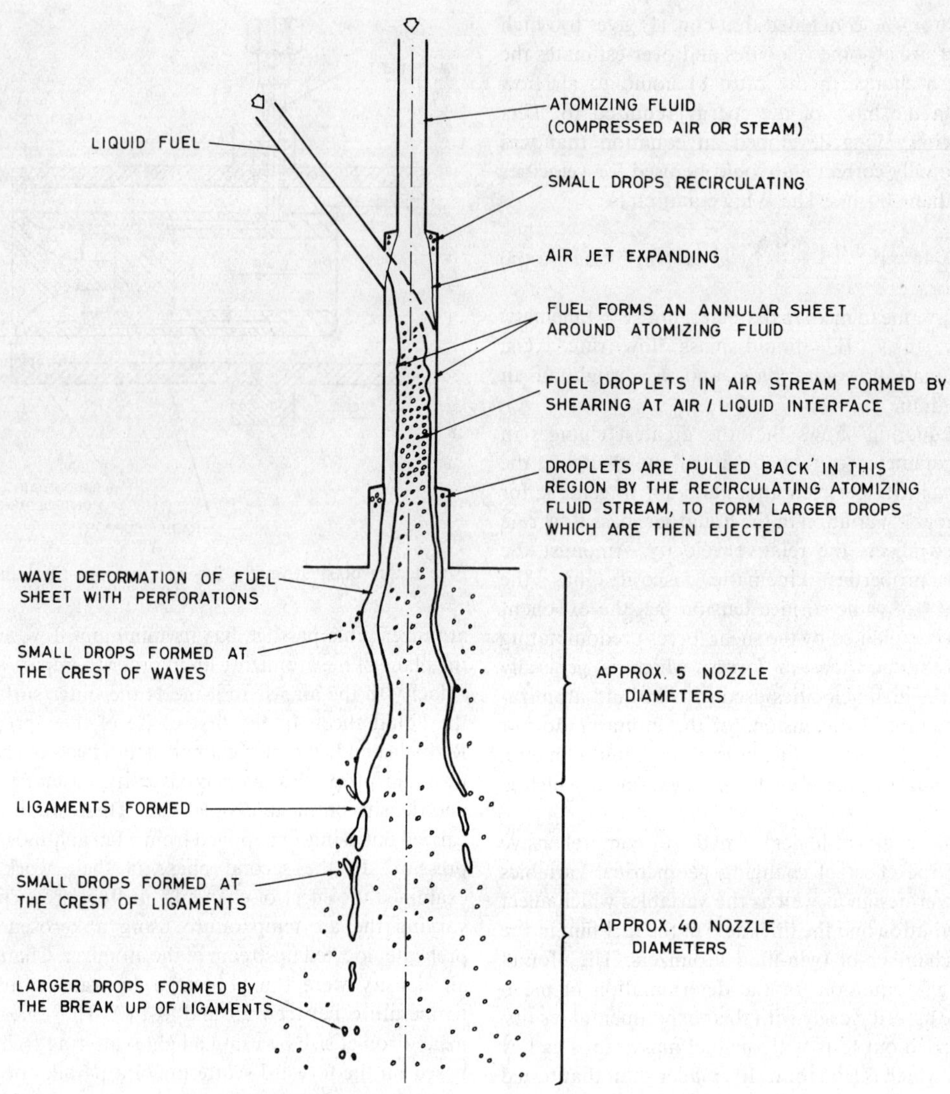

FIG. 5. Liquid atomization in an internal mixing twin-fluid atomizer.[17]

D = diameter of prefilmer, m; η = absolute viscosity of liquid, N s/m^2; σ = liquid surface tension, N/m; and ρ = density, kg/m^3. *Subscripts*, a = air, l = liquid. For liquids of low viscosity such as water and kerosene, the first term predominates and the SMD thus increases with increase in liquid surface tension, density, film thickness and liquid/air ratio, and declines with increase in air velocity and air density. With liquids of high viscosity, the second term acquires greater significance and, in consequence, SMD becomes less sensitive to variations in air velocity and density. The range of variables tested by Rizkalla and Lefebvre[24] was: air velocities from 70 to 125 m/s; air/liquid ratios from 2 to 6; liquid viscosities from 0.001 to 0.044 N s/m^2; and surface tension between 0.026 and 0.073 N/m. For the burning of kerosene, the second term on the right-hand side of eqn. (4) is negligibly small compared with the first term and the SMD is inversely proportional to air pressure. Since gas turbines operate over a wide range of pressures, this finding can be usefully applied in practice.

Both the equations of Wigg,[22] and Rizkalla and Lefebvre[24] were obtained under atmospheric pressure conditions. The effect of increasing pressure on SMD was also examined. The studies of Lewis[21] and Ingebo and Foster[25] suggest that droplet size falls with increasing pressure. In gas turbine combustion chambers any increase in inlet air pressure is always accompanied by an increase in temperature, which increases the rate of evaporation. Air blast atomizers for gas turbine application should, therefore, be designed for the minimum pressure conditions and it can be assumed that, if combustion performance is satisfactory at atmospheric pressure, then atomization quality will be more than adequate at all higher levels of pressure.

The multijet, internal mixing, twin-fluid atomizer is used in oil burners, in boilers and also in gas turbines. Twin-fluid atomizers operate at high combustion efficiencies with low excess air requirements with a wide turn-down ratio. For burners with large flow rates of fuel, multiple jets are used so as to provide flow rates of up to 15×10^3 kg/h in boilers for electricity generation. The physical processes of atomization are shown schematically in Fig. 5. Liquid fuel is injected into the mixing chamber at an angle while the atomizing fluid, compressed air or steam, is introduced centrally to the mixing chamber with sufficient pressure to provide sonic conditions at the jet exit. The liquid fuel forms an annular film around the walls of the mixing chamber, with the high-speed atomizing jet passing centrally through the mixing chamber. Some atomization occurs within the mixing chamber, but the major portion of the liquid emerges from the atomizer in the form of a liquid sheet, which disintegrates into ligaments and, subsequently, into droplets. The secondary atomization occurring outside the atomizer continues for some 50 nozzle diameters downstream. Increasing the air/fuel ratio leads to a reduction in the break-up length. A detailed study of the twin-fluid atomizer, shown in Fig. 5, was made by Mullinger.[17] The internal mixing within the atomizer was determined from photography of a transparent model and Changes in geometry could also affect the distribution and thickness of the fuel film in the mixing chamber, which, in turn, affect the droplet size distribution. The design study showed that, in order to obtain the most effective atomization, the fuel should be supplied so as to form a thin, continuous, steady film in the mixing region. Increases in velocity and density of the atomizing fluid are the most effective means of reducing droplet diameters. A table of the effect of changes on atomizer geometry and performance, together with design recommendations, is given by Mullinger.[17] It can be concluded that minimum droplet sizes are obtained when atomizers are designed to provide maximum physical contact between the air and the fuel. If a low velocity liquid sheet of fuel is "sandwiched" between high velocity air on both sides, this ensures both good atomization and transport of the droplets with the airstream so as to avoid deposition of the fuel on solid surfaces.

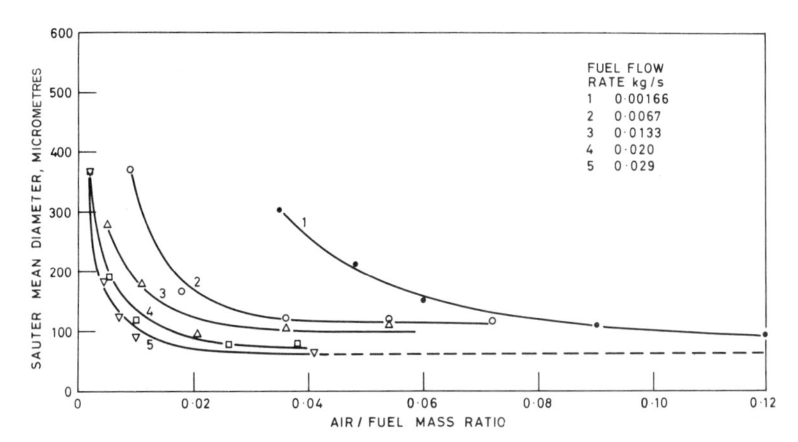

FIG. 6. Variation of Sauter mean diameter with air/fuel mass ratio for various fuel flow rates in a twin-fluid atomizer.[17]

the effects of changes in the atomizer geometry were studied as they affected the size distribution of droplets in the spray. The variation of the Sauter mean diameter of droplets, as measured by the high-speed photographic technique, with changes in air/fuel mass ratio, is shown in Fig. 6 for a range of fuel flow rates. Figure 6 shows the significant reductions in droplet size that can be obtained by increasing the air/fuel mass ratio. For each fuel flow rate, however, there is a limit to the effectiveness of increased air flow rates, beyond which no further significant reduction in droplet diameter is achieved. Since energy considerations and the desirability of limiting the quantity of steam introduced through the atomizer are important, it is useful to establish the minimum air/fuel mass ratio for any given fuel flow rate. It will be noted in Fig. 6 that, for a fixed air/fuel mass ratio, the droplet diameter was reduced by increasing the fuel flow rate. This is explained as being due to the increase in air supply pressures required for the higher fuel flow rates and, hence, leading to an increase in the air density at the mixing point. Variation of atomizer geometry was shown to affect the atomizing fluid velocity and density, which had an important effect on droplet size distribution.

4. IDEALIZED SPRAY FLAME

On the basis of the measurements that have been made in spray flames, we can propose an idealized model of droplet vaporization in spray flames, as shown in Fig. 7. We use the hypothesis that spray flames and gas diffusion flames are similar and that both temperature and oxygen concentrations are low within the spray. In the idealized model, the flame acts as an interface, totally separating the inner fuel vapour from the outer air. All the fuel vapour originates from the vaporization from droplet surfaces. All burning occurs as a consequence of fuel vapour diffusing outwards and air diffusing inwards to the flame front. This interface is convoluted by the turbulence so that, in the time mean, droplets and flame can occur at any one position but never at the same time. The droplet velocity can either be greater or smaller than the gas (fuel vapour) velocity. These velocities may typically have values of the order of 30 m/s with velocity differentials of the order of 5 m/s. The surrounding air velocity will vary from zero for stagnant air surroundings to velocities of the order of 100 m/s and thus higher than droplet velocities. The flame

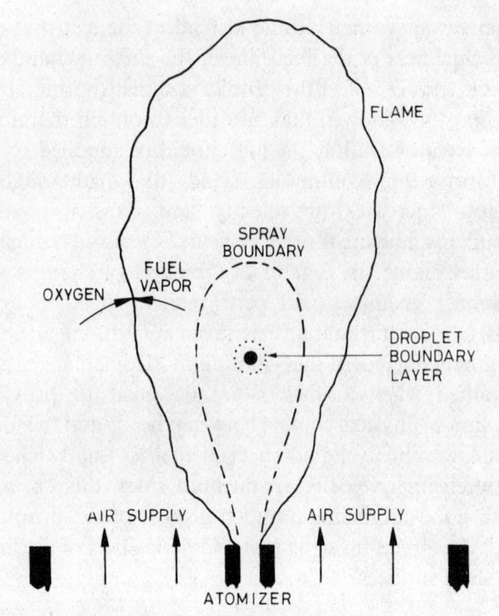

FIG. 7. Vaporization of a typical droplet in an idealized spray flame.

temperature may be assumed to be 1500 K, with both the droplet temperature and the ambient air temperature taken to be 300 K.

All droplets are injected into the spray through the atomizer and vaporization of all these droplets is completed within the flame volume. The major portion, say 99%, of all droplets is contained within the spray boundary so that there is a zone of fuel vapour, without droplets, between the spray boundary and the flame. Droplet sizes will vary from 1 to 500 μm but, particular attention is focused on the largest droplets. No interaction takes place between droplets and the separation between droplets is sufficiently large that each droplet can be considered in isolation. A spherical boundary layer, with a diameter of approximately twice the droplet diameter, envelops the droplet. We ascribe bulk gas temperatures, velocities and concentrations to the gas outside the boundary layer, while restricting the siqnificant gradients of temperature, velocity and concentration to within the boundary layer.

We sub-divide the spray into an initial "cool" zone, where temperatures are maintained at low levels, due to the strong quenching action of the droplets, followed by a "hot" zone, where the temperature rises, due to turbulent convective transfer of heated products from the flame. Within the cool zone, heat transfer is restricted to radiation from the flame front to the droplet surface. In the hot zone, heat transfer takes place both by radiation from the flame front and by turbulent convection. Predictions of the rate of vaporization of the droplets within the spray can be made by using the following equations: the reduction in droplet diameter due to vaporization follows the "d^2 law":

$$-\mathrm{d}(d^2)/\mathrm{d}t = \lambda \tag{5}$$

where λ is the evaporation constant for forced convection.

The evaporation constant in a stagnant atmosphere, k_0, is given by:

$$\lambda_0 = (8k/\rho_l c_p)\ln(1+B), \tag{6}$$

where k is the thermal conductivity of the gas, c_p is the specific heat of the gas, ρ_l is the density of the fuel and $B =$ the transfer number $= (1/L)[\bar{C}_p(T_\infty - T_l) + (QY_{0_\infty}/i)]$, $L =$ latent heat of vaporization per unit mass of fuel, $T_\infty =$ temperature of gas surrounding the droplet, $T_l =$ temperature of drop surface, $Q =$ heat of reaction, $Y_{0_\infty} =$ mass fraction of oxidant in the surrounding atmosphere and $i =$ stoichiometric mixture ratio.

The evaporation constant in forced convection, λ, is:

$$\lambda = \lambda_0(1+0.276\,Re^{1/2}\,Sc^{1/3}), \tag{7}$$

where Re is the Reynolds number and Sc is the Schmidt number. Motion of droplet;

$$m(\mathrm{d}u_d/\mathrm{d}t) = F - mg, \tag{8}$$

where m is droplet mass, u_d is droplet velocity, t is elapsed time, F is the drag force and g is the gravitational constant. Drag force;

$$F = C_d \tfrac{1}{2}\rho_g(u_g - u_d)^2 A, \tag{9}$$

where C_d is the drag coefficient of an evaporating droplet, u_g is the velocity of the gas surrounding the droplet, A is the surface area of the droplet and ρ_g is the gas density.

The drag coefficient of a spherical droplet is:

$$C_d = (24/Re)(1+0.15\,Re^{0.687}). \tag{10}$$

The drag coefficient for an evaporating droplet, $\bar{C}_d$:

$$\bar{C}_d/C_d = 1/(1+B). \tag{11}$$

The heat transfer coefficient is obtained from the Nusselt number, which is given by the relationship:

$$Nu = 2+0.6\,Pr^{1/3}\,Re^{1/2}, \tag{12}$$

where Pr is the Prandtl number.

Onuma and Ogasawara[1] have used the above equations to calculate the reduction in diameter of droplets due to vaporization and also changes in the velocity of the droplets in a spray flame. In their calculations they have used the measured values of local mean gas velocity and temperature and the measured initial size distribution of droplets. They assume that the initial velocity of droplets was equal to the discharge velocity of kerosene from the atomizer. For the case of a total flame length of 0.53 m they found that evaporation of the spray was completed by a distance of 0.3 m from the atomizer. They also determined the distances to completion of evaporation of droplets with sizes from 15 to 155 μm.

The rate of vaporization of droplets is initially low because of the low temperature and high fuel vapour pressure of the gas in the cool zone. This rate of vaporization increases in the hot zone so that, for droplets up to 150 μm in diameter, vaporization is just completed at the tip of the spray. For droplets greater than 200 μm, the possibility exists for isolated droplets to pass through the spray boundary and the flame. When these liquid particles come in contact with cooled chamber walls, deposition and coke formation arise. The rate of vaporization of droplets may, thus, play only a minor role in the combustion process where the rate determining step is the rate of turbulent diffusion of vapour and air to the flame front.

Before this idealized model of spray combustion can be applied to a practical combustor, it is necessary to verify that there is no evidence of envelope flames around individual droplets and that the flame front surrounds the spray. The flame front may not act as a complete separation interface between fuel vapour and air so that air may penetrate into the spray. When air-assist atomizers are used, a proportion of the total air is introduced directly into the spray and this will alter the flame structure. Given the choice between making predictions on the basis of single droplet burning and that of the idealized spray flame model, the experimental studies which have been made on spray flames suggest that the above model is a good approximation to the practical conditions.

5. PHYSICAL MODEL OF SPRAY COMBUSTION IN A GAS TURBINE ENGINE

Mellor[26] has carried out a series of studies in a gas turbine test combustor with air flows up to 2.75 kg/s and pressures up to 15 atm. From the measurement of mass-averaged exhaust emissions from the test combustor, using a water-cooled gas sampling probe with a side-mounted thermocouple, Mellor[26] found that both NO and CO emissions decreased as the differential fuel injection pressure was increased to 15 atm. This demonstrated that the combustion characteristics could not be explained satisfactorily solely on the basis of homogeneous chemical kinetic effects, but rather that some aspect of injector performance was influencing the combustion.

Following Lefebvre,[27] Mellor examined the parameters that could relate injector performance to combustion characteristics and, in particular, to see how changes in fuel pressure could change flame characteristics. A characteristic time for droplet evaporation is given by $\tau = d_0^2/\lambda$, where d_0 is the initial droplet diameter and λ is a function of fuel and ambient gas properties and Reynolds number. It was shown that, since the mean droplet size, $d_m \sim \Delta P^{-0.4}$ and the initial droplet relative velocity $\Delta U \sim \Delta P^{0.5}$, the Reynolds number based on mean droplet size and initial droplet relative velocity is:

$$Re \sim \Delta P^{0.1}. \tag{13}$$

Taking λ to be approximately independent of ΔP, it was concluded that

$$\tau \simeq d_0^2 \simeq d_m^2 \sim \Delta P^{-0.8}.$$

On the basis of the above argument, the pressure drop across the atomizer is shown to be almost inversely proportional to the droplet evaporation time.

In the physical model of spray combustion, as postulated by Mellor, he considers three conditions: (a) $\Delta P < 15$ atm, (b) $\Delta P = 15$ atm, and (c) $\Delta P > 15$ atm. For the first case of low ΔP, the evaporation time is long compared to the residence time, resulting in low combustion efficiency, with long yellow flames. CO emission levels are high, due to incomplete combustion, and the high NO levels are taken to be evidence of envelope flames around individual droplets. At $\Delta P = 15$ atm, Mellor found both a minimum in NO and CO emissions and he concluded that this corresponds to a transition from droplet diffusion flames to wake flames or pure evaporation. For $\Delta P > 15$ atm, the flame was found to be blue and the flame length remained constant. The combustion phenomena resembled that of a gaseous turbulent diffusion flame. NO levels were found to be unaffected by increase in ΔP, which is consistent with the evidence of no increase in flame length. The increase of CO emissions with increasing ΔP is due to the overall equivalence ratio increase with ΔP when the turbulent flame length remains unchanged.

Under conditions of poor atomization, such as those obtained by Azelborn *et al.*,[28] using an air-assist nozzle operating at low air flows, corresponding to idle conditions, it was concluded that the presence of a yellow flame, coupled with high emission levels of both NO and CO, could be ascribed to droplet burning. In a series of measurements made by Sanders *et al.*[29] in a large utility boiler, they found no effect of increasing ΔP from 57.3 to 75.4 atm on NO emissions and they concluded that droplet burning does not occur in a well-atomized spray.

6. AIR BLAST SPRAY FLAMES

In air blast spray flames, the air which passes through the atomizer has the main function of breaking up the liquid sheet into droplets and providing sufficient momentum to transport the droplets. The oxygen in the atomizing air will react, after mixing with the fuel, provided that the temperature is sufficiently high and the mixture ratio is within the limits of flammability. Within the spray sheath, the mixture ratio changes from very lean conditions at the atomizer exit where almost all the fuel is still in liquid form, to the very rich conditions which occur when large proportions of the fuel have vaporized within the spray. Under burning conditions the amount of oxygen entrained into the spray will be negligibly small because the flame surrounding the spray acts as an effective barrier to oxygen entrainment. In the initial regions of the spray, near the atomizer exit, the temperature levels are too low to allow combustion in the dense spray. As we proceed downstream, the spray temperature rises, due to heat transfer from the surrounding flame, and the vapour mixture ratio changes from lean to rich. The vapour within the spray thus passes through the flammability limits, starting from conditions which are too lean and ending with conditions that are too rich. In sprays, where the temperature is sufficiently high to allow ignition, combustion of vapour can take place within the spray. In the sprays which have been examined so far this situation did not occur, with the overall result that no burning takes place within the spray. When combustion does not take place within the spray the characterization of the spray only requires consideration of the vaporization of droplets in a heated fuel–air gas-stream.

Roett[2] carried out an experimental investigation of air blast kerosene spray flames using the same atomizer as Mullinger.[17] The studies were also carried out under unconfined conditions with the spray directed vertically upwards and initially ignited by a pilot gas flame but, subsequently, burning freely in the open atmosphere.

FIG. 8. Photograph of air blast kerosene spray flame in the open atmosphere.

A photograph of this open spray flame is shown in Fig. 8. The central region of the spray is dark, due to the presence of the cloud of droplets, and the flame is seen to be initially confined to an annulus surrounding the spray. Further downstream, flame is seen across the whole cross-section of the jet but, at this stage, the spray region has ended since all the droplets have been vaporized. Roett[2] used the double-image high-speed photographic technique for measuring individual droplet sizes and velocities in the spray. Spray flames with air/fuel mass ratios of 0.2, 0.3 and 0.42 and total jet momenta of 0.061, 0.114 and 0.169 kg m/s^2 were studied. Series of photographs were taken at positions in radial and axial traverses and, following mass averaging, the changes in droplet size through the spray were plotted in terms of iso-mass fraction lines for droplets less than 100 μm (Fig. 9) and for droplets more than 200 μm (Fig. 10). The changes in droplet size within the spray are explained in terms of two separate and apparently distinct phases. An initial

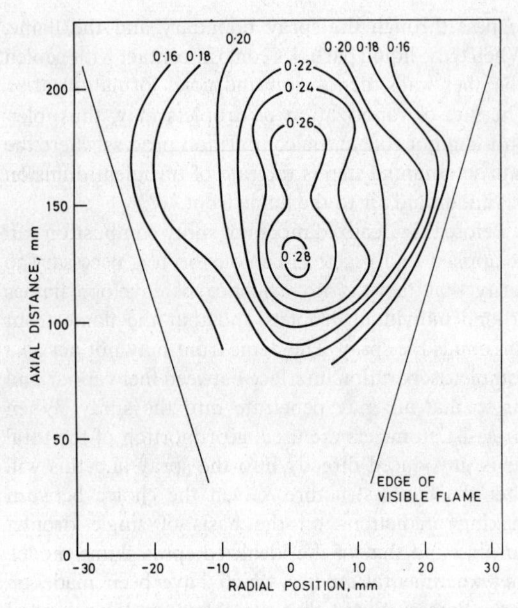

FIG. 9. Iso-mass fraction lines for droplets less than 100 μm in air blast kerosene open spray flame.[2]

phase, up to 150 mm, where larger droplets break up into smaller droplets, leading to an increase in relative proportion of small droplets, with a corresponding decrease in proportion of large droplets. In the second phase, beyond 150 mm, atomization is complete and the effect of vaporization is dominant. The smaller droplets vaporize quickly with a net result that Figs. 9 and 10 show a decrease in the proportion of small droplets. Radial distributions of droplet size show that the small droplets are concentrated near the centre of the spray and that in the outer regions the proportion of small droplets decreases, due to the closer proximity to the flame. There is preferential vaporization of

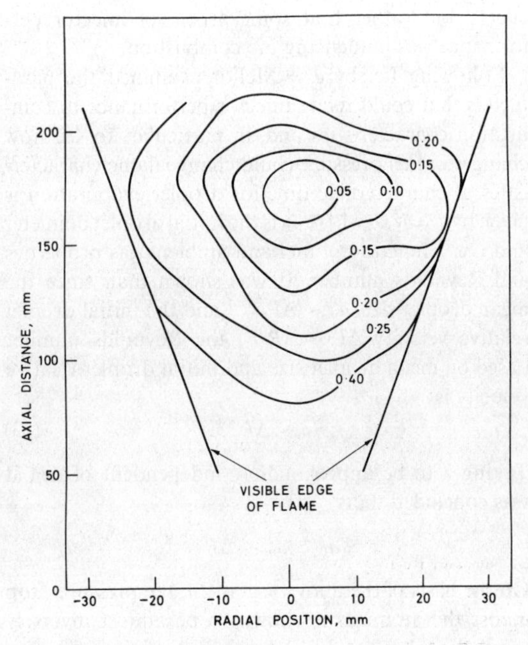

FIG. 10. Iso-mass fraction lines for droplets more than 200 μm in air blast kerosene open spray flame.[2]

smaller droplets in regions of high temperature. If we follow the changes in size of droplets along the axis of the spray, the percentage of droplets less than 100 μm initially increases. Photographs of individual droplets in this region showed some of the larger droplets to be non-spherical and in the process of break-up to smaller droplets.

Changes in the atomizer flow conditions were studied by maintaining the total jet momentum constant and increasing the air/fuel ratio, which led to the production of a finer spray. In studies where the air/fuel ratio was maintained constant but the total jet momentum increased, it was found that this again led to finer sprays but, as found by Mullinger,[17] a point is reached where further increases in momentum do not lead to reduction in average drop size. Experimental techniques are not yet sufficiently advanced to examine the change in size of individual droplets as they move through the spray. Such measurements would allow a direct determination of the vaporization constant and would be useful for prediction of rates of vaporization of droplets in sprays. The high-speed photographic techniques which have been used by Mullinger[17] and Roett[2] have used volume averaging at the "point" at which measurements were made in the axial and radial traverses. The changes in the droplet diameter distributions between one plane of the spray and another plane downstream, are due to the break-up of large droplets and vaporization as well as spreading of the droplets within the spray jet by turbulence and droplet dynamics. Until these phenomena can be separated it is not possible to determine the vaporization constant from the droplet size measurements of the type shown in Figs. 9 and 10.

Using the double-flash high-speed photographic technique, Roett[2] also measured velocity distributions of the droplets in the spray. Droplet velocity profiles were found to be Gaussian in form with velocity maxima on the jet axis. Since the droplets are transported by the atomizing air jet, the droplet velocity profiles are governed by the air jet velocity profiles. The measured drop velocities were in the range 10–30 m/s and the drag to momentum ratio for droplets less than 30 μm was sufficiently high that droplet velocities did not differ significantly from the time average air jet velocities. For the larger droplets, however, slip occurs so that velocities of large droplets are different to that of the surrounding gas-stream. A direct measure of the velocity differential between droplet and gas-streams is required in order to determine the drag forces. The separation of the droplet and surrounding gas-stream velocity could not be achieved by Roett.[2] At the atomizer exit, the airstream has a velocity just below sonic, while the large liquid droplets have relatively low momentum and velocity. Because of this initial velocity difference, the droplets are accelerated by the airstream. As we proceed downstream, gas velocities decrease, both due to the entrainment of air from the surroundings and as a consequence of exchange of momentum with the droplets being accelerated by the gas-stream. Equilibrium conditions, neglecting gravitational forces, will be achieved where droplet and gas-stream velocities equalize. Beyond this position the droplets will retain their momentum to a greater extent than the surrounding gas and Roett concluded that droplets were being decelerated by the gas-stream in the tail region of the spray. Thermal expansion of the gases as a direct consequence of heating also leads to velocity changes and can result in local gas velocity increases.

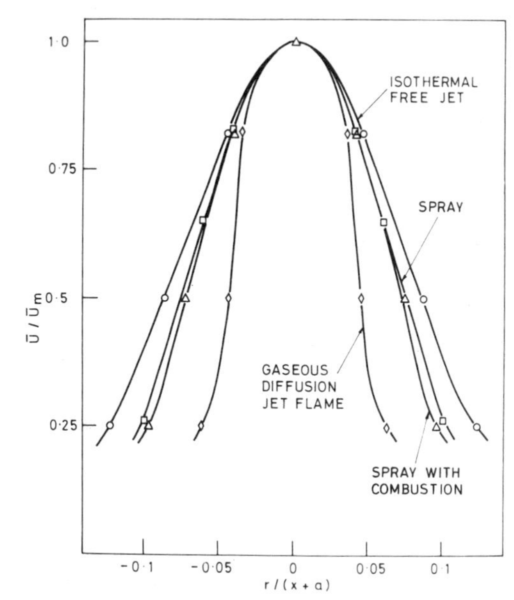

FIG. 11. Radial velocity distributions for various jet systems.[2]

Velocity distributions for a number of jet systems are compared in Fig. 11. In this figure the time average velocity measured at several points in a radial traverse is normalized by dividing by the maximum velocity, which, in each case, is found on the jet axis at $r = 0$. This velocity ratio $\overline{U}/\overline{U}_m$ is plotted against $r/(x+a)$, where x is the distance along the axis from the nozzle exit and a is the distance from the nozzle exit to the effective origin of the jet system. In each of the experiments shown in Fig. 11 the profiles were found to be similar so that they had the same normalized velocity distribution at various cross-sectional planes along the jet. A velocity distribution such as shown in Fig. 11 gives a measure of the spread and the boundaries of the jet. The profile for reference is that of the isothermal free jet, which is represented by an equation of the form

$$\overline{U}/\overline{U}_m = \exp(-k_u \xi^2),$$

where k_u is the velocity spread coefficient and $\xi = r/(x+a)$. The spray flame of Roett[2] is referred to in Fig. 11 as the burning heterogeneous jet. Since, in this spray flame, we can expect changes in the velocity fields, both due to the presence of droplets and due to the presence of flame, profiles are also shown in Fig. 11 of an isothermal liquid spray from the measurements of Hetsroni.[30] This profile, when compared to the reference isothermal free jet profile, shows that the presence of droplets reduces the spread of the jet. The droplets retain their momentum for longer periods of time than the equivalent particles of fluid, thus the exchange of momentum between a spray jet and its surroundings is less than that of a gas jet and, consequently, the spread and the rate of decay of velocity is reduced by the presence of droplets. In order to examine the separate influence of burning, the measurements of Chigier and Chervinsky[31] are shown in Fig. 11. For a turbulent diffusion flame with a cold core, the flame is concentrated in an annulus surrounding the cold core and the net result is that the spread and decay of the jet are reduced when compared to the equivalent nonburning jet.[32] The velocity spread coefficients, as determined from each one of these profiles, are given in Table 1.

TABLE 1. Velocity spread coefficients

	k_u
Isothermal free jet	92
Isothermal spray[30]	130
Spray with combustion[2]	130
Gaseous diffusion flame[31]	360

From Table 1 and Fig. 11 it can be seen that the value of k_u in the burning spray is close to that of the isothermal spray. This indicates that, in the region of the spray where Roett[2] made measurements, the effect of the presence of droplets was dominant and more significant than the presence of flame. On the basis of the measurements of Roett,[2] a physical model has been proposed and is shown schematically in Fig. 12. The twin-fluid atomized jet spray flame is seen to consist of a central core region with a high concentration of droplets and high droplet velocities. In this region no significant reaction can take place as the mixture ratios are too rich and the quenching effect of the liquid is too great. Combustion takes place at the outer periphery of the spray, where, as a consequence of air entrainment, mixture ratios are within the limits of flammability. The flame acts as an effective boundary, confining fuel on the inside and restricting oxygen to the outside. Droplet diameters were seen to reduce rapidly on approaching the flame so that all droplets could be considered to vaporize within the spray and not enter into the flame.

The experimental studies in the air blast spray flame at Sheffield have been taken a stage further by Styles.[33] A laser Doppler anemometer has been used for velocity measurements in the spray flame. Because of the wide sprectrum of particle sizes with maximum droplet diameters up to 400 μm, some special problems arose in the use of the laser anemometer. For general applications in laser anemometry the flow is seeded with particles of the order of 1 μm and the light scattered from these particles is used in order to determine the velocity of the fluid surrounding the particles. In order that the particles' movements will be representative of the fluid flow movements, the drag to momentum ratio requires to be large so that there will be no relative slip between particle and fluid. In the burning spray system differential velocities between particles and gas can be large and the majority of particles are in a state of acceleration or deceleration. The laser anemometer measures the velocity of all particles so that an average velocity at any one "point" does not differentiate between the variation in particle size and the associated variation in velocity. In order to take measurements which can be meaningful in such a system, it becomes necessary to measure simultaneously particle size and velocity.

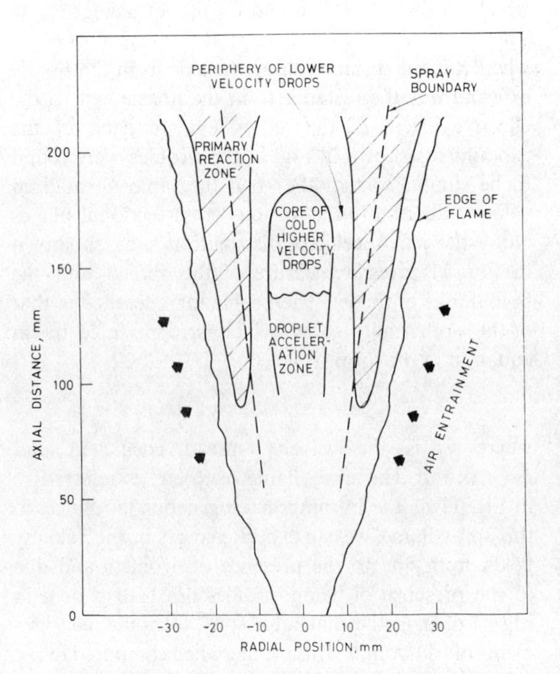

FIG. 12. Schematic diagram of physical model of twin-fluid atomized jet spray flame.[16]

7. PRESSURE JET, SWIRL ATOMIZED, HOLLOW CONE SPRAY FLAMES

The pressure jet, swirl atomized, hollow cone spray is used in many types of combustion chamber where pumping facilities are available for the high liquid pressures required and where high pressure air or steam is not available. Liquid fuel is introduced tangentially at high pressure into a swirl chamber and passes through a diffuser to a circular orifice exit. The liquid is attached to the walls of the diffuser and leaves the atomizer in the form of an annular film with a central air core, generated by the pressure differences within the mixing chamber. This annular film spreads out to form a hollow conical spray, which becomes unstable and disintegrates into ligaments and large drops and, finally, into a spray. A description of the atomizer and the interaction of such sprays with uniform airstreams is given by Mellor.[11,12,32] Following the nonburning study of Mellor, McCreath and Makepeace[4,15,16,18] studied pressure jet spray flames in the wake of a stabilizer disc. The system is described schematically in Fig. 13. Kerosene is introduced at high pressure into a swirl atomizer, which produces a hollow cone

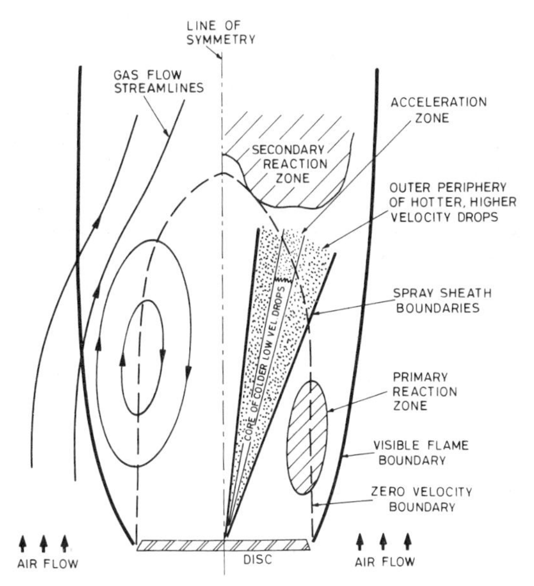

FIG. 13. Hollow cone pressure jet spray burning in the wake of a stabilizer disc.[4]

spray projected vertically upwards. The spray emerges from the centre of a stabilizer disc with a surrounding air flow. A recirculation zone is formed by the air flow in the wake of the disc, as shown by the streamline flow patterns and zero velocity boundary in Fig. 13. A primary reaction zone is formed near the edge of the disc by fuel vapour and small droplets which are transported by the recirculating gas flow towards the outer air flow. Since only a very small fraction of the fuel enters the primary reaction zone, the dimensions of this flame are restricted by the spray and surrounding air flow. This primary flame provides an ignition source, acting as a pilot flame, introducing hot products into the main air flow stream, which ignite the vapour mixture further downstream in the secondary reaction zone.

Figure 14 shows the isotherms within the spray flame as measured by a fine wire Pt/PtRh thermocouple. Temperature levels within the dense spray region are seen to be less than 600°C. The temperature levels within the spray increase with downstream distance as a consequence of entrainment of hot products from the primary reaction zone, which can be seen outside the spray boundary near the edge of the stabilizer disc.

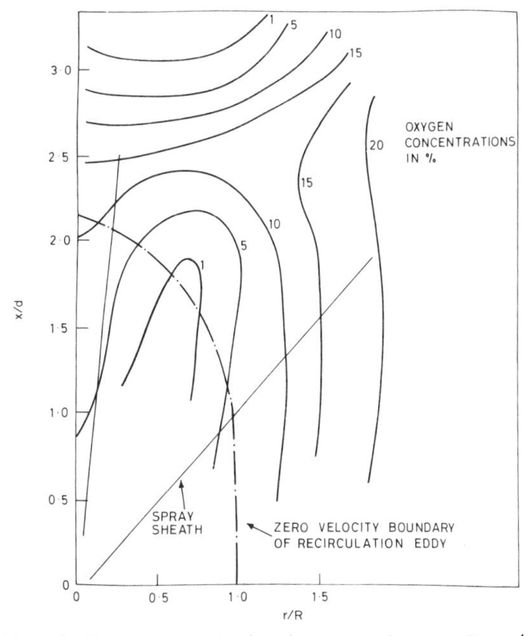

FIG. 15. Oxygen concentrations in pressure jet spray flame.[4]

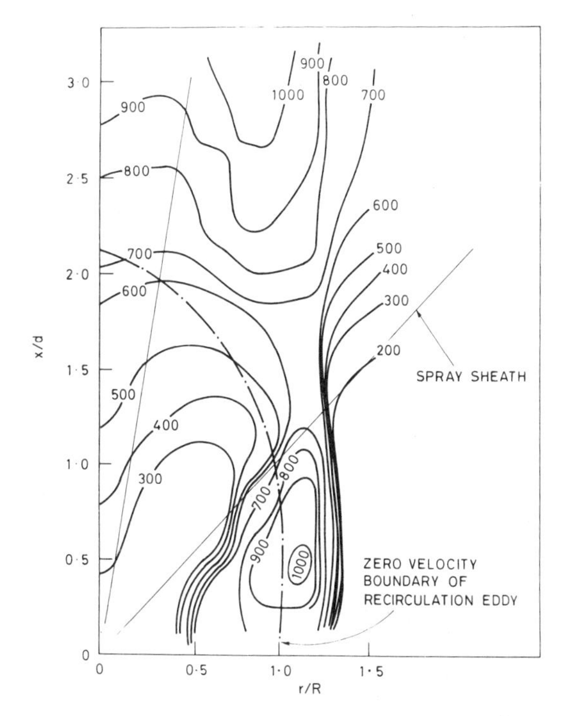

FIG. 14. Isotherms in pressure jet spray flame.[4]

The spray also entrains air from the surrounding cold air flow. Temperatures along the axis of the system are higher than within the spray, due to the transport of hot combustion products from both the primary and secondary reaction zones. The presence of a high concentration of droplets in the spray has a strong quenching action, which inhibits combustion but allows limited vaporization to take place. The spatial distributions of oxygen concentration, as determined by probe sampling, are shown in Fig. 15. Within the dense spray region measured oxygen concentrations were as low as 1%, showing that the mixture ratios are very rich and dominated by fuel vapour. The oxygen concentrations rise with distance downstream as a direct consequence of entrainment and penetration of oxygen from the surrounding airstream. The oxygen concentrations decrease in the secondary reaction zone due to burning. The experimental evidence in Figs. 14 and 15 shows that, in the initial dense spray region, both the temperatures and the oxygen concentrations are too low to allow combustion to take place. Direct photography of this region also showed no evidence of burning. The double-flash high-speed photographic technique was used to measure droplet size and velocity and these measurements showed that large droplets

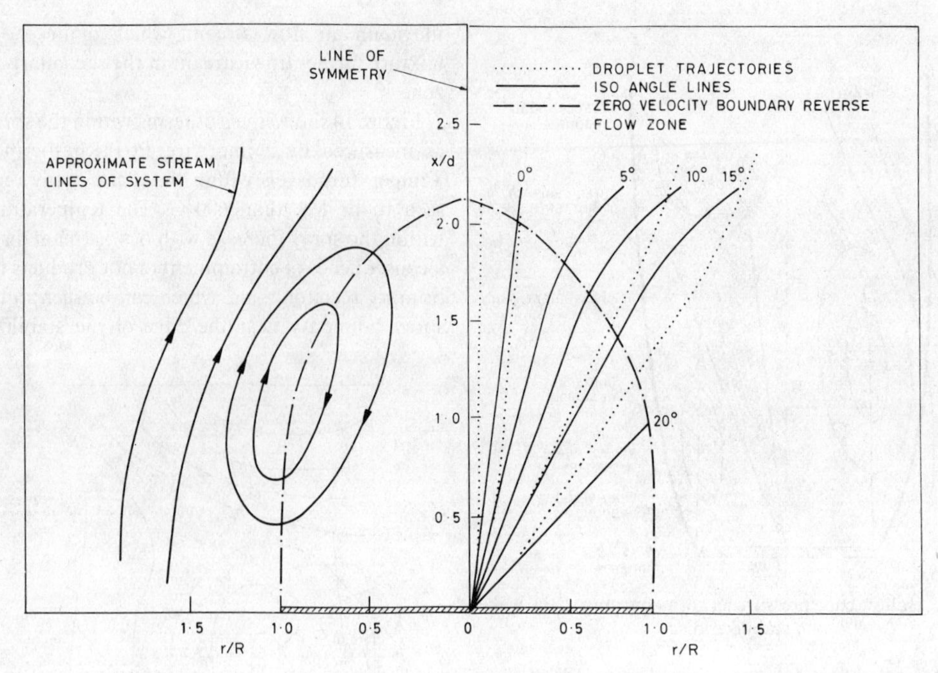

FIG. 16. The influence of recirculating gas stream on trajectories of droplets between 100 and 200 μm in pressure jet spray flame.[16]

were breaking up into smaller droplets in the initial regions of the spray. The size distributions change further downstream as the smaller droplets vaporize rapidly and relatively small changes occur in the size of the larger droplets. The influence of the recirculating gas-stream on the trajectories of large droplets (between 100 and 200 μm) is shown in Fig. 16. These large droplets are seen to have almost linear trajectories and penetrate through the reverse flow zone with very little influence from the recirculating gas-stream. McCreath[4] shows that the smaller droplets, less than 50 μm, are deflected by the gas flow but there is insufficient time for droplets to be decelerated, and subsequently accelerated, in the opposite direction by the reverse flow and little evidence was found of droplets moving in the reverse flow direction towards the stabilizer disc. The measurements of droplet velocities show comparatively small variations and the changes are explained by McCreath[4] as being due to the influence of acceleration and deceleration as droplets interact with gas-streams of varying velocity.

A comparison of trajectories and velocities of droplets, as measured by Makepeace[15,18] in cold sprays and those of McCreath[4] in spray flames, shows that significant changes occur as a consequence of combustion. These differences are partly due to the reduction in drag coefficient of droplets vaporizing in the spray flame. The comparison showed that velocity of both small and large droplets was larger in the flame than in the cold spray. In addition to the effect of reduction in drag coefficient, which may be as high as 80%, the relative velocities between gas and droplets were lower and the density of the hot gas was also lower in the flame as compared to the cold spray. By increasing the velocity of the airstream from 8 to 40 m/s, the drag forces were increased by increasing the relative velocity between droplet and gas. The residence time of droplets within the recirculation zone as well as the rates of mass transfer are changed by increases of the air velocity. The smallest droplets were found to penetrate smaller distances as the reverse flow velocity was increased. For an air velocity of 40 m/s, droplets less than 50 μm were not found beyond 120 mm.

The physical models of spray burning show that sprays are initially dense, surrounded by vapour with rich mixture ratios at low temperatures in which no significant chemical reaction can take place. Reaction is forced to take place at the outer periphery of the sprays, where air/fuel ratios and temperatures are within the limits of flammability. No evidence was found of the classical model of droplets burning with surrounding individual flames. Within the spray core the rate of evaporation of individual droplets plays no significant role in the combustion system, since evaporation is taking place in an atmosphere which is so rich that it is beyond the limits of flammability.

The three prime requirements for flame stabilization, i.e. mixture ratios within the limits of flammability, velocities low enough to match burning velocities, and sufficient supply of heat to retain reaction, are found in the primary reaction zone outside the spray boundary. The main combustion is deferred to distances further downstream where the spray is more dispersed, more oxygen has been entrained from the surrounding air flow and temperature levels and mixture ratios are within the limits of flammability. Liquid spray flames may generally be divided into a number of regions. Within the atomizer nozzle there is bulk liquid flow which interacts with the bulk air. Immediately after break-up of the bulk liquid flow there follows a dense spray region with liquid particles becoming dispersed through the main air flow field. While the concentration of droplets remains high and the concentrations of oxygen within the spray sheath are low, no burning takes place within the spray and the main reaction zones occur at the spray periphery. This type of spray burning can persist for considerable distances from the nozzle and can, at times, be detected by visual observation of dark spray regions within the combustion chambers. Further downstream, liquid particles become more dispersed and reaction zones converge towards the axis. Towards the end of the flame, conditions can arise where droplets are in sufficient isolation, surrounded by air, that envelope flames can form around the droplets. The detailed measurements within spray flames have been made under laboratory conditions but evidence is coming forward that the physical models described above are applicable to combustion chambers in large power station boilers, gas turbines, diesel engines and automobile engines. In each case it is necessary to determine the location of the flame front and establish either the presence of envelope flames around individual droplets or flames confined to the outer periphery of the spray sheath.

8. SINGLE DROPLET COMBUSTION

Many experimental and analytical studies have been made on the burning of single liquid drops. These single drops have usually been held in suspension attached to the end of wires or the process has been simulated by the use of a porous sphere covered with a liquid film. These studies have been reviewed by Beér and Chigier,[32] Williams[6] and Hedley *et al.*[34] The more recent studies of Gollahalli and Brzustowski[35] show that the flame around the porous sphere can either have an envelope flame or a wake flame (Fig. 17). This work follows that of Spalding,[36] who investigated the effect of relative air velocity of the combustion of liquid fuel spheres and observed a critical velocity above which the flame could not be supported at the upstream portion of the sphere. In low velocity airstreams, a flame envelops the leading half of the sphere. Above a critical free-stream velocity (the extinction velocity) the flame on the leading half of the sphere is extinguished and a small flame is stabilized within the wake. A third regime of burning is also sometimes observed, in which the flame is stabilized in the boundary layer at the side of the liquid sphere. In the experiments of Gollahalli,[35] n-pentane was supplied to a porous bronze sphere, 6 mm in diameter, which was suspended in a uniform airstream. Measurements were made of burning rate, temperature and composition profiles in the wake, as determined from micro probe samples. They found that the envelope flame in free convection had a near-wake extending about 5 sphere diameters and a far-wake extending another 15 diameters. For the typical wake flame, these lengths were about 3–6 diameters respectively. The particle track photographs in Fig. 17 show a recirculation zone extending for about 1 diameter in the wake of the sphere, without the presence of a flame. This recirculation zone is not visible in either the envelope or in the wake flame conditions. This is very clear evidence of changes in wake structure under combustion conditions and provides a partial explanation to the observation of reduction in drag coefficient of burning droplets. The composition profiles suggested that the near-wake of the envelope flame is a pyrolysis zone in which n-pentane decomposes to produce lighter hydrocarbons, including acetylene. Combustion takes place at the edges of this zone, similar to a laminar diffusion flame. In the case of a flame totally in the wake of a sphere, the near-wake resembles the flame behind a flame holder. Most of the heat release occurs in this zone and only a small amount of soot burns in the far-wake. The burning rate was found to decrease by a factor of three when the envelope flame was transformed into a wake flame at the critical Reynolds

(a)

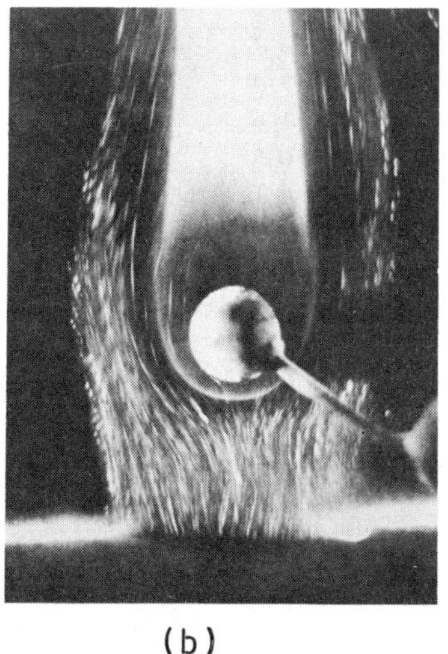
(b)

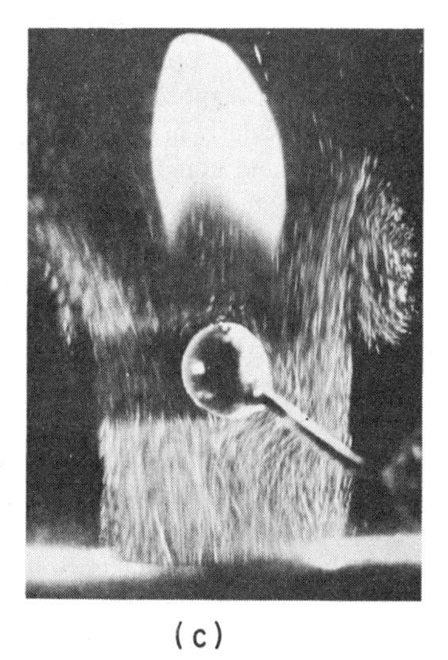
(c)

FIG. 17. Particle track photographs around porous spheres: (a) sphere without flame, $Re = 92$; (b) sphere with envelope flame, $Re = 92$; (c) sphere with wake flame, $Re = 152$ (Gollahalli *et al.*[35]).

number of 138 and envelope flames radiated much more than wake flames. The oxygen concentration of the ambient gas and the intensity of turbulence in the gas-stream can have considerable effects on the extinction velocity.

Several attempts have been made to determine the reduction in drag coefficients of droplets under vaporizing and burning conditions. High-speed movie films show that the velocity with which vapour is emitted from the drop surface is comparable with that of gas flow surrounding the droplet and this results in reduction of skin friction. The particle track evidence of Gollahalli[35] shows substantial reductions in the size and strength of the recirculation zone in the wake of the droplet, which provides additional reduction to the drag coefficient. Khudyakov[37] measured appreciable decreases in drag coefficient of metal spheres wetted with fuel in airstreams at high Reynolds numbers so as to produce wake flames. Spalding[38] in similar work with kerosene-wetted cylinders also found reductions in drag coefficients in cylinders supporting wake flames.

In single droplet combustion a heating-up phase occurs prior to ignition, which is referred to as "ignition delay". This delay comprises the time-interval for the combustible mixture to form around the drop (the physical delay time) and the time-interval between the perceptible chemical reaction and self-ignition (the chemical delay time). The lengths of the chemical and physical delays are similar; the chemical delay is shorter than the physical delay at higher air temperatures and with more volatile fuels.

9. VAPORIZATION OF DROPLETS IN GAS–AIR SYSTEMS

In many of the studies which have been made on single droplet burning it has been assumed that the droplet is surrounded by an infinite quantity of air and that the diffusion of heat and mass can be explained in terms of a boundary layer surrounding the droplet. Mizutani and Nakajima[39] have examined the case of mixing liquid fuel droplets with gaseous fuel and air. These systems can precipitate explosion disasters in industry, particularly in dual fuel-fired systems where both gaseous and liquid fuels are used. In industrial furnaces heavy fuel oil has been added to natural gas in burners as a means of increasing the radiative heat transfer from the flame and, in diesel engines, liquid petroleum gas has been added to diesel oil. In Otto-cycle engines it has been found useful to introduce easily ignitable fuel droplets to act as distributed ignition sources.

Mizutani and Nakajima[39] have reported on the burning velocities and burning characteristics of propane–kerosene droplet–air systems as determined by using an inverted-cone-flame-burner apparatus. An air atomizer was used for atomizing the kerosene and the resultant droplets were added to a stream of propane–air mixture. Size distributions and average diameter of droplets was determined by collection of droplets on magnesium oxide coated glass slides. The burning velocity was determined from the local time average flow velocity and the angle of the flame front. Local flow velocities and intensity of turbulence were measured by a hot wire anemometer in the absence of flame and without injected kerosene. They found that addition of a small amount of kerosene droplets to a lean propane–air mixture increased the burning velocity for a fixed fuel/air ratio and also extended the region of stable burning towards leaner mixtures. They noticed that there was an optimum value of added kerosene droplets and, if kerosene droplets are added so as to exceed this optimum value, the burning velocity fell below that for the propane–air mixture. The effect of droplet addition was found to be greater for lower flow velocity or for weaker intensity of turbulence. The combustion-promoting effects of kerosene droplets were less prominent for higher flow velocities and for larger mean diameters of droplets. By making a separate set of measurements with kerosene mist, the combustion-promoting effects were observed to be directly due to the presence of the liquid droplets and not caused by adding a chemical substance of different thermal or chemical properties. Mizutani and Nakajima[39] give the reasons, in their opinion, why a small amount of added kerosene droplets result in acceleration of the combustion process. These are as follows:

(a) droplets interact with the originally smooth flame surface to make it wrinkled and to expand its surface area, resulting in an increased burning velocity;
(b) burning droplets act as high-temperature heat sources which accelerate the local burning velocities or act as stabilizers for the flame at extremely lean fuel/air ratios;
(c) in the region surrounding evaporating droplets, regions of optimum fuel/air ratio are formed, thereby leading to increase in burning velocities; and
(d) turbulence is generated due to the local thermal expansions of gas associated with randomly located burning droplets, which lead to increase in burning velocities as a result of higher rates of diffusion.

Mizutani and Nakajima[39] demonstrated that the contribution of radiative heat transfer from the flame and burned products to the kerosene droplets upstream of the flame was of little importance because of the low absorptivity of the droplets. They estimated that the radiative heat transfer would increase the temperature of a 60 μm kerosene droplet by only about 4 K above that of the surrounding air when the flame temperature is 1500 K and the temperature of air surrounding the droplet is 300 K.

Mizutani and Nakajima[39] also studied burner flames in a cylindrical combustion vessel with a centrally located spark. This study has particular application to lean mixture operation of Otto-cycle engines. Flame speeds were measured using high-speed photography.

In the cylindrical combustion vessel the conclusions they obtained were as follows:

(a) a small amount of kerosene droplets added to a propane–air mixture intensifies the burning process, raises the maximum pressure for a given overall fuel/air ratio and shortens the time between ignition and maximum pressure. In addition, both the burning and flame propagating velocities are significantly increased by addition of droplets;
(b) whereas a smooth flame sphere of low luminosity is observed in a propane–air mixture, addition of kerosene droplets results in a rough flame sphere with a luminous core region. The zone between the core region and the flame surface remains at low luminosity;
(c) the optimum value of kerosene–air mass ratio was 0.0035, at which the burning and propagating velocities reached a maximum;
(d) the effects of adding fuel droplets are less significant when the flame is turbulent and
(e) there are some differences in the combustion-promoting effects of droplet addition between open burner flames and spherical flames in a vessel, since the acceleration processes of the unburned mixture is different for each case.

10. ATOMIZATION AND EVAPORATION OF FUEL SPRAYS IN DIESEL ENGINES

Diesel fuel is normally injected into diesel engines under pressures of the order of 10^8 N/m^2. The mean droplet diameter of a typical diesel spray is about 15 µm with maximum diameters of approximately 70 µm. Air temperatures in the combustion chamber approach the critical temperature of fuel so that droplets evaporate rapidly and the major portion of the entire spray is in the gaseous form. At low injection pressures, atomization is less effective than at higher pressures and both liquid ligaments and droplets with diameters greater than 100 µm have been found in diesel engines.

Because of the very high pressures and temperatures under which diesel engines operate, atomization and evaporation characteristics of diesel sprays can be substantially different from those of sprays burning under lower pressures and temperatures. Radcliffe[40] and Hiroyasu[41] measured droplet sizes and distributions of diesel sprays under low air temperature conditions. Several attempts have been made to study the specific effects of pressure and temperature on spray characteristics. Photographic observations show that in atmospheric conditions long ligaments of liquid fuel are observed in the central core of the spray. The break-up of the liquid can be seen to take place along the entire length of the spray. The break-up of the spray is not completed for some considerable distance downstream from the nozzle and the number density of droplets is smaller at the periphery of the spray compared with that occurring at a higher gas pressure. The high gas pressures result in higher gas densities, which significantly improve the gas entrainment and the atomization. At the high gas pressures used under normal diesel engine operating conditions, the spray usually breaks up into droplets rapidly, soon after the liquid emerges from the nozzle.

Gas temperatures at the high pressures in diesel engines are typically 200°C higher than the boiling point of diesel fuel and, under these conditions, it has been calculated that droplets would completely evaporate within 10 mm from the nozzle exit. The precise temperature and pressure conditions within diesel sprays have not yet been clearly established and, in the direct-photography studies which have been made, it has so far not been possible to determine clearly the extent to which the fuel spray remains in the liquid phase.

For a number of diesel injection systems, the injection pressure becomes very low towards the end of the injection period and, under these conditions, large droplets are formed. Poor atomization, due to inadequate injection pressure, results in the formation of long liquid ligaments and droplets of the order of 100 µm. This poor, low pressure, atomization is one of the main factors resulting in the formation of smoke and unburned hydrocarbons in diesel engines. Development studies have shown that smoke levels can be reduced by increasing injection pressures and reducing the size of the injector holes—both of which lead to finer spray atomization.

The use of laser optical techniques in diesel engine research has made very substantial progress in recent years and important developments have been made at the Arnold Engineering Development Center.[42,43] By inserting quartz windows in diesel combustion chambers, in order to allow optical access, measurements have been made using the following techniques: high-speed photography, holographic flow visualization, holographic fuel droplet sizing, resonance absorption spectroscopy and laser velocimetry.

High-speed photography, up to 20,000 frame/s, has been used to follow the injection and combustion processes in diesel combustion chambers. The position, time and propagation of combustion have been determined and smoke formation and turbulence in the gas flow have been observed. Recording high resolution images of small fuel droplets by direct photography has not been very successful, except when measurements are made of a very thin plane of droplets.

Holography has become an established technique for the recording and study of three-dimensional particle field distributions. The production of a holograph requires that light reflected from, or scattered by, the object field be mixed with a mutually coherent reference beam and the sum of the two be recorded. The interference patterns resulting from the sum of these two beams constitute the hologram. When the hologram is re-illuminated with the reference beam, the recorded interference fringes scatter it into the form of an image identical to the original object. When pulsed lasers are used in holography, dynamic events in three dimensions can be frozen over very short periods.

REFERENCES

1. ONUMA, Y. and OGASAWARA, M., Studies on the structure of a spray combustion flame, *Fifteenth Symposium (International) on Combustion*, pp. 453–465, The Combustion Institute, Pittsburgh 1975.
2. CHIGIER, N. A. and ROETT, M. F., Twin-fluid atomizer spray combustion, ASME Winter Annual Meeting, New York, Paper No. 79-WA/HT-25 (1972).
3. MELLOR, A. M., Workshop on combustion measurements in jet propulsion systems, Project SQUID, Purdue University (Ed. R. Goulard) (1976).
4. MCCREATH, C. G. and CHIGIER, N. A., Liquid-spray burning in the wake of a stabilizer disc, *Fourteenth Symposium (International) on Combustion*, pp. 1355–1363, The Combustion Institute, Pittsburgh, 1973.
5. BRACCO, F. V., Nitric oxide formation in droplet diffusion flames, *Fourteenth Symposium (International) on Combustion*, pp. 831–842. The Combustion Institute, Pittsburgh, 1973.
6. WILLIAMS, A., Combustion of droplets of liquid fuels, a review, *Combust. Flame* **21**, 1–32 (1973).
7. YORK, J. L. and STUBBS, H. E., Photographic analysis of sprays, *Trans. Am. Soc. mech. Engrs*, **74**, 1157–1162 (1952).
8. DECORSO, S. M., Effect of ambient and fuel pressure on spray drop size, *J. Engng. Pwr.* **82**, 10–18 (1960).
9. BRIFFA, F. E. J. and DOMBROWSKI, N., Entrainment of air into a liquid spray, *A.I.Ch.E. Jl.* **12**, 708–717 (1966).
10. FINLAY, I. C. and WELSH, N., National Engineering Laboratory, England, Report No. 331 (1967).
11. MELLOR, R., CHIGIER, N. A. and BEÉR, J. M., Hollow cone liquid spray in uniform airstream, *Combustion and Heat Transfer in Gas Turbine Systems* (Ed. E. R. Norster), *Cranfield International Symposium Series*, Vol. 11, pp. 291–305. Pergamon Press, Oxford, 1971.
12. MELLOR, R., CHIGIER, N. A. and BEÉR, J. M., Pressure jet spray in airstreams, ASME Gas Turbine Conference, Brussels, Paper No. ASME 70-GT-101 (1970).
13. CHIGIER, N. A., Velocity measurement of particles in sprays, *Flow—Its Measurement and Control in Science in Industry* (Ed. R. B. Dowdell), Vol. 1, pp. 823–832, Instrument Society of America, Pittsburgh, 1974.
14. MCCREATH, C. G., ROETT, M. F. and CHIGIER, N. A., A technique for measurement of velocities and size of particles in flames, *J. Phys. E. Scient. Instrum.* **5**, 601–604 (1972).
15. CHIGIER, N. A., MAKEPEACE, R. W. and MCCREATH, C. G., Aerodynamic interaction between burning sprays and recirculation zones, *Combustion Institute European Symposium* (Ed. F. Weinberg), pp. 577–582, Academic Press, New York, 1973.
16. CHIGIER, N. A. and MCCREATH, C. G., Combustion of droplets in sprays, *Acta Astronautica* **1**, pp. 687–710 (1974).
17. MULLINGER, P. J. and CHIGIER, N. A., The design and performance of internal mixing multijet twin-fluid atomizers, *J. Inst. Fuel*, **47**, 251–261 (1974).
18. CHIGIER, N. A., MCCREATH, C. G. and MAKEPEACE, R. W., Dynamics of droplets in burning and isothermal kerosene sprays, *Combust. Flame* **23**, 11–16 (1974).
19. LEFEBVRE, A. H. and MILLER, D., The development of an air blast atomizer for gas turbine application, The College of Aeronautics, Cranfield, Report Aero No. 193 (1966).
20. NUKIYAMA, S. and TANASAWA, Y., Experiments on the atomization of liquids in an airstream, *Trans. Soc. mech. Engrs. Japan* **5**, 68–75 (1939).
21. LEWIS, H. C., Atomization of liquids in high velocity gas-streams, *Ind. Engng Chem.* **40**, 67 (1948).
22. WIGG, L. D., Drop size prediction for twin-fluid atomizers, *J. Inst. Fuel* **37**, 500–505 (1964).
23. RIZKALLA, A. A. and LEFEBVRE, A. H., Influence of liquid properties on air blast atomizer spray characteristics, ASME Gas Turbine Conference, Zurich, April 1974.
24. RIZKALLA, A. A. and LEFEBVRE, A. H., The influence of air and liquid properties on air blast atomization, ASME-CSME Joint Symposium on Fluid Mechanics of Combustion, Montreal, May 1974.
25. INGEBO, R. D. and FOSTER, H. H., Drop size distribution for cross current break-up of liquid jets in airstreams, NACA Tech. Note 4087 (October 1957).
26. MELLOR, A. M., Simplified physical model of spray combustion in a gas turbine engine, *Combust. Sci. Technol.* **8**, 101–109 (1973).
27. LEFEBVRE, A. H., Factors controlling gas turbine combustor performance at high pressure, *Combustion in Advanced Gas Turbine Systems* (Ed. I. E. Smith), pp. 211–226, Pergamon Press, Oxford 1968.
28. AZELBORN, N. A., WADE, W. R., SECORD, J. R. and MCLEAN, A. F., Low emissions combustion for the regenerative gas turbine, Part 2—Experimental Techniques, Results and Assessment, ASME Paper No. 73-GT-12 (1973).
29. SANDERS, C. F., TEIXEIRA, D. P. and DE VOLO, N. B., The effect of droplet combustion on nitric oxide emissions by oil flames, Western States Section, The Combustion Institute, Paper No. 72-7 (1972).
30. HETSRONI, G. and SOKOLOV, M., Distribution of mass, velocity and intensity of turbulence in a two-phase turbulent jet, *J. Appl. Mech.* **38**, 315–327 (1971).
31. CHIGIER, N. A. and CHERVINSKY, A., Aerodynamic study of turbulent burning free jets with swirl, *Eleventh Symposium (International) on Combustion*, pp. 489–499, The Combustion Institute, Pittsburgh, 1967.
32. BEÉR, J. M. and CHIGIER, N. A., *Combustion Aerodynamics*, Applied Science, London; Wiley, New York, 1972.
33. CHIGIER, N. A. and STYLES, A. C., Laser anemometer measurements in spray flames, *Second European Symposium on Combustion, Orleans, France*, pp. 563–568, 1975.
34. HEDLEY, A. B., NURUZZAMAN, A. S. M. and MARTIN, G. F., Combustion of single droplets and simplified spray systems, *J. Inst. Fuel* **44**, 38–54 (1971).
35. GOLLAHALLI, S. R. and BRZUSTOWSKI, T. A., Experimental studies on the flame structure in the wake of a burning droplet, *Fourteenth Symposium (International) on Combustion*, pp. 1333–1344, The Combustion Institute, Pittsburgh, 1973.
36. SPALDING, D. B., The combustion of liquid fuels, *Fourth Symposium (International) on Combustion*, pp. 847–864, Williams & Wilkins, Baltimore, 1953.
37. KHUDYAKOV, J. N., *Izv. Akad. Nauk. SSSR. Otd. Tekhn. Nauk* **4**, 508–511 (1949).
38. SPALDING, D. B., A theory of the extinction of diffusion flames, *Fuel* **33**, 255–273 (1954).
39. MIZUTANI, Y. and NAKAJIMA, A., Combustion of fuel vapor-drop–air systems, *Combust. Flame* **21**, 343–357 (1973).
40. RADCLIFFE, A., The performance of a type of swirl atomizer, *Proc. Inst. mech. Engrs.* **169**, 93–106 (1955).
41. HIROYASU, H. and KADOTA, T., Study on the removal of the diesel engine smokes, Part II Measurement of the fuel droplet distribution under high pressure, *JARI*, TM2, 51–59 (1971).
42. DOUGHERTY, N. S. and BELZ, R. A., *Holography of Reacting Liquid Sprays*, Arnold Engineering Development Center, Tennessee, 1971.
43. OSGERBY, I. T., *Fuel Evaporation Rate in Intense Recirculation Zones*, Arnold Engineering Development Center, Tennessee, 1974.

Manuscript received, January 1976

Prog. Energy Combust. Sci., Vol. 2, pp. 115–128, 1976. Pergamon Press. Printed in Great Britain

FOSSIL ENERGY RESEARCH AND DEVELOPMENT IN ERDA*

PHILIP C. WHITE

Assistant Administrator for Fossil Energy in the Energy Research and Development Administration, Washington, DC 20545

THE underlying premise for our Fossil Energy Research and Development Program rests on the fact that no matter which projection of future energy demand and supply is examined, it must be concluded that (1) we must reduce demand; (2) increase the utilization of our vast coal and oil shale resources and ensure that technical options to utilize them in an environmentally acceptable manner are available; and (3) we must develop technologies to more efficiently extract oil from known partially depleted reservoirs as well as develop technologies to extract oil and gas from presently uneconomic or unobtainable resources.

The near-term objectives (1975–85) of the Fossil Energy Research and Development Program include the stimulation of petroleum and natural gas production, the development of processes for the direct combustion of coal for electric power generation and for industrial process heat, the development of processes for the conversion of coals to clean liquid and gaseous fuels, the demonstration of *in situ* processes for generating low Btu gases from coal, and the development of *in situ* processes for production of oil and gas from oil shales. Process development includes the construction and operation of demonstration plants which are modules of commercial size plants.

Mid-term objectives (1985–2000) include the commercialization of developed technologies and the development of advanced processes for the combustion of high sulfur coals, the development of advanced electric power generation systems directly utilizing coals, the development and operation of a commercial scale demonstration MHD power plant, and the demonstration and transfer to industry of the technology for commercial scale implementation of significantly improved processes for synthetic fuels from coal.

Long-term objectives (2000+) include the development and demonstration of advanced technologies for producing electric power and process heat at increased efficiency, the development of new synthetic fuels, and the development of *in situ* recovery techniques for fossil energy resources not recoverable by currently available technology.

Our program strategy is based on the fact that we have not until recently been rapidly developing techniques for the conversion of coal to clean, fluid fuels and are not moving fast enough to efficiently extract the fossil fuels represented by much of our oil shale resource, by coal in the thick deep seams of the West and by natural gas in tight rock formations which we must learn to exploit economically. Legislation under which we are now operating clearly instructs us to demonstrate the technical feasibility of new technologies for providing energy from fossil fuels. This requires heavy involvement with various elements of private industry, state and local governments. Our program has been designed and implemented with this in mind. We have a balanced team of systems engineers, research laboratories, architectural/engineering firms, designers, chemical companies, power companies, equipment manufacturers, and coal and oil companies as well as universities providing technical support for our research and development programs. The use of cost-sharing for pilot and demonstration plants assures technology transfer to the industrial sector and total involvement in the scale-up decision by the people who will be responsible for implementing and operating these technologies.

Fossil Energy projects are designed to incorporate appropriate environmental criteria. Environmental monitoring will be conducted at all such plants and installations to insure compliance with applicable environmental regulations. Environmental impact statements and careful assessments will be prepared and issued covering all proposed demonstration sites. It is estimated that Fossil Energy will commit approximately $17 million for environmentally related efforts during Fiscal Year 1977.

CURRENT ORGANIZATION

The Fossil Energy Program is currently organized in the manner shown in Fig. 1. This is a recent organizational configuration and appointments to most of the key positions have been made.

The organization consists of an Assistant Administrator for Fossil Energy, a Deputy Assistant Administrator and five program divisions: Coal Conversion and Utilization; Fossil Energy Research; Fossil Demonstration Plants; Oil, Gas and Shale Technology; and Magnetohydrodynamics (MHD). Staff offices include the Office of Program Planning and Analysis and an Administrative Office. A Senior Staff has been designated to provide special assistance, advice and counsel on program planning and management.

PROGRAM STRATEGY

The Fossil Energy Program seeks to develop a mix of technologies because the complex American economy requires an enormous amount of fuel in

* Statement submitted to the US Congress for the Fiscal Year 1977 Appropriation Hearings.

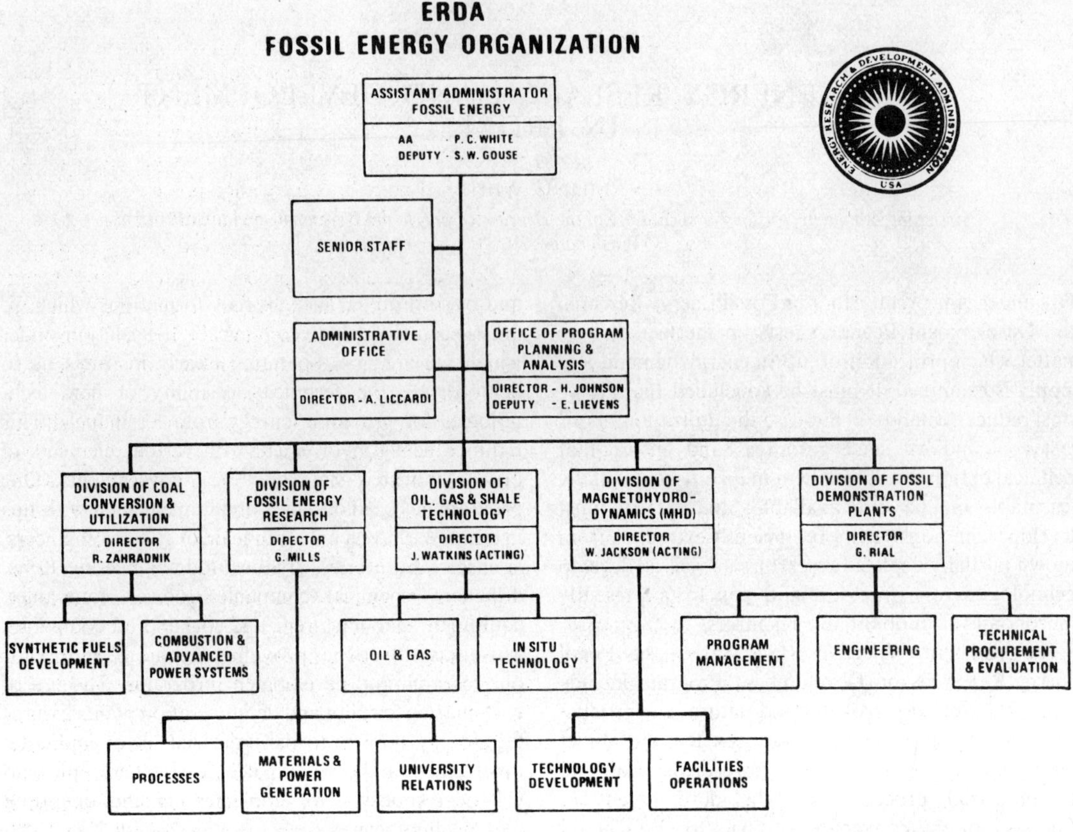

FIG. 1.

various forms: natural and synthetic gas for space heating, electric power and industrial processing, liquid fuels for transportation, electric power, heating and chemical feedstocks; and combustible solids for power generation and industrial application.

The funding levels and activities are different for coal, oil, gas, oil shale, *in situ* coal gasification and MHD. These differences reflect the state of development of conversion and extraction technologies for the different fossil fuels. The state of development has largely been a function of the differing levels of investment by the private sector in research and development for each of the technologies.

There is also recognition of the extent of our resource base for each fossil fuel type. The fact that coal conversion technology requires considerable development to reach maturity, coupled with the fact that coal is our most abundant fossil fuel resource, has resulted in the coal program receiving the largest share of funding.

PROGRAM ACCOMPLISHMENTS

Major accomplishments during Fiscal Year 1976 in the coal conversion and utilization effort include the extended operation of the CO_2 Acceptor Project and the HYGAS process, the near completion of the 30 MW boiler fluidized bed plant, and the successful combustion of coal waste in a fluidized bed combustor.

The Fossil Energy Research program during the past year has produced promising results on new processes and catalysts for new processes for coal derived liquids and gases. Additionally, a new program establishing metal alloys and ceramic materials which will greatly increase component reliability has been initiated.

Our MHD program has completed a 100 h electrode test in the Soviet U-02 installation. Tests have been completed at the University of Tennessee Space Institute (UTSI) demonstrating the feasibility of seed-slag separation under power plant conditions.

The first enhanced oil recovery (EOR) project is presently underway utilizing a micellar–polymer flood technique in El Dorado, Kansas, and is on schedule. Other contracts for fluid injection have been awarded. We also have established this year our Eastern shale gasification program, and are currently evaluating massive hydraulic and chemical fracturing techniques for EOR and stimulated gas recovery programs.

Work is underway on our first demonstration plant, Clean Boiler Fuel, and site selection has been made. Promising responses have been received to our High Btu Demonstration Plant RFP.

THE COAL PROGRAM

The principal objectives of the coal program are to accelerate the development of the environmentally acceptable technology for converting coal to liquid and gaseous fuels, to stimulate improved methods for direct combustion of coal, and to foster the rapid advancement of advanced power conversion systems for generating electricity from coal. The requested Fiscal Year 1977 budget represents an increase of \$76.2 million in budget authority over the amount appropriated in Fiscal Year 1976 to provide for the continuation of a balanced and integrated coal program.

Several significant adjustments in the program structure are reflected in this Fiscal Year 1977 budget request. The coal MHD projects, previously included under the solar, geothermal, and advanced energy systems budget activity for Fiscal Year 1976, are now presented as part of the fossil energy coal program. *In situ* coal gasification projects, previously a component of the low Btu gasification subprogram, are now included under a newly established *in situ* technology budget program. This latter adjustment brings together efforts related to oil shale and *in situ* gasification of coal to permit a more co-ordinated programmatic review of related technology.

To achieve the goal of increased coal utilization, we estimate Fiscal Year 1977 operating expenses will be \$354.5 million and plant and capital equipment obligations will be \$54.5 million. The operating expenses and plant and capital equipment projects described in the budget submission break down in the following way:

Coal Program	Program Estimate Budget Authority (In thousands)		
Subprogram	1976	TQ	1977
Operating expenses			
Liquefaction	\$89,912	\$26,443	\$73,946
High Btu gasification	53,364	9,250	45,054
Low Btu gasification	24,552	6,720	33,052
Advanced power systems	10,001	3,500	22,500
Direct combustion	38,096	13,500	52,416
Advanced research and supporting technology	35,393	8,850	37,085
Demonstration plants	31,900	7,750	53,000
Magnetohydrodynamics	29,544	7,800	37,441
Total	\$312,762	\$83,813	\$354,494
Plant and capital equipment	\$20,000	\$8,000	\$54,450

I would like to briefly describe the objectives of each of these elements, list the major ongoing projects and describe how the requested authorization will be used.

LIQUEFACTION

The general objective of the coal liquefaction subprogram is to provide technology which is economically attractive and environmentally satisfactory to convert coal to a clean liquid for electric power generation or for industrial and residential heating. Simultaneously, technology will also be developed for converting coal into feedstock for production of transportation fuels and of chemicals. Major ongoing projects in this area are given in Fig. 2.

Major categories in the coal liquefaction subprogram include research and development in each of the four methods of converting coal to liquids: direct hydrogenation, solvent extraction, pyrolysis, and indirect liquefaction. A parallel approach has been adopted because different coals will be converted and the optimum product mix from each process differs. Each process under development has uncertainties of scaleup which are reduced by testing the key process steps in a pilot plant. Selected process options are being investigated in the laboratory and bench scale units in order to build a broad technological base. From this base, an efficient process or a combination of processes will be developed.

In Fiscal Year 1976, research and development efforts in direct hydrogenation processes have been considerably accelerated; two projects are moving toward construction of pilot plants. The H-Coal process plant project that was started in Fiscal Year 1975 will continue through Fiscal Year 1977. The design will be evaluated by means of an extended process development unit program and an independent economic feasibility study. In addition, construction of a Synthoil process development unit is planned to be completed in early Fiscal Year 1977.

The operation of a 50-ton per day solvent refined coal plant at Fort Lewis near Tacoma, Washington, construction of which was completed late in Fiscal Year 1974, will be continued through Fiscal Year 1977 with attendant evaluation of operational data. Engineering effort will be directed toward design modification of the pilot plant to recycle unconverted coal, thereby improving the efficiency of the process.

HIGH BTU GASIFICATION

The High Btu Gasification Subprogram is one of the furthest along technically of all our efforts and will be pursued with vigor. Four high Btu gasification pilot plants and one PDU will be in operation by the end of Fiscal Year 1976. The HYGAS plant in Chicago, Illinois, and a Carbon Dioxide Acceptor plant in Rapid City, South Dakota, have been in operation for approximately three years. The Synthane plant in Bruceton, Pennsylvania, was completed in late Fiscal Year 1975 and is presently undergoing shakedown operations. The Self-Agglomerating Ash PDU in Columbus, Ohio, will be completed in Fiscal Year 1976 as well as the Bi-Gas plant in Homer City, Pennsylvania. In addition to the gasification pilot plants, a major auxiliary process project initiated in Fiscal Year 1973 is now in the construction phase adjacent to the HYGAS plant in Chicago. This pilot plant, utilizing the steam-iron process, will produce hydrogen for the hydrogasification of coal in the HYGAS process. It will be completed in late Fiscal Year 1976, and will begin operations in Fiscal Year 1977. Major ongoing projects in this area are shown in Fig. 3.

COAL LIQUEFACTION

MAJOR PROJECTS	*CONTRACT VALUE $M (COST SHARE)	CONTRACTOR	LOCATION	KEY EVENTS
• COAL OIL ENERGY DEVELOPMENT (COED)	$21.0	FMC	PRINCETON, N.J.	PILOT OPERATIONS COMPLETE FY 75
• SOLVENT REFINED COAL (SRC)	$42.0	P & M	TACOMA, WASH.	PLANT MODIFICATIONS NEARING COMPLETION
• H-COAL	$9.5 ($2.7)	HRI	TRENTON, N.J. CATTLETSBURG, KENTUCKY	PDU RUNS FY 75; PILOT PLANT DECISION MID FY 76
• CLEAN COKE	$6.5 ($1.9)	U.S. STEEL	MONROEVILLE, PA.	PDU COMPLETE FY 75; PILOT PLANT DECISION FY 76
• SYNTHOIL	$11.4 ($1.6)	FOSTER WHEELER	PERC BRUCETON, PA.	CONSTRUCTION LARGER PDU IN FY 76

*CONTRACT VALUES AS OF 12/30/75

FIG. 2.

LOW BTU GASIFICATION

The objective of the Low Btu Gasification Subprogram is to develop at the earliest possible date one or more gasification systems which are economically applicable for the use of coal as a substitute for oil or natural gas in power generation, industrial heating applications and the production of chemical feedstocks.

In order to compress the time required to develop a suitable gasification system, several feasible concepts are being studied concurrently. The results of these efforts will permit the evaluation and comparison of techniques and will provide a basis for selection of the most economic process in the shortest time.

The Low Btu Gasification Subprogram was initiated in Fiscal Year 1973, and the projects are still in the early stages of development.

An atmospheric pressure entrained bed PDU of 5 ton/h capacity is under construction. A novel system in which coal will be gasified under pressure in a pool

HIGH BTU GASIFICATION PROGRAM

MAJOR PROJECTS	*CONTRACT VALUE $M (COST SHARE)	CONTRACTOR	LOCATION	KEY EVENTS
• CO_2 ACCEPTOR PROCESS	30.4 (7.8)	CONOCO COAL DEV. CO.	RAPID CITY, S.D.	METHANATION PLANT CONSTRUCTION COMPLETE FY 75
• HYGAS PROCESS	38.7 (11.1)	INSTITUTE OF GAS TECHNOLOGY	CHICAGO, ILL.	STEAM OXYGEN SYSTEM CONSTRUCTION COMPLETE FY 75
• ASH-AGGLOMERATING PROCESS	8.8 (0.6)	BATTELLE COLUMBUS	WEST JEFFERSON, OHIO	COMPLETE PILOT PLANT CONSTRUCTION FY 76
• STEAM-IRON-PROCESS	8.4 (2.3)	INSTITUTE OF GAS TECHNOLOGY	CHICAGO, ILL.	COMPLETE PILOT PLANT CONSTRUCTION FY 76
• BI GAS	66.0 (12.0)	BITUMINOUS COAL RESEARCH/CONOCO COAL DEV. CO.	HOMER CITY, PA.	COMPLETE PILOT PLANT CONSTRUCTION FY 76
• SYNTHANE	9.6	RUST ENGINEERING/ LUMUS CORP.	PERC BRUCETOWN, PA.	COMPLETE CONSTRUCTION FY 75

*CONTRACT VALUES AS OF 12/30/75

FIG. 3.

LOW BTU GASIFICATION

MAJOR PROJECTS	*CONTRACT VALUE $M (COST SHARE)	CONTRACTOR	LOCATION	KEY EVENTS
• FLUIDIZED-BED 2 STAGE FOR ELECTRIC POWER GENERATION	22.4 (6.7)	WESTINGHOUSE ELECTRIC	WALTZ MILL, PA.	PDU OPERATIONAL FY 77
• MOLTEN SALT	6.9 (2.3)	ATOMIC INTERNATIONAL/ ROCKWELL	NORWALK HARBOR, CONN.	PDU COMPLETE FY 77
• ENTRAINED-BED (ATMOSPHERIC)	20.6 (6.9)	COMBUSTION ENGINEERING	WINDSOR, CONN.	PILOT PLANT DESIGN & SYSTEMS FABRICATION FY 77

*CONTRACT VALUES AS OF 12/30/75

FIG. 4.

of molten salt is in the final design phase. Experimental work on a 0.6 ton/h fluidized bed pressurized gasifier is continuing at Waltz Mill, Pennsylvania. This unit is to provide design data for a larger gasifier which will be used to feed a combined cycle power generation system.

Such a combined cycle power plant has the potential of lowering the cost and increasing the overall thermal efficiency of coal conversion to electricity. During Fiscal Year 1977, construction of a plant to explore this concept is planned. It will process from 500 to 800 tons of coal per day, convert the coal to a purified low Btu gas and then utilize that gas to drive a combustion turbine for electrical power generation. Cost-sharing commitments for the project of two-third government and one-third industry have been made.

The Fiscal Year 1977 budget provides for the initiation of efforts for a developmental unit to produce hydrogen from coal. The production of hydrogen from coal is necessary for coal conversion processes which are dependent on the reaction of coal with hydrogen. This developmental unit, for which an RFP is being prepared is being planned jointly with the liquefaction subprogram and may be located at the facilities of an industrial user of hydrogen and cofunding is anticipated. Major ongoing projects in this area are shown in Fig. 4.

ADVANCED POWER SYSTEMS

The objective of the Advanced Power Systems Subprogram is to develop advanced power generation systems to the technology readiness stage in preparation for engineering into prototype electric power generation systems. The primary system under development is the high temperature gas turbine, using low Btu gas, and configured with a steam bottoming cycle. This gasification combined cycle plant has promise for generating electricity at lower cost than is possible from present coal-fired systems. In addition, closed cycle gas turbines, using inert gas as the working fluid, and alkali metal vapor turbines are being evaluated to determine

ADVANCED POWER SYSTEMS

MAJOR PROJECTS	*CONTRACT VALUE $M (COST SHARING)	CONTRACTOR	LOCATION	KEY EVENTS
ECAS	3.8 (2.2)	GE WESTINGHOUSE NASA/LEWIS	VARIOUS	PHASE II PRESENTATION SCHEDULED FEB. 76
OPEN CYCLE GAS TURBINE	7.7	TBD	TBD	RFP ISSUED FEB 75

*CONTRACT VALUE AS OF 12/30/75

FIG. 5.

the economics of their use as coal-fired topping plants in combination with additional power generation from heat recovery boilers and steam turbines. All of these systems are based on the environmentally acceptable use of coal containing appreciable sulfur content or coal derived gaseous or liquid fuels. The major potential economic advantage offered by these advanced power systems results from the fact that high performance combined cycle turbine systems are expected to have lower capital costs and are expected to offer an attractive coal-to-power generation efficiency. Major ongoing projects in this area are shown in Fig. 5.

DIRECT COMBUSTION

The prime purposes of the Direct Coal Combustion Subprogram are first to develop atmospheric systems and then to develop pressurized systems capable of burning combinations of high sulfur coals of all degree, rank and quality in an environmentally acceptable manner. The strategy for combustion research is directed at developing a new mode of clean and economical combustion for power plants and industrial and institutional users, having the potential for substantial improvements over conventional coal firing modes; determining methods to burn synthetic low sulfur fuels derived from coal and chars from coal conversion processes; and minimizing fouling and corrosion of heat transfer surfaces in coal fired boilers. This new mode of coal utilization is that of fluidized bed combustion in which coal is burned in an inert bed of ash and limestone, fluidized by combustion gases. During combustion, the limestone absorbs the SO_2 produced by the coal combustion and enables the emissions to meet air quality requirements without the need for scrubbers.

The Fiscal Year 1977 budget period will see the start of fabrication of several prototype systems in the industrial and institutional fluidized bed combustion applications program. It is anticipated that these prototypes will lead to commercial applications promptly as curtailments of natural gas, the cost of oil, and the uncertainty of oil supply create the immediate need for equipment to burn coal cleanly and economically. It is anticipated that these units will lead to commercial implementation after a year or more of shakedown and testing. It is expected that in the Fiscal Year 1977 period, component test and integration facilities for both atmospheric and pressurized fluidized bed combustion systems will be constructed. These facilities are to provide the capability for component improvement, development testing, and system configuration research. The pressurized fluidized bed combined cycle project will continue with construction of the plant scheduled to begin in Fiscal Year 1978. This plant should lead to a demonstration project in the mid-1980s and later to commercialization. The results will be an increase in power generation efficiency from coal and improvement in power generation costs.

The fluidized bed concept lends itself not only to utilities and to industrial and institutional application, but also to coal conversion processes in that it can utilize the char by-product of coal conversion processes for increased plant efficiency. Major ongoing projects are shown in Fig. 6.

ADVANCED RESEARCH AND SUPPORTING TECHNOLOGY

The Advanced Research and Supporting Technology Subprogram serves as a central research and system studies area for all elements of fossil energy. The major objectives of this subprogram are to develop and maintain the supporting science and technology base; to lay the foundation for innovative, new and improved processes; to assure reliable and efficient operation of synthetic fuel plants; to investigate and develop advanced techniques and support technology for direct

DIRECT COMBUSTION

MAJOR PROJECTS	*CONTRACT VALUE $M (COST SHARING)	CONTRACTOR	LOCATION	KEY EVENTS
• FLUIDIZED BED (ATMOSPHERIC)	14.3	POPE, EVANS, ROBBINS	RIVESVILLE, W. VA.	STARTUP CY 76
• FLUIDIZED BED (ATMOSPHERIC)	3.5	COMBUSTION POWER CO.	MENLO PARK, CALIF.	FY 76 DRY HOT GAS CLEANUP DEVELOPMENT
• FLUIDIZED BED (PRESSURIZED)	24.0 (4.8)	CURTIS-WRIGHT	WOODBRIDGE, N. J.	CONTRACT BEING NEGOTIATED

*CONTRACT VALUES AS OF 12/30/75

FIG. 6.

utilization of coal including gas cleanup and improved high temperature materials for plant construction; to conduct analyses of fossil energy process technologies and systems; to formulate planning policy options for management decision making; to continue efforts to comply with the National Environmental Policy Act; and to assure an adequate supply of trained personnel from the nation's university system.

During Fiscal Year 1976, additional research projects were initiated in the areas of advanced synthetic fuels process research and engineering basic coal science, materials and components, and direct coal utilization. This central research effort is vital to the progress of all program areas of fossil energy. Projects in this area are conducted in ERDA's Energy Research Centers and National Laboratories, other government agencies, private industry and universities.

ADVANCED RESEARCH & SUPPORTING TECHNOLOGY

MAJOR PROJECTS	CONTRACT VALUE $M (COST SHARING)	CONTRACTOR	LOCATION	KEY EVENTS
• FIRESIDE CORROSION	$6.0	TBD	TBD	RESPONSE TO RFP BY 31 JAN
• GASIFICATION MATERIALS	$6.0	NBS ARGONNE MPC/IITRI	WASHINGTON CHICAGO, ILL. CHICAGO, ILL.	---
• VALVES FOR COAL GASIFICATION	$5.0	TBD	TBD	
• ADVANCED SYNTHETIC FUELS PROCESSES	$14.0	45	VARIOUS INDUSTRY ERC'S, & NAT LABS	
• UNIVERSITY PROGRAMS	$9.0/YR (1.0)	44	VARIOUS	---

Fig. 7.

Intensive studies are underway on coal and oil shale structure and reaction characteristics which could lead to development of more efficient conversion processes. Properties of coal, oil shale, and synthetic fuels, combustion chemistry and the chemistry and engineering of conversion to synthetic fuels are likewise the subject of research. Work on the refining of synthetic fuels and testing of fuels which are alternatives to conventional gasoline is underway and is expected to lead to optimization of processes and products and to new fuels for transportation and electrical generation. Major ongoing projects are shown in Fig. 7.

DEMONSTRATION PLANTS

The objective of the Demonstration Plants Subprogram is to translate research and development results into technology that can be transferred to the private sector for use in commercial operations. This objective will be achieved by designing, building, and operating demonstration-scale plants in cooperative efforts with the private sector. Demonstration plants are defined as those plants that utilize either commercial-scale equipment or somewhat smaller units that can be scaled up to full commercial size with normal risk.

Competitive proposal procedures are used to solicit interest from private industry concerning proposed demonstration plants. The plants are all intended to be jointly funded. The present standard for cooperative government/industry joint funding is 50/50 sharing of construction and operating costs. The commitment of private funds indicates the industry's confidence in the process potential, assures the availability of top industrial personnel and continued management attention, and fosters technology transfer.

On January 17, 1975, a contract was awarded to COALCON for the phased design, construction and operation of a 2600 ton/day demonstration plant using a hydrocarbonization process for producing 3900 barrels of liquid product and 22 million cubic feet of pipeline quality gas daily. This plant will support an objective of the coal program which is to provide a fuel substitute from coal for a portion of the low sulfur oil and gas used in the generation of electricity and in industrial heating processes.

This plant is the first demonstration plant that Fossil Energy has undertaken. Initial construction phases will begin in late Fiscal Year 1977. Although the design engineering is being funded by the Government, the industrial partner will share the construction and operating costs on a 50/50 basis. The plant schedule is being accelerated and we hope to begin operation in Fiscal Year 1980.

To meet national needs and continue to accelerate coal conversion technology, we have requested funding to continue conceptual design activities through Fiscal Year 1977 for a high Btu pipeline gas demonstration plant and a low Btu demonstration plant. In addition, the Fiscal Year 1976 appropriations will be utilized in Fiscal Year 1977 to initiate architectural engineering

DEMONSTRATION PLANT PROGRAM

MAJOR PROJECTS	SCHEDULED OPERATIONAL DATE (FY)	TONS/DAY COAL FEED	OUTPUT
• CLEAN BOILER FUEL DEMO PLANT COALCON	1980	3000	3500 BBL BOILER FUEL 22 (10^6) BTU PIPELINE GAS
• HIGH BTU SYNTHETIC PIPELINE GAS DEMO PLANT (RFP 2Q FY 76)	1981	3000-5000	50-80 (10^6) SCF
• LOW BTU FUEL GAS DEMO PLANT (RFP 3Q FY 76)	1981	3000-4000	250-350 MW^e

FIG. 8.

and long-lead procurement activities for these plants. These plants will build on the experience of the coal conversion technology. Products made in demonstration plants will be tested by potential users to ensure compatibility with all elements of the product supply system. Major ongoing projects are shown in Fig. 8.

MAGNETOHYDRODYNAMICS

The objective of the Magnetohydrodynamics (MHD) Subprogram is to achieve successful operation of a commercial facility to generate electricity from coal or a coal derived fuel by the late 1980s. This objective is being pursued in recognition of the potentially higher overall conversion efficiency available in MHD power generation.

Basic feasibility of the open cycle MHD system was established prior to Fiscal Year 1975. Early exploratory work conducted with direct coal combustion was highly encouraging and has led to increasing development activity aimed toward the open cycle, coal burning design concept. A recent experimental channel test, simulating a coal-burning MHD generator, achieved 95 h without failure.

The current US/USSR MHD agreement is contributing useful engineering data on basic "state-of-the-art" performance capability, using clean (natural gas) fuels. Progress in this program is reflected by the recent completion of a 100 h continuous operation test of a 2 kg/s experimental generator. Electrodes used in this successful test were designed and fabricated in the United States and delivered to the USSR for assembly in the test channel.

MAGNETOHYDRODYNAMICS

MAJOR PROJECTS	*CONTRACT VALUE $M (COST SHARING)	CONTRACTOR	LOCATION	KEY EVENTS
• GENERATOR STUDIES	3.5	AVCO	EVERETT, MASS.	FY 76 100 HR. GENERATOR RUN
• HIGH TEMPERATURE AIR HEATERS	8.1	UNIV. TENN. SPACE INSTITUTE	TULLAHOMA, TENN.	FY 78 DIRECT COAL FIRED OPERATION
• HIGH PERFORMANCE GENERATOR DEMONSTRATION EXPERIMENT	3.6	AEDC	AAFS, TENN.	FY 76 15-20% ENTHALPY EXTRACTION
• COMPONENT DEV & INTEGRATION FACILITY	31.5	RFP	TBD	A-E DESIGN AWARDED
• ENGR TEST FACILITY	3.1	RFP	MONTANA	
• OPEN CYCLE PLASMA RESEARCH	2.9	MONTANA ENERGY R&D INSTITUTE	BUTTE, MONTANA	

*CONTRACT VALUE AS OF 12/30/75

FIG. 9.

The MHD effort falls generally into two basic categories, i.e. open and closed cycle systems. Major emphasis is being placed on the open cycle concept because of intrinsic performance advantages and lesser experimental problems. Closed cycle work is being continued, however, mainly as exploratory heat exchanger and generator studies.

A combined cycle (coal burning) generating plant concept in which the open cycle MHD generator serves as the topping cycle offers the dual advantage of high overall thermal efficiency and low emission levels. To provide the necessary technology and design experience to design and build pilot scale facilities, an open cycle MHD program has been structured which progresses through three overlapping phases. The initial (current) phase will develop the technology required to design and test prototypes of all major components. These will be tested separately, and in various combinations, in a fully integrated test facility which has been designed as the Component Development and Integration Facility (CDIF). The second phase carries the program through to a second generation of component design of increased scale and superior performance. These will be tested in the Engineering Test Facility (ETF). The final phase is aimed toward design, construction, and operation of the commercial scale plant.

A recent review of the subprogram's status has resulted in our recognition that the schedule for the Component Development and Integration Facility could be accelerated in Fiscal Year 1976 to permit the more timely evaluation of this technology. We have recently submitted to the Congress a reprogramming request which will provide an additional $20 million in budget authority in Fiscal Year 1976 for this purpose. We also plan to submit an amendment to the Fiscal Year 1977 budget request to identify the Component Development and Integration Facility as a line item construction project. Major MHD Projects are shown in Fig. 9.

PETROLEUM AND NATURAL GAS

Petroleum and natural gas will continue to be the Nation's main fossil energy source for many years. About three-quarters of the Nation's energy is supplied from these energy resources. Very recent estimates indicate that without enhanced recovery, production of domestic petroleum and natural gas will begin to drop off rapidly in the mid-1980s. Enhanced recovery of oil and gas with requisite technological success could extend the supply of domestically available resources by approximately 10 years and is crucially important to the country while alternate sources are being developed.

To these ends we estimate that in Fiscal Year 1977, a $36.9 million for budget authority under operating

PRESENT PETROLEUM AND NATURAL GAS MAJOR CONTRACT PROJECTS

MAJOR PROJECTS	$ (M) FUNDING (COST SHARE)	PERFORMER	LOCATION	STATUS
• MICELLAR - POLYMER	7.1 (4.1)	CITIES SERVICE	ELDORADO, KS.	UNDER INJECTION
• MICELLAR - POLYMER	9.8 (6.4)	PHILLIPS PET	BURBANK, OK	INJECTION TESTS IN MAY '76
• MICELLAR POLYMER	4.4 (2.2)	PENN CRUDE	BRADFORD, PA.	INJECTION TESTS IN MAY '76
• POLYMER FLOOD	3.9 (2.7)	KEWANEE	N. STANLEY, OK.	INJECTION STARTS JUNE '76
• POLYMER FLOOD	7.4 (5.2)	SHELL OIL	E. COALINGA, CA.	INJECTION STARTS JUNE '76
• MASSIVE HYDRAULIC FRACTURE	4.3 (2.2)	COLUMBIA GAS	W. VA. (DEVONIAN SH)	DRILLING 12/75: FRACTURING 3/76
• MASSIVE HYDRAULIC FRACTURE	2.8 (1.5)	CER-GEONUCLEAR	RIO BLANCO, COL.	FINAL TREATM'T & EVAL APR '76 START SEPT '75
• CHEMICAL EXPLOSIVE FRACTURE	.35 (.35)	PHYSICS INTERNAT'L	W. VA.	FIELD TESTS – OCT '75
• CHEMICAL EXPLOSIVE FRACTURE	.13 (.13)	TALLEY FRAC	MINERAL WELLS, TEX.	FINAL EVALUATION BEING MADE
• CHEMICAL EXPLOSIVE FRACTURE	.32 (.32)	PETROLEUM TECH	CAPON FIELD, W. VA.	3RD EXPLOS FRAC IN NOV '75
• CO_2 INJECTION	3.2 (2.0)	GUYAN OIL CO.	GRIFFITHSVILLE W. VA.	CONTRACT SIGNED 9/75
• THERMAL RECOVERY	6.8 (4.1)	HUSKY OIL CO	PARIS VALLEY, CA.	ECON-TECH EVALUATION 12/76
• THERMAL RECOVERY	3.0 (2.0)	HANOVER PET	LITTLE TOM FIELD, TX.	GETTING UNDERWAY 1975
• DEVIATED WELL RECOVERY PROJECT	.22 (.17)	KENT-W. VA. GAS	HAZARD, KY	TO STIMULATE 1ST WELL-FOAM FRAC
• MASSIVE HYDRAULIC FRACTURE	1.6 (0)*	EL PASO NATURAL GAS	EL PASO, TX.	CONTRACT SIGNED 9/75

UNIVERSITY OIL & NAT GAS RESEARCH PROJECTS, NOT INCLUDED

*INDUSTRY HAS ALREADY SPENT $4.6M

FIG. 10.

expenses and $270,000 for plant and capital equipment obligations are required:

Petroleum and Natural Gas Program	Program Estimate Budget authority (In thousands)		
Subprogram	1976	TQ	1977
Operating expenses:			
Oil and gas extraction	$41,423	$8,426	$35,074
Supporting research	1,797	450	1,831
Total operating expenses	$43,220	$8,876	$36,905
Plant and capital equipment total	$100	$100	$270

The oil extraction efforts emphasize the cost shared demonstrations of existing and improved secondary and tertiary recovery techniques. A federal role in enhanced oil and gas recovery is required because of the commercial uncertainties and financial risks associated with economically unproven technologies and the urgent need to accelerate progress beyond normal commercial capability. It is probable that the oil industry would eventually implement competitive advanced recovery techniques; however, this would not be done on the accelerated scale or within the time frame required.

The industry response to the enhanced oil and gas recovery subprogram to date has been outstanding. Of the eight contracts entered into in Fiscal Year 1975 for enhanced oil recovery, the industry is supporting 67% of the total costs; in six existing enhanced gas recovery projects, the industry support is 47% of the total. Conversations with petroleum-industry spokesmen indicate general approval and support of the ERDA enhanced oil and gas recovery program.

An important function of the enhanced oil and gas recovery program is technology transfer. This is being accomplished by a quarterly report that summarizes progress on all contracts, technical presentations and publications and symposia. The first symposium was held in Tulsa in September. Attendance of about 200 was expected; more than 500 persons did attend. It is planned to hold such symposia on at least an annual basis. In addition, we are looking into the possibility of putting all raw data received from our contractors into open files at several strategically located places, so the data can be available to anyone who wishes to consult them.

The natural gas stimulation efforts are designed to stimulate the commercial production of gas from formations containing vast quantities of natural gas but having natural permeabilities so low that commercial production to date has not been feasible. Experimental methods include massive hydraulic fracturing, combinations of hydraulic and chemical explosive fracturing, and wells deviated from normal to intersect natural fractures.

Micellar–polymer flooding is the most promising method for enhanced oil recovery and three field tests are currently underway. Five additional field tests are planned in Fiscal Year 1977 to maintain the accelerated rate of the subprogram. Massive hydraulic fracturing appears to provide the major promise for near-term gas production increases. The Fiscal Year 1977 budget will provide for continuation of this vital technology with emphasis on test of stimulating gas production from Eastern Devonian Shales. Major ongoing projects in this area are shown in Fig. 10.

IN SITU TECHNOLOGY PROGRAM

The *In Situ* Technology Program contains two major subprograms; oil shale and *in situ* coal gasification. Technology for conventional mining and surface retorting of oil shale has advanced to a level believed capable of early commercial application to richer grades of oil shale averaging over 25 gal/ton. However, commercial scale equipment and operations have not been demonstrated. The present ERDA strategy in this area consists of making its Anvil Points facility available for industry sponsored projects to provide for the technology transfer of results through the ERDA observer program, providing general consultation to industry, and conducting laboratory scale supporting research on general problems relevant to oil shale processing, both aboveground and *in situ*.

In order to pursue these important efforts, we request Fiscal Year 1977 operating expenses of $30.6 million and $500,000 for plant and capital equipment. These figures are broken down in the following way:

In Situ Technology Program	Program Estimate Budget authority (In thousands)		
Subprogram	1976	TQ	1977
Operating expenses			
Oil shale	$13,720	$1,950	$21,085
In situ coal gasification	6,137	1,680	8,236
Supporting research	1,265	300	1,310
Total operating expenses	$21,122	$3,930	$30,631
Plant and capital equipment total	$325	$100	$500

The ERDA oil shale subprogram is focused mainly on developing technology capable of improving total resource recovery and lowering environmental impacts and water requirements. This is accomplished through development of various *in situ* extraction techniques capable of exploiting the leaner and deeper oil shale deposits as well as offering an alternative to present technology. Additional advantages associated with *in situ* retorting are reduced requirements for water, manpower and shale disposal. The potential application to lower grade shales is important in that most of the total estimated resource of 1.8 trillion barrels of shale oil in the Western Green River formation is in low grade deposits. Much of this may never be recovered by techniques that involve conventional mining. Laboratory scale research on new process technology includes fundamental research on methods potentially offering similar advantages for above-ground processing.

Eastern oil shales of the Devonian and Mississippian age represent an additional large potential energy resource although their extent is presently poorly defined. These shales are potentially major producers of synthetic gas and oil although they are considered lower grades than the western shale from an oil production standpoint. These are geologically similar to western oil shales and it is believed that similar technology can be used in exploiting these shales by *in situ* methods.

Another major subprogram, *in situ* coal gasification, seeks to further enhance the use of vast coal reserves of the United States which will supply much of our short- and intermediate-term energy needs. More than 85% of our coal resources are too deep, too thin seamed or otherwise unsuitable to be commercially exploited by conventional mining. *In situ* processing, especially by gasification, provides a means to utilize these otherwise unusable resources while providing the benefits of minimal environmental disruptions, lower water usage, relief of mine safety problems and alleviation of the shortage of skilled miners. Major *in situ* and oil shale projects are shown in Fig. 11.

Research on novel extraction processes that may form the basis for improved second-generation technology must be continued. Although the projects are currently oriented largely toward the specific needs of the *in situ* subprogram, the subprogram is considered supportive of oil shale development in general, providing data useful to both above ground and *in situ* processes.

ENERGY RESEARCH CENTERS AND NATIONAL LABORATORIES

The Energy Research Centers and National Laboratories are playing an important role in achieving the ERDA's goal in fossil energy. These facilities are staffed by highly skilled scientists, engineers and technicians with long histories of accomplishments in the production and utilization of fossil fuels. They are looked to for new ideas, new data and new processes to better utilize our fossil energy resources, and project management (Figs. 12 and 13).

CONCLUSION

I would like to note in conclusion that the Fossil Energy Program currently has over 271 contracts outstanding with the total value of nearly $567 million. We hope to maintain the present standard for cooperative joint funding which anticipates one-third private and two-thirds federal funding for pilot plants and 50/50 funding for demonstration plants. Our participants include many aspects of American industry; oil and gas companies, chemical companies, coal companies, architect engineer and manufacturing companies, utilities, universities, research centers and trade associations.

In addition to the contractors performing the major

IN SITU AND OIL SHALE MAJOR PROJECTS

MAJOR PROJECTS	$ (M) FUNDING	PERFORMER	LOCATION	STATUS
• OIL PRODUCTION FIELD TESTS				
TRUE IN SITU	1.62	LERC	LARAMIE, WYO.	TEST EVAL THRU '76
	.60	SANDIA	ALBUQUERQUE, N. M.	INSTRUMENTATION
MODIFIED IN SITU	4.54	CONTRACTS		SIGN-OFF CONTRACT
	.65	SANDIA	ALBUQUERQUE, N. M.	INSTRUMENTATION
	.30	LASL	LOS ALAMOS, N. M.	
• SUPPORT				
EXPER. RETORTING & MODELING	1.80	LLL	LIVERMORE, CA.	EXPER. RETORTING & MODELING
PROCESS EVALUATION	.41	LERC	LARAMIE, WYO.	PROCESS EVALUATION
NEW PROCESS TECHNOLOGY	.25	LERC	LARAMIE, WYO.	NEW PROCESS TECHNOLOGY
CHARACTERIZATION OF OIL SHALE	.31	LERC	LARAMIE, WYO.	CHARACTERIZATION OF OIL SHALE
• GAS PRODUCTION				
EASTERN SHALE	2.50	CONTRACTS		SIGN-OFF CONTRACT
WESTERN SHALE	.60	LERC	LARAMIE, WYO.	
• ENVIRONMENTAL				
	.90	LERC	LARAMIE, WYO.	WATER MONITORING WELLS
	.10	CONTRACTS	VARIOUS UNIV AND LABS	STUDIES
• OIL SHALE SUPPORTING RESEARCH				
PRODUCTION OF CLEAN FUELS	.23	LERC	LARAMIE, WYO.	PRODUCTION OF CLEAN FUELS
ANVIL POINTS MONITORING	.11	LERC	LARAMIE, WYO.	ANVIL POINTS MONITORING

FIG. 11.

RESEARCH CENTER/ NATIONAL LABORATORY	SIGNIFICANT RESEARCH CENTER/NATIONAL LABORATORY SUPPORT OF ERDA/FE ACTIVITY ELEMENTS							
	COAL							
	LIQUEFACTION	GASIFICATION		ADVANCED POWER SYSTEMS	DIRECT COMBUSTION	ADVANCED RESEARCH & SUPPORTING TECHNOLOGY	MHD	DEMONSTRATION PLANTS
		HIGH BTU	LOW BTU					
ARGONNE NATIONAL LABORATORY	SYNTHOIL PHYSICAL PROPERTIES					MATERIALS RESEARCH AND COAL CONVERSION	SUPER CONDUCTING MAGNETS PROGRAM, U25 GENERATOR	INSTRUMENTATION
BARTLESVILLE ENERGY RESEARCH CENTER	CHARACTERIZATION & UTILIZATION OF SYN CRUDE EROM COAL							
BROOKHAVEN NATIONAL LABORATORY								
GRAND FORKS ENERGY RESEARCH CENTER	CO-STEAM PROCESS		SLAGGING FIXED BED GASIFIER			LIGNITE LIQUEFACTION		
LARAMIE ENERGY RESEARCH CENTER						COAL LIQUEFACTION		
LAWRENCE LIVERMORE NATIONAL LABORATORY								
LOS ALAMOS SCIENTIFIC LABORATORY								
MORGANTOWN ENERGY RESEARCH CENTER	LIQUID-SOLID SEPARATION	HYDRANE HOT GAS CLEANUP; H_2 FROM CHAR	FIXED BED PRESSURIZED REACTOR; HOT GAS CLEANUP	COMB. CYCLE; FLUID BED COMBUSTION; FIRESIDE CORROSION	FLUIDIZED BEDS	LIQUEFACTION RESEARCH		
OAK RIDGE NATIONAL LABORATORY	PARTICULATE REMOVAL FROM SYNCRUDE; HYDROCARBONIZATION			ALKALI METAL VAPOR SYSTEMS	FLUIDIZED BED HEATER	CARBONYL CHEMISTRY; MATERIALS RESEARCH		COMMERCIALIZATION; CONCEPTROL DESIGN & PROCESS ENGR STUDIES
PACIFIC NORTHWEST LABORATORY			HOT GAS CLEANUP				MATERIAL CHARACTERIZATION	
PITTSBURGH ENERGY RESEARCH CENTER	SYNTHOIL; CO-STEAM; SRC; FISHER-TROPSCH	SYNTHANE; HYDRANE			SLURRY COMBUSTION AND STACK GAS CLEANING	FUEL PROD. ENHANCEMENT; CATALYST DEV.; COAL SCIENCE	HIGH TEMPERATURE COAL COMBUSTION	
SANDIA NATIONAL LABORATORY	SYNTHOIL CATALYST PERFORMANCE							
SAN FRANCISCO FE OFFICE OF PROGRAM COORDINATION AND MANAGEMENT								

FIG. 12.

RESEARCH CENTER/ NATIONAL LABORATORY	SIGNIFICANT RESEARCH CENTER/NATIONAL LABORATORY SUPPORT OF ERDA/FE ACTIVITY ELEMENTS						
	PETROLEUM & NATURAL GAS				IN SITU TECHNOLOGY		
	GAS AND OIL EXTRACTION			SUPPORTING RESEARCH	OIL SHALE	SUPPORTIVE RESEARCH	IN SITU COAL GASIFICATION
	FLUID INJECTION	NON-NUCLEAR FRACTURING	DRILLING EXPLORATION & OFFSHORE OPERATIONS				
ARGONNE NATIONAL LABORATORY							
BARTLESVILLE ENERGY RESEARCH CENTER	IMPROVED RECOVERY TECHNIQUES; STIMULATION OF HEAVY OIL PRODUCTION	PRODUCTION STIMULATION	RESERVOIR AND THERMODYNAMIC CHARACTERIZATION; OIL UTILIZATION	HEAVY ENDS CHARACTERIZATION, OIL QUALITY IDENTIFICATION FUEL/ENGINE TECHNOLOGY			
BROOKHAVEN NATIONAL LABORATORY							
GRAND FORKS ENERGY RESEARCH CENTER							
ARAMIE ENERGY RESEARCH CENTER	TAR SANDS CHEMISTRY IN SITU RECOVERY FROM TAR SANDS			HEAVY LIQUID CHARACTERIZATION, ASPHALT STUDIES		CHARACTERIZATION, ASPHALT., TAR SANDS AND CHEMICAL STUDIES	LINKED VERTICAL WELLS
AWRENCE LIVERMORE NATIONAL LABORATORY		FRACTURING SUPPORT MASSIVE HYDRAULIC & EXPLOSIVE FRACTURING; RIO BLANCO EXP.				RETORTING IN SITU	PACKED BED PROCESS
LOS ALAMOS SCIENTIFIC LABORATORY						IN SITU RECOVERY FRACTURE DESIGN	
MORGANTOWN ENERGY RESEARCH CENTER	CO_2 RECOVERY	MASSIVE HYDRAULIC FRACTURING, EARTH FRACTURE STUDIES					LONG WALL GENERATOR
OAK RIDGE NATIONAL LABORATORY	CHEMICALS FOR ENHANCED OIL RECOVERY						
PACIFIC NORTHWEST LABORATORY							
PITTSBURGH ENERGY RESEARCH CENTER							
SANDIA NATIONAL LABORATORY		MASSIVE HYDRAULIC FRACTURING SUPPORT; INSTRUMENTATION & FRACTURE DESIGN					
SAN FRANCISCO FE OFFICE OF PROGRAM CO-ORDINATION & MANAGEMENT	SOLVENT RECOVERY OF HEAVY OIL						

FIG. 13.

FOSSIL ENERGY
U.S. ENERGY RESEARCH AND DEVELOPMENT ADMINISTRATION
FY 1977 BUDGET ESTIMATES
(DOLLARS IN THOUSANDS)
SUMMARY TABLE BY SUBPROGRAM

FOSSIL ENERGY DEVELOPMENT OPERATING EXPENSES	FY 1976 ESTIMATE			TRANSITION QUARTER ESTIMATE		FY 1977 ESTIMATE	
	B/A	OBS.	B/O	B/A & OBS.	B/O	B/A & OBS.	B/O
COAL							
1. LIQUEFACTION	$ 89,912	$129,465	$ 92,937	$ 26,443	$ 15,693	$ 73,946	$ 79,546
2. HIGH-BTU GASIFICATION	53,364	57,320	37,338	9,250	7,450	45,054	59,234
3. LOW-BTU GASIFICATION	24,552	49,215	38,026	6,720	4,113	33,052	39,952
4. ADVANCED POWER SYSTEMS	10,001	10,423	7,461	3,500	1,700	22,500	12,800
5. DIRECT COMBUSTION	38,096	65,302	32,645	13,500	5,100	52,416	52,116
6. ADVANCED RESEARCH AND SUPPORTING TECHNOLOGY	35,393	40,087	32,061	8,850	4,600	37,085	36,585
7. DEMONSTRATION PLANTS	31,900	32,650	14,250	7,750	3,700	53,000	50,600
8. MAGNETOHYDRODYNAMICS	29,544	32,593	18,400	7,800	4,200	37,441	27,341
TOTAL COAL PROGRAM	312,762	417,055	271,118	83,813	46,556	354,494	358,194
PETROLEUM AND NATURAL GAS							
1. GAS AND OIL EXTRACTION	41,423	43,546	32,859	8,426	9,000	35,074	30,374
2. SUPPORTING RESEARCH	1,797	1,797	1,582	450	500	1,831	1,831
TOTAL PETROLEUM AND NATURAL GAS PROGRAM	43,220	45,343	34,441	8,876	9,500	36.905	32,205
IN SITU TECHNOLOGY							
1. OIL SHALE	13,720	13,721	9,834	1,950	2,000	21,085	12,085
2. IN SITU COAL GASIFICATION	6,137	7,510	7,560	1,680	1,787	8,236	6,736
3. SUPPORTING RESEARCH	1,265	1,265	1,113	300	300	1,310	1,310
TOTAL IN SITU TECHNOLOGY PROGRAM	21,122	22,496	18,507	3,930	4,087	30,631	20,131
TOTAL OPERATING EXPENSES	$377,104	$484,894	$324,066	$ 96,619	$ 60,143	$422,030	$410,530

FIG. 14.

project work I have described, we have contractors performing engineering evaluations, systems studies and planning studies and environmental analyses. We have several inter-agency agreements including agreements with the Department of Commerce, Corps of Engineers, Department of the Interior as well as a strong intra-agency support. We have international agreements with the United Kingdom, Russia, Poland and Germany. We are constantly advised of work in the private sector and adjust our planning accordingly. We cooperate with both the American industry and utilities through the American Gas Association and Electric Power Research Institute to develop national plans which integrate privately sponsored and federally sponsored research and development.

To assist this Committee in review of the Fossil Energy Program, we have included a summary table in Fig. 14 displaying by program and subprogram, the appropriations required for operating expenses.

We look forward to continuing our relationship with this subcommittee and are pleased to have this opportunity to present our program.

Manuscript received, May 1976

Prog. Energy Combust. Sci., Vol. 2, pp. 129–141, 1976. Pergamon Press. Printed in Great Britain

FLARING IN THE ENERGY INDUSTRY

T. A. BRZUSTOWSKI

Thermal Engineering Group, Department of Mechanical Engineering, University of Waterloo, Waterloo, Ontario, Canada N2L 3G1

Abstract—Flaring is the combustion process used for the safe disposal of large quantities of flammable gases and vapours in the petroleum industry. This paper is a critical review of the technology of flaring and of the state of knowledge on which design information can be based. Discussion includes the length and shape of the flame on an elevated flare, its radiation field, as well as noise and air pollution from flares.

NOTATIONS

C_L = lean limit concentration of flare gas in air, volume fraction
d = diameter
F = fraction of heat release radiated by the flame
$Fr \equiv \frac{u_j^2}{gd_j}\left(\frac{\rho_f}{\rho_\infty - \rho_f}\right)$ Froude number for a flame of constant density
L = flame length measured from the nozzle along the flame axis
$M \equiv \rho u^2$ momentum flux
$\tilde{M}$ = molecular weight
$\dot{m}$ = mass flow
ρ = density
$R \equiv M_j/M_\infty$
s = curvilinear coordinate along flame axis from the flare tip
$\bar{S} \equiv s/(d_j R^{\frac{1}{2}})$
u = velocity
x = horizontal coordinate downwind from the flare tip
X = mole fraction
$\bar{X} \equiv x/(d_j R^{\frac{1}{2}})$
z = vertical coordinate upward from the flare tip
$\bar{Z} = z/(d_j R^{\frac{1}{2}})$

Subscripts

f = flame
i = i'th component of flare gas mixture
j = flare tip discharge conditions
L = lean limit
∞ = ambient air

1. INTRODUCTION

Flaring is the combustion process which has been the traditional method for the safe disposal of large quantities of unwanted flammable gases and vapours in the oil industry. With the advent of air quality standards, flaring has also taken on an added importance as a method of industrial environmental control, since most gases which could previously be vented to the atmosphere must now be burned in a flare.

Flaring has only recently attracted the attention of combustion scientists. It is now recognized that combustion in the common elevated flare occurs in a turbulent diffusion flame in a cross-wind. Such a flame presents a number of fascinating and challenging phenomena for study, including: the effect of cross-wind on its shape and length, its radiation field, the formation and dispersion of smoke and gaseous pollutants, the action of steam in suppressing the formation of smoke, the completeness of combustion of any toxic gases released to the flare, and various kinds of noise—particularly the noise of steam jets and low-frequency vibrations of the flare system.

A thorough understanding of all these phenomena is now required for the design of modern flare systems, which are generally larger and, at the same time, subject to much stricter regulations than their forerunners. The design guidelines which have evolved over the years from experience with smaller flares must be put on a firm basis, and reliable methods for extrapolation in both size and performance must be developed.

The purpose of this paper is to assist this development in two ways. It is written both to acquaint combustion specialists with the particular combustion problems encountered in flaring, and to introduce engineers who specify and design flare systems to the research results which bear on these combustion problems. A critical description of a number of generic and proprietary designs of flares which are available from the principal suppliers in the world market is presented for the benefit of both groups of readers.

2. THE NEED FOR FLARING

The flaring of gases in the petroleum industry occurs in three ways. First, there is a requirement for safe disposal of flammable gases in connection with producing oil fields, in those cases where no provision exists for collecting and processing the gas. There was a time when almost all gas released in this way was flared, but the great value now placed on natural gas has made gas recovery economical in many fields. Nevertheless, if the gas occurs in quantities which are too small for economical processing or if it is so sour that processing it would be very expensive, it may still be flared. Such flaring is usually called production flaring.

Flaring also takes place in petrochemical plants, oil refineries and gas processing plants where the flare system is one of the so-called "offsite" facilities. Process flaring and emergency flaring can occur in this setting. In process flaring, the gas which leaks past the safety valves protecting the various process units is brought to the flare and burned. This gas feeds the small flames which burn almost continuously on refinery flare stacks. Process flaring at much greater rates can occur when process units are evacuated during a shutdown or when off-specification products are produced during

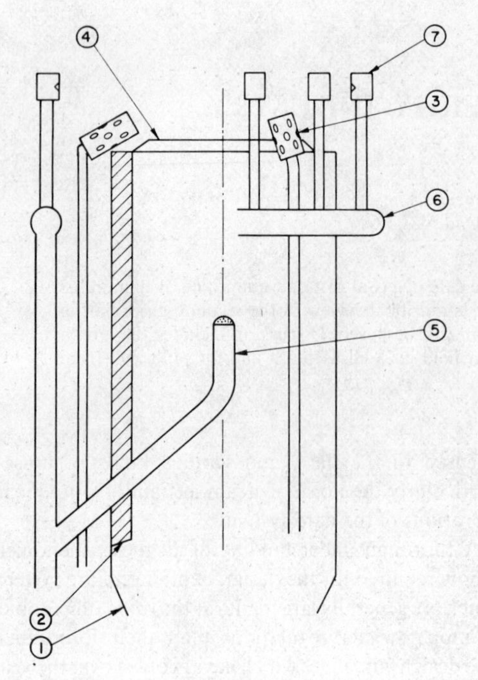

FIG. 1. Principal elements of a 'steam-ring' smokeless flare tip: (1) high-temperature alloy tip, (2) refractory lining, (3) pilot light, (4) nozzle lip or perforated 'flame-holder', (5) centre-steam nozzle, (6) steam-ring header, (7) steam-ring nozzle.

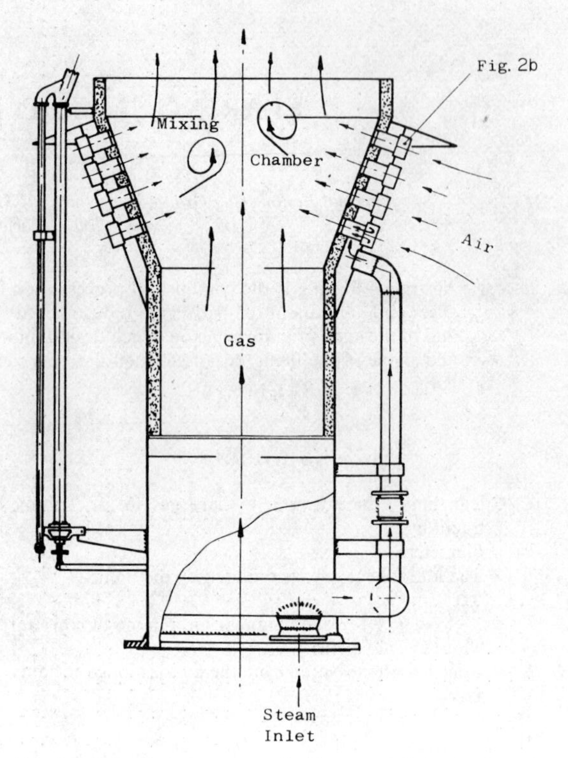

FIG. 2a. Principal elements of the Flaregas FS Antipollutant flare tip.

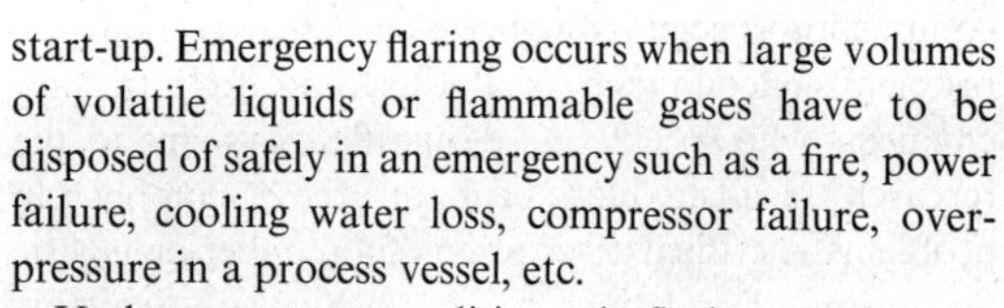

start-up. Emergency flaring occurs when large volumes of volatile liquids or flammable gases have to be disposed of safely in an emergency such as a fire, power failure, cooling water loss, compressor failure, overpressure in a process vessel, etc.

Under emergency conditions, the flaring rate through a single refinery flare might be of the order of 10^5 kg/h, with a heat release rate of the order of 1000 MW for a few minutes. The largest flare installations can reach flaring rates which are even higher, perhaps by an order of magnitude. On the other hand, process flaring rates are usually no more than a few per cent of the emergency flaring rates.

A good introduction to the systems available for process and emergency flaring is given by Jones.[1]

3. THE TECHNOLOGY OF FLARING

The most common types of tips for elevated flares are shown in Figs. 1–5 and are described in Table I. The

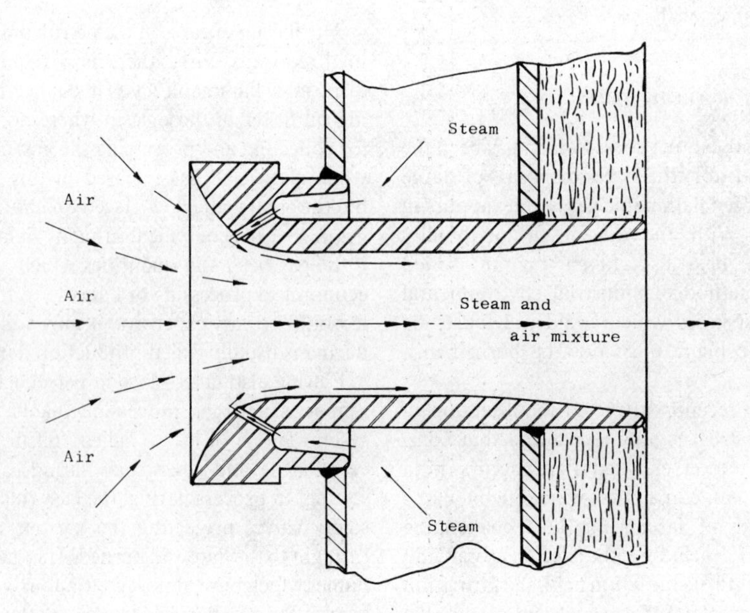

FIG. 2b. Detail of the Coanda nozzle which induces air into the conical mixing section.

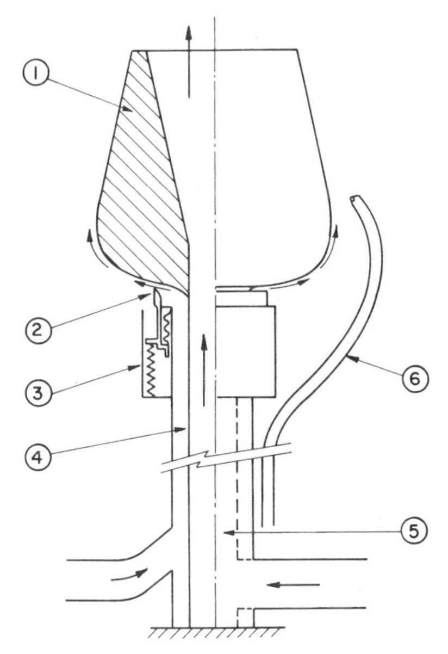

FIG. 3. Principal elements of the Indair Smokeless Flare: (1) conical nozzle inside contoured burner tip, (2) adjustable annular nozzle for h.p. gas, (3) h.p. gas nozzle adjustment mechanism, (4) annular h.p. gas duct, (5) l.p. gas duct, (6) pilot light assembly.

first figure and table refer to three generic designs of flares: the "utility" flare, the "centre-steam" flare and the "steam-ring" flare. Figures 2, 3, 4 and 5 and Table 1-D refers to four proprietary flare designs. The tables contain self-explanatory comments about the construction and the particular advantages and disadvantages of all the flares. Flare tips of these designs are generally available for flare stacks varying between 0.4 and 1.2 m in diameter, but some flare stacks of 1.8 m have also been built. It should be noted that

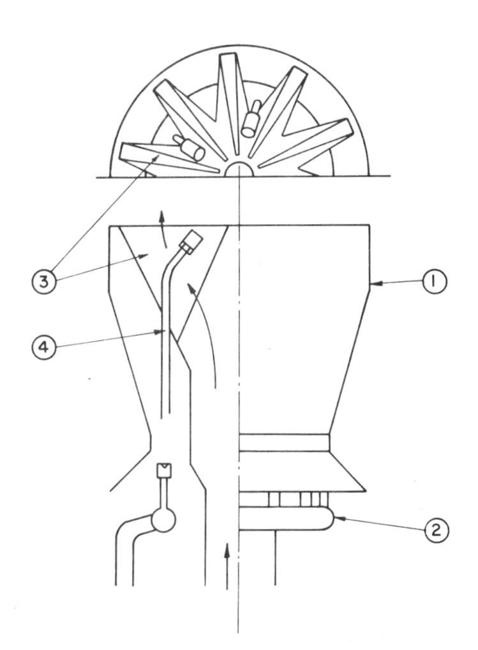

FIG. 4. Principal elements of the Smoke-Ban Model SVL flare tip: (1) shroud, (2) steam-ring header, (3) hollow spokes carrying gas from the central duct to the radial slit nozzles, (4) pilot light.

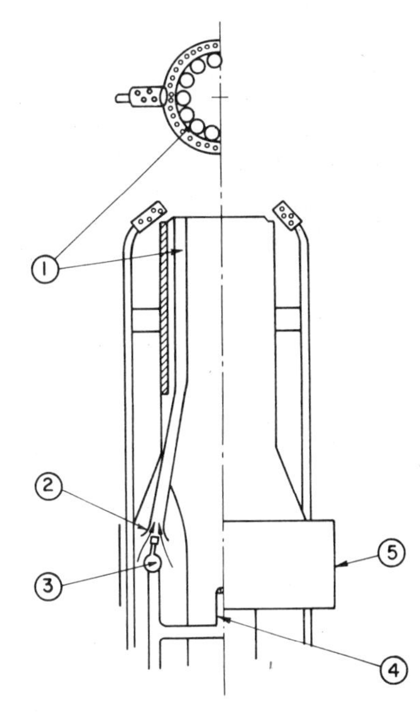

FIG. 5. Principal elements of the Zink Series SA Field Flare tip: (1) ejector tubes carrying steam and induced air into the flame, (2) ejector entrance nozzle, (3) steam-ring header and nozzles, (4) centre-steam nozzle, (5) acoustical shroud.

several of the figures are based on detailed drawings presented in the sales literature of some of the manufacturers; the rest are the author's rendition based on the far less detailed data released by the other manufacturers.

Figure 6 and Table 2 refer to ground flares. Other ground flares, in particular the "Multijet" flare developed by the (now) Exxon Research and Engineering Company, are described in detail by Jones.[1] A somewhat similar design is also offered by Alphons Custodis K.G. of Düsseldorf. Typical design heat release rates for ground flares of the type shown in Fig. 6 lie in the range from less than 1 MW to more than 100 MW, with a flare height of the order of 10–20 m.

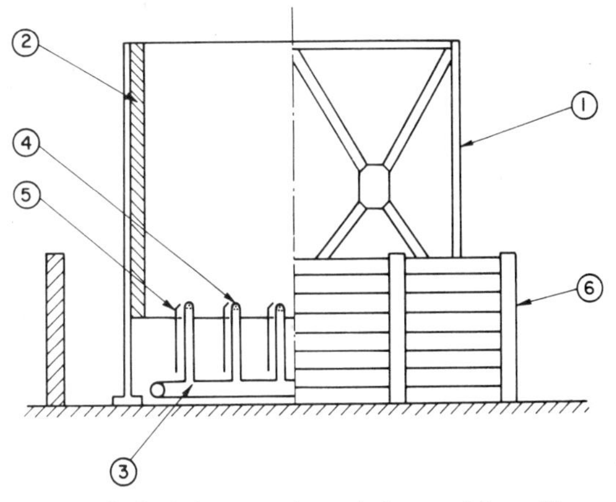

FIG. 6. Principal elements of a typical ground flare: (1) exterior frame, (2) refractory lining, (3) flare gas header, (4) flare gas burners, (5) pilot lights with an independent gas supply, (6) louvred enclosure.

TABLE 1. Tips for elevated flares

A. "Utility" flare tips	
Construction	—tip as in Fig. 1 but without centre steam nozzle 5, steam ring 6 or steam nozzles 7 —may have wind screens surrounding tip of flare
Advantages	—cheap, quiet, requires no utilities except for ignition system and gas for pilots —minimum of parts exposed to the flame
Disadvantages	—no provision for smoke suppression, can be used for flaring H_2, CH_4, H_2S and CO but even small proportions of heavy or unsaturated hydrocarbons cause smoking
Availability	—from most suppliers e.g.—Flaregas series FN —National Airoil NCG —Samia standard flare —Zink series STF-U
B. "Centre steam" flare tips	
Construction	—tip as in Fig. 1 but without steam ring 6 or steam nozzles 7 —may have wind screens surrounding tip of flare
Advantages	—cheapest provision for smoke suppression for installations in which only small quantities of heavy saturates and any unsaturates are flared —hydrocarbons and steam are well mixed at low flaring rates even in a strong wind —steam nozzle not susceptible to flame damage in the absence of steam flow
Disadvantages	—steam is not used to entrain air into flame, as a result a greater proportion of steam to hydrocarbon is required for smoke suppression than in other smokeless flares —steam noise is possible at peak steam flows, depending on location and design of steam nozzle
Availability	—from several suppliers e.g.—Flaregas series FC —National Airoil NCS —Zink (not a designated model)
C. "Steam ring" flare tips	
Construction	—tip as in Fig. 1, some designs do not include the centre steam nozzle 5
Advantages	—cheapest provision for smoke suppression for installations in which large flows of heavy saturates and unsaturates may be flared —steam is used to entrain air into the flame, lowering the proportion of steam to hydrocarbon for suppression of smoke —when fitted with a centre steam nozzle, can be used over wide range of flaring rates —this is the smokeless flare most widely used in refineries and petrochemical plants, proven designs available from several vendors
Disadvantages	—steam ring may be noisy (jet noise) at high smokeless flaring rates —steam ring alone is not very effective for low flaring rates, particularly in a strong wind, excessive steam use at these conditions may produce both jet noise and low-frequency noise (chugging) —steam ring and nozzles are susceptible to damage by flame in the absence of steam flow
Availability	—from several suppliers e.g.—National Airoil NRC —Samia smokeless flare —Zink series STF-S
D. Proprietary and special purpose flare tips	
FLAREGAS SERIES FS ANTIPOLLUTANT FLARE TIP	
Construction	—tip as in Fig. 2a. Principal feature is a perforated conical mixing section into which air is entrained by steam-driven Coanda jet ejectors (detail in Fig. 2b)
Advantages	—confined mixing of hydrocarbon with steam and induced air leads to very efficient steam use —steam wall jets in the Coanda ejectors are much quieter than sonic steam jets from nozzles in typical steam-ring designs
Disadvantages	—expensive construction —openings in mixing section may cause problems with flame stability at low flaring rates in a strong wind
INDAIR SMOKELESS FLARE (GKN Birwelco)	
Construction	—tip as in Fig. 3. Principal feature is a tulip-shaped nozzle with low-pressure gas flowing through it and high-pressure gas flowing over it. The h.p. gas remains in a wall jet and follows the curvature of the nozzle wall (Coanda effect) while entraining air. The l.p. gas burns within h.p. gas flame
Advantages	—special purpose flare for producing fields, off-shore rigs, etc., where h.p. gas is available —entrainment of air into h.p. wall jet is relatively silent and efficient, smoke is suppressed, flame luminosity is low —wall jet of h.p. gas cools nozzle wall, high temperature materials are unnecessary —may be operated vertically or horizontally, with significant liquid carry-over
Disadvantages	—cannot be used unless h.p. gas is available —development to replace h.p. gas by steam seems possible, but it may compromise combustion performance since the l.p. gas jet will not be surrounded by a flame
SMOKE-BAN MODEL SVL	
Construction	—tip as in Fig. 4. Principal feature is a converging-diverging conical shroud. The flared gas flows from the stack into a centre body from which it emerges through slots in radial spokes. Steam nozzles mounted on a steam ring under the shroud induce an air flow over the spokes inside the shroud
Advantages	—large area for mixing the hydrocarbon with air and steam —steam used effectively to induce air flow inside shroud —shroud provides some acoustical shielding for the steam nozzles —large capacity for smokeless flaring
Disadvantages	—expensive construction

TABLE 1. *Continued*

ZINK SERIES SA SMOKELESS FIELD FLARE	
Construction	—tip as in Fig. 5. Principal feature is the location of the steam ring under a shroud near the bottom of the flare tip. The steam nozzles are directed into ejector tubes which carry the induced air flow through the tip up to its exit plane, where the steam-air flow mixes with the hydrocarbon. A centre-steam nozzle is provided below the steam ring
Advantages	—large area for mixing the hydrocarbon with air and steam —steam used effectively in air ejectors —shroud provides some acoustical shielding for the steam nozzles —large maximum capacity for smokeless flaring —centre-steam nozzle gives good performance at low flaring rates
Disadvantages	—expensive construction

TABLE 2. Ground level flares

Construction	—as in Fig. 6. The principal feature is a refractory-lined elevated duct standing inside a louvred enclosure. A number of small burners is mounted on headers and provided with pilot burners using an independent gas supply
Advantages	—cheap and unobtrusive, small ground flares may be built among process units and close to property lines because the flames are not visible, combustion is smokeless and radiant heating is minimum —many well separated small flames are inherently cleaner than a single large diffusion flame —good design of the individual burners can lead to smokeless burning over a wide turn-down ratio —turn-down ratio can be increased by staging the gas flow to headers —utilities such as steam or air can be used to burn difficult materials cleanly
Disadvantages	—small capacity, compared to elevated flares. Ground flares of large capacity cover a large land area —require extreme reliability of flame-out protection when burning toxic gases or gases heavier than air since dispersion by wind at the top of a stack is not available in the event of flame-out —in many designs the burners are set up for clean combustion of specific gases, therefore the flare can be used for routine flaring associated with given process units, but not for plant-wide emergency flaring
Availability	—from most suppliers e.g.—Flaregas Ground Flares —National Airoil NPAC —Zink ZTOF (see note)

Note— The Zink ZTOF is a ground flare of large capacity in the form of a tall cylindrical stack of large diameter standing on grade. With four staged burners mounted at the base, the ZTOF is in many ways similar to an induced-draft furnace. The ZTOF is integrated with an elevated flare through a system of water seals which divert gas to the elevated flare when the ZTOF is operating at capacity. The system offers unobtrusive burning at capacity levels sufficient for routine plant-wide flaring, with the stand-by capacity of an elevated flare for emergency flaring. The one drawback of this system is its high cost.

TABLE 3. Suppliers of flare equipment

Flaregas Engineering Limited, Bentinck House, Bentinck Road, West Drayton, Middlesex, England UB7 7SJ
GKN Birwelco Limited, Mucklow House, Mucklow Hill, Halesowen, West Midlands, England B62 8DG
National Airoil Burner Company, 1284 East Sedgley Ave., Philadelphia, Pa. 19134, USA
Samia, S.p.A., 20145 Milano-Via Guerrazzi, 27, Italy
Smoke-Ban Manufacturing, Inc., 711 E. Curtis St., Pasadena, Texas, 77502, USA
John Zink Company, P. O. Box 7388, Tulsa, Oklahoma, 74105, USA

Table 3 is a listing of the principal suppliers of flare equipment. A number of other companies also build flares, but they generally use the flare tips and control systems provided by the principal suppliers.

Flares are also used in other types of chemical process plants, e.g. heavy water plants using the H_2S cycle. They are also found in steel mills where off gas from the basic oxygen furnace and other gases might have to be flared. In what follows, however, the emphasis will be on flaring in the petroleum industry.

4. CURRENT PROBLEMS IN FLARING

Three kinds of problems are associated with flaring: economic problems, environmental problems and safety problems.

One economic problem is simply the loss of valuable materials. It is generally thought that process flaring leads to a loss of an average of about 0.15–0.5% of feedstock in refineries and petrochemical plants. In a refinery of 200,000 bpd (barrels per day) capacity and with crude oil at $11 per barrel, this could mean a daily loss of as much as $11,000.

The environmental problems associated with flaring are principally those of air pollution and noise. Air pollution is primarily in the form of visible smoke. However, unburned hydrocarbons, SO_2, and NO_x are possible pollutants whose appearance is far less conspicuous. In the case of H_2S flares, the dispersion of this toxic gas as the result of any incompleteness of combustion may be a problem. The potentially far more serious problem of flame-out on an H_2S flare is very unlikely to arise because the pilot light systems used on flares have been developed to a very high level of reliability. For example, it has been shown that pilot lights on a hydrogen flare have provided ignition up to wind speeds of 134 km/h.[2] Once the flame was ignited, winds up to 176 km/h and simulated rain of 114 mm/h failed to extinguish it.

To suppress smoke formation in process flaring, it is customary to inject steam into the flared gas. Various ways of doing this were described in the Tables above. The amount of steam per unit mass of hydrocarbon depends both on the type of hydrocarbon (least for light saturated hydrocarbons, increasing with molecular weight and degree of unsaturation, most for aromatics) and on the design of the flare tips. In practice, these ratios range from some 0.2 kg steam per 1 kg hydrocarbon to over 1 kg steam per kg hydrocarbon. This process of steam addition is the principal source of noise in process flaring.

Steam injection may produce noise in two ways. The first is jet noise associated with the high-velocity steam jets. (Some of the flare tips described in the previous section have been designed to minimize this effect.) The second is a more indirect effect of the steam injector process on the acoustics of the entire flare system, including the stack, the liquid-seal drum in its base, and the flare headers. It is manifest in a sub-audible vibration at low flaring rates and high steam flows, which becomes an environmental noise problem when it rattles windows in a neighbouring community. This appears to be just one of several low-frequency noise phenomena associated with flares.

Combustion noise is not generally a problem in process flaring. However, at emergency flaring conditions when the gas discharges at some 100 m/s the flame could be noisy. It should be noted, though, that in such circumstances the most serious concern is the safety problem caused by thermal radiation from the flame. There may also be massive smoking, since it is entirely impractical to maintain a standby steam capacity of some 10^5 kg/h which can be instantly diverted to the flare during an emergency.

The principal problem in emergency flaring is one of safety. Personnel and sensitive structures must be protected from the intense radiation emitted by a long bent-over flame (flame lengths of the order of 100 m have been reported by eye witnesses, but not documented, in a number of emergencies at refineries and petrochemical plants. Grumer *et al.*[3] have reported a flame 90 m long on a flare 0.75 m in diameter discharging hydrogen at 720 m/s). One way of doing this is to "sterilize" an area surrounding the flare. At this point, the safety problem leads to an economic problem. The sterilized space can be very expensive, particularly if it lies within a plant which is surrounded by built-up areas and which is being expanded to increase capacity.

5. CURRENT STATE OF KNOWLEDGE

This description of the current state of knowledge on flaring is written with reference to the needs of the "offsite" designer who must select the diameter of the flare tip, and then decide on the height of the stack, and on the extent of the restricted area around it.

Several articles on the design of flare systems[4,5,6,7] appeared in the technical literature in the 1960s. All of them suffer to some degree from the same shortcoming, namely, presenting calculations which appear to be very precise but actually involve unproven formulae for basic quantities such as flame length, flame deflection in a wind, etc. Some of these formulae are reasonable guesses based on experience, but some others are questionable. A summary design guide for flare systems, contained in the American Petroleum Institute Guide RP 521,[8] accepts many of these recommendations uncritically. The accuracy of this document itself is questioned by Heitner.[9]

5.1. *Choice of Tip Diameter*

The choice of the flare tip diameter depends on many factors, including the performance of the pilot lights and the flame-holding characteristics of the tip selected, the composition of the gas to be flared, the availability of purge gas to keep air from entering the flare when the flow of flare gas is very low or intermittent,[10,11] and the turndown ratio required of the flare. Of these considerations, the turndown ratio appears now to be the most important factor. Modern flare tips and pilot lights appear capable of stabilizing flames at gas discharge velocities over 100 m/s. Gas seals, such as Zink's "Molecular Seal" and National Airoil's "Fluidic Seal" have reduced the required flows of purge gas to very low values.[12]

It is generally easier to predict the maximum flaring rate at emergency conditions under which particular process vessels have to be depressurized, than it is to predict the minimum rate of process flaring likely to be encountered. Nevertheless, the flare should be designed in such a way that the discharge velocity of the flare gas is as large as possible under all conditions. The reason for this requirement is aerodynamic.

The parameter which governs the behaviour of the flare gas jet in the cross-wind in the region near the flare tip is $R \equiv M_j/M_\infty = \rho_j u_j^2/\rho_\infty u_\infty^2$, the ratio of the momentum flux in the jet to the momentum flux of the wind. At sufficiently high values of R, the jet penetrates relatively far into the cross-wind before it bends over. At the same time, the entrainment of air into the jet near the tip is very rapid. At low values of R, on the other hand, the jet is sheared off at the discharge orifice and is sucked into the wake of the tip where entrainment of air is very slow and a lazy smoky flame is produced. This effect of R is shown in Figs. 7 and 8 reproduced from Reference 13.

Since R varies as $\dot{m}_j^2$ for a fixed nozzle, flare gas, and wind, a turndown ratio of 1000 between the design emergency flaring rate and the minimum expected process flaring rate through the same tip means that the flame will behave somewhat as in Fig. 8 rather than Fig. 7 much of the time. Smoke suppression under such conditions is a problem with at least the steam-ring flare tip (see Table 1-C). On the other hand, R also varies as $1/d_j^4$ at constant $\dot{m}_j$ and gas and wind conditions. This means that if two tips were used in conjunction, one of them having d_j four times smaller than the other, a turndown in flaring rate of 1000 could be accomplished at a decrease of R of only about a factor of four, provided that only the small tip was used

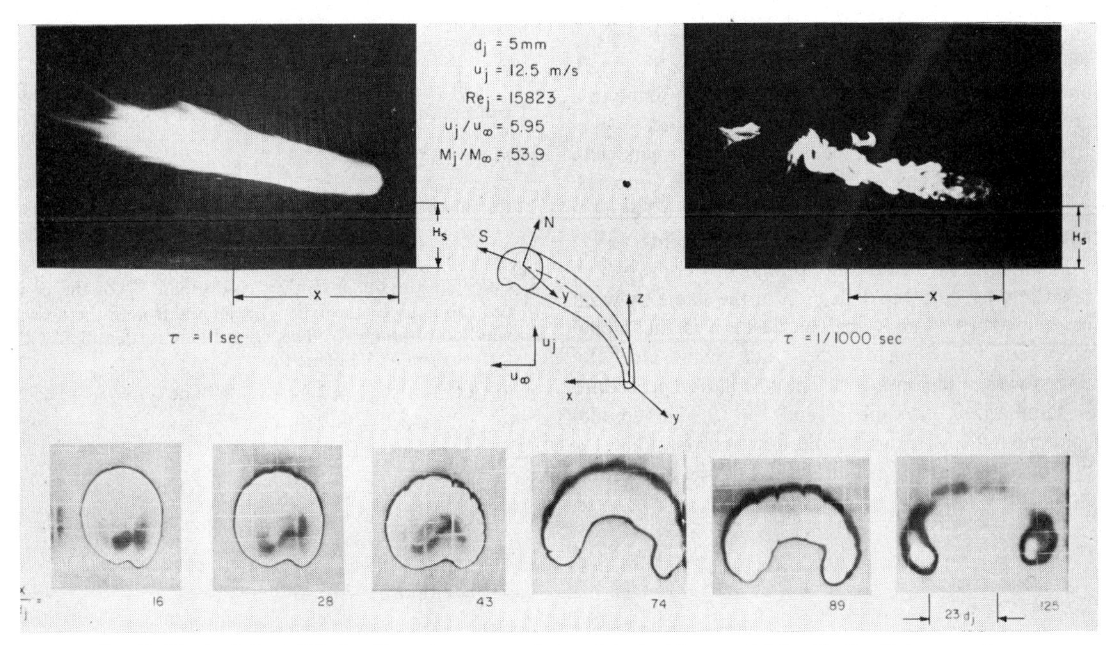

FIG. 7. The turbulent diffusion flame in a cross-wind: a detached flame at high ratio R of jet momentum flux to cross-wind momentum flux. Exposure for the flame fails to show the model flare which is located at the right end of the x interval marked under the photographs. Flame cross-sections are long-time average shapes obtained by immersing painted screens in the flame for several minutes. Temperature increases from the gray colour outside the contour, through the black outline, into the white oxidized region within. (From Reference 13.)

at low flaring rates. It is easy to see that if both tips had the same design discharge velocity, the small tip could handle up to 1/16 of design emergency flaring rate. At higher flaring rates, the excess flow could be directed to the larger flare by a system of suitable water seals. Such systems exist and are not expensive. If two tips were used in this way, the ratio of their diameters would be chosen such that all of the normal process flaring could be handled by the small flare and only a shutdown or emergency depressurization would require the large flare. Both flares could, of course, be mounted on one tower or erected as a single stack, as has often been done with so-called "sour-gas" flares alongside regular flares.

5.2. *Length and Shape of the Flame*

The height of the flare stack and the extent of any restricted area around it depends on the length and shape of the flame and the radiation from it under

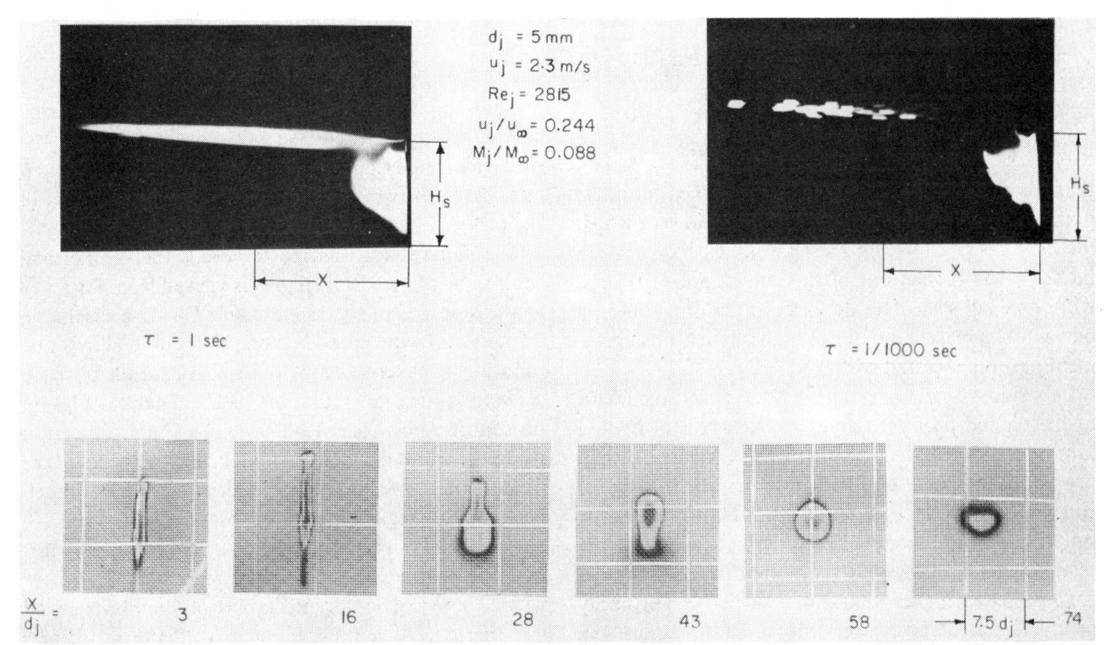

FIG. 8. The turbulent diffusion flame in a cross-wind: an attached flame at a low ratio R of jet momentum flux to cross-wind momentum flux. (From Reference 13.)

emergency flaring conditions. At the moment, there is no completely reliable method of predicting the length and shape of a full-scale turbulent diffusion flame in a cross-wind. Analytical models by Escudier[14] and Brzustowski[15] based on bent-over circular jets with top-hat profiles have yielded some insight into the problem, but much more needs to be done before a reliable predictive procedure emerges from this work.

For the moment, the only design procedure which takes into account the deflection of the flame by wind in even approximately realistic fashion is the model proposed by Brzustowski[16] and elaborated by Brzustowski and Sommer.[17] The calculation procedure is summarized in Table 4 and Fig. 9. This model combines the measured cold-flow correlations for hydrocarbon jets in cross-flow[18] with the suggestion

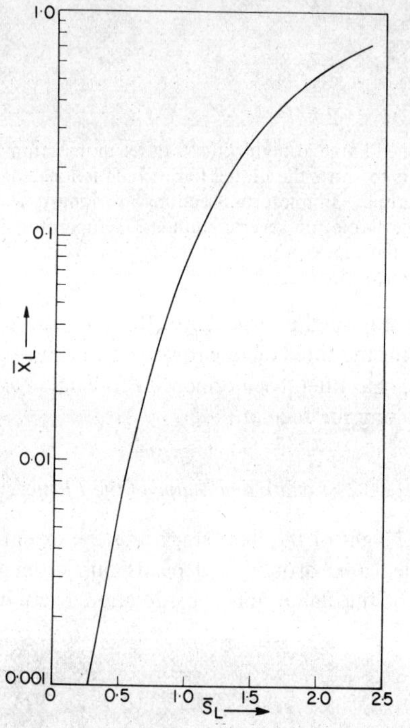

FIG. 9. The relation between $\bar{S}_L$ and $\bar{X}_L$ for $\bar{S}_L < 2.35$. For use with Table 4. (From Reference 16.)

that the end of the flame may be identified with the point on the axis of maximum concentrations at which the flare gas is diluted to the lean limit concentration in air.[19] The predictions of this model are consistent with some full-scale data on flares in the Libyan desert but, since these data are not of high quality, this agreement has to be viewed with caution.

Laboratory-scale studies have been undertaken with hydrogen and propane flames in cross-flow in a wind tunnel to establish the limits of usefulness of this design procedure. In the case of hydrogen, to which even the correlations of Reference 18 had to be extrapolated, the simple procedure seems to be usefully accurate at the relatively low values of R representative of hydrogen flaring. At the higher values of R which might apply to safety valve discharges, the model predicts the correct shape of the flame (which might arise on accidental ignition) but the flame may be too long by

TABLE 4. Calculation of flame shape and length[16]

The method is based on the cold-flow correlations of concentration in wind-blown hydrocarbon jets[18]

1. Required information: $\tilde{M}_j, \rho_j, u_j, d_j, C_L, u_\infty, \rho_\infty, \tilde{M}_\infty$

2. Calculate dimensionless lean-limit concentration of fuel
$$\bar{C}_L = C_L(u_j/u_\infty)(\tilde{M}_j/\tilde{M}_\infty)$$

3. Calculate the dimensionless co-ordinate $\bar{S}_L$ of the concentration $\bar{C}_L$ on the axis of the jet and from it the downwind co-ordinate $\bar{X}_L$. This co-ordinate is identified with the location of the flame tip
   (i) if $\bar{C}_L < 0.5$: $\bar{S}_L = 2.04\,(\bar{C}_L)^{-1.03}$ and $\bar{X}_L = \bar{S}_L - 1.65$
   (ii) if $\bar{C}_L > 0.5$: $\bar{S}_L = 2.51\,(\bar{C}_L)^{-0.625}$
   if $\bar{S}_L > 2.35$ $\bar{X}_L = \bar{S}_L - 1.65$
   if $\bar{S}_L \leqslant 2.35$ $\bar{S}_L \simeq 1.04\,\bar{X}_L^2 + 2.05\,\bar{X}_L^{0.28}$
   In this case $\bar{X}_L$ as a function of $\bar{S}_L$ is shown in Fig. 9

4. Calculate the dimensionless rise $\bar{Z}_L$ of the flame tip above the flare tip $\bar{Z}_L = 2.05\,\bar{X}_L^{0.28}$

5. Calculate the ratio R of the jet momentum flux to the cross-wind momentum flux
$$R = (\rho_j u_j^2)/(\rho_\infty u_\infty^2)$$

6. Calculate the dimensional co-ordinates of the flame tip relative to the flare tip
$$z_L = \bar{Z}_L d_j R^{1/2} \qquad x_L = \bar{X}_L d_j R^{1/2}$$

Notes: (i) This procedure should not be used for
$$u_j/u_\infty > 110$$
(ii) For a mixture of flammable gases, the lean limit can often be approximated by
$$1/C_L = \sum_{\substack{i \\ \text{components}}} (X_i/C_{L,i})$$
This equation is only an approximation. $C_L = \infty$ for inert components. Steam is not strictly an inert and the use of this simple mixing rule for hydrocarbon–steam mixtures may not be accurate.

a factor of two.[20] In the case of propane, the flame shapes are again predicted fairly accurately, but the predicted flame length is systematically low, by as much as a factor of two, except at very high ratios u_j/u_∞ of the order of 20 and greater.[21]

The flame shape referred to here is the geometrical centre-line of the visible flame as viewed from far away at the side, i.e. in the plane defined by the discharge velocity and wind speed vectors. This curve allows one to locate an effective centre of the flame for the purpose of estimating its radiation field. The calculation procedure shown in Table 4 makes no prediction whatever about the lateral extent or the thickness of the flame. The simple analytical models[14,15] are not much better in this regard, since the observed flames (Fig. 7) do not really resemble bent-over circular jets, particularly at large distances downstream from the nozzle where they degenerate into a pair of counter-rotating line thermals. The measurements of flame width and thickness[21] in the wind tunnel illustrate this point further. It should be remembered, however, that these comparisons are between a prediction procedure based on laboratory-scale correlations and other laboratory-scale data.

The use of a cold-flow correlation ignores the effects of buoyancy in a flame. These effects are much more significant in the large scale flame than in the laboratory model. They combine to shorten the flame and to raise its downstream portion.

The analytical studies of Escudier[14] and Brzustowski[15] have both shown that near the nozzle the flame rises no higher than a half-power law, specifically $\bar{Z} = 2.5\bar{X}^{1/2}$.[15] At increasing downwind distances, the flame flattens and then begins to rise again under the accumulated effect of buoyancy. Over the middle portion of the flame there is a region where $\bar{Z}$ varies approximately as $\bar{X}^{1/3}$. The prediction is at least consistent with the wind tunnel experiments,[21] since the line $\bar{Z} = 2.05\bar{X}^{0.28}$ which is used in the cold-flow model correlates with the measured centre-line quite well over a long portion of the flame. This equation was found experimentally in work on air jets[22] and again in Reference 18. It is important to note that, except near the end of the flame, the same dimensionless distances $\bar{X}$ and $\bar{Z}$ correlate the shapes of bent-over cold jets and bent-over flames on the laboratory scale. It would appear that the Froude number is, in fact, the only additional parameter required to model the shape of the flame.[15]

The reason why buoyancy shortens the flame is that it increases entrainment. It is known that the mass flow in a non-buoyant turbulent jet in still air increases in proportion with distance from the nozzle. The rate of entrainment along such a jet is constant.

In a buoyant plume in still air, the flux of buoyancy is conserved and the mass flow is proportional to $z^{5/3}$. The rate of entrainment is therefore proportional to $z^{2/3}$.[22]

In a diffusion flame, however, buoyancy is not conserved. As air is entrained into the flame, heat release occurs and buoyancy is generated. If the rate of heat release is proportional to the rate of entrainment, the average density is constant over the length of the flame. In fact, this appears to be a good model for many diffusion flames. Under these circumstances the mass flow at large distances from the nozzle, and at low Froude numbers in still air, varies as $Fr^{-1/2}z^{5/2}$, from which it follows that the rate of entrainment varies as $Fr^{-1/2}z^{3/2}$ up to the end of the flame.[24] Beyond the end of the flame, the products of combustion flow in a buoyant plume in which the flux of buoyancy is conserved. It is evident from these simple arguments that buoyancy has a significant effect on entrainment at some distance from the nozzle. There is no reason to suppose that the cross-wind obliterates this effect.

It should be pointed out that there is some difficulty in defining the parameters u_j, ρ_j and even d_j for some of the flare tips described in Table 1, e.g. the steam-ring flare or the Flaregas FS tip. Fortunately, in the event of emergency flaring when the shape of the flame needs to be known most accurately, the mass flow of flare gas is so much larger than the largest available steam flow, that the operation of the steam injection system probably does not introduce much uncertainty into the calculations.

5.3. *Flare Radiation*

To predict the thermal radiation field of flares, it is customary to model the flame by a point source located at the centre of the real flame, (i.e. halfway along the flame centre-line from the nozzle to the tip of the flame) and radiating a fraction F of the total rate of heat release. The geometrical aspects of this representation are surprisingly accurate for flare stacks, even in the case of long wind-blown flames,[17] because the flames subtend only a small solid angle as viewed from the receiving element. The errors caused by the use of a point-source model in place of a finite flame are, in fact, likely to be smaller than those arising from the uncertainty in the location of the flame. It has been generally assumed that F is a property of the fuel only,[8] but recent work on laboratory-scale flames in a wind tunnel[25] has shown that F depends on both u_j and u_∞. The behaviour is not simple, but it can be correlated with local temperature measurements. It appears that variations in F can be traced to factors which affect the rates of the competing processes of oxidation and agglomeration of the products of fuel pyrolysis which occurs near the nozzle.

The information on F cannot yet be considered definitive. For example, in the case of methane a maximum value of 0.16 in still air is listed in the American Petroleum Institute RP 521.[8] Reference 25 shows 0.155 for a jet discharging at 30.9 m/s and 0.17 for a jet discharging at 24.5 m/s, both into still air. At a cross-wind velocity of 2 m/s, the value of F has increased to 0.23 and 0.26 respectively. Recent work by Markstein[26] on propane leads to values for F which range from 0.204 to 0.246 to jets discharging into still air from nozzles of increasing diameter. However, for discharge at velocities some two orders of magnitude higher, again into still air, values of 0.17 and 0.18 are reported.[25] A cross-wind increases F for propane but not as dramatically as for methane.

The scaling of these laboratory results to full-scale flares has yet to be undertaken. Among the problems involved is the fact that the length of the pyrolysis zone is probably proportional only to the discharge velocity (multiplied by a pyrolysis time which is probably a constant for a given fuel at a given temperature) and not to the diameter of the flare tip. This is most likely the reason why full-scale flares generally appear luminous over their entire length, even though laboratory scale models discharging the same gas at the same velocity may have a largely non-luminous flame with only a short luminous portion many diameters downstream of the orifice. An added complication is the shielding effect of any mantle of smoke formed around the luminous flame by thermal quenching of the burning soot in the portion of the flame which is convected with the wind. This process is undoubtedly caused by turbulent mixing on a scale characteristic of the wind flowing over the flare tip.

There seem to be enough unanswered questions about flare radiation at this time, that the designer generally has little choice but to be conservative and use the largest value of F that has been documented for any given fuel. The exception, of course, is the rare case when reliable data exist for that fuel when it is burned in the particular flare and at the particular conditions of his design.

There is some latitude, however, with what is to be done with the predicted radiation field. As far as the spacing of equipment around the flare is concerned, a reasonable design criterion seems to be a limiting hot-spot temperature. Such temperatures must, however, be calculated by means of a realistic heat balance, with the recognition that the same wind which blows the flame towards a given structure is also likely to cool it. This point and its consequences are discussed at length in Reference 17. A similar approach is taken by Honda *et al.* in a paper[27] which also contains some measured values of radiant heat flux at grade and of the temperature at some points on the flare tower.

When safe spacing or stack heights are determined with reference to the exposure of personnel to thermal radiation, the design criterion becomes less clear-cut. Design methods found in the literature generally refer to two clinical studies[28,29] in which the experimenters exposed a blackened area of forearm (their own, presumably) to radiation from a focused projector lamp. They measured the time to the onset of pain as a function of the incident radiant heat flux, e.g. 30 s at 3 kW/m^2. The difficulty in applying such clinical data directly to the industrial situation lies in the fact that the industrial worker is generally dressed in heavy denim with a hard hat, heavy boots and gloves, all of which offer considerable protection. Moreover, he is free to take evasive action. It may well be, therefore, that the design criterion should be stated in terms of the maximum permissible temperatures of any objects (e.g. handrails, gate latches, etc.) which personnel must touch in making their escape to safety.

5.4. *Smoke Suppression*

As has already been indicated in Table 1, the most common practical method of suppressing smoke in flares is the addition of steam. This steam has the greatest effect when it is injected into the flame at/or just above the discharge plane of the flare tip, in such a way that it can also entrain some air into this region. It is a fortunate coincidence that steam, which is an effective agent for suppressing smoke in flares, is also the most readily available high-enthalpy fluid in most process plants.

When steam is not available, compressed air or even water can be used. The amounts of water required are enormous, however. Tests on a ground flare in which a water spray was used to suppress smoke in the burning of butane showed that a ratio of 14 kg water per 1 kg butane was required.[30] In the case of small portable flares, and even some larger fixed installations, a motor-driven fan can be used to supply air to the flame, but such a flare depends on the availability of electric power.

The action of steam in suppressing smoke is understood in principle, if not in detail. Obviously, if the steam is used to entrain air into the flame, the flame becomes premixed to some extent. According to Swithenbank,[31] if the amount of air entrained is sufficient to reduce the equivalence ratio from the infinite value in a fuel jet to less than 3, preferably as low as 1.5 for heavy hydrocarbons, there should be no smoke formed. One can also think of this process in terms of the influence exerted on the kinetics of chemical reactions taking place in the flame near the discharge orifice.[25] As the fuel gas pyrolyzes in a region where air and steam are being entrained, the oxidation of pyrolysis fragments proceeds simultaneously with the agglomeration of these fragments and growth of soot particles. Any factor which accelerates the oxidation reactions relative to the agglomeration reactions in the pyrolysis zone will, therefore, have the effect of suppressing smoke formation. The injection of air into the pyrolysis zone is such a factor, as demonstrated dramatically in a recent experiment by Salooja.[32] (This effect is, of course, in marked contrast to the practice of premixing a much smaller amount of oxygen with the fuel gas to increase flame luminosity, and thus radiation heat transfer, from diffusion flames in certain industrial furnaces.) Another such factor is the addition of trace amounts of metals[33] which appear to influence the rate of production of ions among the pyrolysis products. Soot formation is inhibited if the number of such ions is reduced.[34] The latter factor has led to some interesting speculation about smokeless flares using metal additives, but no practical scheme has yet emerged.

The water–gas reaction $C + H_2O \rightleftharpoons CO + H_2$ undoubtedly also occurs in these flames, but its significance in the pyrolysis zone near the nozzle cannot be established without measurements of the temperature and composition fields in that region.

5.5. *Air Pollution from Flares*

It is common practice to evaluate ground-level concentrations (GLC) of pollutants from flares by the same methods as are used for chimneys. These methods are based on turbulent diffusion from a source located at an effective stack height. They are standard and need not be reviewed here.

However, flares are not chimneys and they do behave differently. Some of these differences are sufficiently marked that they should really be incorporated in the prediction of air pollution from flares. Consider process flaring. In this case, the discharge velocity is usually quite low (unless the suggestion in Section 5.1 has been followed) and the plume is sheared off by the wind. The effective stack height is just the same as the actual height, but there are questions about just what is being dispersed. If steam is used to suppress smoke formation to the extent that no visible smoke is produced, it cannot be assumed that combustion is complete. There may be unburned gaseous hydrocarbons released from the flame, and perhaps other unburned components of the flare gas, but information on this point is lacking. If toxic gases are present in the flare gas, the necessary stack height must presumably be based on safe maximum GLC in the unlikely event of flame-out, but whether this design criterion is more severe in process flaring or under emergency conditions depends on the design of the pressure-relief system and the source of the various gas streams which flow into the flare.

In emergency flaring, quite clearly the flame does rise above the stack and it is bent over downwind very significantly. The discharge velocity is generally much higher than for chimneys, but at the same time the discharge may have a very low, zero, or even negative buoyancy if the flare gas stream is heavier than air. The positive buoyancy developed by combustion becomes significant only well along the flame. In the case of flame-out under these conditions, safe dispersion of any toxic, or non-toxic but flammable, gases depends on having the largest possible discharge velocity, which implies the smallest practical diameter of the flare tip, as well as a suitable stack height. This aspect of the design becomes particularly important when the flare gas is heavier than air. Similar considerations enter into the design of vents for discharges from safety valves which, for whatever reason, are not burned in a flare.[35]

When the flame is maintained under emergency conditions, the source of any unburned flare gases is the end of the flame, above and well downwind of the flare tip.

For this purpose, the end of the flame is defined by the relation $d\dot{m}_F/ds = 0$, $s > L$ where L is the value of the curvilinear co-ordinate along the flame axis s at the end of the flame.[36] This definition of the end of the flame may be very different from the end of the visible flame which is probably quite appropriate for the calculations of flame radiation. The end of the flame, as defined here, is the point beyond which there is no more chemical reaction by which the original effluent is consumed. Any unburned flare gas disperses from that point. It does not disperse from within the flame, which sheathes the plume of flare gas, to its environment because the flame itself acts as a sink for it. Flame-generated pollutants such as NO_x might, however, disperse from the flame outwards. To predict the dispersion of these pollutants, it might be necessary to model the flame as a line source, or even a pair of line sources as shown in Fig. 7 for the downstream region.

This discussion has had to be speculative to some extent because detailed measurements of the composition fields inside and outside turbulent diffusion flames in a cross-wind have yet to be made.

5.6. *Flare Noise*

As indicated in Section 4, noise is a problem in flaring. Several kinds of noise are associated with flares. First, one would expect jet noise and combustion roar during emergency flaring. In fact, even though both are undoubtedly present, noise is generally considered a minor problem in emergency flaring. It does not last long and it occurs at a time when the emergency situation, which was the reason for flaring, is a greater concern.

The noise which can arise during process flaring is another matter. It can be of long duration and is generally considered a nuisance. As already pointed out, the steam used to suppress smoke formation is an important source of noise. The noise may be associated with the sonic jets of steam which are ejected from the nozzles on a steam ring, particularly when the flaring rate is low and the steam jets impinge over the flare tip. Such jet noise is well understood by the designers of jet engines, and procedures are available for minimizing it by the proper design of the steam nozzles. The combustion process itself contributes little to the noise of process flaring. The flaring rates and gas discharge velocities are low, with the result that both the combustion roar and the flare jet noise are small contributions.

The most serious, and the least understood, noise of process flaring has its origin in the fact that the flare stack, the liquid-seal drum located at its base, and the flare header form a complicated acoustical system. Few, if any, documented measurements of such noise can be found, but anecdotal evidence abounds, with words like "chugging" and "banging" often used to describe the observed effects. In a useful review article, Seebold[37] describes two possible sources of unsteadiness in the burning process which might produce these effects: a sloshing vibrational mode of the water forming the seal in the seal drum (seal depths are of the order of 100–200 mm, and surface wave amplitudes are apparently of the same order), and a periodic flashback of flame resulting from air entering the stack from the top when the flaring rate is very low, i.e. the low-flow instability limit studied by Grumer *et al.*[3] Perforated baffles mounted just under the water surface in seal drums, which presumably damp out the sloshing vibrations, have been successful in eliminating low-frequency flare noise in certain installations.[37]

One can speculate further that the manometer vibrations of a slug of condensate trapped in a flare header passing under a roadway, or the vibration of a drum acting like a Helmholz resonator, or even the organ-pipe vibrations of the flare stack itself can all become sources of unsteadiness in the burning rate. Since the resonant frequencies of all these modes in many realistic systems are of the order of 1 Hz, intriguing possibilities exist for various couplings among them, which would all lead to similar effects. A possible mechanism for the so-called "chugging" is the periodic extinction and re-ignition of a fluctuating supply of flare gas, which is being diluted by a steady and excessive flow of steam.

It is clear that the elimination of the low-frequency flare noise which can occur in process flaring requires some research, including studies of the interaction between a diffusion flame and an acoustically complex fuel supply system.

5.7. *The State of Scaling Laws*

It has been pointed out in this review that there are many questions connected with flaring which cannot be answered without new information. These deal with the shape and length of the flame, the radiation field, the completeness of combustion, etc. Other questions, which have not yet been raised, deal with the effect of atmospheric stratification and the turbulence characteristics of the wind (e.g. any mechanical turbulence in the wake of a tall distillation column). Normally, one would expect to study these questions theoretically as well as in the laboratory, formulating a set of hypotheses and eventually verifying some of them and rejecting others. From such understanding would follow the scaling laws. If these, in turn, were verified in full-scale tests, one would have a reliable basis for design.

Unfortunately, full-scale testing is almost out of the question in many aspects of flaring. For example, who could seriously expect anyone to offer 50,000 kg of ethylene so that the length and shape of the flame, and its radiation field, etc., might be measured for *one set of conditions?* On the other hand, when massive flaring occurs during an emergency, it is difficult enough to make quantitative observations of any kind, let alone to have any certainty about the values of all the relevant parameters.

The author and his colleagues have attempted to make tests during the well-defined conditions of a shutdown or start-up of a refinery process unit, but even then they have found great difficulty in documenting the necessary parameters.

For such reasons, it is not likely that the design of flares will ever be approached with the kind of certainty with which the designer of a chemical reactor, of a wing, or of a pump approaches his task. Reliance will have to be placed on laboratory work and model studies, and on detailed computer models. Full-scale tests will be exceedingly rare. Pilot-scale tests, if any, will become very significant, but even they will probably continue to be expensive and rare.

5.8. *Future Prospects*

It is likely that, for the foreseeable future, the elevated flare will remain the only reliable means for the safe disposal of large amounts of gases and vapours in an emergency. To keep the flare safe for use on demand, it will be necessary to purge it continuously and keep the pilot burners lit. However, there is every indication that the amount of material to be flared even in an emergency can be significantly reduced by careful design. For example, Klooster *et al.*[38] have recently proposed a series of steps by which the maximum flaring rate in a typical refinery could be reduced from almost 2.5×10^6 kg/h to about 23% of that amount, with the added investment paying for itself in less than 4 years.

As for process flaring, much can already be done with better maintenance of instruments and of safety valves. Indeed, there is no reason to think that a combination of good plant design and good maintenance might not someday reduce process flaring to an obsolete practice, leaving only emergency flaring and its particular problems as a challenge to the engineer.

Acknowledgement—The author's research on flaring has been sponsored by the National Research Council of Canada, the Ontario Ministry of the Environment, and the Petroleum Association for the Conservation of the Canadian Environment. Their support is gratefully acknowledged.

REFERENCES

1. JONES, H. R. *Pollution Control in the Petroleum Industry*, Noyes Data Corporation, Park Ridge, New Jersey (1973).
2. LAPIN, A. Hydrogen vent flare stack performance, *Adv. Cryogen. Engng.* **21**, pp. 198–206 (1967).
3. GRUMER, J., STRASSER, A., SINGER, J. M., GUSSEY, P. M. and ROWE, V. R. Hydrogen flare stack diffusion flames, R. I. 7457, US Dept. of the Interior, Bureau of Mines, Washington (1970).
4. HAJEK, J. D. and LUDWIG, E. E. How to design safe flare stacks, *Petro/Chem. Engr.*, Pt. 1, **XXXII**, C31–38 (1960) and Pt. 2, **XXXII**, C44–51 (1960).
5. KENT, G. R. Practical design of flare stacks, *Hydrocarb. Process. Petrol. Refin.* **43**, 121–125 (1964).
6. TAN, S. H. Flare system design simplified, *Hydrocarb. Process.* **46**, 172–176 (1976).
7. KENT, G. R. Find radiation effect of flares, *Hydrocarb. Process.* **47**, 119–130 (1968).
8. ANONYMOUS. Guide for pressure relief and depressuring systems, R. P. 521, American Petroleum Institute, Div. of Refining (1969).
9. HEITNER, I. A critical look at API RP 521, *Hydrocarb. Process.* **49**, 209–212 (1970).
10. HUSA, H. W. How to compute safe purge rates, *Hydrocarb. Process. Petrol. Refin.* **43**, 179–182 (1964).
11. KILBY, J. L. Flare system explosion, *Chem. Engng. Prog.* **64**, 49–52 (1968).
12. REED, R. D. Design and operation of flare systems, *Chem. Engng. Progr.* **64**, 53–57 (1968).
13. BRZUSTOWSKI, T. A., GOLLAHALLI, S. R. and SULLIVAN, H. F. Experimental studies on hydrocarbon turbulent diffusion flames in a cross-wind, *Proc. Second European Combustion Symposium*, pp. 739–744, The Combustion Institute (1975).
14. ESCUDIER, M. P. Aerodynamics of a burning turbulent gas jet in a crossflow, *Combust. Sci. Technol.* **4**, 293–301 (1972).
15. BRZUSTOWSKI, T. A. Turbulent diffusion flame models—III: The buoyant flame in a cross-wind, *Proc. 5th Canadian Congress of Applied Mechanics*, University of New Brunswick, Fredericton (1975).
16. BRZUSTOWSKI, T. A. A model for predicting the shapes and lengths of turbulent diffusion flames over elevated industrial flares, paper presented at 22nd Canadian Chem. Eng. Conference (1972).
17. BRZUSTOWSKI, T. A. and SOMMER, E. C., JR. Predicting radiant heating from flares, A.P.I. Preprint, No. 64–73, also *Proc. Div. of Refining*, A.P.I., **53**, 865–893 (1973).
18. HOEHNE, V. O. and LUCE, R. G. The effect of velocity temperature and molecular weight on flammability limits in wind-blown jets of hydrocarbon gases, A.P.I. Preprint, No. 56–70 (1970).
19. BRZUSTOWSKI, T. A. A new criterion for the length of a gaseous turbulent diffusion flame, *Combust. Sci. Technol.* **6**, 313–319 (1973).
20. BRZUSTOWSKI, T. A., GOLLAHALLI, S. R. and SULLIVAN, H. F. The turbulent hydrogen diffusion flame in a cross-wind, *Combust. Sci. Technol.* **11**, 29–33 (1975).
21. BRZUSTOWSKI, T. A., GOLLAHALLI, S. R. and SULLIVAN,

H. F. Characteristics of a turbulent propane flame in a cross-wind, *Trans. Can. Soc. of Mech. Engng* (1976) in press.
22. PRATTE, B. D. and BAINES, W. D. Profiles of the round turbulent jet in cross-flow, *Proc. A.S.C.E.*, **HY6**, 53–64 (1967).
23. SCORER, R. S. *Natural Aerodynamics*, Pergamon Press, London (1958).
24. BRZUSTOWSKI, T. A. Turbulent diffusion flame models II: the effects of buoyancy, *Spalanie, Archive of Combustion Processes, Polish Acad. Sci.* **4**, 345–355 (1973).
25. BRZUSTOWSKI, T. A., GOLLAHALLI, S. R., GUPTA, M. P., KAPTEIN, M. and SULLIVAN, H. F. Radiant heating from flares, A.S.M.E. paper 75-HT-4 (1975).
26. MARKSTEIN, G. H. Radiative energy transfer from turbulent diffusion flames, Technical Rept. FMRC Serial No. 22361-2, Factory Mutual Research Corp (1975); also A.S.M.E. paper 75-HT-7 (1975).
27. HONDA, T., MAKIHATA, T., YOGUCHI, M., ITO, T. and TANINAKA, I. Study of large flare stacks: radiation characteristics of flame, *Jap. Petrol. Inst.* **14**, 248–251 (1971).
28. BUETTNER, K. Heat transfer and safe exposure time for man in extreme thermal environment, A.S.M.E. paper 57-SA-20 (1957).
29. STOLL, A. M. and GREENE, L. C., Radiation burns, *Mech. Engng.* **81**, 74 (1959); also The Production of Burns by Thermal Radiation of Medium Intensity, A.S.M.E. paper 58-A-219 (1958).
30. VANDERLINE, L. G. Smokeless flares, *Hydrocarb. Process.* **53**, 99–104 (1974).
31. SWITHENBANK, J. Ecological aspects of combustion devices (with reference to hydrocarbon flaring), *A.I.Ch.E. Jl*, **18**, 553–560 (1972).
32. SALOOJA, K. C. Improved control of smoke, combustion efficiency and flame stability, *J. Inst. Fuel* **XLVII**, 203 (1974).
33. SALOOJA, K. C. Carbon formation in flames: control by novel catalytic means, in *Combustion Institute European Symposium 1973*, pp. 400–405, WEINBERG, F. (ed.); Academic Press, London (1973).
34. BULEWICZ, E. M., EVANS, D. G. and PADLEY, P. J. Effect of metallic additives on soot formation processes in flames, *Fifteenth Symposium (International) on Combustion*, pp. 1461–1470, The Combustion Institute, Pittsburgh (1975).
35. BODURTHA, F. T., PALMER, P. A. and WALSH, W. H. Discharge of heavy gases from relief valves, *Chem. Engng. Prog.* **69**, 37–41 (1973).
36. BRZUSTOWSKI, T. A. Turbulent diffusion flame models I —closed form results, in *Combustion Institute European Symposium 1973*, pp. 595–600, WEINBERG, F. (ed.). Academic Press, London (1973).
37. SEEBOLD, J. G., Flare noise: causes and cures, *Hydrocarb. Process.* **51**, 143–147 (1972).
38. KLOOSTER, H. J., VOGT, G. A. and BRAUN, G. F. Optimizing the design of relief and flare systems, *Chem. Engng. Prog.* **71**, 39–44 (1975).

Manuscript received, 9 March 1976

Prog. Energy Combust. Sci., Vol. 2, pp. 143–165, 1976. Pergamon Press. Printed in Great Britain

TURBULENCE AND TURBULENT COMBUSTION IN SPARK-IGNITION ENGINES

RODNEY J. TABACZYNSKI

Energy Laboratory, Massachusetts Institute of Technology, Cambridge, Massachusetts 02139, USA

1. INTRODUCTION

The turbulent flow field in an engine plays an important role in determining its combustion characteristics and thermal efficiency. Automotive engineers have learned that changes in the combustion chamber shape and inlet system geometry, both of which change the turbulent flow field, influence emissions, fuel economy and the lean operating limit of an engine. Most of this knowledge has been obtained on specific engines through direct experimentation or from global measurements. As a result there exist no general scaling laws to predict the combustion and emission characteristics of an engine.

The empirical method of developing the spark-ignition engine has resulted in a highly refined machine. This method of development proved to be very acceptable prior to the legislation of rigid emission standards and the possibility of legislation of fuel economy standards. Today, the cost of an engine development program can no longer be measured only by the capital outlay in equipment but by the number of months and years before an engine concept can be demonstrated, either successfully or unsuccessfully.

In an effort to shorten development times, researchers have developed analytical techniques to predict the performance and emissions of spark-ignition engines. These techniques show promise for evaluating new design concepts and as research tools for understanding more about engine experiments. As yet these analytical techniques are not totally predictive. A major input into the models is a statement on the turbulence structure in the engine cylinder. In fact, details of the turbulence structure in the engine are needed for determining heat transfer rates, ignition delay times, minimum ignition energy and the rate of mixing and burnup of quench layers.

The importance of the turbulence structure in an engine has been recognized since the early experiments of Clark,[1] in which the intake event was eliminated and the rate of flame propagation decreased. However, the lack of an adequate measuring instrument has made measurements of turbulence quantities in an engine difficult at best. As a result, relatively few investigators have been active in this field and few, if any, general scaling laws governing turbulence characteristics have been developed.

The turbulent field in an engine is produced by high shear flows that occur during the intake process and/or near top dead center (TDC) of the compression stroke for engines that have large squish regions (i.e. a region of small clearance between the cylinder head and the piston crown). Some general features of the turbulent flow in an engine cylinder can be shown with the aid of Fig. 1.[2,3] In this figure, the relative turbulent intensity $u'/\overline{U}$ is plotted versus crank angle. These results show two distinct features of engine turbulence, (1) the flow field is periodic; and (2) the turbulent fluctuations are of the same order as the mean flow. The periodicity of the flow necessitates the use of ensemble averaging techniques where the ensemble is composed of measurements performed at the same point in the engine cycle for many engine revolutions.

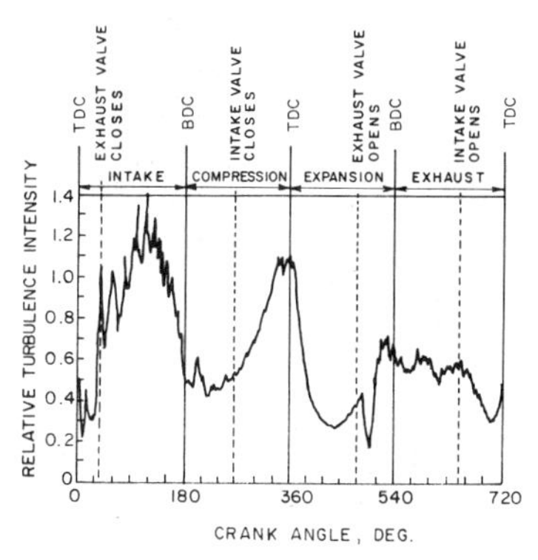

FIG. 1. Relative turbulence intensity, $u'/\overline{U}$ versus crank angle (from Reference 3).

Conventional time averaging techniques are still useful for determining turbulence quantities when the flow can be considered quasi-steady and are the simplest methods for obtaining correlations from which integral time scales and micro time scales can be determined. The high relative turbulent intensities encountered in engine turbulence measurements present a more serious problem in data discrimination and interpretation. This difficulty is primarily due to the use of the hot wire anemometer as the measuring instrument. In fact, it will be shown that a correct interpretation of the hot wire data leads to the conclusion that no mean flow exists, for many engine geometries, near TDC of the compression stroke and into the expansion stroke. Hopefully the use of the laser doppler anemometer, which measures the fluctuating velocity directly and is independent of temperature and pressure, for engine turbulence measurements will quantify the results obtained to data and provide accurate values of the fluctuating velocity.

This paper reviews the features of turbulent flow in an engine during the intake and compression processes and the influence this turbulent field has on the combustion process. An attempt is made to relate conventional turbulence theory to engine turbulence measurements wherever possible. Hence, correlations for certain turbulence characteristics with engine geometry and operating variables can be identified. This review considers only combustion models that have been applied to combustion in closed vessels or engines. For a more complete description of turbulent combustion theories the reader is referred to the recent review by Andrews *et al.*[4]

2. DEFINITION OF TURBULENCE QUANTITIES

To characterize a turbulent flow field in an engine three basic quantities should be measured; the mean velocity $\overline{U}$, the fluctuating velocity u, and the characteristic length or integral scale L. Since an engine is a periodic device which may not be quasi-steady, conventional time averaging is generally not applicable. It is necessary to consider ensemble averaging procedures, where the average is formed from a large number of repeated experiments. Witze[2,3] presents an excellent description of the ensemble averaging techniques and his definitions will be stated here. The instantaneous velocity $U(\phi)$ in the cylinder at any crank angle ϕ is expressed as:

$$U(\phi) = \overline{U}(\phi) + u(\phi) \tag{1}$$

where

$$\overline{U}(\phi) = \frac{1}{N}\sum_{i=0}^{N-1} U(\phi + in\pi) \tag{2}$$

where n is the number of strokes in an engine cyle, $\overline{U}$ the mean velocity and u the fluctuating velocity. Similarly the turbulence intensity u' can be determined by

$$u'(\phi) = \sqrt{\left(\frac{1}{N}\sum_{i=0}^{N-1}\left[U^2(\phi + in\pi) - \overline{U}^2(\phi)\right]\right)}. \tag{3}$$

Determining the integral or macro length, L, scale of the flow in an engine is considerably more difficult. The integral length scale of turbulence is defined as the integral of the autocorrelation coefficient of the fluctuating velocity at two adjacent points in the flow with respect to a variable distance between the points and is a measure of the large scale structure of the flow field.

$$L_x = \int_0^\infty R_x \, dx \tag{4}$$

where

$$R_x = \frac{1}{N}\sum_{i=0}^{N-1} \frac{u(x_0)u(x)}{u'_0 u'_x}. \tag{5}$$

This technique for measuring the integral scale requires the use of two probes and the correlation of large amounts of data. Due to the difficulty of such a technique most investigators who have determined length scales in engines have employed correlations to determine the integral time scale.[2,5–9] The integral time scale of turbulence L_t is defined as a correlation between two velocities at a fixed point in space, but separated in time.

$$L_t = \int_0^\infty R_t \, dt \tag{6}$$

where

$$R_t = \frac{1}{N}\sum_{i=1}^{N} \frac{u(\tau)u(\tau+t)}{u'_\tau u'_{\tau+t}}. \tag{7}$$

This time scale can be determined using one hot wire probe and the non-stationary autocorrelation coefficient R_t. The length scales are then determined via the relationship:

$$L_x = \overline{U} L_t. \tag{8}$$

Equation (8) can be applied to determine integral length scales if the mean flow velocity is constant, the turbulence is homogeneous, and the relative turbulent intensity $u'/\overline{U}$ is small compared to one. An additional assumption that the flow be quasi-steady is also necessary if the integral time scale is determined using the stationary correlation coefficient and conventional time averaging. The relationship between the time scale and the length scale must be further scrutinized since there is a possibility that no mean flow exists in the chamber near TDC of the compression stroke. If no mean flow exists, the relationship given by eqn. (8) is not valid and an expression of the form

$$L_x = Cu'L_t \tag{9}$$

where C is a constant of order one, is more appropriate.

Another scale that is commonly referred to is the Taylor microscale. This scale is determined from the correlation curve produced by expressions (4) and (5). A typical correlation coefficient curve is shown in Fig. 2. The Taylor microscale, λ, is defined by the second

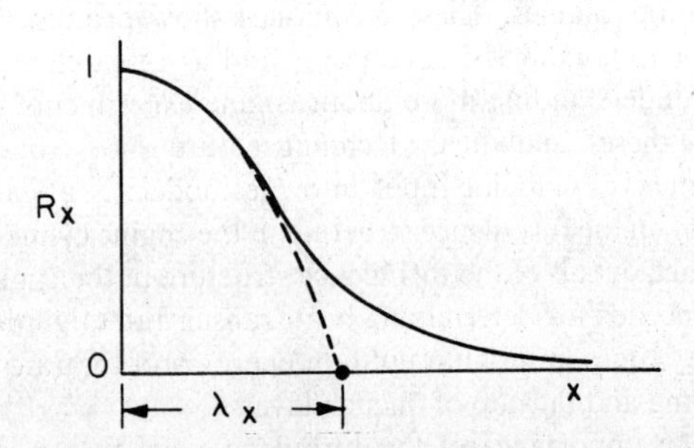

FIG. 2. Correlation coefficient R_x and estimate of λ_x (from Reference 6).

derivative of the correlation coefficient R evaluated at the origin x_0 or t_0. Equations (10) and (11) are the definitions of the micro length scale and time scale respectively.

$$\lambda_x^2 = -2/(\partial^2 R_x/\partial x^2)|x_0 \tag{10}$$

$$\lambda_t^2 = -2/(\partial^2 R_t/\partial t^2)|t_0. \tag{11}$$

This microscale is a measure of the fluctuating strain rate of the turbulent field $[(\partial u/\partial x) \approx (u'/\lambda)]$ and is thought to be important in the ignition delay problem.[5,6,10]

In order to obtain a better understanding of the production and dissipation of turbulent kinetic energy in an engine, it is helpful to determine the kinetic energy as a function of frequency at various times (or crank angles) in the engine cycle. The distribution of turbulent kinetic energy with respect to frequency is called the power spectral density function $E(\kappa)$ where κ is the wave number. If a flow is stationary the power spectrum can be obtained by subtracting the mean velocity from the instantaneous velocity at each crank angle and squaring each remainder. The resulting array is a time-series of data which can be analyzed for its frequency components using Fourier transforms. If the flow is non-stationary the power spectrum can be obtained from a Fourier transform of the autocorrelation coefficient R. A comparison of the experimental power spectrum can then be made with power spectra resulting from classical turbulence theory. Kolmogorov[11] developed an expression for the power spectrum using the concept of local isotropy which implies that small eddies respond quickly to changes in the mean flow. Hence, small eddies are in approximate "equilibrium" with local conditions. The range of wave numbers over which local isotropy is valid is called the "equilibrium range". In this equilibrium range Kolmogorov derived the following expression for the power spectrum:

$$E(\kappa) = \alpha\varepsilon^{2/3}\kappa^{-5/3} \tag{12}$$

where α is a constant, ε a measure of the dissipation of turbulent kinetic energy, and κ the wave number. Since the turbulent flow field is in a state of decay after the inlet valve is closed, the energy spectrum for turbulence behind a grid could also be of interest. A comparison of theoretical and experimental spectra should yield insight into the local isotropy of the flow, the distribution of frequencies as a function of crank angle and the decay rate of the turbulent kinetic energy.

2.1. *Turbulence Measurements*

General features

The general features of the turbulent flow field in an engine are represented by the mean flow velocity and turbulent intensity shown in Fig. 3.[2,3] One important feature of the flow is the large mean velocities and turbulent intensities generated during the intake process. The intake process generates a highly turbulent flow ($u'/\overline{U} > 0.2$) and is the source of the turbulence that persists throughout the compression and expansion phases of the engine cycle. Another general feature of the flow is the rapid decay of both mean flow and turbulent intensity as the piston velocity approaches zero and the intake valve closes. This occurs since there is little or no turbulent production at this point in the cycle and the dissipation time for the turbulent kinetic energy is of the order of a millisecond.

The compression and expansion phases are highly dependent on combustion chamber shape. For disk chambers with no squish or swirl (a large scale motion produced by intake system geometry) the dissipation process generally continues but at a much slower rate than during intake. For chambers with squish or swirl the structure of the turbulent field is similar to that shown in Fig. 3. The squish region generates high

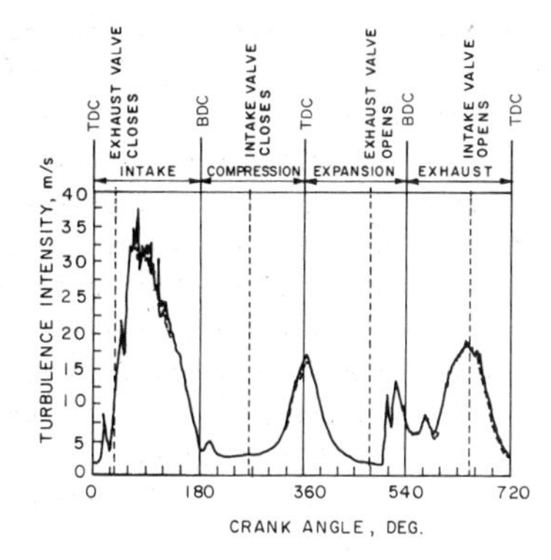

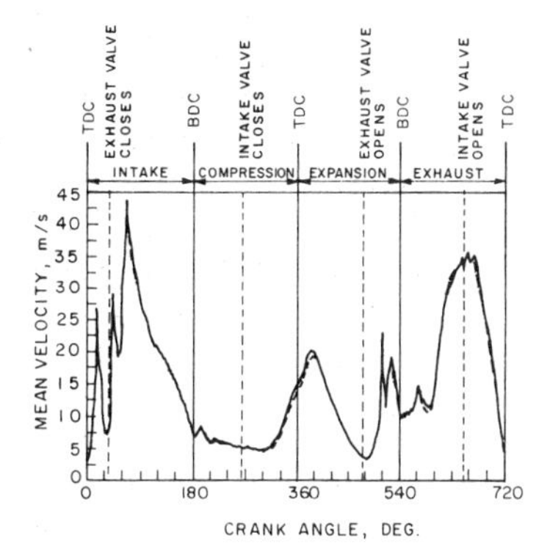

FIG. 3. Turbulence intensity u' and mean velocity $\overline{U}$ averaged over 1000 cycles, dashed curve represents separate run demonstrating repeatability of results (from Reference 3).

velocities when the piston approaches TDC and there is turbulence production resulting in an increase in the turbulence intensity.

During the expansion phase the decay process continues up to the time of exhaust valve opening. As the exhaust valve opens a boundary layer sink flow develops and high mean flows and turbulence levels are recorded. Two peaks in velocity occur during exhaust; an initial peak as the exhaust valve opens, and a second peak near maximum piston velocity (630° crank angle).

Although the flow field in an engine is highly turbulent, the averaged data for 1000 cycles is extremely repeatable. This implies that models of the turbulent field and combustion process that represent the averaged process should yield good agreement with engine performance data.

The following paragraphs will describe the intake and compression processes in greater detail. There will be no further discussion of the expansion and exhaust phases of the cycle. Obviously knowledge of the turbulence characteristics in these regimes is necessary if the turbulent mixing and oxidation of hydrocarbon rich boundary layers is to be quantified. Since combustion is essential to these studies, laser doppler techniques must be used for these measurements and the relevance of hot wire measurements in motored engines (the only studies performed to date) is questionable.

Intake process

The jet flow of gas through the inlet valve of an engine during the intake process is generally thought to be responsible for the production of the turbulent field in the engine cylinder. To test this hypothesis, Clark[1] performed an experiment where he observed the rapid decrease in the rate of flame propagation in an engine after the intake and exhaust processes were eliminated. Semenov[12] repeated this experiment in a motored engine and showed the rapid decay in the instantaneous velocity as the intake and exhaust process were eliminated, Fig. 4. Very few investigators have probed velocity gradients across the chambers and the definite jet nature of the intake flow process. The existence of these velocity gradients also implies that the intake process is highly anisotropic and cannot be studied by using single point measurements.

Since the mean velocity and turbulent intensity are such strong functions of probe position during intake it is not meaningful to compare absolute values of these quantities from investigator to investigator. Instead the general shape of the mean velocity and turbulent intensity versus crank angle should be compared. A more interesting quantity to discuss is the rate of energy dissipation during intake. Before this can be done an estimate or measurement of the turbulent scale must be made.

Several investigators have made measurements of the

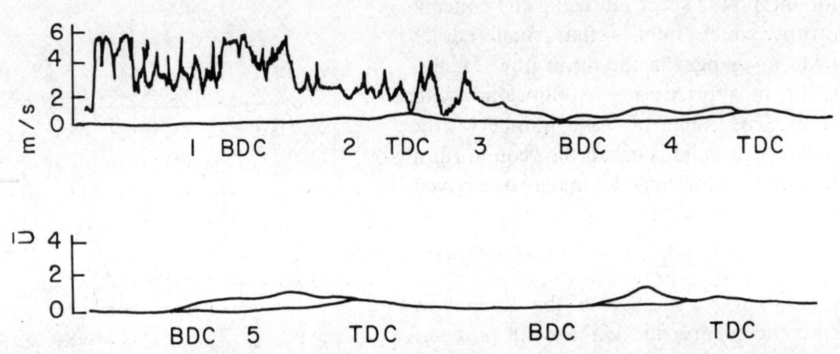

FIG. 4. Oscillogram trace of velocity during transition from normal cycles to a cycle in which intake and exhaust are absent (from Reference 12). (1) Intake; (2) compression; (3) expansion; (4) second compression; (5) third compression.

an engine cylinder to determine jet profiles. Semenov[12] is the only investigator that has made extensive measurements at different points in the chamber during intake. His measurements of the mean velocity during intake at various locations in the cylinder are shown in Fig. 5. These results show the presence of large turbulent scale in an engine during intake.[5-9,12] Semenov[12] used the measurements of the velocity gradient and the turbulent intensity to determine a length via the Prandtl mixing length theory. The resultant integral scale was of the order of 2 mm. Lancaster[7,8] and Dent and Salama,[5,6] used integral

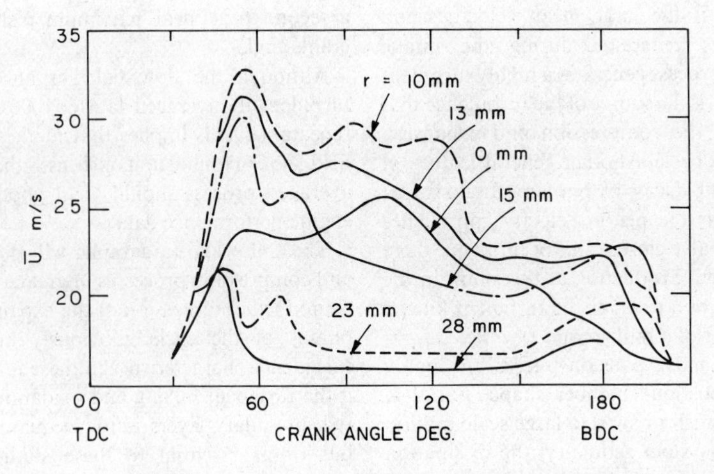

FIG. 5. Variation of mean velocity during intake at various points in the chamber (from Reference 12).

time scale measurements determined via eqn. (6) to infer a length scale. Lancaster[7,8] shows the integral scale to be from 3 to 5 mm during intake for a non-shrouded valve and 10–20 mm for a shrouded valve, Fig. 6. Dent and Salama only report values of the

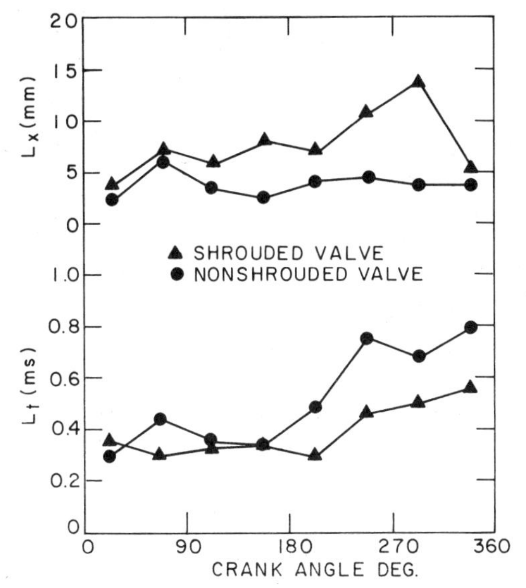

FIG. 6. Turbulent scale during intake and compression (from Reference 7).

Taylor microscale which are of the order of 0.9 mm. If the assumption of equilibrium turbulence is made this results in a value of 9 mm for the integral scale during intake. Similarly, Nanopolous[13] reports values of the integral scale on the order of the valve lift which was 5 mm for his steady flow experiment. These results imply that the integral length scale during intake is the order of the valve lift. Although the lift is a physical scale in the problem, correlations based on valve lift may be fortuitous. This is discussed further in Section 2.2.

Having determined the order of magnitude of the turbulent scale during intake it is now possible to make estimates of the dissipation rate ε where

$$\varepsilon = u'^3/L$$

and the time for an eddy to decay

$$\tau_e = L/u'.$$

Obviously such a calculation will depend on probe position, engine speed, and engine throttle position or volumetric efficiency. Typical values for ε and τ_e are $2.0 \times 10^3\ \mathrm{m^2/s^3}$ and 1 ms respectively. This implies that the anisotropic intake process is decaying rapidly relative to engine time scales.

In order to understand the decay process in more detail, Tsuge *et al.*[14,15] studied the decay of an initial anisotropic turbulent field similar to that which exists in an engine during and just after the intake process. In this work a perforated disk was pulled across a cylinder to create the initial turbulent field simulating the intake process. The turbulent intensities and time scales were analyzed as the flow relaxed. Dent and Salama[5,6] compared engine results for the micro time scale with the results of Tsuge *et al.*[14,15] These results, shown in Fig. 7, illustrate the similarity of the experiments. Tsuge *et al.*[14,15] classified the flow into

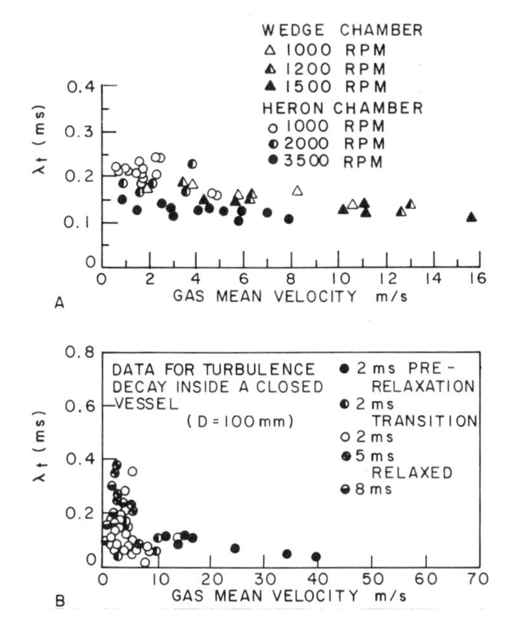

FIG. 7. Variation of micro time scale with gas mean velocity. (A) Engine data; (B) closed vessel data (from Reference 6).

three regimes, (1) the pre-relaxed region where the influence of the jet motion is predominant; (2) a transition region; and (3) a relaxed region where the turbulent structure is influenced by the dimensions of the chamber. The energy spectra for the relaxed and pre-relaxed regions are shown in Fig. 8. Typically the energy is contained in eddies of the higher frequency components (1000 Hz) during the pre-relaxed stage and low frequency (250 Hz) eddies for the relaxed portion of the flow process. Similar results have been obtained for engines. Figure 9 shows engine data from Dent and Salama.[5] These data show the same trend from intake

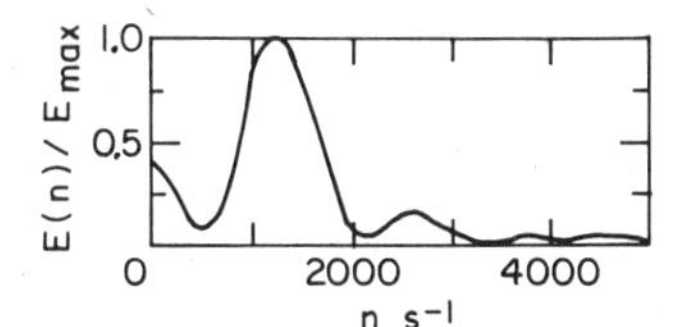

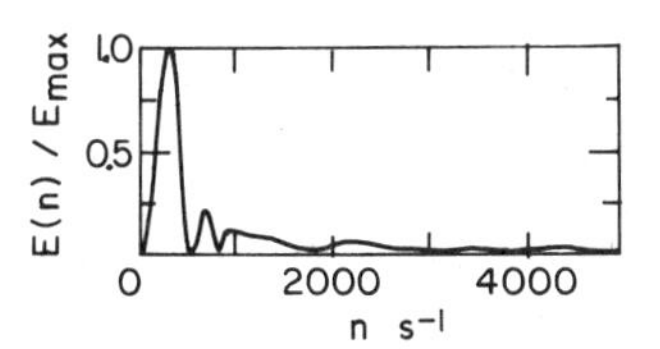

FIG. 8. Typical examples of energy spectrum function for a closed vessel. (a) Pre-relaxed stage; (b) relaxed stage (from Reference 5).

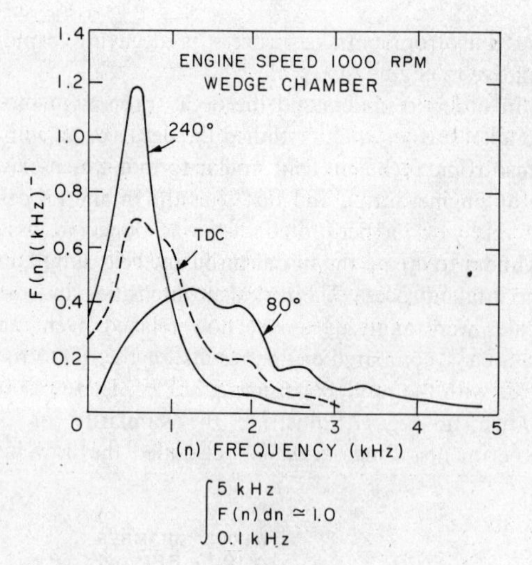

FIG. 9. Spectral density distribution at 3 crank angle positions in the engine cycle (from Reference 6).

to compression as Tsuge *et al*,[14,15] and indicate that the flow is relaxing to the shape of the chamber during the compression process.

Compression process

Since combustion occurs in the vicinity of TDC during the compression process, it is not surprising that most of the measurements of the turbulent flow field have been concentrated on the compression process. As was mentioned previously, the flow field in the vicinity of TDC is highly dependent on combustion chamber shape and inlet swirl. Most of the early work[12,16–18] on turbulence measurements was for low turbulence chambers such as the disk chamber of the CFR engine. Some of the more recent works[5–8,19] have studied the effect of squish and inlet swirl on the turbulent field.

During the early portions of the compression process the turbulent flow field is still in a pre-relaxed state and the influence of the intake process is dominant. The mean velocity is decaying rapidly since little energy is being supplied to the flow. Similarly the turbulent intensity decreases. These features are depicted in Fig. 10 which shows some typical results for the compression process. These results show that, during the early portions of compression, combustion chamber shape, inlet swirl and valve effective area are only important in establishing the initial levels of turbulence from which the decay process begins.

As TDC is approached the flow field is influenced by the inlet swirl and the squish region in the combustion chamber. If no swirl or squish is present then the mean velocity tends to level off and then rise slightly near TDC, Fig. 11, since there is no turbulent production except due to the piston motion. The results of a calculation for the contribution of the piston motion to the turbulent kinetic energy are presented in a later section. For geometries where either large squish regions or high swirl rates exist, the turbulence produced by these techniques is much greater than that due to mean flow produced by piston motion.

The effects of swirl and squish on the mean velocity and turbulence intensity are shown in Figs. 10 and 12. Near TDC the effects of both parameters are to increase the mean flow and turbulence levels. Although the production of mean velocities and turbulence

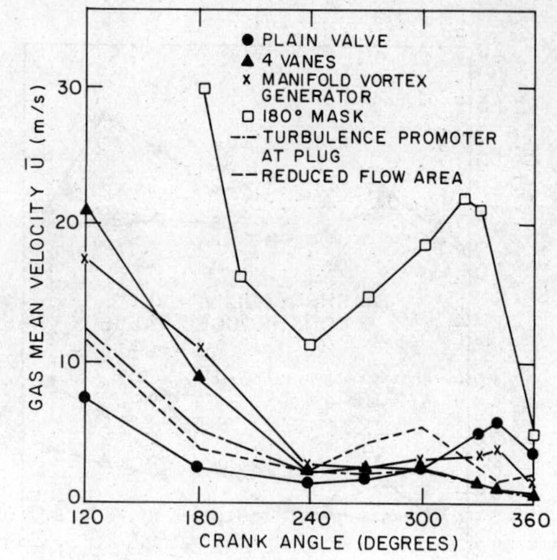

FIG. 10. Effects of various turbulence promoters on velocity variations for a wedge chamber at W.O.T. 1000 rpm (from Reference 5).

intensities near TDC due to squish is obvious, this result for swirling flow is not straightforward. If the swirling motion was similar to solid body rotation no variations in the mean velocity should occur except for

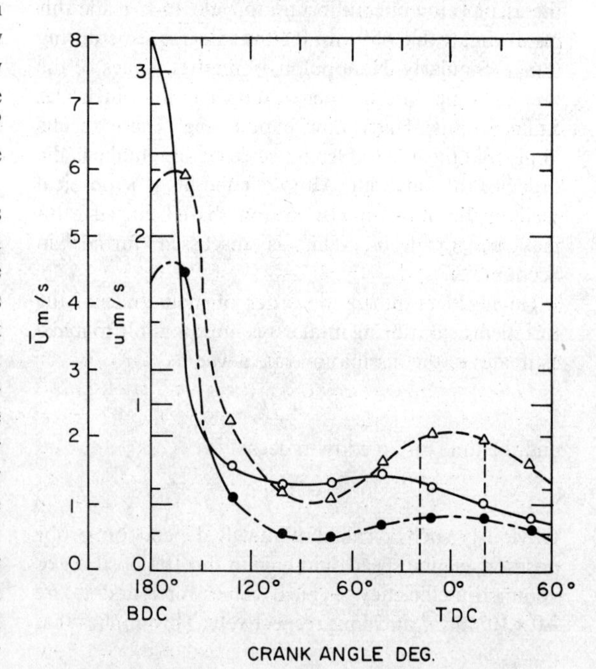

FIG. 11. Average and fluctuating velocities during intake: ● = u'; △ = ku'; ○ = $\bar{U}$ (from Reference 12).

the effects of viscous dissipation. It is evident from the results of Dent and Salama[5] and Lancaster[7] that solid body rotation does not exist and the swirling motion must be highly non-uniform. The compression process should reduce this non-uniformity. The appearance of

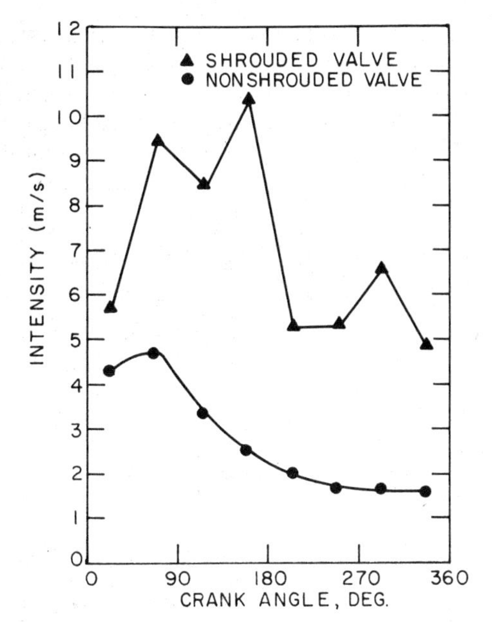

FIG. 12. Turbulence intensity during intake and compression for shrouded and nonshrouded valve geometries (from Reference 7).

an increase in the swirl component late in the compression process may be a result of a non-homogeneous velocity field. The rapid decay just prior to TDC can be due to viscous dissipation, turbulent dissipation, or non-uniformities. The most plausible explanation is; turbulent dissipation which occurs in the order of 1–3 ms or 7–21 crank angles whereas viscous times are of the order of minutes.

In addition to the magnitude of the turbulent intensity as a function of crank angle and engine geometry, the uniformity and isotropy of the turbulent flow field throughout the combustion space is important. The directional behavior of the flow field is important in determining if a mean flow exists, whereas the uniformity of u' is necessary for the isotropy and homogeneity of the flow field. The most extensive work on the distribution of turbulent intensities and mean flows across the combustion chamber is the work of Semenov[12] who determined distributions of $\overline{U}$ and u' as a function of radius. In his experiment, the mean velocity gradient across the chamber late in the compression process did not exceed $20\,s^{-1}$ compared with $600\,s^{-1}$ during intake. Lancaster[7,8] and Winsor *et al.*[18] demonstrated that near TDC a directional behavior of the flow did not exist, Fig. 13. Winsor demonstrated this for mean velocities with no swirl and Lancaster for both the shrouded and unshrouded valve geometries. These results indicate that no mean flow existed near TDC and hence the mean flows registered by the hot wire anemometer are essentially fluctuating velocities. This measurement of a mean flow where no mean flow exists is a consequence of the interpretation of the results for this measuring technique since the hot wire only measures absolute values of velocity. Dent and Salama[5] and James and Lucas[19] have also investigated the directional sensitivity of the flow field by varying the hot wire probe angles and have showed similar results for disk and wedge type chambers. For the chambers with squish a gradient in velocity across the chamber was evident. Winsor[18] varied the depth of his hot wire probe in the combustion chamber and determined no significant variation in mean velocity. Investigators have taken these results to imply that in chambers with no squish the turbulence is homogeneous and isotropic. The experimental evidence tends to support the condition of isotropy. However, there is insufficient data to imply that the turbulence is homogeneous, since a uniformity of length scale must also be demonstrated.

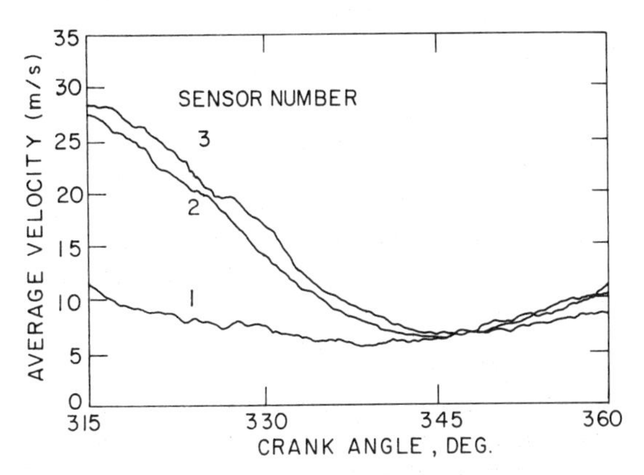

FIG. 13. Average velocity near TDC of compression stroke for 3 sensors orientation, shrouded valve, 6.84 compression ratio (from Reference 7).

Several investigators have measured turbulent length scales during the compression process[2,6,7,8] and typical results are presented in Figs. 6 and 14. The turbulent

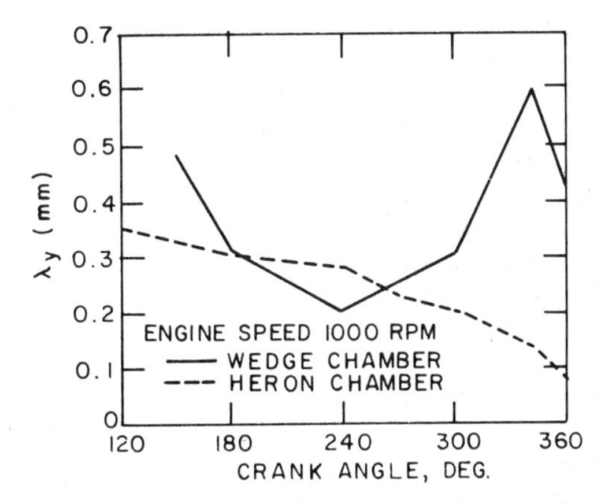

FIG. 14. Variation of spacial microscale with crank angle (from Reference 6).

scale was inferred from measurements of the correlation time and related to the length scale via eqns. (6) and (8). Dent and Salama[6] state that the integral length scale that they have determined at TDC is 4–5 times greater than the reported microscales. This is in agreement with the scales reported by other investigators.

The length scales do not change appreciably during the intake and compression process. This may imply that the time for the flow to relax to the shape of the chamber is at least of the order of the time from intake closing to TDC (see Section 2.2). Also, the correlation times determined via eqn. 6 for the various chamber shapes and inlet swirl are approximately equal. This implies that the variations in turbulent scale are directly related to variations in the turbulent velocity. Care must be exercised in comparing turbulent scales since no mean flow may exist and for this condition the relations given by eqn. (8) may not hold.

An indication of the local isotropy of the flow field can be obtained by comparing the experimental energy spectrum with the equilibrium relationship developed by Kolmogorov.[11] Lancaster[7,8] presents his results for the energy spectrum and compares them to the Kolmogorov law which has a minus 5/3 slope, Fig. 15.

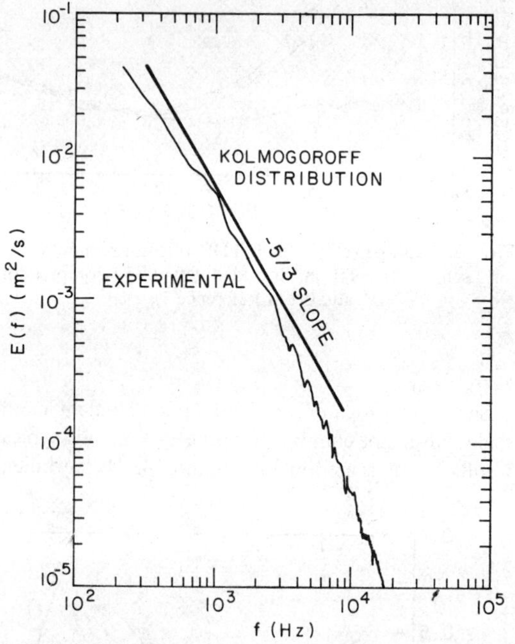

FIG. 15. Energy spectrum with shrouded intake valve compared with Kolmogorov distribution (from Reference 7).

These results do not lend convincing evidence to the concept that the flow is in local equilibrium near TDC. However, the Kolmogorov spectrum may not be the proper spectrum for comparisons.

Another method for determining if the turbulence is isotropic and the integral scale correlates with the chamber height is demonstrated in Fig. 16. In this figure the isotropic relationship between the integral scale and the Kolmogorov scale is used to correlate the experimental data. The integral scale has been replaced by the chamber height so the isotropic relationship becomes

$$h/\eta = A\left(\frac{u'h}{\nu}\right)^{0.75} \tag{13}$$

Where A is a constant order one, ν the kinematic

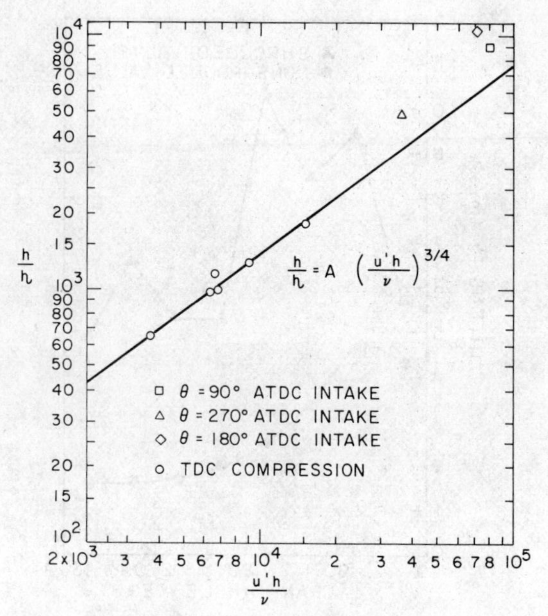

FIG. 16. Ratio of chamber height to Kolmogorov scale versus the turbulent Reynolds number.

viscosity, h the chamber height and η the Kolmogorov scale. The data of Lancaster was used to develop this figure and for the experimental data at TDC the data takes the form of eqn. (13). This implies that the integral scale is being governed by the shape of the chamber (i.e. the chamber height) and that the flow is isotropic. Tsuge[14,15] presented data in a similar format for a constant volume experiment and showed that the turbulent flow field had relaxed to the state where the chamber governed the integral scale when eqn. (13) was valid. From Fig. 16 the constant A can be determined and has the value of 1.43. This can be compared to the value of 1.73 which Tsuge determined for his geometry. Although this analysis does not prove that the flow is isotropic and the integral scale is governed by the chamber height, this is evidence supporting these statements. In the next section a derivation for the Taylor microscale will be presented using similar arguments.

Effects of engine variables

Since an engine operates over a wide range of conditions—engine speed, load and compression ratio—it is important to understand how turbulence quantities vary with these operating variables. Most investigators have concentrated their studies on engine operating variables at TDC of the compression stroke.

The most universal result is the linear variation of turbulent intensity and mean velocity with engine speed, which is independent of combustion chamber shape and inlet swirl, Fig. 17a,b. This behavior can be predicted using dimensional arguments for the rate of change of turbulent kinetic energy in the cylinder. If only one velocity and length scale were present in the cylinder then

$$\frac{dE}{dt} = \frac{-E}{\tau} \tag{14}$$

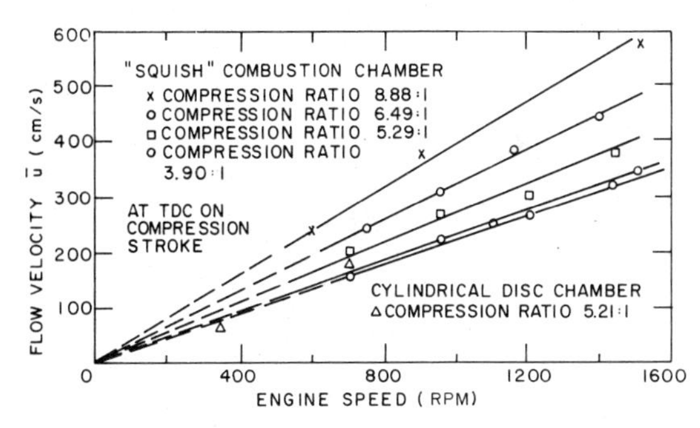

FIG. 17a. Mean flow velocity versus engine speed (from Reference 19).

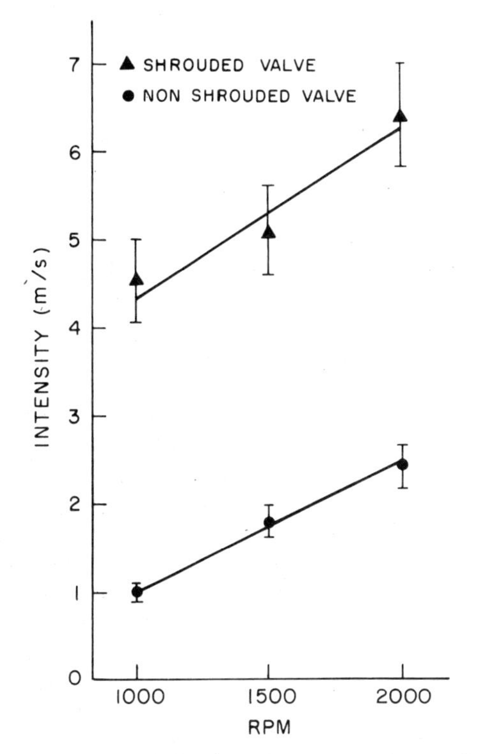

FIG. 17b. Turbulent intensity versus engine speed (from Reference 7).

where

$$E = \tfrac{1}{2}u'^2 \quad \text{and} \quad \tau = L/u' \tag{15}$$

therefore

$$-\frac{du'}{u'^2} = \frac{dt}{2L} \tag{16}$$

and

$$\frac{u'_f}{u'_i} = \left(-1 + u'_i \int_0^t \frac{dt}{2L}\right)^{-1}. \tag{17}$$

Since u'_i scales with engine speed ($u'_i = C_1$ rpm), and

$$t = C_2\,\theta/\text{rpm} \tag{18}$$

we obtain the expression for L constant

$$u'_f = C_1\,\text{rpm}/(-1 + C_1 C_2\,\theta/L) \tag{19}$$

where C_1 and C_2 are constants, θ the crank angle in degrees and i and f denote initial and final states. This expression shows how the turbulent intensity scales with engine speed. This scaling should hold even if piston motion is important for turbulent production since the velocities produced by the piston will scale with engine speed.

The variation of the turbulent length scale with engine speed is shown in Fig. 18. For the non-shrouded valve the integral length scale is not dependent on engine speed; for the shrouded valve the integral length scale decreases with engine speed, although the dependence is small. If the turbulent flow field is relaxed then the integral scale should be of the order of the height of the chamber and should not change with engine speed. These results indicate that the flow has relaxed since the dimensions of the chamber are controlling the turbulent scale. The integral time scales used to obtain the integral length scales are also shown in Fig. 18, and are essentially the same for the shrouded and non-shrouded valve geometry.

The next operating variable to consider is the compression ratio. From the previous discussions, one would expect the integral scale to decrease as compression ratio is increased if the flow is relaxed. Dent and Salama[6] show such a decrease in the Taylor microscale as compression ratio is increased. In a study of ignition delay, Blizard and Keck[10] presented a correlation for the change in turbulent scale as a function of distance from the cylinder head. This correlation can be written in terms of a compression ratio and is given as

$$\lambda = 0.17\,(\text{valve lift})/(\text{compression ratio}). \tag{20}$$

Although Keck and Blizard did not identify this scale as the Taylor microscale, Dent and Salama's[6] results for the Taylor microscale as a function of compression ratio show good agreement with eqn. (20), Fig. 19. Blizard and Keck's[10] results can be derived from standard turbulence theory if we assume that the flow is relaxed and the integral scale is the height of the chamber. Then for isotropic turbulence

$$\lambda/L = (15/C)^{1/2}(u'L/\nu)^{-1/2} \tag{21}$$

where C is a constant of order one. It remains to determine the dependence of the kinematic viscosity with compression ratio. If we assume that

$$\mu \approx T^{0.62} \tag{22}$$

and that the compression is adiabatic then

$$\nu = k_1\,CR^{-0.75} \tag{23}$$

where CR is the compression ratio and k_1 is a constant.

If we further assume that L is of the order of the chamber height h where

$$h = \frac{S}{CR-1} \tag{24}$$

and S is the stroke, then eqn. (21) can now be written as

$$\lambda = \frac{Ku'^{-1/2}CR^{-7/8}}{(1-1/CR)} \tag{25}$$

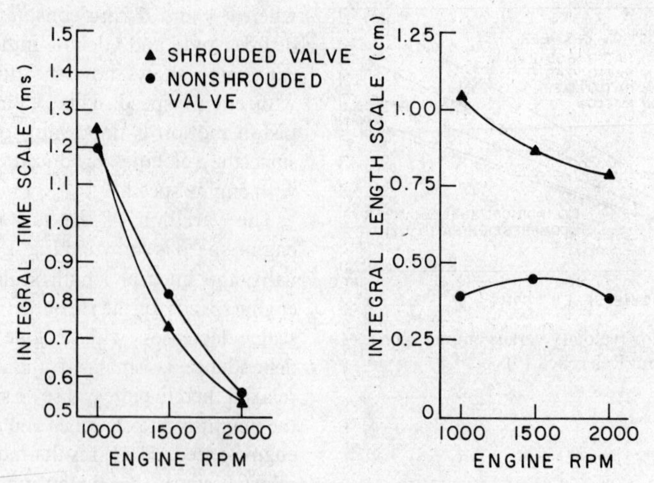

FIG. 18. Variation of turbulent integral time and length scales with engine rpm (from Reference 7).

for $CR \gg 1$

$$\lambda = Ku'^{-1/2}CR^{-7/8} \quad (26)$$

which is good agreement with Keck and Blizard's results (see Fig. 19) and provides another verification that the integral scale is of the order of the chamber height and that the turbulence is in equilibrium. Note that the constant C is related to the constant A determined using Fig. 16 and has the value of 4.

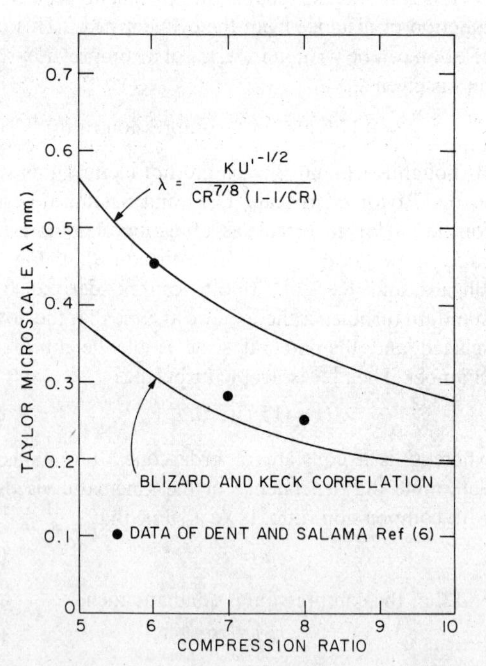

FIG. 19. Variation of spatial microscale with compression ratio (cylindrical combustion chamber) compared with Blizard and Keck correlation and the relationship derived from isotropic turbulence.

The final operating variable considered is the volumetric efficiency or load varying variable. When the intake pressure and density are reduced the flow velocities through the intake valve decrease and the volumetric efficiency of the engine decreases. This essentially reduces the amount of kinetic energy supplied to the flow. One would expect that the turbulent intensity would increase as load increases, as is shown in Fig. 20. This effect is not large for the

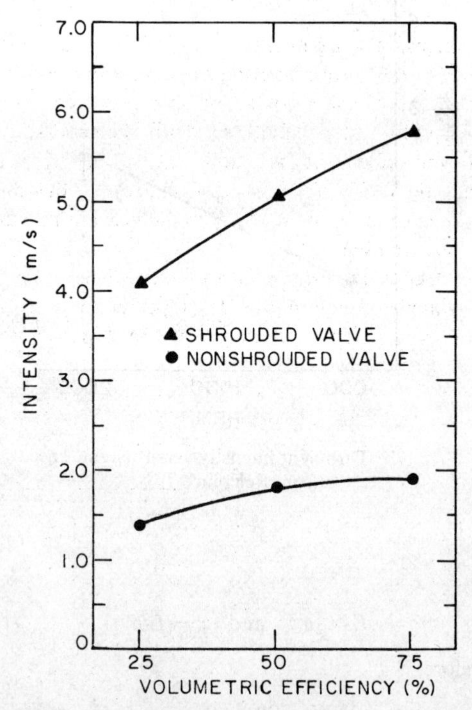

FIG. 20. Variation of turbulent intensity with volumetric efficiency (from Reference 7).

non-shrouded valve, low-turbulence chamber since there is ample time for dissipation of the kinetic energy before TDC of the compression stroke. The organized motion of swirl decays much more slowly and hence shows a much larger dependence on volumetric efficiency. The turbulent time scales and length scales are shown in Fig. 21. The time scales are nearly independent of the volumetric efficiency, whereas the

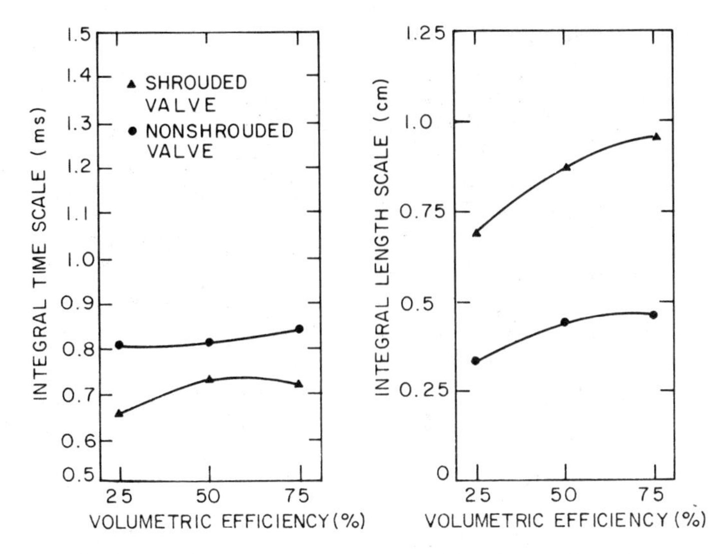

FIG. 21. Variation of turbulent scale with volumetric efficiency (from Reference 7).

length scales show a slight increase as volumetric efficiency increases.

2.2. *Analysis of Intake and Compression Process*

Dimensional arguments are used to analyze the turbulent flow process during the intake and compression process. The purpose of this discussion is to develop physical explanations for the behavior of the turbulent kinetic energy and integral length scale during the intake and compression process. Basically, the dimensional arguments presented in the previous sections are used.

First consider the turbulent kinetic energy versus crank angle. Equation (17), developed in the previous section, can be rephrased in terms of the kinetic energy E and has the form

$$E_f^{-1/2} - E_i^{-1/2} = \int_0^t \frac{dt}{\sqrt{2}\,L(t)} \tag{27}$$

In order to use this equation it is necessary to choose some value for the integral scale L. A schematic of the flow process and the possible choices for the length scale are shown in Fig. 22. The possible choices for the integral length scale are the width of the jet, b, the diameter of the intake valve, D_i, and the distance from the cylinder head to the piston face h. These length scales are used to analyze Semenov's[12] experimental data for the kinetic energy of the mean flow. The mean flow was chosen since large values of the relative turbulent intensity have been observed for turbulent flows in engines and there is evidence that the measured

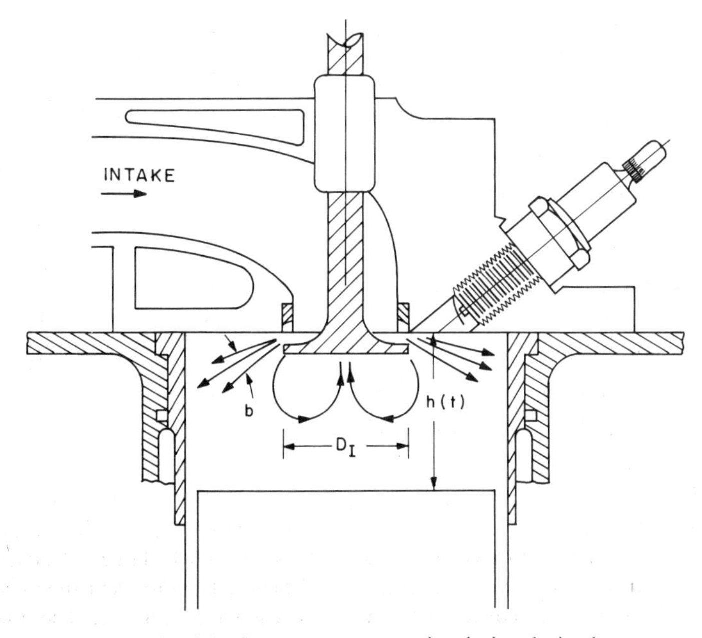

FIG. 22. Schematic of the flow processes occurring during the intake process.

mean flows are related to the actual turbulent intensity by the relationship $\overline{U} = 1.41\ u'$ (see the following section for further discussion). Semenov's[12] results and the rate of turbulent kinetic energy decay based on the various length scales are presented in Fig. 23. The initial decay process is best modeled by the length scale associated with the width of the intake jet. The decay of the kinetic energy based on the intake jet width is very rapid and no turbulence would exist at TDC of the compression process if the kinetic energy continued to decay at this rate.

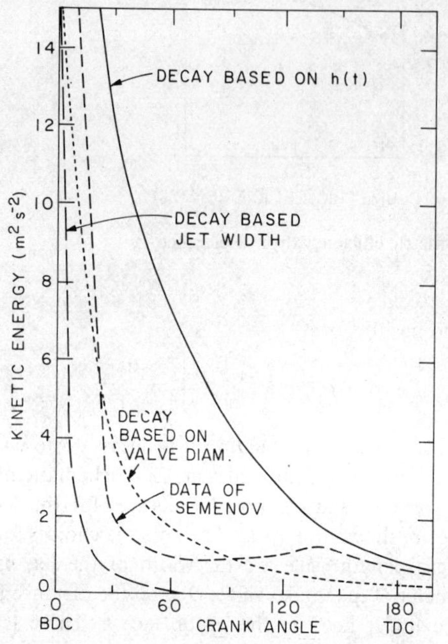

FIG. 23. Predicted decay of turbulent kinetic energy during the combustion process compared with the experimental data of Semenov (Reference 12).

There are several plausible explanations for the observed levels of turbulence at TDC. One explanation is the existence of multiple length scales which cause the kinetic energy to decay at a slower rate. Figure 23 shows that the length scales associated with the geometry of the engine result in a slower decay rate for the kinetic energy. During the intake process it is very plausible that the kinetic energy is distributed in at least the three scales mentioned. Another explanation for the observed kinetic energy history is the production of turbulence due to compression. If the eddies have a definite structure (u' and L) then as the compression process occurs the characteristic length of the eddy decreases proportionally with the volume change in order to conserve angular momentum (i.e. the rim speed of the eddy increases as L decreases). An example of this effect is shown in Fig. 24. This compression process gives values for the turbulent kinetic energy consistent with the experimental data. The mean velocity induced by the piston motion is another source of turbulence production for engines. In Fig. 25, the mean gas velocity produced by the piston motion is plotted. This mean

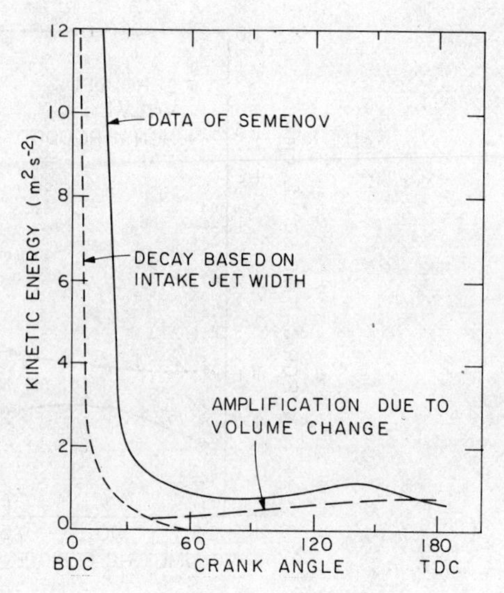

FIG. 24. Predicted turbulent kinetic energy due to compression via a volume change compared to data of Semenov (Reference 12).

velocity is obtained by assuming bulk flow for the gases in the chamber with a zero velocity at the head and the piston velocity at the face of the piston. Since the relative turbulent intensity is large it is plausible that most of the energy put into the flow by the piston motion is directly translated into turbulent kinetic energy. This process also yields turbulent kinetic energies of the correct magnitude.

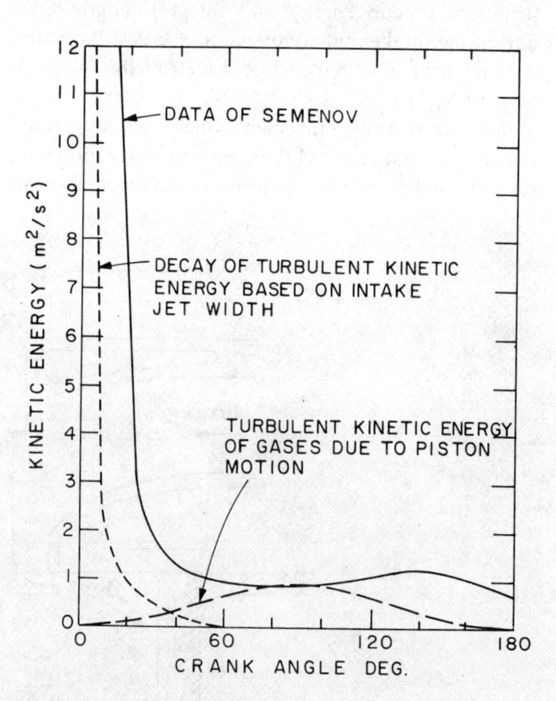

FIG. 25. Turbulent kinetic energy due to piston motion versus crank angle compared with data of Semenov (Reference 12).

The above discussion describes three different mechanisms for the production of turbulence in an engine. The critical experiments to sort out the dominant mechanisms for maintaining and producing turbulence during the compression process are yet to be performed.

Turbulent length scales

In Section 2.1, correlations for the integral length scale and Taylor microscale based on the assumption of isotropic turbulence were presented. These correlations indicated that near TDC of the compression process the integral length scale was governed by the height of the chamber. However, these correlations were not valid for earlier times in the intake and compression process. The following discussion is concerned with the observed behavior of the integral length scale versus crank angle (see Fig. 6).

During the intake process the gas flows into the engine in the form of a turbulent jet and the intake valve lift is a function of time. Although the experimental results indicate that during intake the integral scale is of the order of the maximum intake valve lift, the lift is not the governing variable for the width of the intake jet. Turbulent entrainment theory for a plane jet yields the relationship

$$b = 4\alpha X \tag{28}$$

for the jet width, b, where α is the entrainment parameter (generally $\alpha \approx 0.1$) and X the distance from the orifice. Hence it is a consequence of measurement location that the observed integral scale is of the order of the valve lift. If the hot wire probe was allowed to move with the piston or was positioned in the center of the chamber, different length scales would probably be measured.

The experimental values of the integral scale do not change appreciably throughout the intake and compression process. After the intake flow ceases, the turbulent boundary layer begins to grow in the chamber. Since the measuring location is a fixed distance from the wall, the measured integral scale is dependent on the distance from the wall as in turbulent pipe flow. Hence it is not surprising that the integral scale does not change with crank angle for the measuring locations that were selected. It will be necessary to vary the measuring location to learn more about the distribution of turbulence scales in an engine. These measurements will be necessary to determine the scales that govern the turbulent kinetic energy decay and relaxation time of the flow field.

2.3. *Hot Wire Anemometer Measuring Technique*

The hot wire anemometer is one of the standard instruments used to measure turbulence quantities in a flow field, and it is the instrument that has been used for all the turbulence measurements in engines to date. Most hot wire measurement applications are in flows that have a definite mean flow and low relative turbulent intensities. This allows the use of linearized equations for the data reduction process. Engine turbulence, as we have observed, may not have a mean flow component. Also, when a mean flow component does exist, the relative turbulent intensity is high (i.e. $u'/\bar{U} > 0.25$).

A very important consideration for measuring turbulence in an engine is probe position. Several investigators[5,7,8] have reported results where the hot wire signal was influenced by the cavity in which the hot wire probe was mounted. These investigators did not observe the rapid decay of turbulence after the intake valve closed since the recirculating flows in the cavity persisted much longer than the lifetime of a characteristic eddy. A basic experiment in which probe position in the engine is varied is necessary to check for anomalous behavior.

Few investigators have recognized the fact that no mean flow may be present in an engine and the "fictitious" mean flow deduced from the hot wire anemometer measurements was actually a result of the inability to determine flow direction from one hot wire probe. Semenov[12] is the only investigator that states that the mean flow registered in his experiments is actually a low frequency turbulent velocity. Many investigators[6–8,12,19] have made measurements in which the probe angle was varied and showed equal values for the velocity, such as in Fig. 14. These investigators used this information to claim the flow is isotropic but only Semenov[12] recognized that such a measurement implies that no mean flow exists. In a treatment of turbulence in a bomb, Semenov[20] develops a series of equations that can be used to analyze hot wire results for conditions similar to engine operation.

Several conclusions concerning the use of hot wires to measure engine turbulence can be drawn from his work. They are:

(1) Equality of hot wire anemometer measurements regardless of wire orientation implies that no mean flow exists and there is isotropic turbulence.
(2) The approximate value of the turbulent intensity if the turbulence is almost isotropic and no mean flow exists is 0.6–0.7 of the recorded value of the mean velocity.
(3) Errors in the Euler turbulence scale may result due to the low frequency hot wire response which is essentially recorded as a d.c. level. For Semenov's experiments the measured Eulerian scale was low by a factor of 3–4 as a result of these spectral distortions.

Hopefully the use of laser doppler anemometers in engines will add more quantitative information on the turbulent field in engines.

3. TURBULENT COMBUSTION AND FLAME PROPAGATION

The turbulent combustion process in an SI engine can be characterized by a plot of flame front position as a function of time. Figure 26 shows such a plot depicting the various regimes of the combustion process. The initial phase of the combustion process is characterized by its dependence on the local turbulent field at the spark plug at the time of ignition. This

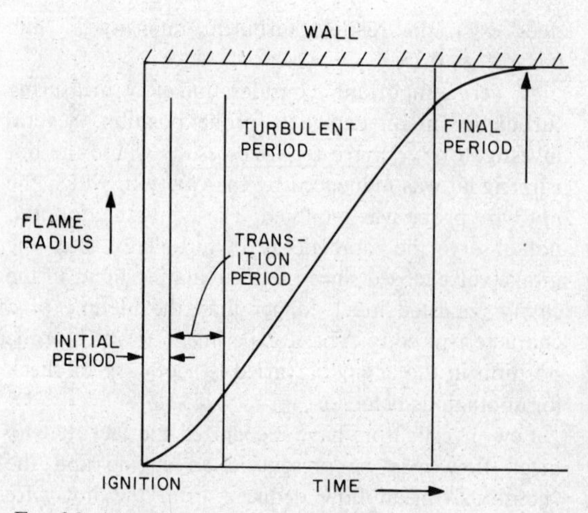

FIG. 26. Average flame front position as a function of time.

initial phase continues until several eddies, the size of the chamber height, are fully burned. Variations in this initial burning process are responsible for much of the cyclic variation observed in SI engines. The fully developed turbulent phase of the combustion process takes place over most of the chamber and is the phase in which most of the charge is burned. This phase of the combustion process can be analyzed by the average turbulence quantities since the flame front is entraining many eddies at a time as it progresses across the chamber. For this phase, the turbulent fluid mechanics is the governing mechanism for flame propagation. A final combustion phase after most of the charge is burned is also present and is dependent both on the local turbulence structure just ahead of the flame front and also on the rate of chemical reaction especially at lean conditions.

The following sections will consider the combustion process from spark initiation to completion of the combustion process. In many instances, the critical experiments have not yet been performed in an engine. In these cases the results from continuous combustion research are used to infer trends.

3.1. *The Influence of Turbulence on Ignition*

Before the turbulent flame propagation process can begin, the existence of a viable flame kernel must be established. For this reason it is of interest to know the effect of turbulent intensity and scale on the ignition process. Although the process of ignition has been studied widely, the effects of turbulence on ignition has been examined by relatively few investigators.[21–24] In general, theoretical analyses of the ignition process are a result of balancing the heat generated by ignition and combustion against the heat loss due to heat conduction to the unburned gases. Also, the experimental work that has been performed is restricted to low pressure, flowing gas streams, or combustion bombs, with approximately isotropic turbulence. Although the flow processes that are studied do not simulate those in an engine, some insight into trend behavior of the ignition process is possible.

Experimental observations

The experimental results of Ballal and Lefebvre[21] illustrate the influence of turbulence on the minimum ignition energy (Fig. 27). These results show the

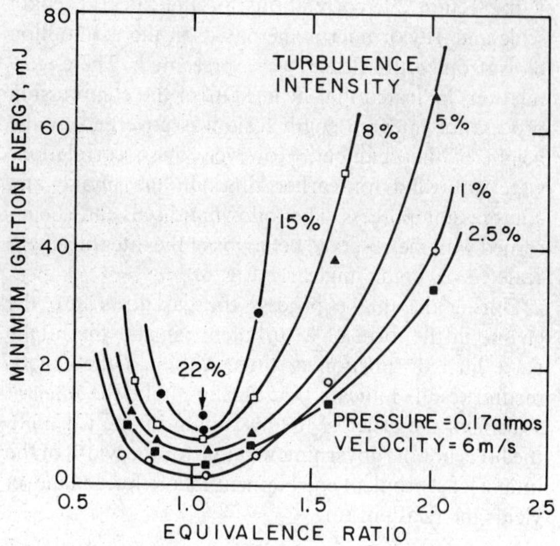

FIG. 27. Graph illustrating the effect of relative turbulence intensity on minimum ignition energy (from Reference 23).

minimum ignition energy is a strong function of turbulent intensity and equivalence ratio. The influence of turbulence is increased as the stoichiometry deviates from the chemically correct value. Ballal and Lefebvre also studied the effect of turbulent scale on the minimum ignition energy. Figure 28 shows the

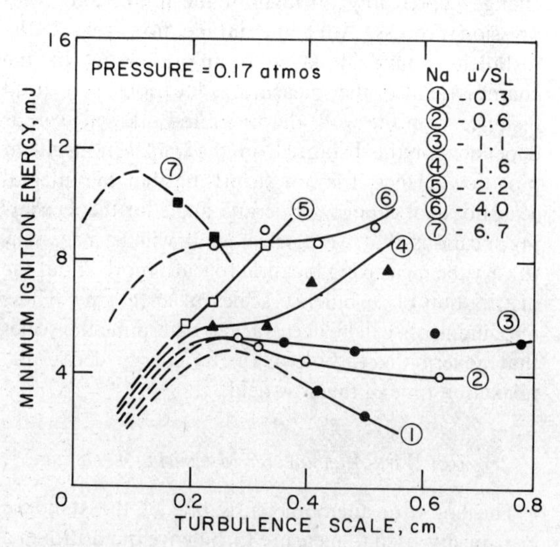

FIG. 28. Influence of turbulent scale on minimum ignition energy (Reference 23).

results of this investigation for a range of values for u'/S_l, where S_l is the laminar flame speed. These results are for stoichiometric mixtures and show mixed trends for the behavior of minimum ignition energy versus turbulent scale, which depend on the level of turbulence.

Typical values of u'/S_l for engines are of the order of 2–4 and hence one would expect the minimum ignition energy to increase as turbulent scale increases.

Theoretical consideration

The minimum ignition energy is usually determined by balancing the heat produced by chemical reaction and the heat dissipated via turbulence. Of critical importance is the size of the flame kernel when its temperature has fallen to the adiabatic flame temperature of the mixture. At this point, the criterion for successful ignition is that the rate of heat release from the reaction zone be greater than the heat loss from the volume. Several qualitative theories have been developed for the ignition process in an isotropic turbulent field.[21–23] Most of these theories are semi-empirical and contain the turbulence as an implicit function. The work of DeSoete[22] is the most detailed and physical of the models and it is outlined below.

The major contributions of DeSoete[22] are the introduction of the turbulent diffusivity as an increase in the overall transport phenomena and the concept that only a fraction of the total released energy, ζE_c, due to spark discharge is efficient for the ignition phenomenon. DeSoete derived the expression

$$\zeta E_c = \rho C_p (T_f - T_i) \pi d e^2 (u'^2 \theta / D_m + 1) \tag{29}$$

for the critical ignition energy of a line source where E_c is the critical ignition energy, e the laminar flame thickness, ρ the unburned gas density, T_f the flame temperature, T_i the initial mixture temperature, d the width of the spark gap, u' the turbulent intensity, θ the spark duration, D_m the molecular diffusivity and C_p the specific heat. The basic assumptions behind this derivation are (1) the turbulent flow is isotropic; (2) the influence of large scale fluctuations is negligible for small flames like those involved in ignition; and (3) the turbulent diffusivity is a function of time when diffusion times are small compared with turbulent fluctuating times. Figure 29 shows comparisons of DeSoete's[22] theory with experimental values of the critical ignition energy. These results show quantitative agreement between theory and experiment.

Several general conclusions on the influence of turbulence on ignition can be stated.

(1) The presence of turbulence increases the reaction zone thickness which in turn increases the critical volume into which the ignition energy must be released.
(2) The presence of turbulence increases the rate of heat dissipation which results in a decrease of the energy available for ignition.
(3) Spark duration and the distribution of the energy input strongly influence the critical ignition energy.
(4) A change in turbulent scale can either increase or decrease the critical ignition energy. However, for engine turbulence where u'/S_l is of the order of 2 or greater, an increase in scale should result in an increase in the minimum ignition energy.

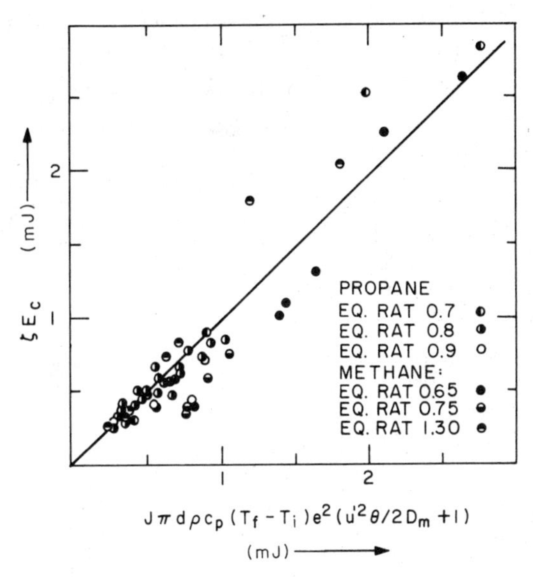

FIG. 29. Comparison of experimental values of critical ignition energy with theoretically estimated values under turbulent conditions (from Reference 22).

3.2. *Influence of Turbulence on Ignition Delay*

Experimental observations

Ignition delay in a spark-ignition engine is generally defined as the time from spark initiation to the time when 10% of the mass is burned. The phenomenon of ignition delay has been the subject of many investigations and it has been shown that variations in the ignition delay time are the major cause of cycle to cycle variations in the cylinder pressure.[25] The variations in the ignition delay time can be caused by variations in the velocity near the spark plug and mixture non-homogeneities.[17] Although investigators do not fully agree on the relative importance of these two phenomena, it is well accepted that cycle to cycle variations in the turbulent field near the spark plug result in cycle to cycle variations in the cylinder pressure. Several investigators[16–18] have sought to quantify this effect by measuring the velocity field at the spark plug and correlating standard deviations of the velocity with peak pressure and the angle of occurrence of peak pressure. The correlation coefficients were of the order of 0.97, showing a high degree of correlation between cycle to cycle variations of velocity and pressure.

Schlieren photographs of the combustion process can also show the variations from cycle to cycle for a given initial volume to burn. Iinuma and Iba[26,27] have studied the flame initiation process using the Schlieren technique. These authors state that the cyclic variations seem to have no regularity and no completely identical diagrams were observed in twenty consecutive cycles. Figure 30 shows a plot of initial versus maximum flame velocities. Little correlation exists, which is not surprising since the initial burning process is highly dependent on the local structure of the turbulence (i.e. neighboring eddies) whereas the major portion of

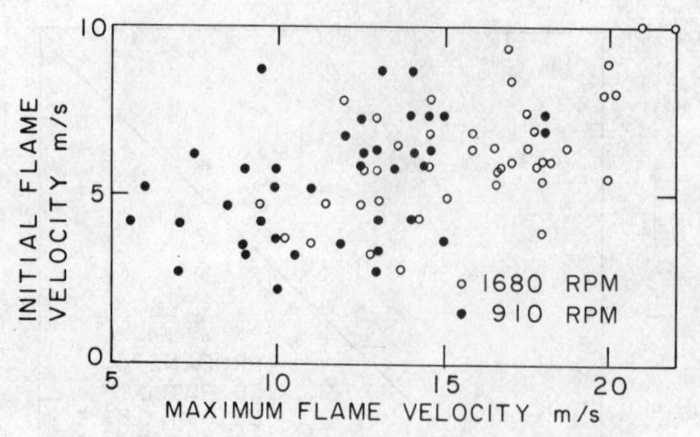

FIG. 30. Initial versus maximum flame velocity (from Reference 26).

the turbulent flame propagation will be better represented by an average turbulent velocity since the flame front is large compared to the eddy size. Another interesting observation of Iinuma and Iba[26] is illustrated in Fig. 31. This is a Schlieren photograph of the motion of a flame kernel that has detached from the spark electrode under conditions of severe turbulence. This figure shows that the flame kernel undergoes a random walk process during the early stages of combustion.

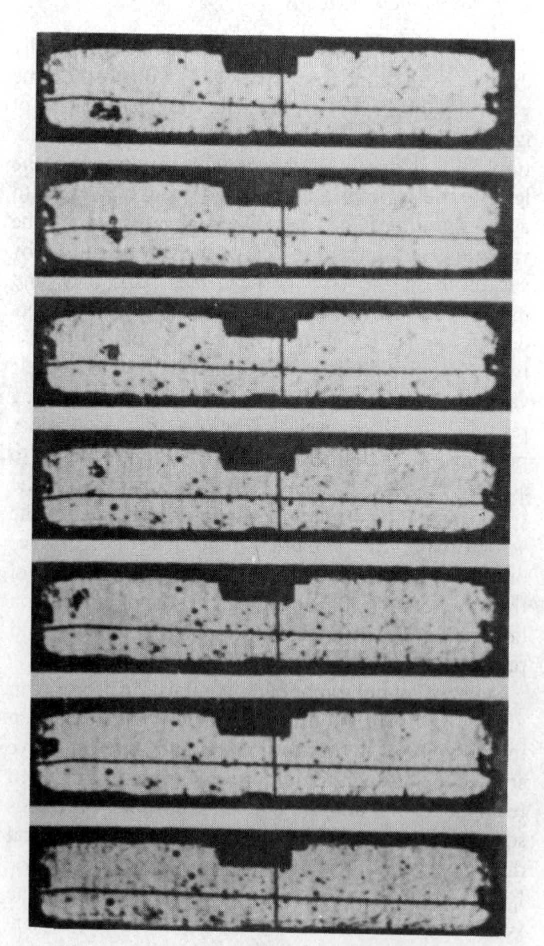

FIG. 31. Motion of the flame kernel under conditions of intense turbulence. Spark plug is located at right hand side of the chamber (from Reference 26).

Another experimental observation on ignition delay is the increase in the percent dispersion as longer mean flame travel times are encountered (Fig. 32). At the same engine speed generally longer flame travel times are associated with lower turbulence levels or leaner mixture strengths, either phenomena can change the ignition delay time through slower interactions or smaller expansion velocities. In each case the time to reach a critical volume would be longer. However,

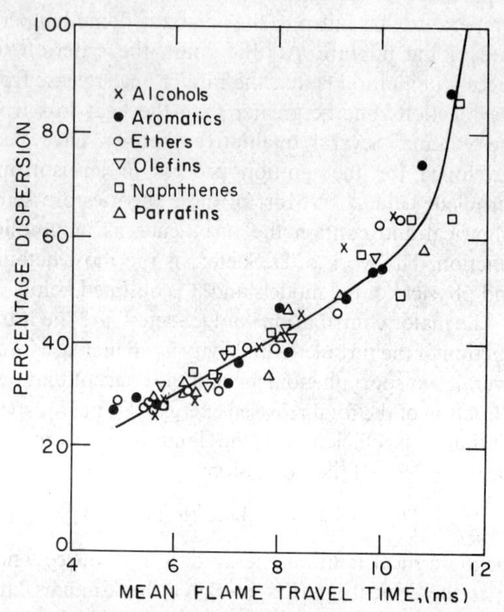

FIG. 32. Relationship between mean flame travel times and percent dispersion for various fuels (from Reference 31).

since the percent dispersion increases as the mixture is made leaner and decreases as engine speed decreases, an increase in the characteristic reaction time for the leaner mixture is probably the cause for the increase in dispersion.

Theoretical considerations

In an attempt to quantify the influence of turbulence on ignition delay, investigators[16,18] have developed models of the initial flame propagation process. These models have generally been developed so that correlations between velocity fluctuations and pressure fluctuations could be interpreted. These models generally assume that the volume burned is a function of the area of the flame front, the laminar flame speed and a turbulent entrainment speed, for example

$$\frac{dV_b}{dt} = A_f S_l f(U_e) E \tag{30}$$

where V_b is the volume burned, A_f the area of the flame front, S_l the laminar flame speed, $f(U_e)$ the increase in S_l due to turbulence and E the expansion ratio. Barton *et al.*[16] used eqn. (30) to develop their correlation between initial burn duration and velocity variations shown in Fig. 33 where F(o.v.) is a function of air fuel ratio and charge density. Winsor and Patterson[18] developed an expression for the critical distance (i.e.

the size of the reacted volume over which cyclic variations in velocity are important) which was dependent on the turbulent flame speed S_t and S_l using similar assumptions as Barton *et al.* Their critical distance was of the order of the chamber height which is consistent with the Schlieren pictures of Iinuma and Iba.[26,27] This observation is also in agreement with

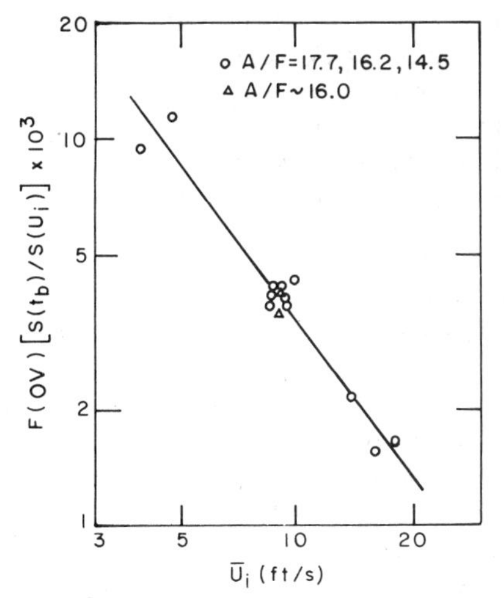

FIG. 33. Correlation of initial burn duration and velocity variations using model presented in Reference 16. F(o.v.) is function of air-fuel ratio and density, $S(t_b)/S(U_i)$ is the ratio of the standard deviations of initial burn time and velocity, and $\bar{U}_i$ is the mean velocity at ignition.

standard turbulence theory if one accepts the hypothesis that the integral scale is of the order of the chamber height. Models of the rate of flame propagation such as those proposed by Blizard and Keck[10] and Sokolik *et al.*[28] should be able to give values for average ignition delay times, since the reaction rates are treated separately from the entrainment rate. The model proposed by Blizard and Keck[10] has been used to correlate ignition delay times by assuming the ignition delay is proportional to a characteristic reaction time

$$\tau_i \sim \tau = l_e/S_l. \tag{31}$$

The characteristic eddy size, l_e, in this equation is of the order of the Taylor microscale which is consistent with the work of Corrsin[29] on the mixing of eddies of different concentrations, if we consider the burned and unburned gases as the two species of different concentrations.

To date there are no complete models of the ignition delay process which account for the statistical non-uniformities in eddy size, velocity, and composition. However, models such as those proposed by Blizard and Keck[10] and Sokolik *et al.*[28] can be modified to account for such non-uniformities. It also may be possible to describe a stochastic model of the ignition delay process similar to that which has been developed for continuous combustion.[30]

Flame propagation in engines

The fully developed portion of the turbulent flame propagation process in an engine (i.e. 10–90% burned) has been observed to scale almost linearly with engine speed.[31] This fact, combined with the observation that the turbulence intensity is also a linear function of engine speed, implies that the turbulent flame speed is approximately a linear function of the turbulent intensity. This linear relationship between turbulent flame speed and intensity has been verified in studies of turbulent combustion in bombs,[28,32] and engines.[7,9] The most extensive combustion bomb work is that of Sokolik *et al.*[28,33–39] Figure 34 shows the approxi-

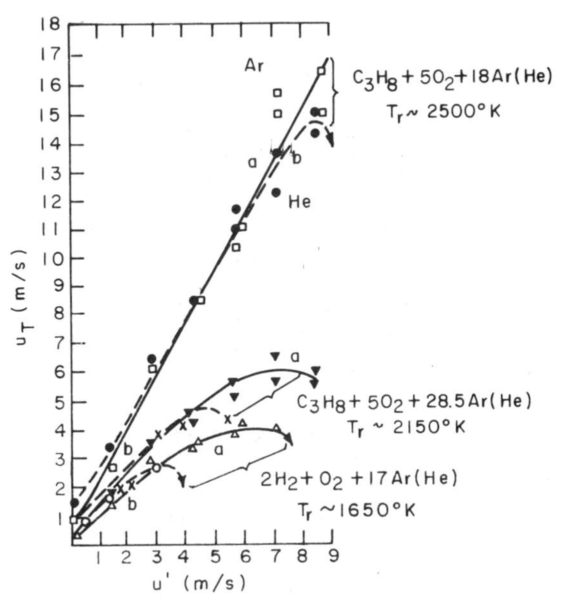

FIG. 34. Turbulent flame speeds for various fuels as a function of turbulent intensity and adiabatic flame temperature T_r (from Reference 28).

mately linear relationship which exists between the turbulent flame speed and the turbulent intensity. The sudden drop in turbulent flame speed for large values of the turbulent intensity is attributed to the extinction of the flame due to the high turbulence levels. Lancaster *et al.*[9] present their results for a CFR engine in terms of a flame speed ratio defined as the ratio of the turbulent flame speed to the laminar flame speed, Fig. 35. These researchers measured the flame speed ratio as a function of compression ratio and showed no dependence existed. This implies that the turbulent scale is not important in determining the turbulent flame velocity under normal operating conditions. Perhaps a better parameter for correlating data of this sort is the turbulent Reynolds number $(u'\lambda/\nu)$ which Andrews *et al.*[4] used in their review paper. The turbulent Reynolds number does contain the added parameter of scale and should provide a wider range of correlation.

Ohigashi[32] and Lancaster[9] also measured the effect of equivalence ratio on the turbulent flame speed. Their

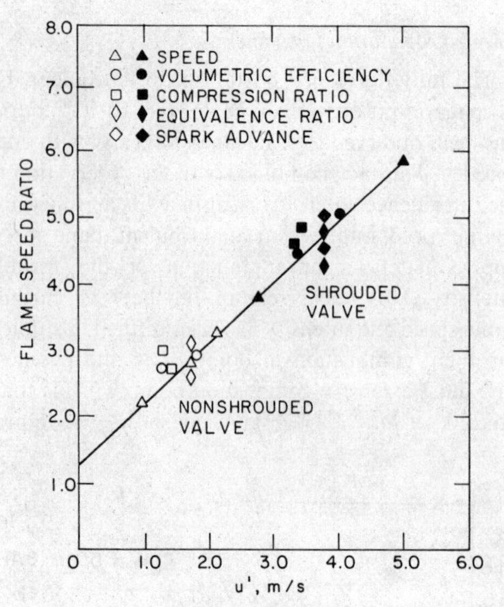

FIG. 35. Correlation of flame speed ratio with turbulent intensity (from Reference 9).

results imply that under normal operating conditions the flame speed is only a function of the turbulent intensity. However, as the flammability limits are approached, the turbulence scale should play an important role since it is the shearing of the reaction zone that is of primary importance for flame extinction.[40] Systematic studies of the effects of turbulent scale on the flammability limits in engines have not been performed. However, the similarity between the phenomena of ignition and blowout implies that turbulence scale is important and can have various effects depending on the intensity of the turbulent field (see Fig. 28).

In order to gain more insight into the process of flame propagation in an engine it is of importance to understand more about the structure of a turbulent flame. Iinuma and Iba[26,27] have studied flame propagation in an engine using a Schlieren technique. In this experiment special attention was given to obtain one dimensional flame propagation. Figure 36 is a reproduction of one of their photographs. The flame propagates from right to left. Several interesting features are apparent. During the initial phase of combustion the reaction zone develops slowly but soon engulfs a large volume of the chamber even though only a small percentage of the mass is burned. Once the reaction zone dimensions become the order of those of the chamber, it propagates in a regular manner until it reaches the far wall of the chamber. Note the reaction zone thickness is of the order of the height of the chamber which is consistent with measurements of the integral scale. The small scale turbulence is representative of the shearing and is of the order of the Taylor microscale. After the turbulent flame has traversed the chamber burning occurs for some time until combustion is complete.

Several other investigators have observed the turbulent reaction zone in combustion bombs.[32,35] Figure 37 shows a typical ionization probe signal as the flame front passes the measuring location. The irregularities in this signal are associated with reacted volumes of gases which interact with the probe confirming the observations that the reaction zone is composed of unburned, burned and burning gases. Also shown in this figure is the variation of the average period of pulsation as a function of mixture composition. A merger of this type of data with laser doppler velocity results could yield information on both the length and velocity scales in the reactive zone. These details of the combustion process must be adequately described by any proposed model of turbulent combustion for an engine. In the next section various models of turbulent combustion in spark-ignition engines will be reviewed.

Theories of turbulent flame propagation

Various models of turbulent combustion in spark-ignition engines have been proposed. Included in these models are the wrinkled flame front model, turbulent diffusion models and entrainment models. This review is not intended to describe all the various turbulent combustion models and the interested reader is referred to the recent review by Andrews *et al.*[4] for a detailed summary. The combustion models that have been applied to engine combustion can be divided into two categories; (1) models that assumed the turbulent burning velocity are related to the laminar flame speed; and (2) models that use overall or global reaction rates.

Laminar flame speed models

The majority of models describing the combustion process in an engine use the laminar flame speed in their analyses. One of the most common techniques involves the description of the turbulent flame speed via the laminar flame speed.[7,9,41–44] The expression

$$S_t = kS_l \qquad (32)$$

is generally used to calculate the turbulent flame speed where k is found empirically. The constant k is related to the turbulent intensity u' and a correlation for k is necessary to make such a description useful. Figure 35 showed the flame speed ratio S_t/S_l versus u' for a CFR engine.[7] A good correlation exists. However, more extensive measurements on different engines must be made before such a correlation could be called universal. In fact, as previously stated, a better correlating factor may be the turbulent Reynolds number based on the integral scale or Taylor microscale and the kinematic viscosity of the unburned mixture. Theories that use expression (32), do not include the effects of turbulent scale on flame propagation. This may not be important for predicting engine performance or even NO_x emissions since the major portion of the burning does not depend on the turbulent scale. However, if some insight is to be gained into ignition delay or flammability limits, techniques for including the turbulent scale are necessary.

Another model that uses the laminar flame speed is the wrinkled flame front model. This model assumes that laminar flame propagation proceeds normal to the

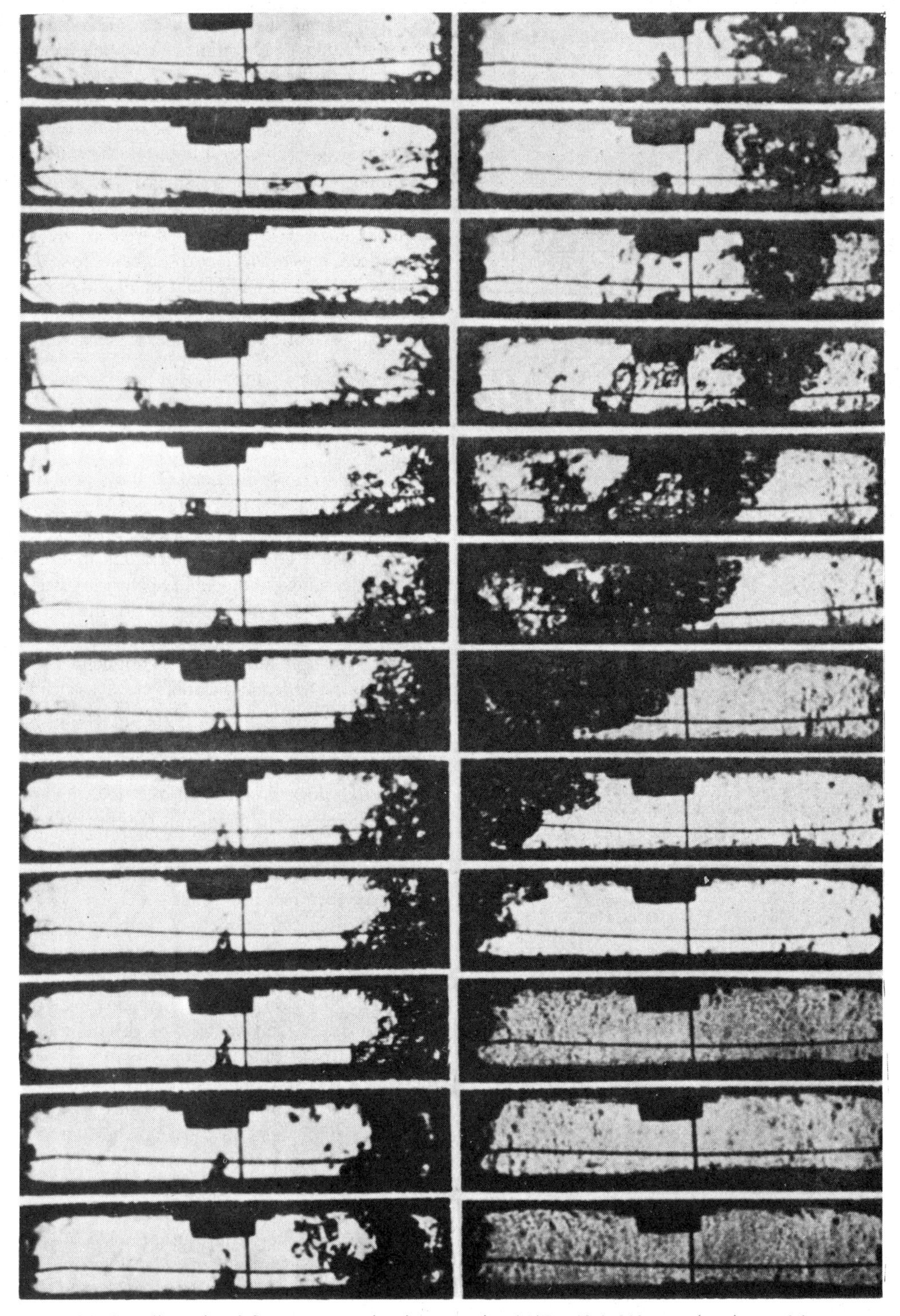

FIG. 36. One-dimensional flame propagation in an engine ($A/F = 19.4$, 910 rpm, time interval between frames 1.5 ms) spark occurs on right hand side of chamber (Reference 26).

flame front but the flame front area is greatly increased due to turbulence. This model was first proposed by Damkohler,[45] and many other investigators[44–68] have adopted this approach. There is a wealth of experimental evidence supporting such models which claim the identification of isolated laminar reaction zones in a turbulent flame.[69–76] Recently two concepts of this model have been applied to engine combustion. One model[77,78] is an extension of the work of Shchelkin[46] to an engine. The model assumes that

$$S_t = (S_l^2 + k_2 \, \mathrm{Re}^{k_3})^{1/2} \tag{33}$$

where

$$\mathrm{Re} = \frac{V_i D}{\nu_u} \tag{34}$$

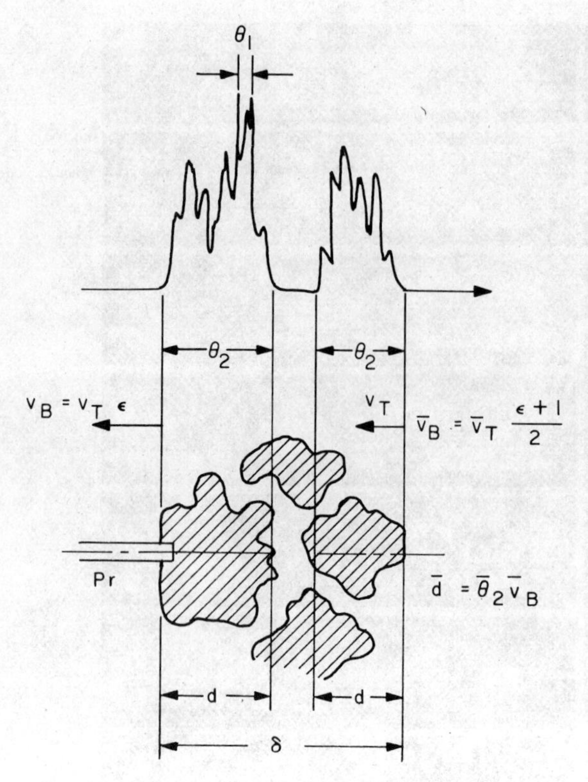

FIG. 37a. Diagram illustrating the movement of the combustion zone past the ionization probe and the record of the ionization current. v_t is the turbulent flame speed, v_b the velocity of the burning front, θ_2 the frequency associated with the large reacting volumes, θ_1 frequencies of oscillation in the turbulent reacting volume and δ the thickness of the flame.

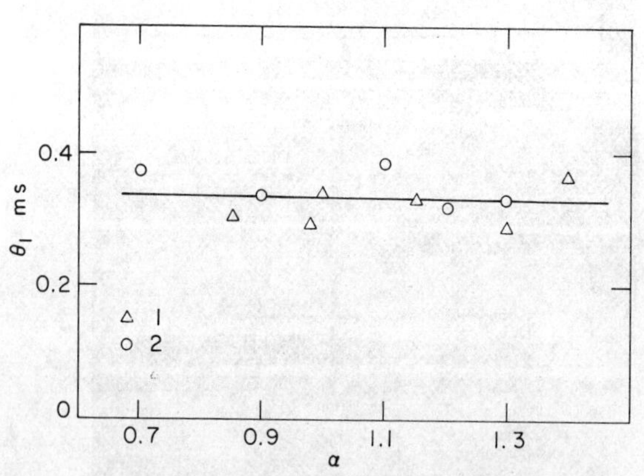

FIG. 37b. Variation of the small scale oscillation characterized by θ_1 with mixture composition for: (1) methane; (2) propane initial bomb pressure 0.8 atm, $u' = 3\,\text{ms}^{-1}$.

and V_i is the average inlet velocity, D the diameter of the chamber, v_u the kinematic viscosity of the unburned mixture and k_2 and k_3 constants. The flame front surface is assumed to consist of fluctuating flame cones. Longwell[79] points out that such a model may not have enough surface to correctly predict turbulent burning rates. Also, the large surface areas necessary for predicting turbulent flame speeds would necessitate cones with high aspect ratios which would be unstable in a turbulent flow field. Another disadvantage of the model of Shchelkin[46] and that employed by Samaga and Murthy[77] is that it does not include the effect of turbulent scale.

A model proposed by Blizard and Keck[10] for turbulent flame propagation in an engine includes the turbulent intensity and scale as well as the laminar flame speed. This model incorporates some of the ideas of Sokolik[60] who proposed that small eddies gradually burn up in laminar flames. The Blizard and Keck model is basically an entrainment model which assumes

$$\dot{m}_e = \rho_u A_f u' \tag{35}$$

and

$$\tau = l_e/S_l \tag{36}$$

where $\dot{m}_e$ is the mass entrained, A_f the area of the flame front and l_e a characteristic eddy size. Using expressions (35) and (36) the rate of mass burned can be calculated. The model overcomes the objections that Longwell raised concerning the model of Shchelkin by assuming that the large eddies are responsible for the entrainment of fresh charge and that smaller eddies burn up laminarly. Blizard and Keck[10] do not define the length scale l_e but Dent and Salama[6] have associated the Taylor microscale λ with l_e. The applicability of the Blizard and Keck[10] model to engine combustion has undergone only preliminary verifications. However, this model does contain all the relevant physical variables necessary for predicting the entire combustion process. In fact, the correlations for the length scale l_e were developed via the ignition delay time and the expression for τ (see Section 2).

Before discussing models of turbulent combustion that do not depend on the laminar flame speed, it is useful to determine the scale of turbulence over which laminar combustion occurs in an engine. The following derivation uses the assumptions of equilibrium and isotropic turbulence to develop a relationship between the scale over which laminar combustion occurs and the various turbulent length scales. Basically the inequality that must hold true if the laminar flame speed is to be important is

$$\tau_l < \tau_e \tag{37}$$

where τ_l is the time to burn up an eddy of size r and τ_e is the lifetime of an eddy of size r. Rewriting eqn. (37) in terms of the known quantities and assuming isotropic equilibrium turbulence we obtain;

$$\frac{r}{S_l} < \frac{r}{V} \tag{38}$$

and

$$V = u'\left(\frac{r}{L}\right)^{1/3} \tag{39}$$

Substituting eqn. (39) into (38), we obtain the result

$$\frac{r}{L} < \left(\frac{S_l}{u'}\right)^3 \tag{40}$$

Table 1 shows a comparison of the various turbulent length scales and the scale for laminar burning when both sides of expression (37) are set equal.

TABLE 1.

u'	S_l	L	λ/K	η/L	r/L	ν
ms^{-1}	ms^{-1}	m	—	—	—	$m^2 s^{-1}$
2	0.5	0.01	7×10^{-2}	2×10^{-3}	2×10^{-2}	6×10^{-6}

Although laminar flame speeds can be influenced by curvature, diluents, pressure and temperature, the results of Table 1 indicate that the laminar flame mechanisms may not be adequate for describing engine combustion. The ratio r/L in this table represents the scale over which laminar burning is possible. Since the Kolmogorov scale, η, is smaller than r turbulence interactions with the laminar flame are possible. This implies that the diffusion processes that govern a laminar flame will be altered by the small scale turbulence and the reaction times for laminar flames may not be applicable for turbulent combustion in engines.

Non-laminar flame speed models

Only a few investigators have used models of turbulent combustion in engines that do not involve laminar flame speeds. The studies of Sirignano and Bracco[80–84] are the only published papers that directly relate to engine combustion. These authors introduce the effect of turbulence by using an overall turbulent diffusivity to calculate the rate at which fresh charge is brought into the reaction zone and an overall global reaction rate to describe the rate of energy release. The turbulent diffusivity used is based on the dimensions of the chamber and a characteristic velocity based on engine speed. Their results indicate that a thick overall reaction zone exists.

The Russian researchers, Sokolik *et al.*,[28] propose that combustion occurs in microvolumes at a rate that is independent of the laminar flame speed. Their model is similar to that of Blizard and Keck[10] except the characteristic reaction time τ is dependent on the turbulent transport in the small microvolumes. To support their theory, Sokolik *et al.*[28] performed extensive investigations in a combustion bomb with known turbulence levels. A mixture of gases having varying laminar flame speeds but equal flame temperature were studied. Their results indicate that the turbulent flame speed is dependent only on the combustion temperature and the turbulent intensity. The authors claimed that their experimental evidence implied that turbulent combustion was not related to laminar combustion. As yet this model has not been applied to spark-ignition engine combustion problems and there has been some question as to the interpretation of their experimental data.

To date there has been no experimental verification of the non-laminar combustion models in engines. The validation of any combustion model for spark-ignition engines is complicated by the lack of information on laminar flame speeds and global reaction mechanisms for typical fuels at engine conditions.

4. SUMMARY

Many questions still remain unanswered with regard to the turbulence structure of the flow field in the cylinder of a spark-ignition engine and influence of turbulence on combustion. Few experts in the fields of turbulence and turbulent combustion have applied their knowledge to this area of application. Nevertheless, a number of useful conclusions concerning the turbulence characteristics in an engine cylinder can be stated.

(1) The high shear flows generated during intake are the major turbulence production source for combustion chambers with small squish regions and no swirl.
(2) The turbulent flow field rapidly decays so that the shape of the chamber becomes a dominant factor governing the turbulence structure near TDC on the compression stroke.
(3) Little or no mean flow exists near TDC for engine combustion chamber geometries with no swirl and squish.
(4) There is strong evidence that the integral scale of turbulence near TDC is of the order of the chamber height.
(5) Conventional theories for isotropic and equilibrium turbulence, boundary layer turbulence and turbulent flow behind a grid can yield correlating variables for the flow processes that are occurring in spark-ignition engines.

Turbulent combustion in spark-ignition engines has been described using models that employ laminar flame speeds and global kinetics. There is no conclusive evidence validating either one of these approaches. Also there has been little or no work on the effects of turbulence on ignition and flammability limits in engines. However, several statements on the structure of the turbulent flame front and its rate of propagation can be made.

(1) The turbulent flame has a finite thickness which is on the order of the height of the chamber.
(2) The turbulent burning process takes place on a scale the size of the Taylor microscale. There is evidence that this scale is a dominant variable for the ignition delay process.
(3) Turbulent flame speeds correlate well with the turbulent intensity and show a nearly linear dependence with this variable.
(4) Under normal operating conditions the fluid mechanics (turbulent intensity) govern the rate of flame propagation. The turbulent scale is important during ignition and extinction.

Acknowledgement—The author thanks Dr. David P. Hoult and Dr. Richard C. Flagan for the many informative and stimulating discussions on the nature of turbulence and turbulent combustion. I also thank Drs. Peter O. Witze and

David Lancaster for their comments on the manuscript. This work was supported in part by the Department of Transportation under Contract Number DOT-TSC-1034.

5. REFERENCES

1. BONE, W. A. and TOWNEND, D. T. A., *Flame and Combustion in Gases*, London, 1927.
2. WITZE, P. O., Hot wire measurements of the turbulent structure in a motored spark ignition engine, 14th AIAA Aerospace Sciences Conference, January 26–28, 1976.
3. WITZE, P. O., Hot wire turbulence measurements in a motored internal combustion engine, presented at the 2nd European Symposium on Combustion, Orleans, France, September 5, 1975.
4. ANDREWS, G. E., BRADLEY, D. and LWOKABAMBA, S. B., Turbulence and turbulent flame propagation—a critical appraisal, *Combust. Flame* **24**, 285–304 (1975).
5. DENT, J. C. and SALAMA, N. S., The measurement of the turbulence characteristics in an internal combustion engine cylinder, Paper no. C83/75, I. Mech. E. Conference on Combustion in Engines, Cranfield, England, 1975.
6. DENT, J. C. and SALAMA, N. S., The measurement of the turbulence characteristics in an internal combustion engine cylinder, SAE Paper no. 750886, Automobile Engineering Meeting, Detroit, October, 1975.
7. LANCASTER, D. R., Effects of turbulence on spark ignition engine combustion, Ph.D. Thesis, University of Illinois, 1975.
8. LANCASTER, D. R., Effects of engine variables on turbulence in a spark-ignition engine, SAE Paper no. 760159.
9. LANCASTER, D. R. and KRIEGER, R. B., Effects of turbulence on spark-ignition engine combustion, SAE Paper no. 760160.
10. BLIZARD, N. S. and KECK, J. C., Experimental and theoretical investigation of turbulent burning model for internal combustion engines, SAE Paper no. 740191, 1974.
11. KOLMOGOROV, A. N., *Dokl. Akad. Nauk. SSSR* **30**, 301 (1941).
12. SEMENOV, E. S., Studies of turbulent gas flow in piston engines, NASA Technical Translation F-97, 1963.
13. NANOPOLOUS, N. C., Study of turbulence in a steady flow for an engine cylinder geometry, M.S. Thesis, M.I.T., 1975.
14. TSUGE, M., KIDO, H., KATO, K. and NOMIYAMA, Y., Decay of turbulence in a closed vessel, *Bull. J.S.M.E.* **16** (1973).
15. TSUGE, M., KIDO, H., KATO, K., and NOMIYAMA, Y., Decay of turbulence in a closed vessel, *Bull. J.S.M.E.* **17** (1974).
16. BARTON, R. K., LESTZ, S. S. and MEYER, W. E., An empirical model for correlation cycle-by-cycle cylinder gas motion and combustion variations of an SI engine, *SAE Trans.* **80**. Paper 710163, 1971.
17. MATSUOKA, S., YAMAGUCHI, T. and UMEMURA, Y., Factors influencing the cyclic variation of combustion spark ignition engine, Paper 710586 presented at SAE Mid-Year Meeting, Montreal, June, 1971.
18. WINSOR, R. E. and PATTERSON, D. J., Mixture turbulence, a key to cyclic combustion variation, SAE Paper no. 730086, 1973.
19. JAMES, E. H. and LUCAS, G. G., Turbulent flow in spark ignition engine combustion chambers, SAE Paper no. 750885.
20. SEMENOV, E. S., *Combustion Explosion and Shock Waves*, Vol. 2, p. 83 (1965).
21. SWEET, C. C., Spark ignition of flowing gases, NACA Report no. 1287, June, 1956.
22. DESOETE, G. G., The influence of isotropic turbulence on the critical ignition energy, 13th Symposium (International) on Combustion, p. 735, The Combustion Institute, 1973.
23. BALLAL, D. R. and LEFEBVRE, A. H., The influence of flow parameters on minimum ignition energy and quenching distance, 15th Symposium on Combustion, p. 1473, The Combustion Institute, 1975.
24. KARPOV, V. P. and SOKOLIK, A. S., Ignition limits in turbulent gas mixtures, Institute of Chemical Physics, Academy of Sciences, USSR. Translated from *Dokl. Akad. Nauk. SSSR* **141**, no. 2, 393–396 (1961).
25. PATTERSON, D. J., Cylinder pressure variations, a fundamental combustion problem, *SAE Trans.* **75**, Paper 660129, 1967.
26. IINUMA, K. and IBA, Y., Studies of flame propagation process, JARI Tech. Memo. no. 10, 1972.
27. IINUMA, K. and IBA, Y., Studies of flame propagation, JARI Tech. Memo. no. 15, 1973.
28. SOKOLIK, A. S., KARPOV, V. P. and SEMENOV, E. S., Turbulent combustion of gases, *Fizika Goreniya Vzryva* **3**, no. 1, 61–76 (1967).
29. CORRSIN, S. A., *A.I.Ch.E. Jl.* **3**, 329 (1957).
30. APPLETON, J. P. and FLAGAN, R. C., A stochastic model of turbulent mixing with chemical reaction: nitric oxide formation in a plug-flow burner, *Combst. Flame* **23**, 249–267 (1974).
31. HARROW, G. A. and ORMAN, P. L., A study of flame propagation and cyclic dispersion in a spark-ignition engine, *Advances in Automotive Engineering*, Part IV, Pergamon Press, Oxford, 1966.
32. OHIGOSHI, S., HAMAMOTO, Y. and KIZIMA, A., Effects of turbulence on flame propagation in closed vessel, *Bull. J.S.M.E.* **14**, no. 74 (1971).
33. KARPOV, V. P. and SOKOLIK, A. S., Laminar and turbulent hydrazine-decomposition flames, *Russ. J. Phys. Chem.* **38**, no. 6 (1964).
34. KARPOV, V. P. and SOKOLIK, A. S., Relation between spontaneous ignition and laminar and turbulent burning velocities of parafin hydrocarbons, *Dokl. Akad. Nauk. SSSR* **138**, no. 4, 874–876 (1961).
35. SOKOLIK, A. S. and SEMENOV, E. S., Nature of chemical ionization in flames, *Russ. J. Phys. Chem.* **38**, no. 7 (1964).
36. KARPOV, V. P. and SOKOLIK, A. S., The effect of pressure on the rate of laminar and turbulent combustion, Institute of Chemical Physics Academy of Sciences, USSR. Translated from *Dokl. Akad. Nauk. SSSR* **132**, no. 6, 1341–1343 (1960).
37. SOKOLIK, A. S. and SEMENOV, E. S., An ionization current study of the macrokinetic characteristics of turbulent propane flames, *Russ. J. Phys. Chem.* **39**, no. 9 (1969).
38. SOKOLIK, A. S. and KARPOV, V. P., The effects of temperature and laminar velocity on the rate of turbulent combustion, *Dokl. Akad. Nauk. SSSR* **129**, 168 (1959).
39. KARPOV, V. P., SEMENOV, E. S. and SOKOLIK, A. S., Turbulent combustion in an enclosed space, *Dokl. Akad. Nauk. SSSR* **128**, 1220 (1959).
40. LEWIS, B. and VON ELBE, G., *Combustion, Flames and Explosions of Gases*, Academic Press, New York and London, 1961.
41. BENSON, R. S., ANNAND, W. J. D. and BARUAH, P. C., A simulation model including intake and exhaust systems for a single cylinder four-stroke cycle spark-ignition engine, *Int. J. mech. Sci.* **17**, 97–124 (1975).
42. PHILIPS, R. A. and ORMAN, P. L., Simulation of combustion in a gasoline engine using a digital computer, *Advances in Automotive Technology*, Part IV, Pergamon Press, Oxford, 1966.
43. HIROYASU, H. and KADOTA, T., Computer simulation for combustion and exhaust emissions in spark ignition engine, 15th Symposium (International) on Combustion, The Combustion Institute, 1975.
44. LUCAS, G. G. and JAMES, E. H., A computer simulation of a spark ignition engine, SAE Paper no. 730053.
45. DAMKÖHLER, G., *Z. Elektrochem.* **46**, 601 (1940); and NACA TM 1112, 1947.

46. SHCHELKIN, K. I., *Zh. éksp. teor. Fiz.* **13**, 520 (1943). (English translation, NACA TM 1110, 1947.)
47. FRANK-KAMENETSKII, D. A., Contribution to the theory of micro-diffusive turbulent combustion, *Trudy nauchno-Issled. Inst.* **7**, Oborongiz (1946).
48. LEASON, D. B., *Fuel* **30**, 233 (1951).
49. KARLOVITZ, B., DENNISTON, D. W. and WELLS, F. E., *J. Chem. Phys.* **19**, 541 (1951).
50. KARLOVITZ, B., *Fourth Symposium (International) on Combustion*, p. 60, Williams and Wilkins, Baltimore, 1953.
51. KARLOVITZ, B., *Selected Combustion Problems*, p. 248, Butterworths, London, 1954.
52. KARLOVITZ, B., *Chem. Eng. Prog.* **61**, 56 (1965).
53. SCURLOCK, A. L. and GROVER, J. H., *Fourth Symposium (International) on Combustion*, p. 645, Williams and Wilkins, Baltimore, 1953.
54. WOHL, K., SHORE, L., ROSENBERG, H. and WEIL, C. W., *Fourth Symposium (International) on Combustion*, p. 620, Williams and Wilkins, Baltimore, 1953.
55. WOHL, K., *Ind. Engng Chem.* **47**, 825 (1955).
56. TALANTOV, A. V., A study of combustions in turbulent flow, *Trudy Nauchno-Issled Inst.* **8**, Oborongiz (1955).
57. TALANTOV, A. V., Velocity of flame propagation and length of combustion zone in a turbulent flow, *Trudy kazan. aviats. Inst.* **31** (1956).
58. TALANTOV, A. V., Research on combustions in flow, *IVUZ aviats. Teknika*, 1967. (English translation, FTD-MT-24-209-68.)
59. TALANTOV, A. V., ERMOLAEV, V. M., ZOTIN, V. K. and PETROV, E. A., *Combustion, Explosion and Shock Waves* **5**, 73 (1969).
60. SOKOLIK, A. S., The experimental basis of the theory of turbulent combustion, in: *Combustion in Turbulent Flow*, L. N. KHITRIN, (ed.) Moscow, 1959. (English translation, IPST, p. 54, 1963.)
61. POVINELLI, L. A. and FUHS, A. E., *Eighth Symposium (International) on Combustion*, p. 554. Williams and Wilkins, Baltimore, 1962.
62. KOZACHENKO, L. S., *Izv. Akad. Nauk SSSR*, Otd. Tek. Nauk, Energetika Automatika **2**, 21 (1959). (English translation, *ARS J.* **29**, 761 (1959).
63. KOZACHENKO, L. S., The combustion of gasoline-air mixtures in turbulent flow, *The Third All Union Congress on Combustion Theory*, Vol. 1, p. 126. Flame Propagation and Detonation in Gas Mixtures, Moscow, 1960.
64. KOZACHENKO, L. S., *Izv. Akad. Nauk SSSR*, Otd. Khim Nauk, **1**, 45 (1960). (English translation, *Bull. Acad. Sci. USSR*, Div. Chem. Sci. No. 1, 37, 1960.)
65. KOZACHENKO, L. S., *Eighth Symposium (International) on Combustion*, p. 567, Williams and Wilkins, Baltimore, 1961.
66. KOZACHENKO, L. S. and KUZNETSOV, I. L., *Combustion, Explosion and Shock Waves* **1**, 22 (1965).
67. TUCKER, M., Interaction of a free flame front with a turbulence field, NACA TN 3407, 1955.
68. RICHARDSON, J. M., Proc. Gas Dynamics Symposium on Aerothermochemistry, Northwestern University, Illinois, 1956.
69. BURGESS, D., Structure and propagation of turbulent bunsen flames, Bull. 604, U.S. Bureau of Mines, 1962.
70. GRUMER, J., SINGER, J. M., RICHMOND, K. and OXENDINE, J. R., *Ind. Engng chem.* **49**, 305 (1957).
71. FOX, M. D. and WEINBERG, F. J., *Br. J. App. Phys.* **11**, 269 (1960).
72. BERL, W. G., RICE, J. L. and ROSEN, P. *Jet Propul.* **25**, 341 (1955).
73. GROVER, J. H., FALES, E. N. and SCURLOCK, A. C., *ARS J.* **29**, 275 (1959).
74. GROVER, J. H., FALES, E. N. and SCURLOCK, A. C., *Ninth Symposium (International) on Combustion*, p. 21, Reinhold, New York, 1963.
75. KOKUSHKIN, V. V. *Izv. Akad. Nauk SSSR*, Otd. Tekh. Nauk **8**, 3 (1958). (English translation, Johns Hopkins University APL Translation TG 230-T56, OTS 5911929.)
76. KOKUSHKIN, N. V., *The Third All Union Congress on Combustion Theory* Vol. 1, p. 109, Flame Propagation and Detonation in Gas Mixtures, Moscow, 1960.
77. SAMAGA, B. S. and MURTHY, B. S., On the problem of predicting burning rates in a spark-ignition engine, SAE Paper no. 750688.
78. GAJENDRA BABU, M. K. and MURTHY, B. S., Simulation and evaluation of a 4-stroke single-cylinder spark ignition engine, SAE Paper no. 750687.
79. LONGWELL, J. P. and WEISS, M. A., *Fourth International Symposium on Combustion*, pp. 615–620, Butterworth, London, 1959.
80. BRACCO, F. V. (ed.) Stratified charge engines, *Combust. Sci. Tech.* **8**, nos. 1 and 2, 1–100 (1973).
81. SIRIGNANO, W. A., One-dimensional analysis of combustion in a spark-ignition engine, *Combust. Sci. Tech.*, **7**, no. 3, 99–108 (1973).
82. BRACCO, F. V. and SIRIGNANO, W. A., Theoretical analysis of wankel engine combustion, *Combust. Sci. Tech.* **7**, no. 3, 109–123 (1973).
83. ROSENTWEIG-BELLAN, J. and SIRIGNANO, W. A., A theory of turbulent flame development and nitric oxide formation in stratified charge internal combustion engines, *Combust. Sci. Tech.* **8**, 51–68 (1973).
84. BRACCO, F. V., Theoretical analysis of stratified, two-phase wankel engine combustion, *Combust. Sci. Tech.* **8**, nos. 1 and 2, 69–84, 1973.

Manuscript received, 1 June 1976

Prog. Energy Combust. Sci., Vol. 2, pp. 167–179, 1976. Pergamon Press. Printed in Great Britain

FUNDAMENTALS OF OIL COMBUSTION

ALAN WILLIAMS
Department of Fuel and Combustion Science, The University of Leeds, Leeds LS2 9JT

NOMENCLATURE

A	Reaction rate pre-exponential factor
B	Transfer number
B_{ev}	Transfer number for evaporation
C_p	Specific heat
C_d	Drag coefficient
D	Diffusion coefficient
E	Activation energy
Gr	Grashof number
H	Enthalpy
I	Combustion intensity
K	Burning rate constant
Le	Lewis number
M_f	Molecular weight
M_0	Mass flow
Pr	Prandtl number
Q	Heat of combustion
Re	Reynolds number
Sc	Schmidt number
S_u	Burning velocity
T	Temperature
V_c	Extinction velocity
Y	Weight fraction
F_A	Reaction rate correction factor
F_B	Natural convection correction factor
F_C	Forced convection correction factor
$d_{3,2}$	Sauter mean diameter
i	Stoichiometric mixture ratio
m	Mass
q	Chemical rate of formation
r	Radial distance
t	Time
v	Radial velocity
ϕ	Fuel-air ratio
λ	Thermal conductivity
ν	R.M.S. velocity fluctuation
ρ	Density

Subscripts

F	Fuel
L	Liquid surface
O	Oxygen
i	Species i

1. THE NATURE OF THE SPRAY COMBUSTION OF FUEL OIL

The use of the combustion of sprays of liquid hydrocarbon fuels commenced first over a century ago when it was discovered that the atomization of fuel oils provided a convenient means of obtaining self-sustaining flames.[1–3] Prior to that time of course, only the light hydrocarbon fuels were burned as diffusion flames by means of wick burners.

Whilst the technology of spray combustion of fuel oils in domestic and industrial furnaces advanced quite considerably during the early part of this century little effort was devoted to the understanding of the process and sprays were considered to burn in a way analogous to coal. The whole emphasis changed however with the advent of the aircraft gas turbine and liquid-fuelled rockets and very extensive research programmes were devoted to the combustion of the lighter fuels, such as the aviation gasolines and kerosines. Consequently the most commonly considered case, as far as the literature is concerned, is that of droplet combustion involving a volatile liquid fuel burning in a surrounding oxidizing atmosphere. The droplet evaporates and acts as a source of fuel vapour which burns with the surrounding oxidant (which is usually air), as a diffusion flame around the droplet. This classical case[3–11] of droplet burning is illustrated in Fig. 1.

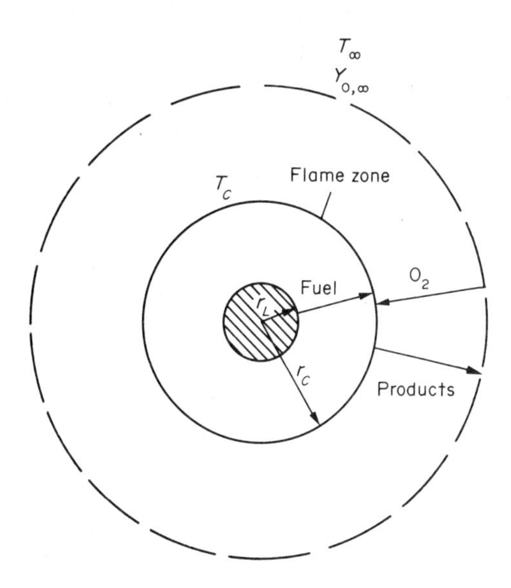

FIG. 1. Typical model of the combustion of a droplet of a volatile liquid fuel.

However this type of combustion only represents one extreme form of the combustion of fuel oils. The other extreme is illustrated by the combustion of heavier fuel oils as shown in Fig. 2. In this type of combustion the initial stage of devolatilization of the lighter components is followed by liquid cracking of the high molecular weight components to carbon (or coke) and its subsequent burn-out by an essentially heterogeneous (i.e. surface) combustion reaction.

The way in which the fuel spray and the combustion air are mixed is also crucial in determining the way in which the overall combustion proceeds. Thus if the fuel spray is premixed or initially well mixed with the combustion air two cases of combustion can be identified, these are "heterogeneous" combustion and "homogeneous" combustion. In the former case, as indicated in Fig. 3a the droplets burn either as individual droplets each with a surrounding flame or as droplets in a sea of flame, but in either event the system is essentially heterogeneous in that two phases

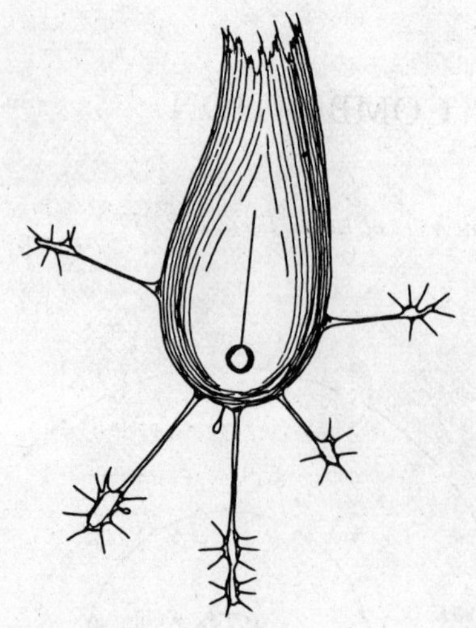

FIG. 2. Typical model of the combustion of a droplet of heavy fuel oil.

are present. In homogeneous spray combustion the droplets evaporate prior to their arrival in the flame zone as indicated in Fig. 3b and the flame front is supported essentially by fuel vapour and air. In the case of jet flames, in which the spray and combustion air are initially separated as in the case of a pressure jet,

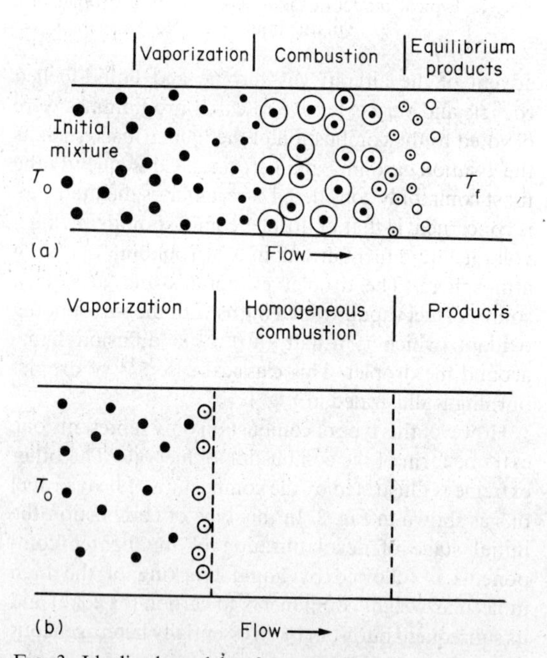

FIG. 3. Idealized models of premixed 1–D spray flames: (a) "heterogeneous" flame in which droplet burning occurs throughout the reaction zone, and (b) "homogeneous" flame in which droplet evaporation occurs prior to the flame zone. Many actual flames may involve both cases.

two extreme cases can again be identified, heterogeneous jet flames and homogeneous jet flames. Here the latter is analogous to a gaseous fuel jet flame in that the droplets evaporate prior to the flame zone whilst in the heterogeneous jet flame discrete droplets burn in the flame zone. These two cases are illustrated diagrammatically in Figs. 4a and 4b. Homogeneous combustion is favoured in systems containing small droplets (*ca* $< 20\,\mu m$) particularly of volatile fuels.

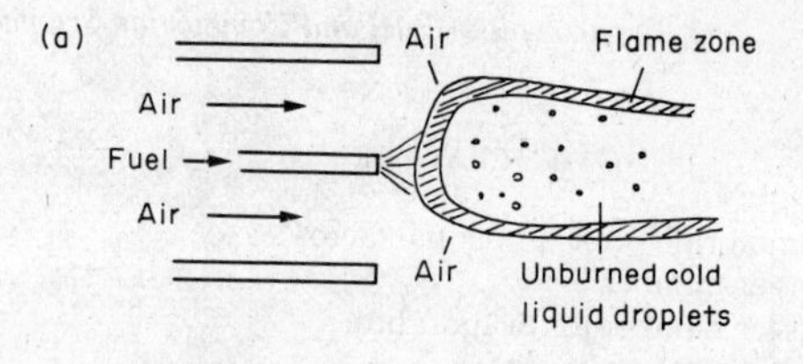

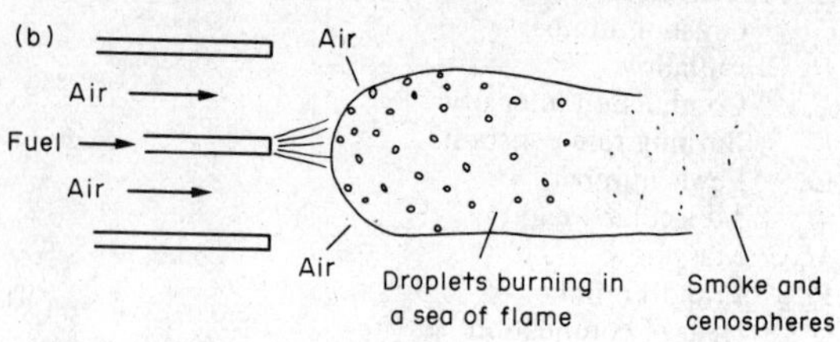

FIG. 4. Idealized models of jet spray flames: (a) "homogeneous" spray combustion in which droplets on the periphery of the core of the spray are vaporized and the vapour maintains the flame, and (b) "heterogeneous" spray combustion in which the evaporating droplets burn in a sea of flame.

Clearly, in real systems which will usually involve a wide range of droplet sizes both the homogeneous and heterogeneous modes of combustion can occur simultaneously and this is frequently advantageous.

In this review, recent work concerning the mechanism of combustion of sprays of fuel oils in these various systems will be outlined.

2. THEORETICAL AND EXPERIMENTAL STUDIES ON THE RATE OF COMBUSTION OF SINGLE DROPLETS OF LIGHT FUEL OILS

2.1. *Theoretical Analyses of Droplet Burning Rates*

The development of theoretical models of droplet combustion has been largely determined by the needs of the jet aircraft and liquid-fuelled rocket research programmes. Consequently only volatile liquid fuels, such as aviation fuels, have been considered and this work is only relevant to the lighter fuels, such as kerosines, domestic heating fuels and the light fuel oils. The state of the art is now such that single droplet models can be used to model combustion processes in combustion chambers in a fairly accurate realistic way.

Generally the approach used is to assume a sphericosymmetric model of the type shown in Fig. 1 based on a vaporizing droplet in which the rate controlling process is the physical step of molecular diffusion and the rate of combustion is not controlled in any way by chemical kinetics. So that analytical solutions may be obtained, a series of simplifying assumptions are made as follows:

(1) The fuel is a pure liquid having a definite boiling point.
(2) The combustion system has spherical symmetry as shown in Fig. 1. The spherical droplet of liquid fuel is surrounded by a concentric flame zone.
(3) Combustion takes place between the fuel and the air which diffuse into the flame zone and which react at a rate which is normally assumed to be infinitely fast in which case the flame is infinitesimally thin. This is the infinite reaction rate or thin flame approximation. In some analyses finite chemical kinetics are assumed.
(4) Heat is transferred from the flame zone to the droplet by convection to provide the latent heat of vaporization. Radiant heat transfer from combustion gases, particles or combustion chamber walls are normally neglected.

Various analytical approaches have been undertaken but they must all be based on the use of the continuity equations which may be expressed in the following form:

Global mass conservation

$$\frac{\partial \rho}{\partial t}+\frac{1}{r^2}\frac{\partial r^2 v\rho}{\partial r}=0 \quad (1)$$

which, in the case of steady-state combustion when $\partial\rho/\partial t=0$, reduces on integration to

$$\dot{m}_f = 4\pi r^2 \rho v = \text{constant} \quad (2)$$

where $\dot{m}_f$ is the mass of fuel vapour leaving the droplet surface, i.e. the mass burning rate, r is the radial coordinate, ρ the gas density and v the radial velocity.

Species mass conservation

$$\frac{\partial}{\partial t}(\rho Y_i)+\frac{1}{r^2}\frac{\partial r^2}{\partial r}\left\{\rho v Y_i-\rho D_i\frac{\partial Y_i}{\partial r}\right\}=q_i \quad (3)$$

where Y_i is the weight fraction, D_i the diffusion coefficient and q_i the chemical rate of formation of species i.

Conservation of energy

For an adiabatic system this takes the form

$$\frac{\partial}{\partial t}p\sum_i(Y_i H_i)+\frac{1}{r^2}\frac{\partial r^2}{\partial r}\left\{\rho v\Sigma(Y_i H_i)-\frac{\lambda\partial T}{\partial r}\right\}=-\Sigma q_i H_i \quad (4)$$

where H_i is the enthalpy of species i, λ the thermal conductivity of the mixture and T the temperature.

The equation of motion

$$\frac{\partial(\rho v)}{\partial t}+\frac{1}{r^2}\frac{\partial}{\partial r}(r^2\rho v^2)=-\frac{\partial p}{\partial r} \quad (5)$$

where p is the pressure which is normally assumed to be constant. In steady-state analyses the time dependent terms become zero and no liquid-phase reactions are assumed to take place in the droplet, this latter assumption being responsible for light oils. Thus at $r=r_L$, $T=T_L$ and $Y_F=Y_{F,L}$ and at $r=r_\infty$, $T=T_\infty$ and $Y_O=Y_{O\infty}$ where Y_F and Y_O are the mass fraction of fuel vapour and oxygen respectively.

Various methods of manipulation of the equations have been adopted.[7–8] If the combustion reaction is simply assumed to be

Fuel + Oxygen → Products

and if the reaction rate is assumed to be very fast then all steady-state solutions for the infinite reaction rate kinetic case are of the type given by Wise and Agoston,[7] namely

$$\dot{m}_F = 4\pi r_L \bar{\rho}\bar{D}\ln(1+B) \quad (6)$$

where $\dot{m}_F$ is the mass of the droplet, r_L its radius, $\bar{\rho}$ and $\bar{D}$ the averaged gas density and diffusivity respectively and B is the transfer number given by

$$B\equiv\frac{1}{L}\left\{\bar{C}_p(T_\infty-T_L)+\frac{QY_{O,\infty}}{i}\right\} \quad (7)$$

where $\bar{C}_p$ is the average specific heat, T_∞ the ambient temperature and T_L the temperature at the liquid surface which is set at its boiling point, T_B, or better as 0.9 T_B. Q is the heat of combustion, $Y_{O,\infty}$ the oxidant mass fraction, L the latent heat of vaporization and i the stoichiometric mixture ratio.

Generally the terms $\bar{\rho}\bar{D}$ in eqn. (6) are substituted by $\bar{\lambda}\bar{C}_p$ on the basis of a Lewis Number of unity where λ and $\bar{C}_p$ are the mean thermal conductivity and specific heat respectively.[8,9] Now

$$\dot{m}_F = -\frac{d}{dt}(4/3\pi r_L^3\rho_L) \quad (8)$$

where ρ_L is the density of the liquid at the appropriate temperature and r_L the radius of the liquid droplet. Equation (8) can be rewritten thus:

$$\frac{d(d_L)^2}{dt}=\frac{2\dot{m}_F}{\pi\rho_L r_L} \quad (9)$$

where $d_L=2r_L$. Since experimentally it has been observed that the experimentally determinable burning rate constant, K, is given by eqn. (10)

$$K=\frac{d(d_L)^2}{dt} \quad (10)$$

then eqn. (6) can be rewritten as:

$$K+\frac{8\bar{\lambda}}{\bar{C}_p\rho_L}\ln(1+B). \quad (11)$$

The application of this equation is very much dependent upon the choice of the values for $\bar{\lambda}$ and $\bar{C}_p$, and precise methods should be used for their calculation.[12,13]

The major assumptions upon which eqn. (11) is based are (a) infinite chemical (combustion) kinetics; and (b) steady-state combustion. The analysis is also restricted by the fact that only a pure liquid is considered and it is burning in a weak convective field.

The validity of the assumptions (a) and (b) have been

considered by a number of workers and these arguments are discussed in the next two sections.

Finite chemical kinetics for the combustion reactions

A number of analyses have been made in which various assumptions limiting the rate of the chemical reactions have been made.[9,12–14] However the form of the reaction rate expression has been limited to the form:

$$\text{Rate} = A[\text{Fuel}]:[\text{Oxygen}]\exp(-E/RT)$$

where A and E are the reaction rate parameters. The

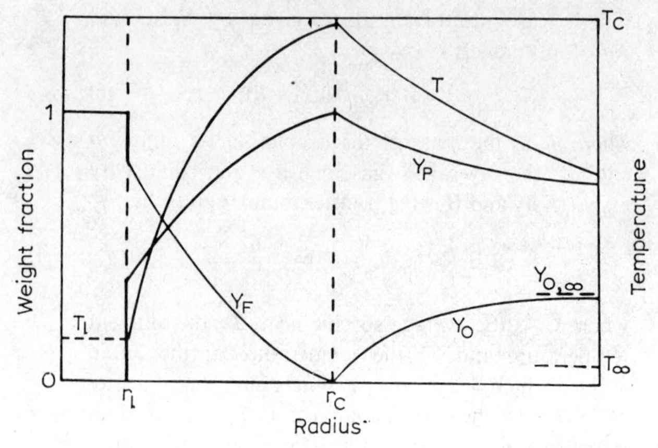

FIG. 5. Theoretical temperature (T) and composition profiles for a burning fuel droplet. Y_F, Y_P and Y_O are the mass fractions of the fuel vapour, combustion products and oxidant respectively, r_c is the radius of the combustion (flame) zone and T_c its temperature.

effect of this type of analysis is to illustrate that unless a zero activation energy is assumed then the reaction zone has a finite thickness. It is thus of interest to compare the "thin-flame" case as shown in Fig. 5 with the reaction zone of a n-heptane flame,[15] as in Fig. 6.

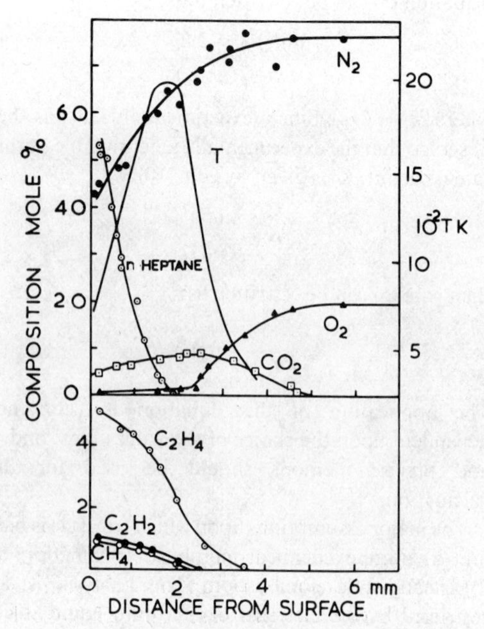

FIG. 6. Experimental temperature and composition profiles for a droplet of n-heptane.[15]

This clearly shows that in addition to oxidation there is significant cracking of the fuel vapour between the flame zone and the liquid surface. This cracking, which results in the formation of carbon particles in the region between the liquid surface is thus largely responsible for the characteristic yellow colour of flames of oil sprays.

The net outcome of these studies has been to show that except under extremes of pressure the assumption of infinite chemical kinetics is reasonable although the chemical detail within the reaction zone cannot be predicted.

2.2. *Experimental Studies of Rates of Droplet Burning in Weak Convective Fields*

The experimental techniques that have been used to obtain data on the rate of combustion of single droplets are (1) the suspended droplet method; (2) the supporting sphere technique; and (3) the free droplet

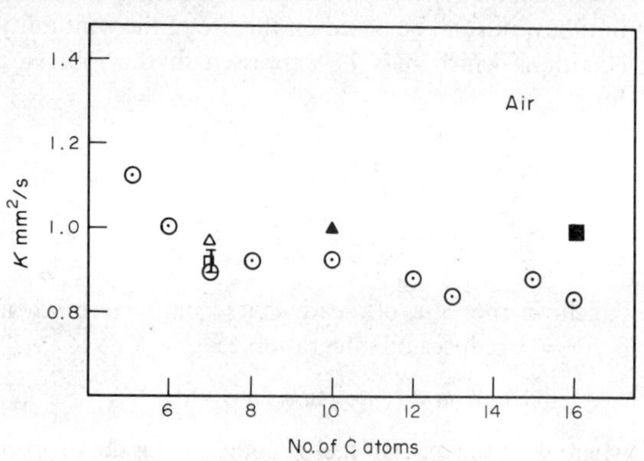

FIG. 7. Burning-rate constants for n-alkanes burning in air as a function of molecular weight. $T = 25°C$, $P = 1$ bar (after ref. 10).

technique. The data obtained by these techniques have been summarized by a number of authors, one of the most recent being a review by the present author which covers work up to early 1973.[10]

The suspended droplet work generally yields data on the burning rate coefficients for small (*ca* 1 mm diameter) droplets burning under conditions of natural convection and the data that has been obtained are given in Figs. 7, 8 and 9. The data are taken from references 7 and 10. It is clear that the influence of the molecular weight is of little significance for molecules larger than n-heptane.

Interpretation of data obtained from this type of experiment is not straightforward since the rates are affected by natural convection effects as well as effects from non-steady state combustion.[16–21] A number of studies have been undertaken in which the data has been correlated by equations of the form

$$K = K_0\{1+f(\text{Gr, Sc, Le})\}$$

where K_0 is the quasi-steady state burning rate coefficient under zero convection conditions and Gr, Sc and Le are the Greenhof, Schmidt and Lewis Numbers respectively.

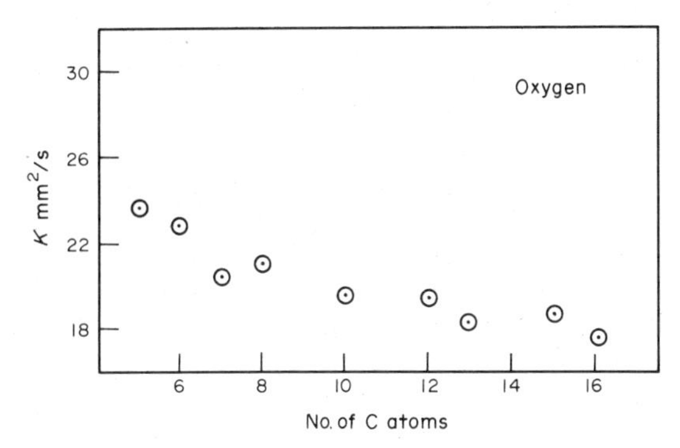

FIG. 8. Burning rate constants for n-alkanes burning in oxygen. $T = 25°C$, $P = 1$ bar (after ref. 10).

Law and Williams[11] have correlated the observed burning rate coefficients by the expression

$$K = K_0(1-f_A)(1+f_B)(1+f_C) \tag{12}$$

where

$$f_A = A(d^2_{L(av)}\rho^l)^{-1/m-1} \tag{13}$$

$$f_B = a(\mathrm{Gr})^n \tag{14}$$

and

$$f_C = 0.276\,\mathrm{Re}^{1/2} \tag{15}$$

The variation of the reaction rate with pressure is included in the convection factor, f_A, whilst the effect

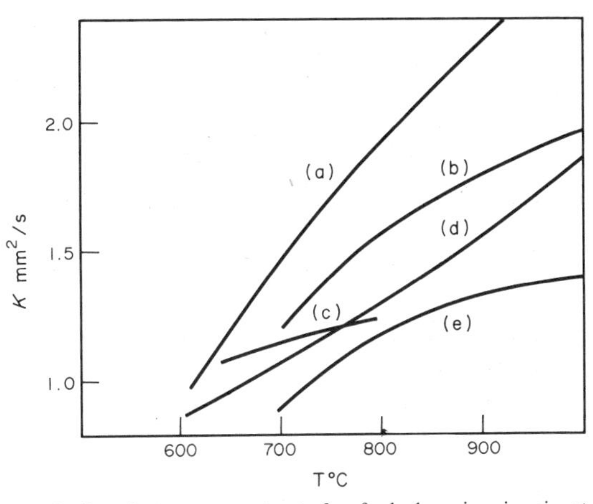

FIG. 9. Burning rate constants for fuels burning in air at elevated temperatures: (a) cetane, (b) light diesel oil, (c) heptane, (d) aviation kerosine and (e) benzene. Heavy fuel oils are similar to the light diesel oils up to 700°C but measurements at higher temperatures are not possible because of droplet swelling.

of natural and forced convection are included in f_B and f_C respectively. The effect of the initial droplet diameter, which influences the approach to steady-state burning is included in the average droplet diameter $d_{L(ar)}$. The terms l, m, n and a are determined empirically.

2.3. *Combustion of Droplets Under Forced Convection*

In droplets burning under conditions of natural convection, or in low velocity air streams or, as is the main practical case, the droplet is travelling through air at a low relative velocity, the flame envelopes the leading hemisphere as a stagnation point flame. A typical example of a simulated droplet is shown in Fig. 10.

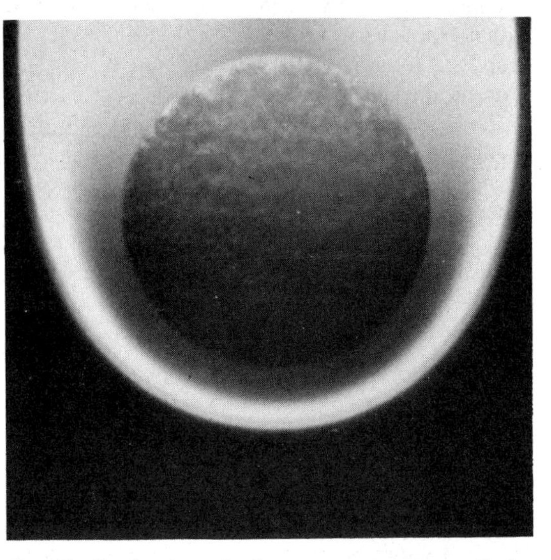

FIG. 10. Combustion of n-heptane on a porous cylinder. Carbon formation in the zone between the flame and the liquid surface is clearly visible.

A number of studies have been made of theoretical and experimental aspects of droplet combustion under forced convection conditions. Thus Brzustowski and Natarajan[22] have carried out a quasi-steady state analysis in which the convective mass transfer of the oxidant in the usual analysis is replaced by a modified oxidizer flux. This leads to the expression analogous to the classical solution but including terms for convection:

$$K = \frac{8M_f\lambda}{\rho L\bar{C}_p}(\beta\mathscr{D}^*_0 + \ln(1+B_{ev})) \tag{16}$$

where M_f is the molecular weight of the fuel and $\mathscr{D}^*_0$ is the dimensionless flux given by

$$\mathscr{D}^*_0 = \frac{Y_{O,\infty}}{2\,\mathrm{Le}\,i}\{2+0.60\,\mathrm{Gr}^{1/4}\,\mathrm{Sc}^{1/3}\} \tag{17}$$

Recently more detailed theoretical analyses have been formulated by Fendell *et al.*,[23] Buckmaster,[24] Law[25] and by Birchley.[26] The work by Birchley, which is essentially an extension of the earlier work by Fendell *et al.*,[23] produces a solution as a perturbation of the steady-state spherically symmetric droplet combustion case. This leads to detail about the influence of the Peclet number on the boundary layer region near the droplet, the rear stagnation region and the downstream wake region and such studies will be of great assistance in clarifying the effects of droplet drag.

The work by Buckmaster[24] and Law[25] again deal with perturbation techniques based on the Damkohler number (D), which is defined as the ratio of the characteristic flow time to the chemical time. Such analyses enable stability criteria to be examined and leads to information on the ignition and extinction of droplets.

Non-steady state analysis

Whilst the "d^2 law" (eqn. 5) is reasonably obeyed within experimental error during the combustion of droplets at (or near) atmospheric pressure the burning droplets are actually in a transient state during most of the life of the droplet.[16–21] This is best shown by changes in r_c/r_L during droplets, although when this phenomenon was first observed it was attributed to other causes.

Kotake and Okazaki[17] have undertaken a numerical analysis of these effects whilst Krier and Wronkiewicz[18,19] have shown that the variation of d^2 may be expressed in terms of the ratio of r_c/r_L. Thus

$$d^2 = d_0^2 - K\left\{t + \frac{1}{b}\ln\left(1 = \frac{b}{(a-1)}t\right)\right\} \tag{18}$$

where the variation in r_c/r_L is given by $(a+bt)$ in which a and b are constants and d_0 is the initial droplet diameter. However in practical terms, and particularly so in the combustion of light fuel oils under the majority of operating conditions, the "d^2 law" is sufficiently accurate for design purposes. A number of analyses have also been undertaken involving the non-steady state heating-up of the liquid droplet and its subsequent behaviour. Commercial liquid fuels are of course multi-component mixtures and during droplet heat-up and combustion the liquid droplet undergoes fractional distillation. Wood *et al.*[20] have given the following expression for the burning rate coefficients of binary mixtures.

$$K_{1,2} = \frac{8\lambda}{\bar{C}_p(Y_1\rho_{L,1}+Y_2\rho_{L,2})}\ln\left\{1+\frac{1}{Y_1L_1+Y_2L_2}\times\left[\bar{C}_p(T_\infty-T_L)+Y_{O,\infty}\left(\frac{Q_1Y_1+Q_2Y_2}{i}\right)\right]\right\} \tag{19}$$

where K_1, K_2, etc. refer to the components 1 and 2.

If the boiling points of the components differ greatly then disrupted boiling occurs. This does not actually happen with the lighter fuels but is a major feature of the combustion of heavy fuel oils together with liquid phase cracking as outlined in Section 3.

Experimental results have been correlated by a number of empirical formulations, which emanate basically from Frossling and Ranz and Marshall, as given by Agoston *et al.*[27] Perhaps the best form of these types of equations is given by Faeth and Lazar,[28] namely

$$\underset{\text{(forced convection)}}{K} = K_0\{1+0.278\,\mathrm{Re}^{1/2}\,\mathrm{Pr}^{1/3} \times(1+1.237/\mathrm{Re}\,\mathrm{Pr}^{4/3})^{-1/2}\} \tag{20}$$

this being consistent with experimental data in the range of $10 < \mathrm{Re} < 800$. Under higher gas velocities the mode of combustion of the droplet can change from "envelope" combustion to that of "wake" combustion, this occurring at the "extinction" velocity.* In this mode, combustion occurs only in the wake of the droplet and this greatly influences the drag characteristics of many droplets and the nature of the pollutants formed, this latter aspect is discussed in Section 5.

Data has been given by Agoston *et al.*[27] and Sjognen[29] and others[9,10] on the variation of the extinction velocity, V_c with droplet diameter. Whilst there is still some uncertainty about the interpretation of the data available at present, it does seem that $V_c \propto d_L^{1/2}$. The data obtained by Sjognen[29] show that 1 mm droplets have extinction velocities of about 50 cm/s in air and this implies that extinction velocities for droplets burning with air in a spray having, say, diameter of about 100 μm, would have an extinction velocity in the range of 5–10 cm/s. The extinction velocities also drop to zero when the oxygen content drops to 14–16 mole %.

A considerable number of studies have been undertaken to obtain information on the drag coefficients of evaporating and burning droplets. These have however been restricted to pure, relatively volatile fuels, in which case it has been found that the data can be summarized[30] as follows:

$$0 < \mathrm{Re} \leqslant 80 \qquad C_D = 27\,\mathrm{Re}^{-0.84}$$

$$80 < \mathrm{Re} \leqslant 10^4 \qquad C_D = 0.271\,\mathrm{Re}^{0.217}$$

$$10^4 < \mathrm{Re} \qquad C_D = 2$$

Natarajan and Ghosh[31] have recently found however that over the range $2 < \mathrm{Re} \leqslant 10^3$ that $C_D \simeq 1$. Since there is some divergence[10] between experimental data it seems that additional experimental data is desirable.

3. THEORETICAL AND EXPERIMENTAL STUDIES ON THE RATE OF COMBUSTION OF SINGLE DROPLETS OF HEAVY FUEL OILS

3.1. *The Nature of Combustion of Heavy Fuel Oils*

During the combustion of droplets of heavy fuel oils the composition of the liquid droplet changes because of fractional distillation which often takes place in a violent way as shown in Fig. 2. The high boiling point components initially present in the heavier oils can reach sufficiently high temperatures so that liquid phase cracking can take place leading to the formation of coke residues. Because of these phenomena the combustion of heavy fuel oils can only be adequately described by non-steady state analyses except for the early stages of combustion when the more volatile components are being evaporated in a similar way to the light oils. In view of the difficulties of handling both fractional distillation and liquid phase cracking mathematically very few analytical studies have been made and most of the information about medium and heavy fuel oils has been derived from experimental studies.

3.2. *Experimental Studies*

A number of studies of the combustion of medium and heavy fuel oils have been undertaken and it has

* This is sometimes termed the "transition" velocity.

been shown that combustion is complicated by the occurrence of disruptive boiling,[20] swelling and the formation of a carbonaceous residue. The processes involved are shown[32] in Fig. 11. The first major studies

(5) slow heterogeneous combustion of the carbonaceous residue at a rate of about one tenth that of the initial droplet (in terms of burning rate constants).

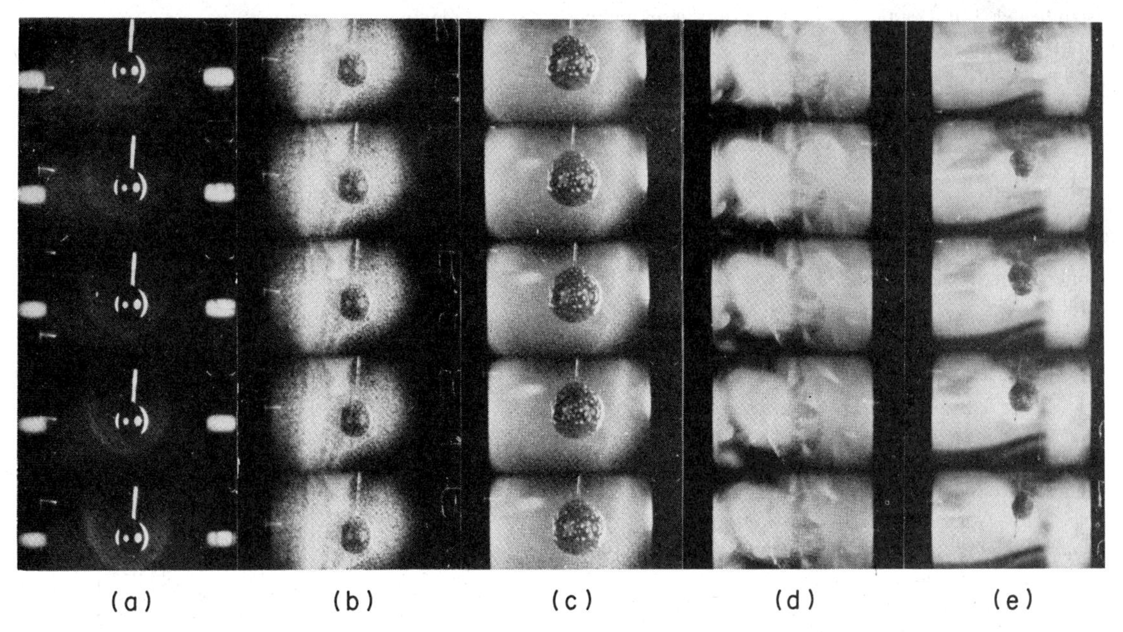

FIG. 11. Combustion of residual fuel oil in air, 650 μm dia droplet at 100 frames/s: (a) initial evaporation; (b) advanced evaporation; (c) flame initiated and bubbling commences; (d) maximum swelling, and (e) coke button formed (after ref. 32).

of the nature of the combustion of heavy fuel oils were undertaken by the MIT school[33] and by Topps,[34] Kobayasi[35] and Wood *et al.*[20] Whilst little work was undertaken during the 1960s on the combustion of heavy fuels a number of recent investigations[36–39] have been made prompted by pollution and energy conservation considerations.

Clearly the behaviour of medium and heavy fuel oil droplets is markedly different to that of the lighter fuel oils. Firstly, particularly in the case of residual oils, radiation plays a much greater role than in the case of light fuels. Secondly, since the droplet undergoes extensive swelling the "d^2 law" does not hold although equivalent burning rate constants can be defined.[35,36] The course of combustion of a medium to heavy fuel can be summarized as follows:

(1) heating up of the droplet and vaporization of the low boiling point components;
(2) self-ignition and combustion with slight thermal decomposition and continued vaporization of the volatile components;
(3) extensive disruptive boiling and swelling of the droplet together with considerable thermal decomposition to give a heavy tar;
(4) combustion of remaining volatile liquids and gases from the decomposition of the heavy tar which collapses and forms a carbonaceous residue having an open structure called a cenosphere;

The course of combustion of a typical heavy fuel oil is shown in Fig. 12a and the equivalent burning rate is also defined. When disruptive boiling is enhanced by emulsification of water with the fuel oil, extensive swelling of the droplet can take place during the

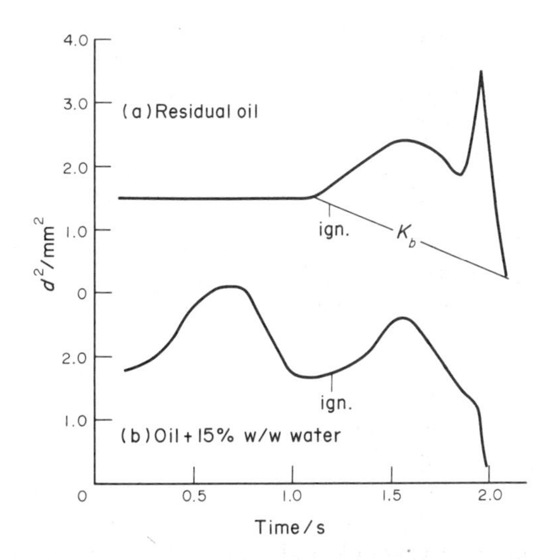

FIG. 12. Plot of d^2 against t for droplets at 800°C of, (a) fuel oil; and (b) fuel oil + 15% wt water (after ref. 36). The burning-rate constant (K_b) based on the burning time is also defined.

preheat stage prior to ignition as indicated in Fig. 12b.[36,40] This can result in extensive disruption of the droplet effectively resulting in finer atomization. In some instances the small satellite droplets thrown off from the parent droplet can ignite resulting in premature ignition. In many applications this is of little significance but it can cause problems in diesel engine applications.

The nature of the carbonaceous button, or cenosphere that is obtained at the completion of combustion is a function of the time-temperature history of the fuel, the nature of the fuel and particularly its asphaltene and ash content and whether it has been subject to an oxidizing or reducing environment. The sizes and quantity of the particulates produced is very much a function of droplet size and in the case of single droplet studies the size of the residual carbon button is dependent upon the supporting fibre.

3.3. *Theoretical Studies*

Analyses of the rate of combustion of heavy fuel oils generally follow the pattern outlined earlier for volatile fuels. However since the rate of the endothermic liquid phase cracking is a function of the internal temperature of the liquid droplet it must be calculated throughout the course of combustion. A number of such analyses have been undertaken by El-Wakil *et al.* to estimate the extent of liquid phase cracking.[37,38] Since the liquid phase decomposition process involves the overall reaction:

$$\text{High boiling point liquid} \rightarrow (\text{Low boiling point liquid}) + (\text{gas}) + (\text{coke})$$

the analysis has to take account of the contribution of the gas produced to the combustion process.

4. THE MECHANISM OF COMBUSTION OF DROPLETS AND SPRAYS OF FUEL OILS

Combustion of fuel oil sprays in practical systems is very complicated aerodynamically and consequently most detailed analyses have been applied to relatively simple geometrical systems.

4.1. *One-Dimensional Premixed Laminar Flames*

A number of theoretical studies of one-dimensional premixed flames of the type shown in Fig. 3 have been undertaken.[9,10] These have been principally by Tanasawa and Tesima,[41] Williams[8] and Nuruzzaman *et al.*[42] and such analyses present few problems. In contrast, analyses leading to the burning velocity of such a mixture are much more complicated. Williams[8] and Paterson and Fairbanks[43] have made such studies and generally these lead to theoretical burning velocities of about 40 cm/s, which are consistent with experimental data for most hydrocarbon fuels. These theories however are dependent upon assumptions of the ignition temperature of droplets.

Mizutani and Ogasawara[44] have determined burning velocities and low flammability limits for tetralin. They have proposed a "network" flame propagation theory which is based on the ignition delays of adjacent fuel droplets being the controlling mechanism. This theory gives good agreement with experimental data, at least over the range studied.

Very few measurements of burning velocities of premixed spray flames have been undertaken. Burgoyne and Cohen[45] have presented data for tetralin flames stabilized on a burner. Aldred and Williams[46] have also made burning velocity measurements for kerosine flames using 60 μm droplets and over a range of oxygen concentrations. Mizutani and Ogasawara[44] have given extensive data for tetralin sprays burning in a tube and this is given in Fig. 13 together with values computed

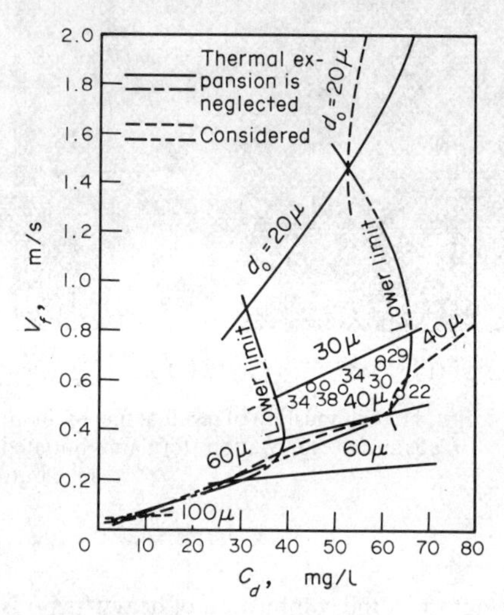

FIG. 13. Flame propagation velocities for tetralin/air mixtures measured by Mizutani and Ogasawara (ref. 44); 0, data of Burgoyne and Cohen (ref. 45).

from a spray combustion model. The experimental data obtained by Burgoyne and Cohen[45] are also plotted in this diagram and it is clear that the burning velocity is a function of the mass concentration (C_d) and droplet size. Furthermore the flame propagation is a function of the direction of propagation of the flame because of sedimentation effects. Thermal expansion effects due to combustion are also significant as indicated by Fig. 13. Data on the influence of oxygen concentration was also given by these authors.[44]

4.2. *Turbulent Premixed Flames*

A number of experimental investigations have been made of the burning velocities of turbulent premixed flames. The early studies were mainly concerned with droplet combustion rates in turbulent flames and the data obtained was limited because of the complexity of such systems.

Basevich and Kogarko[47] investigated the rate of flame propagation in turbulent fuel/air mixtures and found that in heterosized sprays the burning velocity decreased with increasing the drop size. Cekalin[48] obtained extensive data on turbulent flame velocities for various fuel/air ratios, mean drop size, droplet concentrations and turbulence intensities. It was shown that when the droplet diameters were greater than 30 μm then the "network" or relay mechanism of propagation dominated. The most detailed studies have been by Mizutani and Nakajima[49] in two publications which also summarize the earlier work. This work is primarily concerned with the burning velocities of propane–kerosine drop-air systems on open burners whilst the earlier work is concerned with kerosine drop clouds. Their results, which are summarized in Fig. 14,

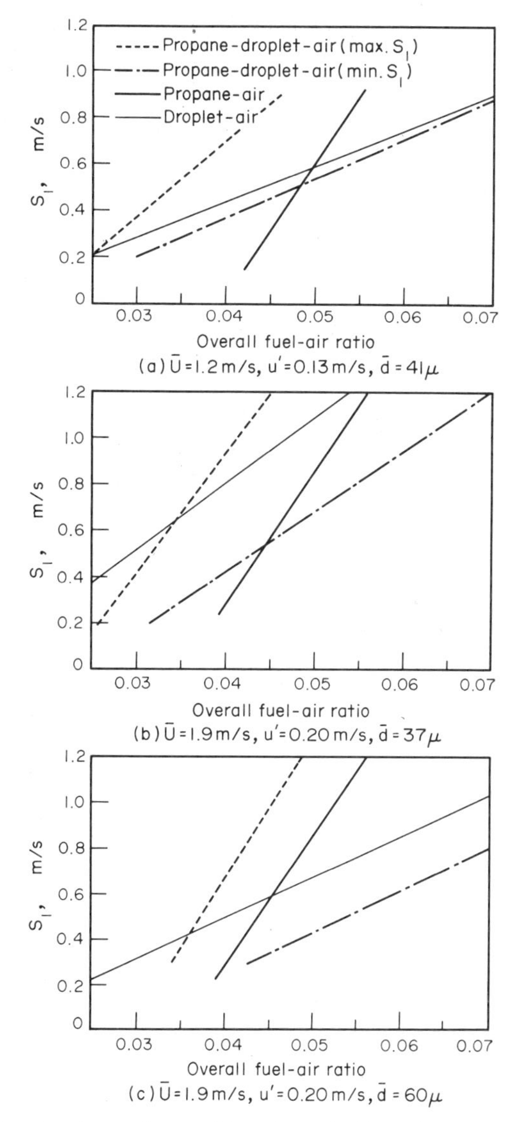

FIG. 14. Comparisons of burning velocities of propane–kerosine drop-air systems, propane–air mixtures and kerosine sprays (Mizutani and Nakajima[49]).

show that small amounts of kerosine drops added to a propane–air mixture markedly accelerate the burning velocity of the overall fuel–air ratio and extends the region of stable combustion towards the lean level. It was also found that the weaker the turbulence intensity in the flame zone, or the finer the droplets the greater are the effects observed. Kerosine mists containing sub-micron droplets do not exhibit such combustion-promoting effects. The same authors also studied combustion in a closed bomb and essentially the same features were observed.

Polymeropoulas and Das[50] studied the effect of droplet size on the burning velocity, S_u, of kerosine–air sprays but using mixtures containing a distribution of droplet sizes. Their results showed that as the degree of atomization of the spray increases, the burning velocity increases to a maximum value, and then decreases to the burning velocity approaching that of a premixed gas flow. The results obtained[50] are consistent for droplets in the 30–100 μm range with the expression obtained by Mizutani and Nakajima,[49] namely

$$S_u = \frac{6.800}{d_{3,2}}(\phi - 0.012)(v')^{1.15}\,\mathrm{ms}^{-1}$$

where ϕ is the fuel–air ratio, $d_{3,2}$ the Sauter mean diameter (μm) and v' is the *rms* velocity fluctuations in m/s.

However the form of this equation in which $S_u \alpha 1/\bar{d}$ does not appear to hold for sprays with $d < 30$ μm.

4.3. *Stirred Reactor Flames*

In this idealized type of reactor a spray is fed into a well-stirred combustion chamber. This system approximates to a number of industrial high intensity combustion chambers such as gas turbines or certain heaters although apparently there have been no studies using reactor designs equivalent to those used for gaseous fuel combustion research. Courtney[51] has considered the combustion of monosized propellants and shown that the combustion intensity, I, is given by

$$I \simeq \frac{M_0}{1 + 2r_0^2 M_0 / 5K\rho_g} \tag{21}$$

where M_0 is the mass flow of gas from the chamber, r_0 the initial droplet size, ρ_g the gas density and K the burning-rate coefficient appropriate to the turbulent conditions within the reactor.

Grout and Eisenklam[52] have studied theoretically the gasification of a spray in a well-stirred reactor followed by complete combustion in a second-stage furnace or convertor. The system was optimized with regard to minimum solid emission using a mechanistic model based on evaporation control in the gasifier and carbon burn-out in the converter.

4.4. *Jet Flames of Volatile Fuels*

Very many spray combustion studies have been undertaken in which turbulent jets of fuel sprays burn with air. The combustion of sprays of volatile fuels, such as aviation kerosines, has been the subject of a number of reviews[8,9,53] and is discussed by Chigier[54] elsewhere; consequently it is only dealt with briefly here. The experimental studies on such systems by

Chigier and co-workers[51,54] and by Onuma and Agasawara[55] have shown that these volatile fuels burn in much the same way as an equivalent gaseous jet. Essentially these flames are of the homogeneous jet type as shown in Fig. 4b and droplet burning involving envelope flames plays only a minor role and this being restricted to systems which contain large drops.

Fewer studies have been made of laminar flames. Briffa and Dombrowski[56] and Dombrowski *et al.*[57] have made a number of investigations of flat and swirling flames. One such typical flame is shown in Fig. 15.

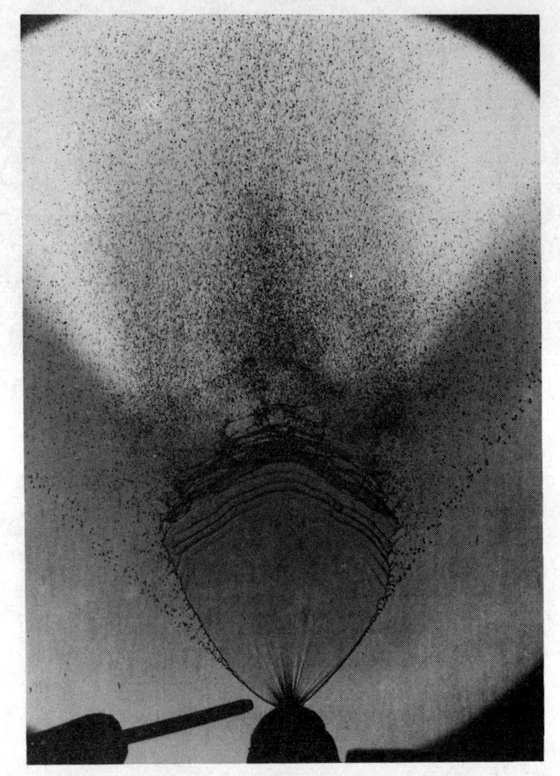

FIG. 15. Flat spray flame.

Recently Moore *et al.*[58] have undertaken studies of *n*-heptane burning in such flames and has made a mathematical analysis based upon the approach developed by Spalding and co-workers.[59] Good agreement appears to have been achieved indicating the great potential for this modelling approach. The technique can also be extended to turbulent systems.

Whilst most work reported has been concerned with unconfined jets of sprays one investigation, by Bracco,[60] has been concerned with the combustion of a layered spray confined in a tank. The spray layer is initially in the upper part of the tank and the droplets fall under gravitational forces to a lower hot air layer.

A number of investigations related to gas turbine applications have also been undertaken. One of the central problems arises from the fact that under many operational conditions the droplets are injected into a high temperature environment under what are essentially critical conditions. Consequently there are no latent heat of vaporization requirements and the droplet, at least in theory, is replaced by a puff of gas.[10] However, the liquid must still be heated for the critical conditions to be reached, which represents an energy requirement similar to a heat of vaporization. A number of studies[10,61,62] of single droplets have been under combined high temperature and high pressure environments but it is not at all clear that instantaneous gasification of the droplet occurs. Consequently the standard laws of droplet burning apparently apply even under these conditions and modelling in say, gas turbines, has been carried out apparently successfully on this basis.[63]

4.5. *Jet Flames of Heavy Fuel Oils*

Most studies of the heavy fuel oil flames are made in combustion plant and the data obtained is not generally amenable to fundamental interpretation of the combustion mechanism. Most data capable of interpretation has been obtained in the furnace facility at the International Flame Research Foundation in Ijmuiden and the work is summarized in their annual reports[64] and by Beér and Chigier.[65]

Experiments involving heavy fuel oils are extremely difficult to undertake because of the large quantities of particulate solids and liquids that are present. However extensive data is available on gross flame properties such as combustion intensities and flame radiative properties so that modelling of spray combustion involving heavy fuel oils can be undertaken.[66]

A considerable amount of work on the combustion of diesel fuel oils has been carried out. Again critical temperatures are exceeded in the combustion chambers but in practice it seems that standard droplet vaporization and combustion laws are obeyed. The effect of droplet size distribution on combustion under swirling conditions in diesels has been investigated by Henein.[67] Under high swirl conditions the very small droplets are concentrated in the leading edge of the swirled spray resulting in a considerable variation in the fuel/air ratio across the spray.

A number of studies have been undertaken[68–71] in which the spray injected into the combustion chamber is considered to react as an equivalent gaseous jet. This approximation is sufficient to permit modelling of smoke and NO_x but since arbitrary parameters are employed the assumption of this particular model is not indicative that this is what actually physically occurs.

5. THE MECHANISM OF FORMATION OF POLLUTANTS DURING THE COMBUSTION OF OIL SPRAYS

Since the combustion of sprays mainly involves the gas phase combustion of hydrocarbon vapour with air the predominant pollutants are typical of such systems, namely carbon monoxide, NO_x, SO_x, unburned hydrocarbons and smoke produced via gas phase reactions. In addition, as outlined earlier, particulate products also result from reactions in the liquid droplets as a consequence of cracking reactions. Any inorganic material present in the oil is also concentrated in this product as ash. The actual amounts of these pollutants in the combustion products is very dependent upon the combustion conditions and is therefore a function of the plant or engine used.[72] These practical aspects are not discussed here but rather the recent work relating to the mechanism of their formation is outlined.

5.1. *The Formation of Particulate Pollutants*

The formation of smoke and cenospheres from the combustion of sprays is one of the dominant present-day pollutant considerations. There are two sources of particulates, smoke from gas phase reactions of the fuel vapour and cenospheres or coke (carbonaceous) particles from the liquid phase cracking of the liquid.

If droplets are burning in a heterogeneous mode then the mechanism and extent of soot formation is dependent upon whether the droplet is burning as an envelope flame or a wake flame. In the case of envelope combustion, the soot is formed via gas phase cracking of the hydrocarbons[15,73] (c.f. Fig. 6) in the region between the flame zone and the surface. In the case of wake combustion, the soot is only formed in the stagnation region of the flame and several studies have been undertaken of the behaviour of wake flames.[74,75]

Gollahalli and Brzustowski[74] have studied this effect in detail and made measurements of the composition profiles across the flame burning in the wake of both "envelope" and "wake" flames. Their structures are quite different, the former has a near-wake of several diameters similar to a gaseous flame with a long soot-producing trail. The wake flames are similar to the flame stabilized behind a flame holder and the soot-producing region is small. Cernez and Dobovisek[39] investigated overall soot emissions from diesel fuel oil again using the porous sphere technique. Sphere diameters ranged from 1 to 5 mm and both envelope and wake flames were investigated. It was concluded that the porous sphere technique does not give realistic values of blow-off velocities although they found that the extinction velocity is a function of $d_L^{1/2}$ (c.f. Section 2.3). They also found that soot emissions increase with droplet diameter and approach air velocity and reach their maximum just at the transition point. Wake flames do not produce soot.

Considerable interest has recently been expressed in the formation of solid residues from the liquid phase cracking of droplets of heavy fuel oils. During the ignition delay period prior to ignition, the liquid droplet loses the lighter components by fractional distillation. Once ignition occurs the temperature in the central region of the droplet increases resulting in the thermal decomposition of the liquid hydrocarbon. This is a function of the aromatic content of the fuel which is particularly a characteristic of the high asphaltene content oils. It is also a function of the organo-sulphur and nitrogen compounds present in the oil. Little is known at the present time about the actual mechanism leading to the carbonaceous product but it is markedly controlled by the rate at which the centre of the droplet attains a temperature high enough for cracking to occur and the length of time it remains at this temperature. For this reason the formation of carbonaceous residues, or cenospheres, from the combustion of heavy fuel oils particularly in highly asphaltic oils is a function of droplet size and its extent is greatly minimized by finer atomization or water emulsification.

A number of studies have been made of the nature of the carbon residues produced by the combustion of various fuels and extensive work has been undertaken by Hottel,[33] Topps[34] and the CEGB.[32]

Other studies have been concerned with the theoretical modelling of the liquid phase cracking reactions and whilst chemically simple mechanisms have been adopted they nevertheless give a reasonable insight into the controlling features.

In the case of combustion of sprays in the homogeneous mode then only gas phase smoke is produced and the mechanism is similar to that in an equivalent hydrocarbon gaseous flame.

5.2. *The Formation of CO and NO_x*

The formation of carbon monoxide in spray combustion is not very dependent upon the combustion of individual droplets but upon local fuel/air concentrations generally and the subsequent conditions which affect its burnout. Thus to all intents and purposes sprays can be treated as equivalent gaseous fuel flames for modelling purposes. Considerable data are available on carbon monoxide yields from spray combustion systems in diesels, gas turbines as well as furnaces and boilers but these are all dependent upon the individual configurations.

The mechanism by which CO emissions result from droplet burning is well understood. In all spray systems CO is initially formed in the hydrocarbon combustion zone and then "burns out" by the reaction

$$CO + OH \rightleftharpoons CO_2 + H$$

until equilibrium levels of CO are reached. The level of CO actually reached is thus frequently dependent upon the reaction time available and the degree of mixedness rather than any feature of spray combustion itself.

Few studies have been made of CO formation during the combustion of single droplets. Aldred *et al.*[15] studied the formation of CO in the region between the flame zone and the liquid surface in an envelope flame using a porous sphere (Fig. 6). Whilst the CO levels are high in this region the CO is reacted before leaving the flame zone.

A number of studies have been undertaken of NO_x formation from a variety of spray equipment and the general features of the mechanism of formation have been clarified.

Essentially there are three basic routes that may be followed, the first of which is the thermal or Zeldovich route involving the reactions:

$$O + N_2 \rightleftharpoons NO + N$$

$$N + O_2 \rightleftharpoons NO + O$$

and in rich systems the following reaction also becomes important

$$N + OH \rightleftharpoons NO + H$$

Under fuel-rich conditions there is now fairly comprehensive evidence that another route can be invoked. This is the "prompt"-NO reaction which involves the following steps

$$\text{Fuel, } RH \rightarrow \rightarrow CH,\ C_2$$

$$CH + N_2 \rightarrow CN + HN$$

$$C_2 + N_2 \rightarrow 2CN$$

$$CH + H_2 \rightleftharpoons HCN$$

$$HCN \rightarrow \rightarrow NO$$

$$HN \rightarrow \rightarrow NO$$

Since this reaction occurs so rapidly that it is completed in the flame zone, it implies that NO_x yields from systems in which this occurs will be higher for a given residence time. A further consequence is that under very rich conditions HCN may be emitted.

The third route involves the oxidation of any fuel-nitrogen that may be present in the oil and this is of course a function of the origin of the oil and whether it is a light fuel oil or a heavy fuel oil since fuel–nitrogen compounds tend to predominate in the latter. The mechanism of this reaction is still the subject of some speculation but essentially the fuel–nitrogen compounds are converted into NO by means of a reaction which is not very temperature dependent. However, the extent of conversion is a function of the fuel–nitrogen content itself being about 100% for low levels (0.1%) and reducing to about 50% for higher levels (0.5%). This implies that this contribution to the total NO_x yield, which may be as much as 30% in some cases, cannot be reduced by a reduction in the combustor temperature.

Finally the conditions in the burning spray will determine the ratio of NO_2/NO. The extent of NO_2 formation is dependent upon the rate of the reaction

$$NO + HO_2 \rightleftharpoons NO_2 + OH$$

where the HO_2 is produced by the reactions

$$OH + CO \rightleftharpoons CO_2 + H$$

$$H + O_2 \rightleftharpoons HO_2$$

and is a function of the excess air in the system and the pressure. Thus the highest NO_2/NO ratios are observed in diesel engines and gas turbines.

A number of studies have been made of NO_x formation using the porous sphere technique. Hart *et al.*[76] studied the formation of NO and NO_2 in the flame zone for *n*-heptane and methanol flames and also investigated the influence of a fuel–nitrogen compound, pyridine, on the NO_x produced. Their conclusions are consistent with the mechanism outlined earlier. Bracco[77] has also investigated NO_x formation in methanol flames and pyridine flames. Whilst there are certain differences between these two studies the general conclusions are in accord. Bracco[78] has also undertaken a theoretical analysis of NO formation around single droplets.

Cernez and Dobovisek[39] have studied specific NO_x emissions from porous spheres and found that it is dependent upon ambient conditions and type of fuel. The emission increases in the region of the envelope flame and reaches its maximum at the flame transition velocity. In the case of wake flames, the NO_x emission is smaller than that of the envelope flames presumably because the size of the flame is smaller.

The application of these analyses are of course dependent upon the mode of combustion of the spray, heterogeneous spray combustion behaving in a significantly different way to homogeneous spray combustion.

REFERENCES

1. BAILEY, T. C., *J. Inst. Fuel* **32**, 202 (1959).
2. ROMP, H. A., *Oil Burning*, Martinus Nighoff, The Hague, 1937.
3. GODSAVE, G. A. E., *Nature* **164**, 708 (1949); **171**, 86 (1953).
4. GODSAVE, G. A. E., *Fourth Symposium on Combustion*, p. 818, Williams and Wilkins, Baltimore, 1953.
5. SPALDING, D. B., *Fourth Symposium on Combustion*, p. 847, Williams and Wilkins, Baltimore, 1953.
6. AGAFANOVA, F. A., GUREVICH, M. A. and PALEER, I. I., *Sov. Phys. tech. Phys.* **2**, 1689 (1958).
7. WISE, H. and AGOSTON, G. A., *Literature of the Combustion of Petroleum*, p. 125, American Chemical Society, Washington, D.C., 1958.
8. WILLIAMS, F. A., *Combustion Theory*, Addison Wesley, Reading, Mass, 1965.
9. WILLIAMS, A., *Oxidation and Combustion Reviews* **1**, 1 1968.
10. WILLIAMS, A., *Combust. Flame* **21**, 1 (1973).
11. LAW, C. K. and WILLIAMS, F. A., *Combust. Flame* **19**, 393 (1972).
12. ANNAMALAR, K., KUPPU RAO, V. and SREENATH, A. V., *Combust. Flame* **16**, 287 (1971).
13. LOVELL, J., WISE, H. and CARR, R. E., *J. Chem. Phys.* **25**, 325 (1956).
14. TARIFA, C. S., PEREZ DEL NOTARIO, P. and MORENO, F. G., *Eighth Symposium on Combustion*, p. 1035, Williams and Wilkins, Baltimore, 1962.
15. ALDRED, J. W., PATEL, J. C. and WILLIAMS, A., *Combust. Flame* **17**, 139 (1971).
16. KUMAGAI, S. and ISODA, H., *Sixth Symposium on Combustion*, p. 726, Reinhold, New York, 1957.
17. KOTAKE, S. and OKAZAKI, T., *Int. J. Heat Mass Transfer* **12**, 595 (1969).
18. KRIER, H. and FOO, C. L., *Oxidation and Combustion Reviews* **6**, 111 (1972).
19. KRIER, H. and WRONKIEWIEZ, J., *Combust. Flame* **18**, 159 (1972).
20. WOOD, B. J., WISE, H. and INAMI, S. H., *Combust. Flame* **4**, 235 (1960).
21. WILLIAMS, A., *The Combustion of Sprays of Liquid Fuels*, Paul Elek, 1976.
22. BRZUSTOWSKI, T. A. and NATARAJAN, R., *Can. J. Chem. Eng.* **194** (1966).
23. FENDELL, F. E., *Astronautica Acta* **11**, 419 (1965).
24. BUCKMASTER, J. D., *Combust. Flame* **24**, 79 (1975).
25. LAW, C. K., *Combust. Flame* **24**, 89 (1975).
26. BIRCHLEY, J. C., *Proc. Roy. Soc. Lond.* **A347**, 195 (1975).
27. AGOSTON, G. A., WISE, H. and ROSSER, W. A., *Sixth Symposium (International) on Combustion*, p. 708, Reinhold, New York, 1957.
28. FAETH, G. M. and LAZAR, R. S., *A.I.A.A.J.* **9**, 2165 (1971).

29. SJOGNEN, A., *Fourteenth Symposium (International) on Combustion*, p. 919, The Combustion Institute, Pittsburgh, 1973.
30. DICKERSON, R. A. and SCHUMAN, M. D., *J. Spacecraft* **2**, 99 (1956).
31. NATARAJAN, R. and GHOSH, A. K., *Fuel (Lond.)* **54** (1975).
32. ANSON, D. and STREET, P. J., C.E.G.B. Report No. RD/M/M56.
33. HOTTEL, H. C., WILLIAMS, G. C. and SIMPSON, H. C., *Fifth (International) Symposium on Combustion*, p. 101, Reinhold, New York, 1955.
34. TOPPS, J. E. C., *J. Inst. Pet.* **37**, 535 (1951).
35. KOBAYASI, J., *Fifth Symposium on Combustion*, p. 141, Reinhold, New York, 1955.
36. JACQUES, M. J., JORDAN, J. B. and WILLIAMS, A., *Deuxieme Symposium European Sur La Combustion*, p. 397, The Combustion Institute, 1975.
37. CHEN, C. S. and EL-WAKIL, M. M., Diesel engine combustion, *Proc. Inst. Mech. Eng.* **184**, Part 3J, 95 (1969–1970).
38. SHYU, R. R., CHEN, C. S., GOUDIE, G. D. and EL-WAKIL, M. M., *Fuel* **51**, 135 (1972).
39. CERNEJ, A. and DOBOISEK, A., Private communication, 1975.
40. IVANOV, V. M., KANTOROVICH, B. V., RAPROVETS, L. S. and KHOTUNTSEV, L. L., *Vestn. Akad. Nauk S.S.S.R.* 56 (1957); IVANOV, V. M., KANTOROVICH, B. V., RAPROVETS, L. S. and KHOTUNTSEV, L. L., *Trudy Inst. Gorychikh Iskopaenykh, Akad. Nauk. U.S.S.R.* **11**, 156 (1959); IVANOV, V. M. and NEFEDOV, P. I., NASA Tech. Transl. TTF-258, 23 (1965).
41. TANASAWA, Y. and TESIMA, T., *Bull. J.S.M.E.* **1**, 36 (1958).
42. NURUZZAMAN, A. S. M., SIDDALL, R. G. and BEÉR, J. M., *Chem. Engng Sci* **26**, 1635 (1971).
43. PATERSON, R. E. and FAIRBANKS, D. F., A.P.I. Research Conference on Distillation Fuel Composition, Chicago, 1961.
44. MIZUTANI, Y. and OGASAWARA, M., *J. Inst. Heat Mass Transfer* **8**, 921 (1965).
45. BURGOYNE, J. H. and COHEN, L., *Proc. Roy. Soc. Lond.* **A225**, 375 (1954).
46. ALDRED, J. W. and WILLIAMS, A., To be published.
47. BASEVICH, V. YA. and KOGARKO, S. M., *Eighth Symposium on Combustion*, p. 1113, Williams and Wilkins, Baltimore, 1962.
48. CEKALIN, E. K., *Eighth Symposium on Combustion*, p. 1120, Williams and Wilkins, Baltimore, 1962.
49. MIZUTANI, Y. and NAKAJIMA, A., *Combust. Flame* **20**, 343, 351 (1973).
50. POLYMEROPOULAS, C. E. and DAS, S., *Combust. Flame* **25**, 247 (1975).
51. COURTNEY, W. G., *A.R.S.J.* **30**, 356 (1960).
52. GROUT, P. D. and EISENKLAM, P., *High Temperature Chemical Reaction Engineering*, The Institution of Chemical Engineers Symposium Series No. 43, 1975.
53. CHIGIER, N. A. and MCCREATH, C. G., *Astronautica Acta* **1**, 687 (1974).
54. CHIGIER, N. A., *Prog. Energy Combust. Sci.* **2** (1976).
55. ONUMA, Y. and OGASAWARA, M., *Fifteenth Symposium (International) on Combustion*, p. 453, The Combustion Institute, 1975.
56. BRIFFA, F. E. J. and DOMBROWSKI, N., *Proc. Roy. Soc.* **A320**, 309 (1970).
57. DOMBROWSKI, *N.*, HORNE, W. and WILLIAMS, A., *Combust. Sci. Tech.* (1975).
58. MOORE, J. G., JELLINGS, W. and MOORE, J., *Sixteenth Symposium (International) on Combustion.* To be published.
59. PATANKAR, S. V. and SPALDING, D. B., *Heat and Mass Transfer in Boundary Layers* (2nd edn.), Int. Textbook, London, 1970.
60. BRACCO, F. V., *Unsteady Combustion of a Confined Spray*, National Heat Transfer Conference, Atlanta, Georgia, 1973.
61. NATARAJAN, R. and BRZUSTOWSKI, T. A., *Combust. Sci. Tech.* **2**, 259 (1970).
62. CANADA, G. C. and FAETH, G. M., *Fourteenth Symposium (International) on Combustion*, p. 1345, The Combustion Institute, 1973.
63. ODGERS, J., *Fifteenth Symposium (International) on Combustion*, p. 1321, The Combustion Institute, 1975.
64. International Flame Research Foundation, Annual Reports, Ijmuiden, Holland.
65. BEÉR, J. M. and CHIGIER, N. A., *Combustion Aerodynamics*, Applied Science Publishers, London, 1972.
66. BUETERS, K. A., COGOLI, J. G. and HABELT, W. W., *Fifteenth Symposium (International) on Combustion*, p. 1245, The Combustion Institute, 1974.
67. HENEIN, N. A., Combustion and emission formation in fuel sprays injected in swirling air, SAE Paper No. 710220, 1971.
68. ADLER, D. and LYNN, W. T., *Proc. I. Mech. Eng.* **184**, Part 3J, 171 (1969–1970).
69. KHAN, I. M., *Proc. I. Mech. Eng.* **184**, Part 3J, 36 (1969–1970).
70. KOMIAYAMA, K., *Deuxieme Symposium European Sur La Combustion*, p. 641, The Combustion Institute, 1975.
71. CROOKES, R. J., JANOTA, M. S. and TAN, K. J., *Combustion Institute European Symposium*, p. 427, 1973.
72. HELION, VON R., DELATRONCHETTE, C., SUNDERMANN, P. and BRANDEL, A., *V.G.B. Kraftwerkstechnik 55* **2**, 88 (1975).
73. ABDEL-KHALIK, S., TAMARU, T. and EL-WAKIL, M. M., *Fifteenth Symposium (International) on Combustion*, p. 389, The Combustion Institute, 1974.
74. GOLLAHALLI, S. R. and BRZUSTOWSKI, T. A., *Fourteenth Symposium (International) on Combustion*, p. 1333, The Combustion Institute, Pittsburgh, 1973.
75. GOLLAHALLI, S. R. and BRZUSTOWSKI, T. A., *Combust. Flame* **22**, 313 (1974).
76. HART, R., NASRALLA, M. and WILLIAMS, A., *Combust. Sci. Tech.* **11**, 57 (1975).
77. LUDWIG, D. E., BRACCO, F. V. and HARRJE, D. T., *Combust. Flame* **25**, 107 (1975).
78. BRACCO, F. V., *Fourteenth Symposium (International) on Combustion*, p. 831, The Combustion Institute, 1973.

Manuscript received, 25 June 1976

Prog. Energy Combust. Sci., Vol. 2, pp. 181–238, 1977. Pergamon Press. Printed in Great Britain

GEOTHERMAL ENERGY DEVELOPMENT

H. Christopher H. Armstead
Rock House, Ridge Hill, Dartmouth, South Devon, England

1. THE SPECTRE OF AN ENERGY FAMINE

Ever since the dawn of the industrial revolution Man's appetite for energy has grown at a steadily increasing rate. The human way of life has become such that dislocation or even disaster can threaten if that growing appetite is not satisfied. The political events of 1973 in the Middle East have shown that dislocation, at any rate in the highly industrialized countries, can occur almost overnight at the crack of the Arab whip, and there is no lack of prophets who foretell disaster within a few decades or even sooner. It is important to examine the historical pattern of this energy appetite and to speculate about its future trends, so that we may better understand the urgent rôle which geothermal energy may be expected to play.

From the year 1900 to 1950 the total world production of energy grew at an average annual compound rate of 2.6%. From 1950 to 1960 this growth rate rose to an average of 4%. From 1960 to 1970 it rose further to about 5.5% per annum.[1] These figures show that the energy demand has not merely been continuously growing, but that the *rate of growth* has been accelerating; that is to say, energy growth has been more rapid than in accordance with simple exponential law.

The implications of this profligacy are alarming. Even allowing for the fact that the world population rose during the first 70 years of this century by a factor of more than 2.31[1] the figures are still disturbing; for the energy consumption during the same interval of time increased more than nine-fold. Thus the energy consumption *per capita* rose nearly four-fold in 70 years.

Until 1973 no difficulty had been experienced in matching the supply of energy to the demand: the world growth in energy consumption until that year could be regarded as uninhibited and subject to normal market forces. In 1973, however, the Middle Eastern oil crisis occurred and a very marked, politically inspired, rise in the price of oil fuel, which has not been without influence upon the prices of alternative fuels such as coal. The pattern of energy growth suddenly became disturbed, and it is still too early to say with any certainty whether that disturbance was temporary or permanent: (it takes about 3 years or so for world statistics to be collected and published). For the purpose of the argument which follows, 1972 will be taken as the last year of "natural" energy growth.

About 98% of the world's energy demands were, until the 1950s, satisfied by fossil fuels. The remainder was mainly accounted for by hydro power. By 1972, partly owing to the advent of nuclear power and a miniscule contribution from geothermal energy, the share of fossil fuels had very slightly fallen to about 97.5%. We still rely almost entirely upon fossil fuels.

The World Power Conference estimates of world energy resources published in 1968 showed a total believed stock of fossil fuels—proved and inferred—of about 8.8×10^{12} tonnes of coal equivalent. The world energy demand in 1972 was 7566×10^6 tonnes of coal equivalent.[2] The estimated fossil fuel resources would therefore have sufficed to keep the world supplied with its energy needs for 1163 years at the 1972 level of demand. On the face of it, this would seem to give no cause for immediate alarm, especially when it is remembered that fossil fuels are not the only available sources of energy; but the figure of 1163 years makes no allowance for *growth.*

In order to see the effect of growth upon this preliminary evidence it is first necessary to assume some growth law. Now despite the accelerating growth in energy consumption during the first 70 years of this century it was a fact that the average annual growth from 1970 to 1972 was only about 4%. This could well be due to inevitable short-term fluctuations lacking long-term significance, or perhaps to the first rumblings of the economic troubles that were to assail the world only a year or two later; on the other hand it could be indicative of the beginning of an end to accelerating growth—perhaps of the establishment of something nearer to simple exponential growth. Endless speculation could be devoted to such considerations; but just in order to illustrate the alarming effects even of simple exponential growth, let it here be assumed that from 1972 onwards energy growth will be maintained at a steady rate of 5% *per annum* (compound).

Allowing for 5% p.a. exponential growth as from 1972, it can be shown that the 1968 estimated world stock of fossil fuels, unsupported by other energy sources, would last only 82 years until the year 2054—little more than three-quarters of a century from now. Our 1163 years have shrunk to about 1/14th of the time! Now of course the 1968 World Power Conference estimates were *only* estimates, which could have been wildly inaccurate. But let it be supposed that the estimates were ten times too low—a rather improbable assumption—then all we should gain would be a further 47 years, i.e. until just after the year 2100, or about four or five generations from now. Even if the estimated fuel stocks were 100 times too low we should only gain a further 47 years.

Now these periods assume that fossil fuels would remain available, as though at the turn of a tap, until they finally became exhausted. In practice, as they become scarcer they would become less accessible, prices would soar, there would be keener international struggles to acquire the remaining fossil fuels, and something like market chaos would set in long before stocks became depleted. Moreover, fuels will become increasingly in demand for the chemical industries and not merely for burning. So the position would be even far more alarming than these figures would suggest.

Two conclusions can be drawn from these revealing figures:

(1) We cannot, as in the past, rely mainly upon fossil fuels as our chief source of energy for very much longer.

(2) We cannot sustain an exponential growth in energy demand at anything approaching 5% annual rate. In fact we cannot sustain exponential growth at *any* rate indefinitely because we live on a finite planet with large, but finite, energy resources, while exponential growth is a divergent series that runs towards infinity at an alarming rate.

Figure 1 shows the likely broad pattern of future energy supply and demand. Curve "A" represents actual world energy demand from 1900 to 1972 and extrapolated at 5% p.a. exponential growth (to scale) starting at 1972 with a relative demand of 1. The shaded area "D" represents the entire world energy

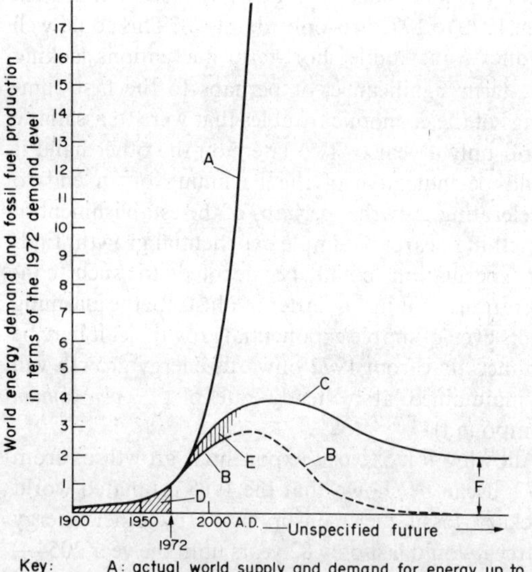

Key:
A: actual world supply and demand for energy up to 1972, and the extrapolated future demand assuming 5% annual exponential growth.
B: probable shape (not to scale) of future fossil fuel production
C: probable shape (not to scale) of future world energy demand if disaster is to be avoided.
D: total world energy consumption from 1900 to 1972 (about 2% of the WPC 1968 estimates of world's fossil fuel reserves
E: deficit between world energy demand and fossil fuel availability, to be met from new energy sources and economics.
F: ultimate stability when worlds energy demand can be met from renewable sources alone.

FIG. 1. The probable pattern of future fossil fuel supplies and World energy demand. (*N.B.* On the vertical scale "1" represents the World demand for energy—7566 million tonnes coal equivalent—in 1972, the last year of uninhibited energy supply and demand before the Middle Eastern crisis of 1973.)

consumption since the year 1900 (which would scarcely differ from the total consumption since the dawn of history). All but a tiny fraction of this area "D" represents fossil fuels, and it seems likely that the whole area "D" accounts for only about 2% of the total world stocks of such fuels. But, as has been shown above, the remaining 98% could serve the needs of Curve "A" only until shortly after the middle of the next century if we have to continue to rely almost entirely on fossil fuels. Curve "B" represents the probable *shape* (not to scale) of the available supplies of fossil fuels. Sheer necessity will probably cause the supplies of fossil fuels to rise for some years, because too violent changes in long established trends could wreak havoc upon the world economy; but it is highly probable that the rate of growth will be slower than in past years and that it will continue to decline, partly because it is now the avowed policy of some of the oil-producing countries to conserve more of their oil in the ground against future needs, partly because of the declining coal production in some countries, partly because of the rising prices of fuels in terms of real money (which will force consumers to economize) and partly because new energy forms will undoubtedly be developed to an increasing extent. The discovery of new oil fields and the exploitation of less commercial fuels such as oil shales will to some extent offset these influences, but on the whole it seems likely that the production of fossil fuels will rise at a decreasing rate to a peak—possibly in the early part of the next century—and then gradually start declining somewhat in the manner shown in Fig. 1, approaching zero asymptotically at some very remote future date when stocks become virtually depleted. Against the background of history, by far the bulk of fossil fuels will have been squandered within perhaps 100–150 years—a mere instant in the life of Mankind, with very attenuated production over longer periods on either side of the hump.

To maintain the demand curve "A", enclosing ever-increasing areas beneath it that represent cumulative consumption, will clearly be impossible. The real enemy is not so much demand as growth. We could cope with the 1972 level of demand from fossil fuels alone for 1100 or 1200 years but, as already shown, only for 78 years (from 1976) if we are to feed a 5% annual exponential growth.

Even by making supreme efforts to economize and to develop new energy sources as rapidly as possible, we could not for long hope to bridge the rapidly widening gap between Curves "A" and "B". Apart from the problem of discovering new energy sources there must also be a limit to the rate at which we can construct the necessary equipment for harnessing them and converting them into the form needed. For instance, in 50 years' time the energy demand at a sustained exponential growth of 5% p.a. would have increased about eleven and a half times, whereas the population (which for some time has been growing at about 2% p.a.) would have increased about 2.7 times. Hence the production of exploitation equipment *per capita* would have to increase by about 4.25 times.

After 100 years this factor would have risen to more than 18.

Curve "A" must, therefore, somehow be bent into some shape such as that of curve "C" (also not to scale) which may be permitted to lie above curve "B" only to the extent that we can develop new energy sources to supplement fossil fuels. Growth will have to be checked, halted and perhaps even reversed if disaster is to be avoided, regardless of the inevitable impact upon our way of life. It is in the area "E" that the mission of geothermal energy lies. The maximum level to which curve "C" may be permitted to rise may well be influenced not only by our ability to develop new energy sources but also perhaps by environmental considerations. Ultimate equilibrium in the remote future can be established only when the demand curve "C" can be brought to coincide with the availability of *renewable* energy sources, after all accessible capital stocks of stored energy have been depleted.

Although less than 15% of the estimated total world hydro power resources have hitherto been developed, even if these resources were now fully developed they could contribute only about one-seventh part of our present annual energy needs; and as the total world energy demand rises, the proportional contribution that could ever be made by hydro power will of course dwindle. Likewise, wind, wave and tidal power, though sometimes locally useful, could provide no more than small scale palliatives. There remain only the following large scale potential sources of energy with which fossil fuels may be substantially supplemented:

(1) *Nuclear fission*, already accomplished technically and providing a small fraction of the world's energy needs. But the resources of fissile matter are by no means unlimited. Breeder reactors would effectively multiply these resources by a factor of about sixty, and the successful exploitation of thorium would further enlarge the potentialities of nuclear fission very considerably. But there are undoubted pollution hazards, from wars, earthquakes, sabotage and mere handling accidents which have created a very powerful anti-nuclear lobby of world opinion. The generation of nuclear heat and power will doubtless grow substantially, but it is too problematical for us to stake our future entirely upon it. It is likely, however, to give us a much needed breathing space.

(2) *Solar energy*. Vast though this is in quantity, and useful though it already is for small scale exploitation, it is too diffuse to harness (by means at present conceivable) on the immense scale necessary for industrial communities. An old proposal, advanced in 1881 by D'Arsonval, and put into practice on a very small scale in the 1920s by Claude, has recently been revived. This is to exploit the temperature differences between the surface waters and depths in tropical seas by means of inefficient heat engines: it is a form of solar energy development. But here again, though locally impressive plants may perhaps be installed in the near future, development on the vast scale required would seem to be most improbable.

(3) *Controlled nuclear fusion*. If this could be mastered, then the flood gates of available energy could indeed be opened and our problems would recede from our present field of concern. But despite many years of intense research the key to this source of energy has so far eluded us.

(4) *Geothermal energy*. It is in this source of energy that our best hopes would seem to lie. The resources are gigantic and it is one of the least pollutive energy sources. It has already been successfully and cheaply exploited, and although at present we can win it only in limited quantities and in restricted zones, it is now thought to be far more prevalent than was formerly believed. New techniques are being pursued which hold out hope that before long we may be able to win earth heat in large quantities at economic cost in many—and perhaps ultimately in all—parts of the world. Earth heat is also an extremely versatile form of energy, capable of exploitation for a wide variety of purposes.

2. PRESENT LIMITED EXTENT OF GEOTHERMAL DEVELOPMENT

Despite some impressive local projects—for example more than half a million kilowatts of power plant in California—geothermal energy is now (1976) contributing only about one-thousandth part of the world's electricity needs and probably a much lower fraction of its low grade heat requirements. Considering that serious geothermal development began at the turn of the twentieth century, this does not seem to be much of an achievement, especially as geothermal power and heat, wherever developed, has always proved to be substantially cheaper than power or heat from competitive energy sources. It is of interest to examine the four-fold reasons for this slow rate of development:

(1) Nature has made geothermal energy accessible (with the techniques now at our disposal) in only a limited number of places, some of which are topographically remote from potential markets.

(2) Cheap alternative energy sources that can be exploited by well established techniques involving no more than conventional engineering methods are sometimes available in countries possessing geothermal resources. New Zealand and Iceland are cases in point. Both are rich in hydro power, while New Zealand also has coal and natural gas. It is true that New Zealand has nevertheless developed quite a large geothermal power plant and that Iceland has undertaken extensive district heating and domestic hot water supply schemes. But had either of these countries been poor in alternative energy sources, it is highly probable that they would have

developed their geothermal resources far more rapidly and extensively by now.

(3) Fuels discovered anywhere in the world can find a ready market and can be easily shipped to any other part of the world for consumption. A geothermal field on the other hand must be exploited more or less locally, where the market may sometimes be very limited.

(4) Geothermal exploitation must be preceded by exploration which requires a fairly large outlay of "risk" capital which, if successful, can be amply repaid by results; but which, if unsuccessful, represents just so much money wasted (except perhaps for some purely scientific knowledge gained as a "spin-off"). Governments, particularly those of developing countries, are often reluctant to invest risk capital if an alternative source of energy is available without risk—even at a higher price than that at which geothermal energy could perhaps, but by no means certainly, be won if the outcome of the risk were successful. In short, governments will seldom be willing to take a gamble. Private enterprise, on the other hand (e.g. the oil companies) is often willing to invest risk capital, but only if it can foresee a fairly generous return on its money if the risk should pay off. The commonest applications for geothermal energy have hitherto been electricity supply and district heating, both of which, being public services, are apt to be intolerant of high profits. So the private investor too has had little incentive to take the risks inherent in geothermal exploration. Examples can of course be cited both of governments and of private enterprises which *have* taken the risks and whose investments have paid off handsomely; but the philosophical gap between "risk" and "security" has undoubtedly acted as a deterrent to rapid geothermal development in the past.

In the field of power it took more than half a century—from 1904 to 1958—for less than 300 MW of geothermal electric power to be developed; but in the subsequent 15 years the capacity of geothermal power plant throughout the world was approximately quadrupled. This sounds quite impressive, and it represents an average annual growth rate of about 9.5%. But the growth of world electricity consumption (as distinct from total energy consumption) over the same 15 years was about 8% p.a. average; so it would seem that geothermal development kept only slightly ahead of conventional power development during those years. In point of fact, these figures are not quite fair to geothermal energy; for whereas the world growth of about 8% p.a. referred to kilowatt-hours produced, the geothermal growth of about 9.5% p.a. referred to kilowatts capacity. Now the average world load factor during those 15 years was about 48% annual, whereas the average load factor at which the geothermal plants generated power was probably something between 85 and 90% annual; so that in terms of energy the geothermal contribution was probably rather better than the growth figures suggest, because geothermal power plants can operate throughout their lives at higher load factors than any alternative type of thermal power plant—even super-modern "giant" installations and nuclear plants. Too much should not, however, be made of this point because much would depend upon the types of alternative plants that would have been installed if the geothermal plants had not been adopted; but it would be generally true to say that a geothermal kilowatt is worth more in terms of energy than any other kind of kilowatt—except perhaps from run-of-river hydro plants.

Taking all this into consideration it must be admitted that the average growth rate of geothermal power development from 1958 to 1973 was not very much more rapid than that of conventional power development. It is clear that we must do much better than that before we can claim that earth heat is a major contributor to the world's electrical energy needs.

On closer examination of the available figures, however, grounds for moderate optimism may be detected. Using as milestones the years in which United Nations conventions have been held, there is evidence of a substantial acceleration in geothermal growth in recent years, as the following figures will show:

1961 (Rome conference)—420 MW of geothermal power installed	} 5.4%
1970 (Pisa symposium)—675 MW of geothermal power installed	} 14.2%
1975 (San Francisco symposium)—1310 MW of geothermal power installed	

The figures on the right represent the average annual growth rates in geothermal plant capacity per annum over the periods between the conventions. It will be seen that the progress from 1970 to 1975 was far more rapid than from 1961 to 1970, and moreover that it greatly exceeded the world electricity growth rate. If geothermal power plans "in the pipeline" be taken into consideration it would seem that this acceleration is still being maintained and that an important corner has been turned. Parallel evidence of non-power geothermal applications is difficult to obtain quantitatively, but there are signs of a growing interest in low grade geothermal heat. It would generally seem that geothermal energy is now rapidly taking a higher proportional share of Man's energy burden, although that share is still lamentably small. The reasons for this recent acceleration are probably as follows:

(1) Improved exploration techniques are reducing the risk element of geothermal development.

(2) Rocketing fuel prices have made the risks much more worthwhile, because of the relatively greater prizes if the (reduced) risks pay off.

(3) A growing consciousness is becoming apparent of the need for abundant, cheap low-grade heat for purposes other than power generation.

(4) More large low-grade thermal fields are being discovered outside the Seismic Belt.
(5) Growing public concern in environmental pollution is directing greater attention towards geothermal energy which is far less pollutive than fuel combustion.
(6) Every successful geothermal enterprise engenders confidence, and helps to convince people that geothermal energy is not merely a "freak" development.

The important question is whether, and for how long, this geothermal "acceleration" can be sustained. The honest answer to that question is probably "not for very long while we are restricted to the capabilities of our present techniques for winning earth heat". The reason for this lies in point (1) p. 3, col. 2, namely, that earth heat is accessible *with the techniques now available* only in a limited number of places. It is for this reason that many writers and commentators on geothermal energy have tended to write it off as little more than a resource, sometimes of great local value to certain countries, but certainly not as a major contributor to the solution of the world's energy problems. This, in the author's opinion, is an unwarranted defeatist attitude. Great and exciting developments are even now the subject of experiment at Los Alamos, New Mexico, USA; and although much remains to be done there are reasonable grounds for optimism and for believing that within a decade or thereabouts we may possess the means for vastly extending the areas of the world where geothermal energy may be won, and that in the not too distant future it may even be possible to win earth heat at commercially acceptable cost at any point on the Earth's surface. The author sincerely believes that we are on the eve of an "energy renaissance" and that earth heat will fairly soon assume an importance at least comparable with, and perhaps exceeding, that of fuels in present times.

3. HISTORICAL NOTE

The oldest adaptations of earth heat were "balneology.. and "crenotherapy", or (in common parlance) hot baths and "taking the waters" for alleged medicinal reasons. The ancients of Greece and Rome made great use of hot and warm springs which, apart from the comfort afforded by hot baths, were alleged to possess healing and prophylactic properties when applied externally, as in bathing, and sometimes when taken internally or used for douches. The Roman bath became an institution—not merely as a health centre, but also as a focus for social intercourse; and the custom was revived in the eighteenth and nineteenth centuries when a multiplicity of "spas" and "watering places" appeared in many countries as gathering points for invalids and for the world of fashion. The waters at many of these spas were hot; that is to say they were of geothermal origin. In the Ottoman Empire, Japan, Mexico and in Maori New Zealand thermal waters were used for centuries for purposes of health and hygiene. Balneology and crenotherapy flourish to this day. In 1971 in Italy alone fifteen million patients were treated at more than two hundred thermal clinics, while in the USSR more than ten million people are treated annually with thermal waters.[3] The medical profession, in general, has endorsed the therapeutic value of many "waters". The "hamams" of Turkey are much frequented to this day by sufferers from various ailments and by women hoping for fertility. The Maoris used geothermal fluids for a variety of domestic purposes—cooking, bathing, laundering—ever since they settled in New Zealand in the fourteenth century. The intrusion of geothermal exploitation into their traditional way of life may be seen to this day in a village near Rotorua.

The next oldest practical application of geothermal fluids was the extraction of chemicals. The Etruscans obtained boric acid from the hot springs near Velatri (modern Volterra) and used it for the splendid enamels for the decoration of their vases.[4] From the thirteenth to the sixteenth century sulphur, vitriol and alum were extracted from the same springs by the Italians. In the early sixteenth century, Hernan Cortez used the sulphur deposited near the summit of the volcano Popocatepetl to make his own gunpowder during his conquest of Mexico. Three hundred years later, in 1818, Francesco Larderel, Count of Montecerboli, started to build up a prosperous boric acid industry in Tuscany, using the hot springs that had been exploited by the Etruscans about two and a half millennia previously, which survived until quite recently.[4] The count's name is commemorated by Larderello, the centre where the modern Italian geothermal power industry has been built up.

Geothermal energy attracted the attention of Lord Kelvin in the late nineteenth century when he advanced various theories to account for the hot interior of the Earth. Although he outlived by about 10 years the discovery of radioactivity by Becquerel in 1896, Kelvin's theories to account for earth heat and to deduce therefrom estimates of the age of the Earth were mainly based on classical nineteenth century physical conceptions. He never succeeded in satisfactorily explaining the origin of geothermal energy but he certainly directed much thought towards the search for a solution of the mysteries of subterranean heat.

In 1904 Sir Charles Parsons, the eminent engineer whose name is chiefly associated with the development of the steam turbine and the astronomical telescope, advocated the sinking of a shaft 12 miles deep into the Earth at an estimated cost of £5 millions—a very considerable sum in those days. He believed that the work could be achieved in about 85 years and that it would greatly contribute to our knowledge of the Earth's structure and thermal characteristics. He light-heartedly referred to his proposal as the "Hellfire exploration project".

It was in the same year, 1904, that the geothermal age was really initiated when Prince Piero Ginori Conti first promoted electric power generation in the Larderello area of Tuscany. After some ill-fated attempts to use conventional reciprocating engines fed with natural steam, which failed on account of rapid corrosion, a 250 kW power plant was successfully commissioned in 1913, using clean steam raised by means of a heat exchanger from "dirty" natural steam. Gradually, by introducing materials of improved quality, the chemical problems were overcome so that natural steam could be fed directly to turbines without the intermediary of heat exchangers and their attendant losses. By the early 1940s about 130 MW of power were being generated from geothermal steam in the Lardarello neighbourhood.[4] This plant, the first of its kind in the world, was totally destroyed by enemy action during World War II, but shortly after the coming of peace to Europe the plant was reconstructed and development extended to other nearby areas. Now, a complex of several geothermal power plants in Tuscany is supplying more than 400 MW to the integrated Italian electricity system of "ENEL". Although there are now larger geothermal power plants in service, it is to the eternal credit of the Italians that geothermal power generation first became an accomplished fact.

In the 1950s New Zealand followed the Italian example by developing a 192 MW geothermal power plant at Wairakei in North Island.[5,6] The technical problems in New Zealand differed somewhat from those encountered in Italy in that the Wairakei steam was "wet" while the Larderello steam was "dry". A few years later California, after a cautious start, built up by stages what has now become the largest group of geothermal power plants in the world.[7] These, like the Italian plants, have to deal with "dry" steam. More than half a million kilowatts of plant are already in service and big extensions are planned. Japan,[8] the USSR,[9] Mexico,[10] El Salvador followed suit from about the mid-1960s. Several other countries are now starting to develop their geothermal resources for power generation; and electricity production from earth heat has now become one of the established facts of life.

But electric power has not been the only modern application of earth heat. Even before power was first generated in Italy the spasmodic exploitation of geothermal energy for domestic space heating and hot water supplies was being practised in Iceland, Japan, the States of Idaho and Oregon (USA) and probably elsewhere in the latter years of the nineteenth century. The first large scale break-through in this form of geothermal development occurred in about 1930 when a public hot water reticulation system, fed from natural subterranean waters, was established in the city of Reykjavik, Iceland for domestic space-heating and washing supplies.[11] This system grew steadily until now all the houses in Reykjavik, except about 1%, are provided with this service which was economically very competitive with fuel-heated domestic systems even before the fuel price increases of 1973 and after. Several other districts in Iceland are now served with domestic heat from geothermal sources and it is hoped that quite soon more than half the homes in Iceland will enjoy this amenity and that the proportion will continue to rise thereafter. Meanwhile, other domestic heating schemes have been or are being initiated in Hungary,[12] Japan, New Zealand, the USSR, the USA, France and even in the Himalayan districts of India.

In parallel with domestic heating, the use of earth heat for agriculture, horticulture, fish-breeding and animal husbandry has been steadily developing during the twentieth century. In Iceland, by 1975, 140,000 m^2 of cultivation was being effected in geothermally heated greenhouses[11] in which fruit, vegetables and flowers are successfully raised in a country afflicted with a climate so rigorous that such an achievement would otherwise be impossible. The benefits of this upon the trade balance of a country normally dependent upon imports for so much of its food can be substantial. In Hungary no less than 1.7 million m^2 of geothermally heated greenhouses are now in use. Similar agricultural and horticultural applications of earth heat are being practised in the USSR, Japan, the USA and probably elsewhere, while fish-breeding and the raising of exotic zoological creatures such as alligators in a climate differing widely from that of their natural habitat is being practised in Japan as a tourist attraction. The washing and drying of wool and the biodegradation of organic wastes are other farming activities practised in certain countries with the aid of earth heat. Air-conditioning, as a corollary to space-heating, was successfully developed in New Zealand[13] in the late 1960s while geothermal refrigeration as an adjunct to farming for the preservation of produce is now being developed in various countries.

The application of earth heat to industry has not yet been so rapid as could have been wished, but the very large mill of the Tasman Pulp and Paper Co. at Kawerau, New Zealand, was established in the 1950s and consumes about 200 ton/h of geothermal steam for process purposes and for the generation of small scale power for the factory's needs.[14] At Namafjall in Iceland a diatomite recovery and upgrading plant, using about 50 ton/h of natural steam was set up in 1967.[14] Salt and sulphuric acid production and a brewery using geothermal heat are in service in Japan and other geothermally activated industries which need not here be enumerated are to be found in several parts of the world.

In the exploratory sciences, geology may be said to date from Goethe to whom may reasonably be given the credit for developing the study of rocks into a serious science worthy of systematic treatment. It developed rapidly in the nineteenth century, particularly after Charles Darwin gave an evolutionary perspective to the whole of pre-history; and it made further advances in the twentieth century when isotropic studies became possible. Geophysics and geochemistry are really the products of the twentieth century; while plate tectonics—a subject that has contributed so much to the theory of the natural occurrence of geothermal fields—dates from 1915 when Alfred Wegener first mooted the theory of continental drift.

4. APPLICATIONS OF GEOTHERMAL ENERGY

One of the greatest assets of geothermal energy is its versatility.

The generation of electric power has until now been by far the most important application of geothermal energy and is likely to take pride of place also in the future. This is because electricity is readily marketable, has for long been in demand at a growth rate exceeding that of primary forms of energy and is easily transported over fairly long distances. Moreover, the geothermal fields that first attracted the attention and interest of the developers were all capable of producing fluids at fairly high temperatures and pressures at which the steam turbine could operate at acceptable (though not high) efficiencies. Furthermore, electricity is itself so versatile that it can be adapted to almost any application—motive power, heating, lighting, etc.—so that its generation was analogous to the minting of coin—a commodity having the great merit of convertibility into whatever form of energy was needed. There has consequently been a tendency to concentrate upon the development of geothermal fields capable of producing fairly high temperatures and pressures suitable for power production. Nevertheless, the use of geothermal heat to generate electricity which may subsequently be used for heating purposes is a most inefficient procedure, even if it may happen to be convenient in certain circumstances. Where possible, if the required end product is heat, then it would be far more sensible to use geothermal heat directly without the intermediary of electricity; but obviously the flexibility and ease of transmission and distribution of electricity will often make it more practicable to use electricity for widespread small scale heating requirements despite the thermal inefficiency of so doing.

Where low grade earth heat has been available, it has primarily been hitherto used for balneology and crenotherapy or, if sufficiently hot, for district heating and domestic hot water supply. But district heating and domestic hot water supply become commercially attractive only for sizeable communities which are soldom located close to low grade geothermal fields—and the transmission of heat over long distances is not economic. Also domestic heating is needed only in cold climates, which further restricts the availability of suitable thermal fields for such purposes. This does not imply that district heating and domestic hot water supply are not an important application for earth heat; in fact it has been shown that nearly 20% of the total energy consumed in the USA is accounted for by space heating.[15] But it does show that this particular application lacks some of the ease of marketability possessed by electricity.

Before considering other possible applications of earth heat, it is necessary to recognize three very important facts:

(1) High enthalpy geothermal fields suitable for power generation are now believed to be far less abundant in nature than low enthalpy fields. It is thought[15] that what low temperature fluids lack in enthalpy they more than make up for in quantity. As utilization techniques improve, lower temperature fluids are likely to become of increasing economic importance.
(2) Geothermal power generation is essentially an inefficient process owing to inescapable thermodynamic constraints, and the lower the temperature of the heat source the lower will be the efficiency of power generation.
(3) On the other hand the direct use of geothermal heat for space heating, industrial processes or husbandry can be highly efficient; the incurred losses being imposed only by such imperfections as must inevitably arise from insulation losses, drains and terminal temperature differences in heat exchangers. These losses can be controlled within economic constraints: they are not dictated by thermodynamic laws.

Clear recognition of these facts (the second two of which are rather obvious) has only come about within the last two or three decades and it is but recently that the full importance of non-power applications (other than district heating, domestic hot water supply, balneology and crenotherapy) have begun to be fully appreciated.

The advantages of low grade heat for farming purposes of all kinds—greenhouse heating, soil warming, soil sterilization, heating of farm buildings, biodegradation of farm wastes, etc.—have been recognized for some time; but, as with district heating, this particular group of applications is generally confined to cold climates. Nevertheless there is one interesting example of geothermal aid to farming in a temperate climate, namely, at Agamemnon near the Mediterranean coast of Turkey. Here there is a hot spring from which hot water at 70°C issues. The water is first used in a "haman", or bathing establishment, and is then discharged through open channels for irrigating the fields. By the time the water reaches the crops it is warm rather than hot, but the local farmers claim that their crop yields are substantially higher than from other fields in the neighbourhood which rely upon normal rain water. This could possibly have a chemical explanation, but it could be simply be due to the warmth: it is a point that would seem to warrant further investigation, as there could perhaps be similar applications for discharged thermal waters in other parts of the world.

It is only since the middle of the twentieth century that there has been any real awareness of the immense industrial potentialities of earth heat, and this awareness has been very slow in developing. The great success of the Tasman Pulp and Paper Mill should have attracted wider attention to these potentialities than it did; and it cannot be denied that the industrial applications of geothermal energy have developed very slowly despite the fact that many industries are highly heat-intensive. Clearly geothermal heat would be of no value for industries requiring high grade or "fire heat"—such as smelting, potteries, glass making, etc.—but there are innumerable industries that need only low grade heat. Figure 2, presented by Lindal in an interesting paper,[14] shows that there are uses for heat over the whole temperature spectrum from 20 to 200°C, and it is probable that there are plenty of other industries and applications that are not specifically shown in that figure. Reistad[16] has convincingly demonstrated that in the USA, which are probably fairly representative of other industrialized countries, about 41% of the country's total energy requirements could be served by geothermal energy *if it were available.* This proportion includes all energy-consuming processes requiring temperatures up to 250°C. Reistad further points out that if the upper temperature limit were dropped from 250 to 200°C, the proportion that could be served by geothermal energy (if available) would be scarcely affected—dropping only from 41 to 40%—because very few processes use heat in the 200–250°C range. If the temperature limitation were further dropped to 150° and 100°C the percentage of national energy requirements that could theoretically be satisfied by means of earth heat would drop to about 30% and 20% respectively. These figures are most revealing, and show that the salient problem is one of availability rather than of applicability. Reistad's figures exclude not only high temperature processes but also intermittent processes like traction and cooking, so his figures are on the conservative side. They illuminate the urgency of rectifying the imbalance between availability and applicability.

To define all heat-intensive industries would be a formidable task, but the following list is indicative of industries that could benefit from the availability of low grade (i.e. 200–250°C or less) heat:

- Desalination[17]
- Many chemical industries, including:
  - extraction of valuable minerals from geothermal fluids;
  - salt production from sea water;
  - heavy water production;[18]
  - protein and vitamin production;
  - sulphur and sulphuric acid production.
- Textile industry
- The dye industry
- Paper manufacture.[14]
- Mining, including:
  - recovery and upgrading of minerals;[14]
  - mining in areas of intense frost (e.g. Siberia);
  - peat drying.
- Plastic manufacture
- Synthetic rubber manufacture
- Alumina from bauxite
- Rayon production
- Total gasification of coal (Lurgi process)
- Timber seasoning
- Refrigeration
- Air conditioning[13]
- Mushroom culture
- Powdered coffee production
- Dried milk production
- Fruit and juice canning and bottling
- Rice parboiling
- Cattle meal from Bermuda grass
- Sugar processing (with the use of bagasse as an industrial material, and not as a fuel)
- Sugar beet industry
- The bottling of mineral waters
- Fish drying and fish meal production
- Other food processing and canning
- Crop drying (seaweed, grass, etc.)
- Molasses fermentation
- Road heating in frosty climates
- Anti-freeze for fire-fighting
- Sewerage heat treatment

This list, which makes no claim to being complete, omits power generation, space heating, farming and balneology which have already been discussed. For any heat-intensive industry geothermal heat could prove to be invaluable provided the grade of temperature required is not too high. Geothermal heat, moreover, has proved generally to be vastly cheaper than fuel heat and, when available, is usually very abundant. If the grade of available geothermal heat is rather too low for the purpose required, it could sometimes be economic to boost it by using modest quantities of supplementary fuel, either directly or (conceivably) by means of heat pumps. Heat pumps, which would enable the geothermal fluids to be rejected at a lower temperature, would however seldom be economic.

The use of dual or multi-purpose projects could sometimes offer very promising and economic applications for geothermal energy; e.g. a combined power plant and chemical factory, a combined power and desalination plant, a combined power or industrial plant discharging warm water to a balneological institution—there are endless possibilities for combinations of high grade/low grade heat processes that could be coupled or even arranged in groups of three or more processes. Such dual or multi-purpose projects would enable the costs of exploration, drilling and certain other common items to be shared between the diverse end products. Combined plants would not be without their problems—the chief of which is the matching of the variable demand patterns of the different end products; but such problems are not necessarily insuperable.

A common fallacy concerning the usefulness of geothermal heat is that of assuming that the *location* of a field must often rule out its exploitability. Of what use, it may be asked, would be a geothermal field in the middle of the Sahara, in the Northern Arctic wastes or in other off-the-map locations? The answer to this question is that for an energy-intensive industry, if the available heat is cheap enough and abundant enough it may sometimes pay to ship the raw materials of an industry and the necessary labour to a remote geothermal field, and to transport the end products thence to the markets. This has already been demonstrated to be economic in the aluminium industry, in which it may pay to ship the ore half-way across the world to a site where cheap energy is available, to establish a local labour colony of expatriates and to ship the product to any point in the world where it may be needed. Such developments might even sometimes enable the rehabilitation of underpopulated areas of the world to be socially advantageous.

In closing these remarks on "applications" it may be said that high grade heat is generally more versatile in application: it can be used for many purposes including power generation; whereas low grade heat is more suited to district heating, certain industries and husbandry, etc., as shown in Fig. 2. Later, however, it will be shown that the use of binary fluids could perhaps enable even fairly low grade heat to be used for power generation; but this is still subject to some controversy.

(Also of relevance to geothermal applications are bibliography references nos. 3, 8, 11, 19 and 20.)

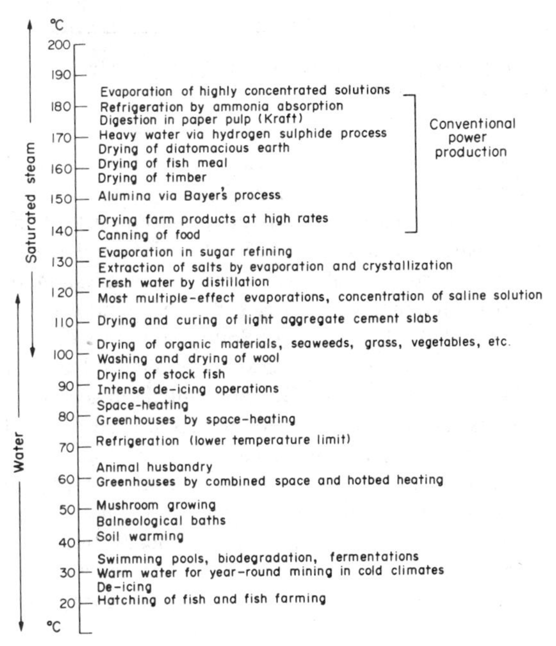

FIG. 2. The required temperature of geothermal fluids (approximate) (Lindal, p. 146, ref. 14).

5. THE STRUCTURE OF THE EARTH

It is surprising that despite our growing familiarity with outer space we are still relatively ignorant of what is going on only a mile or two beneath our feet, though that ignorance is rapidly being dispelled. The only parts of the Earth with which we are directly familiar are the atmosphere and the crust (which includes the oceans); but these together account for less than half of one percent of the total mass of our planet. The remaining 99.5% and more lies hidden beneath the crust and has not yet become accessible to Man. Our knowledge of its nature is indirect, largely deduced from the study of earthquake waves and of lavas, and from measurements of the outward flow of heat from

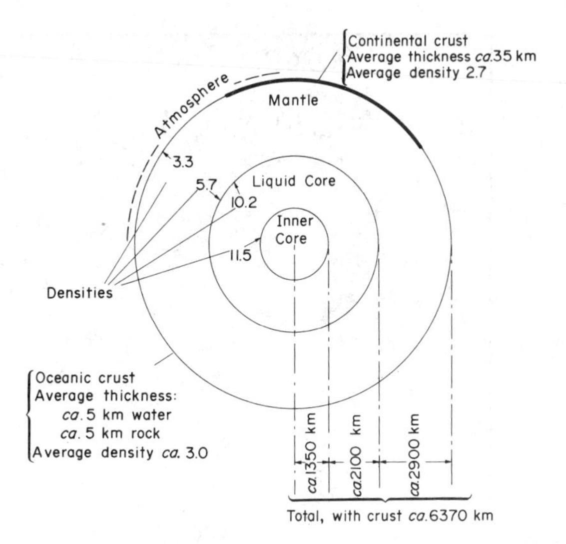

FIG. 3. Concentric layers of the Earth (after Bullard, p. 20, ref. 21).

below-ground to the surface. Nevertheless, this indirect knowledge has enabled us to build up a fairly clear picture of the structure of the Earth, which is now believed to consist of five distinct concentric spheres, assembled "onion fashion" (Fig. 3). Passing from the outside towards the centre of the Earth, these consist of the *atmosphere* (itself a complexity of various layers which need not here concern us), the *crust* (including the land masses, the seas and oceans and the polar ice caps), the *mantle*, the *liquid core* and finally the *innermost core* (which is believed to be solid). Both temperatures and densities rise rapidly as the centre of the Earth is approached.

Thus we have a picture of a very hot planet, efficiently lagged by a thin crust of low thermal conductivity. Thanks to this insulating crust, the conditions on the land surfaces and in the seas are sufficiently temperate to support life. Although the vast quantities of heat stored within the entire planet are, of course, "geothermal", it is only with the crust and the upper layers of the mantle that we need here concern ourselves for practical purposes.

The crust bears to the rest of the Earth a relationship not unlike that of an eggshell to an egg, except that its thickness is far from uniform. Figure 4 shows a section (not to scale) through the Earth's crust at an "inactive" coastline such as the Eastern seaboard of the North American continent. (The meaning of the word "inactive" will later become apparent). The crust forming

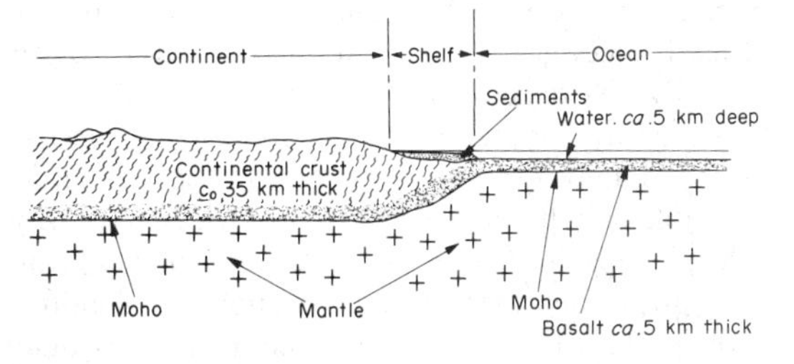

FIG. 4. Simplified section (not to scale) through the Earth's crust at an "inactive" coastline—e.g. the Eastern seaboard of North America (after Bullard, p. 20, ref. 21).

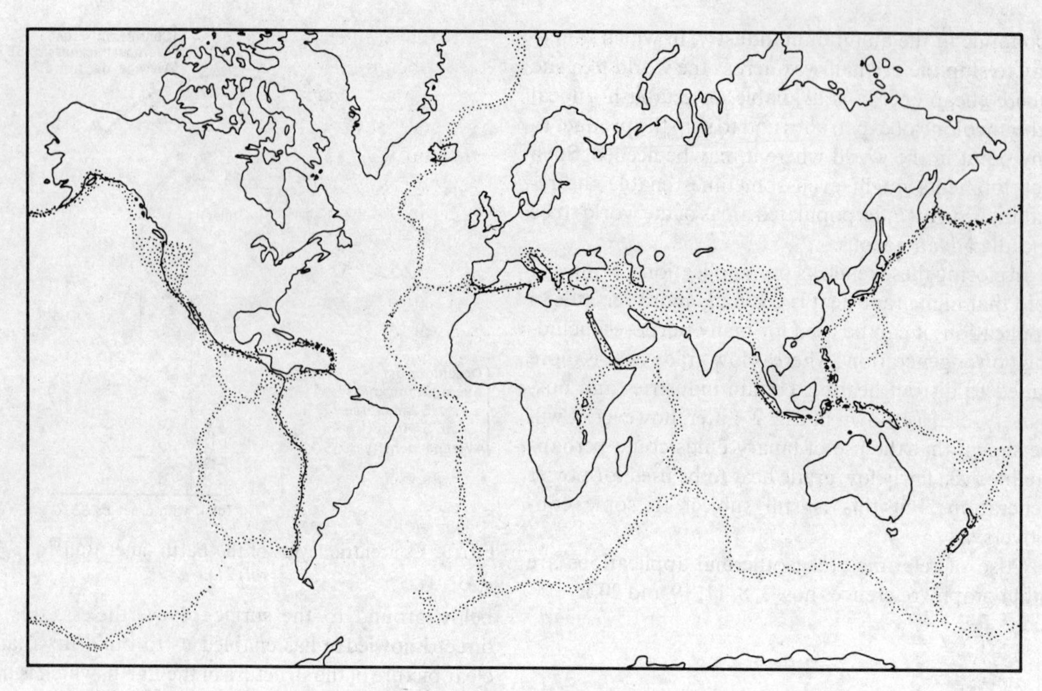

FIG. 5. Principal earthquake zones of the World. (From map of World seismicity, 1961–1969, published by the National Earthquake Centre, Washington, DC.)

the bed of the ocean is very thin (about 5 km) while the continental crust is about seven times as thick—even more in mountainous districts—and is fringed with a sedimental "shelf" of varying width. It is in this shelf that offshore oil and natural gas may sometimes be found (as in the North Sea). Relatively shallow waters (up to about 200 m) cover the shelf but the depth of water in mid-ocean averages about 5 km and reaches about 8 km in certain "trenches". The boundary that separates the crust from the mantle is known as the Mohorovičić Discontinuity—after its Yugoslavian discoverer—or "Moho" for short. At this boundary there is a sudden change in the velocity of earthquake waves which indicates a corresponding change in material composition and physical state. The temperature at the Moho is believed to be of the order of 600°C, but far higher temperatures are thought to occur at greater depth—about 4000°C at the base of the mantle and about 6000°C at the centre of the Earth.

The outer layers of the Earth, as pictured in this simplified model, are not everywhere in a state of mechanical equilibrium. High stresses are built up in parts of the crust and upper mantle, and when these stresses are relieved by material movements the released forces give rise to earthquakes. Seismologists are now able to detect many thousands of earthquakes every year, varying in intensity from the lightest of tremors, imperceptible to the human senses and detectable only by the most delicate devices, to violent shocks capable of devastating cities. It is now possible to map precisely the intensities, locations and depths at which all detectable earthquakes occur, and it is found that with relatively few exceptions they are confined to ribbon-like zones which spread over the surface of the globe in the manner illustrated in Fig. 5. These zones are collectively known as the Seismic Belt, much of which lies beneath the oceans.

The explanation of this zone pattern baffled investigators for a long time. To describe the Belt as a "zone of crustal weakness" was to beg the question. In 1915 a German, Albert Wegener, drew attention to the fact that the East coast of South America would fit snugly, like pieces of a jigsaw puzzle, against the West coast of Africa (Fig. 6). He also pointed out geological similarities found along the two coasts and deduced that the two continents had at one time in the remote past been joined together and had later—for some reason not yet understood—split and drifted apart. His theory was at first received sceptically, but in the course of time it has since been verified that the Americas are in fact moving away from Europe and

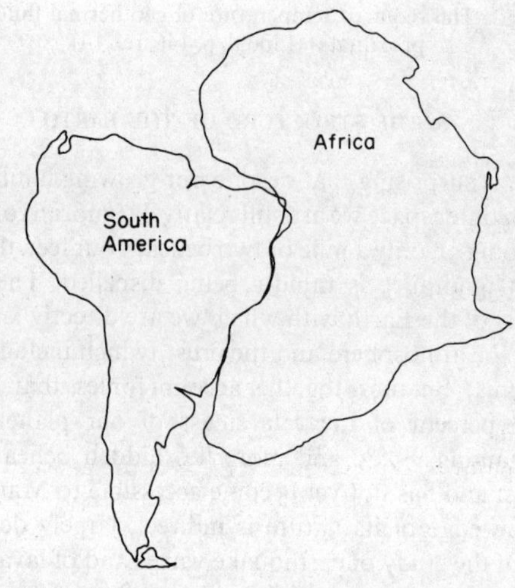

FIG. 6. Wegener's observation of continental "fit".

Africa at a slow, but detectable, rate of about 4 inches a year. Similar relative movements have been detected between other crustal masses. Since the late 1950s a body of very impressive evidence has been assembled by Sir Edward Bullard and others, and it now seems that Wegener's theory was basically correct. The modern belief, which is not without its challengers, is that the crust of the Earth consists of several discrete "plates"—six major and a few smaller ones—that are in a state of continuous relative motion at rates of a few inches a year. These relative motions imply that at some places, where plates are moving apart, the crust must be splitting; while at others, where the plates are colliding, the crust must be crumpling. In at least one place (e.g. the San Andreas fault in California) the plates are sliding past one another.

frictional and tensile resistances to motion. The motive power behind the plate movements is not fully understood, but could well arise from convection currents in the mantle which, though acting as a solid when transmitting earthquake waves, behaves as an extremely viscous fluid under the influence of strong forces maintained over very long periods: that is to say, the magma *creeps* as do other materials under the combined influences of high temperatures and stresses.

A feature common to all types of earthquake zone—crust-splitting, plate-colliding or plate-sliding—is the presence of volcanoes—active, dormant or extinct. The mid-Atlantic Seismic Belt coincides with a long range of submarine volcanoes, while the presence of visible volcanoes round the periphery of the Pacific Ocean has earned for that zone the name "The Belt of

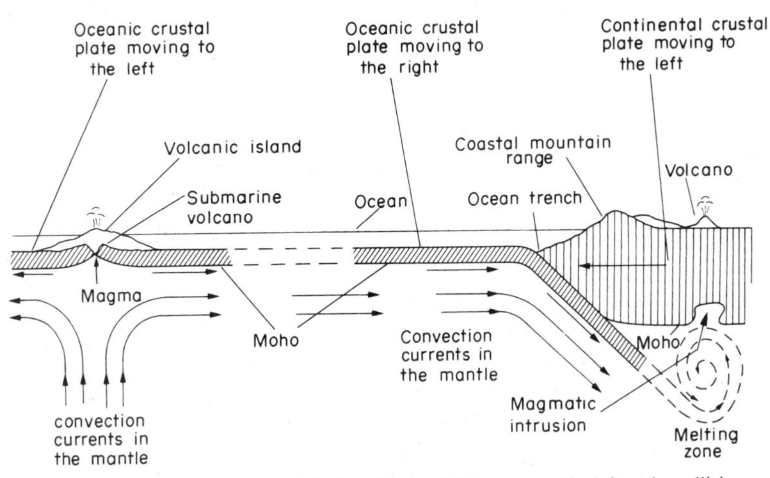

FIG. 7. Crustal plate movements (not to scale), showing a splitting zone to the left and a collision zone to the right.

Figure 7 gives a rough idea of the crust-splitting and plate-colliding processes. That part of the Seismic Belt which runs down the middle of the Atlantic Ocean is typical of the crust-splitting process. Here, the gap formed between the separating plates is continuously being replenished by the upward flow of hot "magma" from the mantle, to form a steadily spreading ocean floor. This process is illustrated at the left of Fig. 7. The Andes, on the other hand, are typical of a plate-colliding zone, where the crumpling has thrown up great mountain ranges. This crumpling effect is modified by the fact that one of the colliding plates yields to the other and slides beneath it at a downward angle, ultimately merging into the mantle whence it originated millions of years ago. This collision process is illustrated at the right hand side of Fig. 7. Such a coastline is described as "active" as distinct from the "inactive" coastline of Fig. 4.

According to the theory of continental drift, or plate tectonics as it is sometimes called, the Seismic Belt follows the lines along which the plates are either separating, colliding or sliding past one another. For the motion of the plates is believed to occur not steadily, but in jerks—large or small—which give rise to major or minor earthquakes respectively. These jerks occur whenever the built-up stresses exceed the

Fire". In a few places the mid-Atlantic Rift rises above ocean floor level to appear as volcanic islands such as Iceland, the Azores, St. Helena, Tristan da Cunha and others. Crust-splitting zones also occur in a few places in the continental land masses—notably the Great African Rift Valley. As a corollary to volcanoes there must be, or have been, magmatic penetrations of hot material from the mantle into the crust as a result of the disturbances at the edges of the plates. Such penetrations may be complete, when lava issues from a volcano; or partial, when they take the form of magmatic intrusions—pockets of mantle material bulging into the crustal rocks at depths of perhaps a few kilometres. Partial intrusions may be volcanoes in embryo. It is of interest to note that a line of volcanoes occurs to the East of the main range of the Andes (and similar formations may be found at other plate-collision zones). It is believed that the intense frictional heat generated by the forcing of the descending plate into a reluctant resisting mantle causes the upper mantle to become relatively fluid at a certain depth (see the right hand part of Fig. 7) and it is this semi-fluid mantle material that gives rise to the magmatic intrusions and volcanoes behind the main mountain range.

The modern theory of plate tectonics, as roughly

outlined above, is supported by impressive evidence,[21] for the more detailed pursuit of which the reader is referred to the bibliography. Wegener's theory included the suggestion that in the remote past all the land masses formed a single continent, which he named "Pangeia", and that at some time in pre-history a cataclysmal event took place which caused this great land mass to disintegrate and disperse to form the present configuration of continents and islands which we know today. The separation of the Americas from Europe and Africa would seem to have taken about 50 million years to attain their present relative positions, while that of the Tonga Isles from the Eastern Pacific Rift would have taken about twice as long. But geologists tell us that the Earth's age is about 4500 million years. Why, after surviving more than 4000 million years, should Pangeia suddenly have begun to disintegrate? It is now believed to be more probable that continental drift has been a continuous process since an early stage in the Earth's history, with plates being simultaneously created and destroyed at their edges. The plates are believed to "float" on the mantle, and in the course of the ages they have re-formed and changed their directions of motion under the influence of magmatic forces not yet properly understood. There may never have been a Pangeia. It seems that the Earth functions as a gigantic heat engine whose precise method of working is still unknown, but whose internal energy motivates the crustal creep and the changing pattern of magmatic convection.

Although of low thermal conductivity, the crustal rocks do in fact carry a certain amount of heat from below-ground to the surface, and in doing so a temperature gradient is formed. It has long been known that as we penetrate deeply into the Earth, by drilling or by mining, temperatures steadily rise. This observation gives positive proof of outward heat flow. The *rate* of this heat flow varies widely from place to place, as also does the temperature gradient. In some places the temperature gradient may be as low as 10°C/km, but a fair average in the non-seismic parts of the world would be of the order of 25–30°C/km. In a few favoured parts of the world the temperature gradient may be very much greater.

The rate of outward heat flow is the product of the temperature gradient and the thermal conductivity of the rock, and since both factors vary from place to place it follows that the heat flow too is far from constant. Accurate heat flow measurements have been made in many parts of the world, even in the ocean bed, and are found to vary widely; being lowest in the ocean bed far from submarine rifts and highest in parts of the Seismic Belt. The worldwide average is about 1.5 $\mu cal/cm^2\,s$. The escape of heat through volcanoes, though locally impressive, forms a negligible fraction of the total outflow of heat from the Earth. Geysers, fumaroles, hot springs, etc., are even less significant on a world scale.

It might be thought that this outward heat flow originated entirely from the stored heat in the mantle and cores of the Earth, but in non-seismic areas (which form by far the greater part of the Earth's surface) this stored heat accounts for only part of the heat flow: the balance originates in radioactive rocks contained within the crust. Variations in the degree of radioactivity of different rocks may account to some extent for the observed variations in the outward heat flow at different places on the surface of the Earth.

The theory of plate tectonics here broadly described gives meaning to the bald statement that the Seismic Belt marks a zone of crustal weakness. It also accounts for the fact that all the known high temperature geothermal fields have been located within, or close to, this belt. It is true that lower temperature fields have been found outside the belt—e.g. in Hungary, in the Russian Steppes, in the Arzacq Basin of France, in Switzerland, Australia and doubtless elsewhere. The existence of these low grade fields could perhaps be attributed to any or all of the following conditions:

(1) There could be parts of the crust that bear abnormally high proportions of radioactive rocks.
(2) The crust could well be relatively thin in places, so that the Moho is at shallower depth than in more normal "non-thermal" areas. (This emphasizes that Fig. 4 is somewhat stylized: there is no reason to suppose that the Moho is uniformly horizontal beneath the continental masses.)
(3) The temperature of the mantle just beneath the Moho is not necessarily the same everywhere: there could be local magmatic disturbances that give rise to hot spots in certain places.

The fact that the known high temperature fields are all found in or close to the Seismic Belt does not imply that such fields are to be found anywhere and everywhere within that belt. They occur only at certain places within the belt where the geological and hydrological conditions favour the formation of high temperature fields: hence their rarity.

(Also of relevance to the structure of the Earth are bibliography references nos. 22–24.)

6. ESTIMATED HEAT RESERVES OF THE EARTH

Any attempt to estimate the heat reserves of the Earth must necessarily be extremely tentative, for it involves the making of assumptions as to the specific heat of materials which form the deeper layers of the mantle and the cores at densities and temperatures of which we have no direct experience. Moreover, if the expression "heat reserves" is to have any practical meaning it must be interpreted as *exploitable* heat reserves; and who is to say how much heat may safely be extracted from our planet for the use of Man without risking seismic upheavals, intolerable climatic changes or excessive pollution? Nevertheless, it is possible to deduce certain figures that are at least indicative of the order of magnitude of the quantities of heat we are talking about.

If it be assumed that the *average* specific heat of all the materials of which the Earth is composed be

0.15 cal/g °C, then it can be calculated quite simply that by lowering the average temperature of our planet by 1°C, approximately 10^{21} kWh(thermal) of energy would be released. The United Nations adopt the practice of expressing their energy statistics in terms of metric tonnes of "coal equivalent" (tce), defining 1 tce as 8000 kWh(thermal). Hence, the lowering of the Earth's average temperature by 1°C would release about $10^{21}/8000$, or 1.25×10^{17} tce.

Now, as already mentioned, the total world consumption of energy in 1972 (the last year of uninhibited growth) was estimated at 7.566×10^{9} tce. Hence the heat released by cooling the planet through 1°C would suffice to supply the world with its energy requirements at the 1972 level for $1.25 \times 10^{17}/7.566 \times 10^{9}$, or about 16.5 million years.

Another way of viewing the problem would be to assume that all the heat released by the 1°C cooling process were converted into electricity at an average efficiency of 16%—a low figure deliberately chosen on account of the modest temperatures and pressures normally encountered in geothermal fields. The number of kilowatt-hours that could thus be generated would be about 0.16×10^{21}, or 1.6×10^{20}. The total world production of electricity in 1972[2] has been estimated at 5.6287×10^{12} kWh. Hence the one degree cooling process would supply the world with all its electrical needs at the 1972 level for $1.6 \times 10^{20}/5.6287 \times 10^{12}$, or about 28.5 million years.

The dramatic effects of *growth* upon figures of this nature have already been demonstrated. Hence, if energy demands and electricity demands are likely to grow appreciably in future years these periods of about 16.5 and 28.5 million years will be very greatly reduced. Nevertheless, even if the effects of growth were such as to reduce the periods of sustained energy supply to hundreds of thousands, or even tens of thousands of years (rather than tens of millions), it is clear that we are talking of energy in quantities that represent periods of super-historical dimensions with which it would be unrealistic to concern ourselves now. At all costs we must avoid trying to extrapolate into the realms of absurdity! Who can possibly foresee how Mankind will have developed in 10, 20 or 50 thousand years? He may by then have learned to harness energy from sources as yet not dreamed of. He may have learned so to modify his way of life as to be able to do with far less energy—*per capita* or *in toto*—than at present. There may have occurred tectonic or solar cataclysms that will have destroyed or drastically reduced human and animal life. He may even (not altogether improbably) have destroyed himself by continuing to follow his foolish policies, perhaps at the same time annihilating all other forms of life on this planet, or perhaps leaving other species to survive, thus putting back the evolutionary clock by aeons.

From all this it is clear that we need not think in terms of cooling the Earth through an average temperature drop of anything approaching one degree Celsius, particularly as the rather crude calculations here ignore a factor of immense importance, namely, that no allowance has been made for the continued generation of *new* heat on a vast scale by radioactive decay in the crustal rocks.

The upshot of these calculations and observations is that for practical purposes the reserves of earth heat may be regarded as virtually infinite and thus treated as a form of *renewable* energy. This, however, does not mean that we yet have access to this rich store. At present we can only nibble at the edges of it, but the knowledge of its immensity provides us with an incentive to persevere in our efforts to tap it on a far larger scale than has hitherto been possible.

One important point must be borne in mind. Even though our energy needs could perhaps be satisfied for centuries or even millennia by reducing the average temperature of the Earth by a miniscule fraction of a degree, quite substantial *local* temperature reductions would be inevitable, and we must be on our guard not to set up intolerably high seismic stresses at the points of large scale exploitation.

7. THE NATURE OF GEOTHERMAL FIELDS

It is convenient to classify all the areas on the Earth's surface into three broad groups, as follows:

Non-thermal areas having temperature gradients ranging from about 10 to 40°C/km of depth.

Semi-thermal areas having temperature gradients up to about 70°C/km of depth.

Hyper-thermal areas having temperature gradients many times as great as those found in non-thermal areas.

Semi- and hyper-thermal areas may loosely and collectively be described as "thermal", or "geothermal". It would perhaps be more logical to classify areas in terms of heat flow rather than temperature gradient; but as the latter can be measured much more quickly and easily, and as it gives a better indication of the depths to be drilled in order to reach a required temperature, it can serve as a basis for preliminary classification.

The existence of a thermal *area*, however, does not necessarily imply the presence of a geothermal *field*. A geothermal field may be defined as a thermal area (either hyper- or semi-) where the presence of permeable rock formations allows the containment of a working fluid without which the area, with the techniques at present available, would be unexploitable. Thus a thermal area associated with impermeable rock formations does not constitute a field. The working fluid in a field—water and/or steam, usually associated with certain gases—serves as a medium for conveying deep-seated heat to the surface for exploitation.

Geothermal fields may also be classified conveniently into three types:

(1) *Hyper-thermal fields*:

(i) *Wet fields* producing pressurized water at temperatures exceeding 100°C; so that when the fluid is brought to the surface and its pressure reduced, a fraction will be flashed into steam while the greater part remains as boiling water.

(ii) *Dry fields* producing dry saturated or superheated steam at pressures above atmospheric.

(2) *Semi-thermal fields* capable of producing hot water at temperatures of up to 100°C.

By some kind of perverse "law" of nature, the world distribution of fields and areas seems to be more or less inversely proportional to their actual or potential economic worth. Thus non-thermal areas, at present valueless, are by far the most common while hyper-thermal dry fields, economically the most valuable, are the rarest.

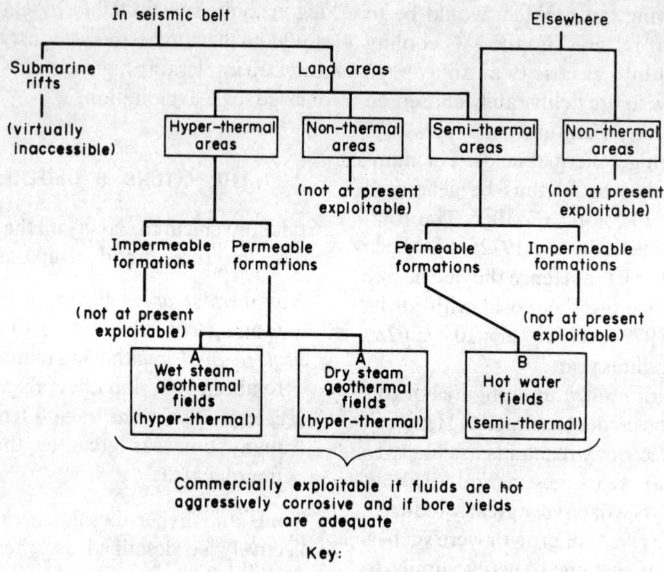

FIG. 8. Classification of "areas" and "geothermal fields".

It is emphasized that this classification of areas and fields has no official or classic standing: it is purely the author's proposal for aiding a general geothermal perspective.

As already mentioned, hyper-thermal fields appear to be confined to the Seismic Belt. The same may be said of hyper-thermal areas. Semi-thermal areas and fields may occur more or less anywhere, while by far the greater part of the Earth's surface is non-thermal. Figure 8 sets out "areas" and "fields" in a manner which shows that the comparative rarity of hyper-thermal fields, both wet and dry, is only to be expected.

(There is a curious type of "freak" thermal field which does not fall within the classification given above nor appear in Fig. 8, known as a geo-pressurized field. Such fields are rare, but are found round the northern fringe of the Gulf of Mexico. They occur in what has here been defined as "non-thermal areas" where the temperature gradients are only about 27–30°C/km. It is only on account of their great depth (up to about 6 km in places) that temperatures of up to 150°C or more are encountered. These curious fields consist of isolated pockets of water trapped in the geological formation as a result of growth faults, so that the water bears a share of the weight of the overburden and its pressure exceeds the hydrostatic value corresponding to the depth by anything from 40 to 90%. The trapped water has prevented the full compaction of the formation. Geo-pressurized fields can have a three-fold value by virtue of their hydraulic pressure, their heat and their content of dissolved natural gas.[25])

The presence of a hyper-thermal field is usually, but not always, accompanied by surface manifestations such as geysers, hot springs, etc.; but the converse is not necessarily true. Surface manifestations may sometimes simply be heat leakages through occasional cracks in a generally impervious formation. Evidence of semi-thermal fields is seldom betrayed at all at the surface, unless perhaps by warm springs; hence these fields may perhaps be more widespread than hitherto suspected.

The structure of geothermal fields and the mechanism of the upward transfer of deep-seated heat to higher levels are still the subject of much speculation, but it is usually possible for a hypothetical "model" to be built up from geological, hydrogeological, geophysical and geochemical observations to explain the behaviour of a field. In fact the postulation of such models is an essential step in the exploration process; for without such a model it would be difficult to know where to sink bores with a reasonably high probability of successfully winning hot geothermal fluids.

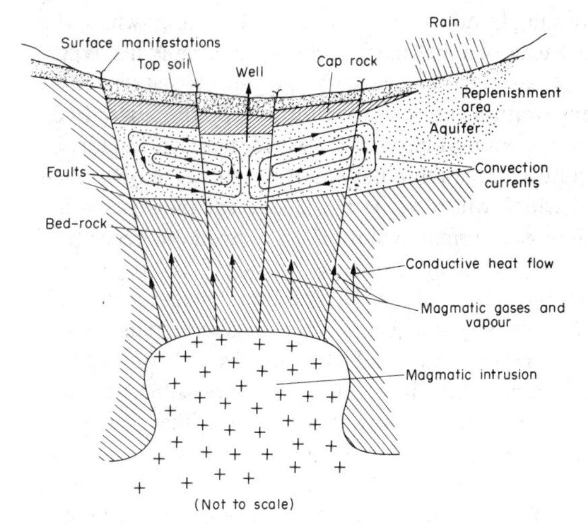

FIG. 9. Stylised hypothetical model of a "wet" hyper-thermal field.

Figure 9 (not to scale) shows a stylised hypothetical model of a wet hyperthermal field, for the functioning of which five main features are required. First there must be a source of heat, and this is generally agreed to be a magmatic intrusion into the Earth's crust, having a temperature of some 600–900°C at a depth of perhaps 7–15 km (see also Fig. 7). Secondly there will be a generally impermeable bedrock, probably fractured here and there by faults, interposed between the magmatic intrusion and the field proper. Thirdly there must be an adequate supply of water. Fourthly there must be a permeable "aquifer" to contain that water. Lastly there must be an impermeable "cap-rock" to act as "the lid of the kettle" to prevent the wholesale dissipation of the underground heat into the atmosphere.

The water in the aquifer may originate from one or both of two possible sources. "Meteoric waters" (an academic term for rain water!) may enter the aquifer at ground level, sometimes at a considerable distance away from the field, where the permeable aquifer outcrops at the surface. At the same time "magmatic waters" may enter the aquifer from beneath through faults in the bedrock, up which will pass vapours and gases released from solution out of the cooling and crystallizing magma below. The vapours will largely consist of very hot superheated steam which will condense on coming into contact with the cooler meteoric waters in the aquifer. At one time it was believed that magmatic waters formed the bulk of the fluid contained in the aquifer and that they, with their associated gases, accounted for most of the heat transfer from depth. Isotopic analysis, however, has now established that magmatic waters seldom account for more than perhaps 5% of the field fluid (for magmatic waters have a rather higher deuterium content than meteoric waters, and the proportions can thus be quite simply deduced). By the same token it is now believed that the greater part of the deep-seated heat enters the aquifer by conduction through the bedrock, rather than by the admixture of magmatic fluids.

It is highly probable that convection cells will be set up within the aquifer in the manner shown in Fig. 9, bringing the hotter waters from below to higher levels. In this process there will be a tendency for steam to flash off as the hot water rises. Some of this steam may find its way to the surface in fumaroles, while some will be condensed by cross-flows of cooler ground waters. Except at the outer edges of the convection cells, where the water is sinking only because it has become cooled, the bulk of the water in the aquifer will tend to be at about the boiling temperature corresponding with the hydrostatic pressure at the appropriate depth, though there will be local cool spots caused by subsurface ground waters, and sometimes local hot spots resulting from short-circuiting of the convection loops through fissures of low flow resistance which may bring hot fluids from depth more quickly to higher levels. Bores sunk through the cap-rock into the aquifer will tap this boiling water which, on rising to the surface and being subjected to a reduction of pressure, will flash off a fraction into steam which may be passed through a turbine to generate power, or alternatively used for some industrial or other purpose. The residual water may also be used to produce additional steam if its pressure is further reduced, or it may be exploited for some other purpose.

It might be thought that the presence at one place of all five of the essential features of a hyper-thermal field of this type would be too coincidental to be credible; but it has been shown that impermeable cap-rock can be formed by a self-sealing process when geothermal fluids pass through the fissures of an initially permeable formation.[26] The impermeability of the bed-rock, apart from fault fissures kept narrowly open by jets of magmatic gases at extremely high pressure, can be accounted for by the high lithostatic pressures at the great depths involved, which would effectively compact the rock so as to be free of general voids and fissures; while the permeability of the aquifer can be explained by the lower compacting lithostatic pressures at shallower depths, coupled with the action of the high tectonic forces experienced in volcanic zones. Hence the hypothetical field illustrated in Fig. 9 would not really be such a "miraculous" phenomenon as might at first be thought.

The field shown in Fig. 9 would almost certainly be "wet", partly because there is nowhere for steam to accumulate in large quantities (and such pockets as might tend to exist could well be cooled by ground waters), and partly because the replenishment area at a relatively high level would provide hydrostatic pressure to prevent the formation of much free steam. But sometimes the formation of the cap-rock may be dome-shaped, in which case the steam flashed off from the rising plumes of the convection cells could accumulate in large quantities, and the field would be "dry", or steam dominated. It may even happen that a field is too hot to retain water at any level, and will be filled with superheated steam. Dry fields are rarer than wet fields. The Tuscan and Californian fields are the best known dry fields, while the New Zealand, Mexican,

Icelandic, El Salvadorian and some of the Japanese fields are all wet.

Semi-thermal fields, being at temperatures no higher than 100°C, need no cap-rock to retain the heat; but they too must have a heat source, which could be accounted for by any of the three propositions set out earlier. These fields may be at any depth—even quite close to the surface—though they usually lie between 1 and 2 km deep or thereabouts.

Hyper-thermal fields, both wet and dry, may be exploited for almost any purpose (other than very high temperature processes) including the generation of power, unless the fluids are too hostile chemically. Semi-thermal fields, on the other hand, are more suited to district heating and domestic hot water supply, to a wide range of heat-intensive industries requiring fairly low grade heat and for farming in all its aspects. Nevertheless, it is technically possible to generate power even from semi-thermal fields yielding fairly low grade heat, but not too low, by using binary fluids.

(Also of relevance to the nature of geothermal fields are bibliography references nos. 12, 27–31.)

8. EXPLORATION

Geothermal exploration is essentially a team-work operation, involving a close partnership between the geologist, the hydrogeologist, the geophysicist, the geochemist and the drilling engineer. Though each is a specialist in one particular discipline he will largely rely upon the evidence of his colleagues when deciding how to proceed. The "mystery" posed to the exploratory team has several facets. Does a geothermal field exist? If so, what is its location, extent in area, depth and probable range of temperatures? Is it steam dominated or hot water dominated (i.e. dry or wet)? What is the order of magnitude of the field's thermal capacity?

A clue to the possible existence of a geothermal field is the presence of surface thermal manifestations such as geysers, hot springs, fumaroles and pools of boiling mud; but there are fields which totally lack such phenomena and which can be detected only in the light of known geologic environment and of other evidence detected by the team members.

The exploration specialist must be on his guard against false clues. For example, hot springs may be indicative of a field quite a long distance away, since the escaping hot fluids may have been deflected along inclined faults or fissures to a point far removed from the source of heat. The cross-flow of ground waters may also sometimes displace laterally the evidence of a field so as to suggest that its location is a fair distance "downstream" from its true position.

At present, the task of the geothermal explorer is to detect and evaluate geothermal *fields* within the meanings defined earlier. Before very long it is quite probable that he may be required to detect thermal *areas* lacking natural permeability but indicative of the presence of hot dry rocks, the future exploitation of which is becoming increasingly probable.

The first task of the exploring team is to assemble and digest all available records of the area where the presence of a thermal field is suspected; and if a hyper-thermal field is being sought it would be well to confine exploration activities to within the Seismic Belt. Even in the remotest parts of the world there will usually be quite a lot of preliminary information on record, much of which will have been amassed for purposes other than geothermal exploration. All available records of topography, meteorology, geology, hydrogeology, observations of surface thermal manifestations, geochemistry and geophysical measurements should be carefully examined and analyzed. From all the assembled data it will usually be possible to select certain promising localities that are likely to repay closer investigation. Collected data should not be confined solely to areas that appear to be "thermal"; for the more information that is available concerning non-thermal areas surrounding a thermal area, the more effectively can thermal anomalies be characterized.

The broad task of the *geologist* is to build up a three-dimensional "model" of the geological structure of the allegedly thermal region to as great a depth as may be practicable. Since all but a tiny fraction of his model will be concealed from the eye, he must rely more upon deduction than upon direct observation. Supporting evidence from his exploration colleagues of other disciplines will be of great help to him in constructing his model. His own tools for exploration will consist of surface geological mapping, the study of the tilt of outcrops, the results of cored soundings, if any, obtained by previous investigators (perhaps in pursuit of ground water or minerals), his own exploratory drillings where he may consider them to be justified, and the observation of faults and of thermal manifestations. He should be on the look-out for permeable aquifers and impermeable cap-rocks.

The *hydrogeologist's* task is to deduce the probable paths by way of which water may flow underground through and between the various geological strata and boundaries of the geologist's model. He should seek to explain where meteoric waters can enter any detected aquifer, how they escape therefrom to the sites of surface thermal manifestations, how they are contained so as to prevent them from escaping elsewhere, and how they may be expected to behave when provided with artificial escape paths in the form of drilled boreholes. He must study gradients, porosities and permeabilities of the constituent geological formations and, with the help of the geochemist, he may be required to detect the relative proportions of meteoric and magmatic waters in an aquifer.

The function of the *geophysicist* is to detect "anomalies" of all kinds and to attempt to interpret them. He may be regarded as the "finger-print expert" of the team and he has a number of tools with which to work. The most important—and fortunately the cheapest—is the thermometer, which may take various forms (the geothermograph, the Amerada gauge, the thermocouple, the thermistor, the platinum resistance thermometer and even the common mercury thermometer) each of which has its own particular application. By embedding thermometers at various depths over wide areas he can deduce temperature gradients and (in conjunction with thermal conductivity tests) heat flow patterns, thus obtaining circumstantial evidence of more deeply seated hot spots. When deducing the *total* natural heat flow from an area, he must allow not only for conducted heat through the ground but also for heat carried away by field seepages and by cross-flowing ground waters. This calls for careful flow measurements as well as temperature measurements, and it is here that close collaboration with the hydrogeologist is needed. Another valuable weapon in the geophysicist's armoury is the measurement of the electrical resistivity of the ground at different depths. By plotting isotherms and iso-resistivity lines on a map, the geophysicist can detect anomalies which he must try to interpret. Hot water, especially if containing large quantities of dissolved salts, will show up as low resistivity: steam as high resistivity. But resistivity can be affected also by the presence of minerals and by differences of rock formations, so the geophysicist must constantly be on his guard against false clues. Clay formations, for example, can masquerade as a water-filled aquifer because they too give rise to low electrical resistance. The geophysicist also has several other tools at his disposal such as gravity intensity measurements, seismic (reflective or refractive) measurements, micro-wave techniques, electromagnetic techniques, radio-frequency interference, ground noise measurements, microseismicity, aerial scanning for infrared radiation and audio-magnetotellurics. Some of these techniques are expensive and should only be invoked in particularly baffling circumstances, e.g. to resolve irreconcilable interpretations or apparently conflicting evidence from the cheaper techniques.

The geochemist's observations are relatively inexpensive to obtain and he may sometimes be instrumental in saving heavy exploration expenditure by demonstrating at the outset, by means of fairly simple investigations, that some particular area is likely to be geothermally worthless. Certain features of the analysis of thermal fluids discharged from hot springs are fairly reliable indicators of temperatures at depth; so the geochemist may be regarded as a highly sophisticated thermometrician who can aid the geophysicist. He too must beware of false clues occasioned by the dilution of hot spring waters with near-surface ground waters. The concentrations of silica and magnesium, and the ion ratios of sodium and potassium are all indicative of temperatures at depth, and if the evidence of all these "indicators" should fortunately coincide, then this evidence may reasonably be regarded as fairly sound.

The work of all these specialists is of course far more complex than can be described within the short space here available, but it is hoped that some broad idea of their work will have been conveyed to the reader.

As soon as the four earth scientists are broadly agreed that a field exists and have by common consent postulated a plausible model showing how it functions, the most critical moment of exploration will have been reached. A decision must be made on where to sink the first exploratory bore. It is at this point that the drilling engineer joins the team. If the deduced model closely approximates to the true structure of the field, it should be possible to choose a site at which a productive bore may be sunk. If the model is inaccurate, the chosen drill site may well result in an unproductive bore. It is most improbable that the preliminary model will be perfect, and the sinking of a non-productive bore should not necessarily give rise to despair. Every bore sunk—including unproductive bores—adds greatly to the knowledge of a field and so makes the choosing of subsequent drill sites progressively less risky. Until the first hole has been drilled, all the evidence is superficial or circumstantial. Once a hole has been sunk, factual evidence of the geology can be obtained by coring, temperature gradients may be confirmed or modified by downhole measurements, and the chemistry of subterranean fluids can be directly measured by sampling at depth. A new and reliable dimension will have been added to the available evidence.

It is of course always possible that the considered opinion of the earth scientists is that no exploitable field exists. In that case no drilling need be undertaken and the cost of exploration, abortive though it will have been, will have been limited; for by far the most costly item of an exploration programme is the drilling. For this reason, no drilling should be initiated until the earth scientists are reasonably confident that a field exists, and the greatest possible care should be exercised in the choice of drilling sites so that a minimum of unproductive bores will be sunk. It is unlikely that all the first few bores will be productive; but if a true field has been detected, the success ratio of drilling should steadily improve as more and more data are collected from each successive bore—whether successful or unsuccessful—until the process of *exploratory* drilling merges gradually into that of *productive* drilling.

Very possibly some of the first few bores may be deliberately sunk to greater depths than are necessary to penetrate to a good yielding "horizon" in the aquifer, in order to gain knowledge of the more profound depths of a field.

As soon as three of four good productive bores have been won, the work of exploration is virtually complete. A plausible model of the field will have been conceived and the broad pattern of productive drilling will have been laid down, though both the model and the choice of drill sites may have to be modified to some extent as the evidence of each bore is added to the sum total knowledge of the field. Exploration work may of course continue for some time after serious production drilling has been initiated, so as to carry investigations beyond the presumed boundaries of the field until those boundaries have been positively proven.

(Of relevance to the exploratory sciences are bibliography references 32–45.)

9. DRILLING

The depth of most geothermal bores ranges from about 500 to 1000 m. Some are rather shallower than the lower figure, while a few may extend down to 2000 m, or even more in rare instances.

Drilling for steam is a process that differs little, fundamentally, from that of drilling for oil or natural gas; but as harder and hotter rocks are usually encountered in geothermal drilling, and as corrosive liquids may sometimes be struck, rather special techniques, materials and equipment may have to be used. As with oil drilling, the rock is abraded away by means of "bits" of toughened steel—usually of tri-cone form, but sometimes annular if cores are to be recovered for geological information. The bits are fixed to the bottom of a hollow stem, consisting of jointed sections of steel tubing, which is rotated by means of a power unit at the surface—usually a diesel engine. As the bit descends more deeply, the stem is lengthened by coupling on further sections of tubing; and a tall power-operated derrick is provided for positioning these sections and for lowering or raising the drill stem into or out of the bore.

To cool and lubricate the bit, to wash away the rock chippings from the hole, to prevent the bore walls from caving in and to cool the stem and the surrounding ground, mud is forcibly circulated down the drill stem and up the annular space surrounding it. The mud becomes heated in this process and has to be cooled by passing it through a cooling tower, which dissipates the heat either by natural or forced draught as may be necessary before being pumped again down the drill stem in recirculation (Fig. 10). The muds used are always heavier than water, and usually consist basically of bentonite or other clay-based slurries, with various admixtures of special alkaline ingredients when high temperatures are encountered: without these additives the mud is apt to "gel" and lose its fluidity if it becomes too hot. Sometimes compressed air is used instead of mud; and although air drilling is cheaper and quicker than mud drilling, it is of no use in formations that are very wet or that tend to slough. Often mud is used down to the bottom of the production casing (see below), while air is used when penetrating the production zone lower down.

It is important that bores be neither too big nor too small. A small bore will restrict the upward flow of the geothermal fluids by offering too high a resistance; whereas a large bore is expensive, takes a long time to sink and may contain such a large weight of water (in a wet field) that a flow of thermal fluids may not be possible to sustain. It is usually considered wise to start drilling a new field, or suspected field, with rather small bores and to increase their diameter only if experience should show that the fluid yields are large enough to justify doing so. The bottom bore diameters generally range from about 160 mm for yields of 10–25 tonnes/h to about 215 mm for yields of 50–80 tonnes/h.

An important feature of geothermal bores is the sequence of concentric steel casings which should extend from the surface down to the top of the production zone of the aquifer. The uppermost and largest is known as the *surface* casing: it should be firmly fixed to the "cellar" (see Fig. 10) and should extend sufficiently far below-ground to pass through the upper weak formations. The next, or *anchor* casing, is perhaps 100–125 mm less in diameter and should extend as far as the cap-rock, where such exists, which serves to hold fast the lower end. Within these two casings is placed a smaller and longer *production* casing which should extend right down to the top of the production zone. Below that, the bore diameter is reduced where it penetrates as far into the production zone as may be necessary to ensure a good yield of thermal fluids. This bottom part of the bore is sometimes left unsupported, but more usually it will contain a slotted liner, standing at least half an inch

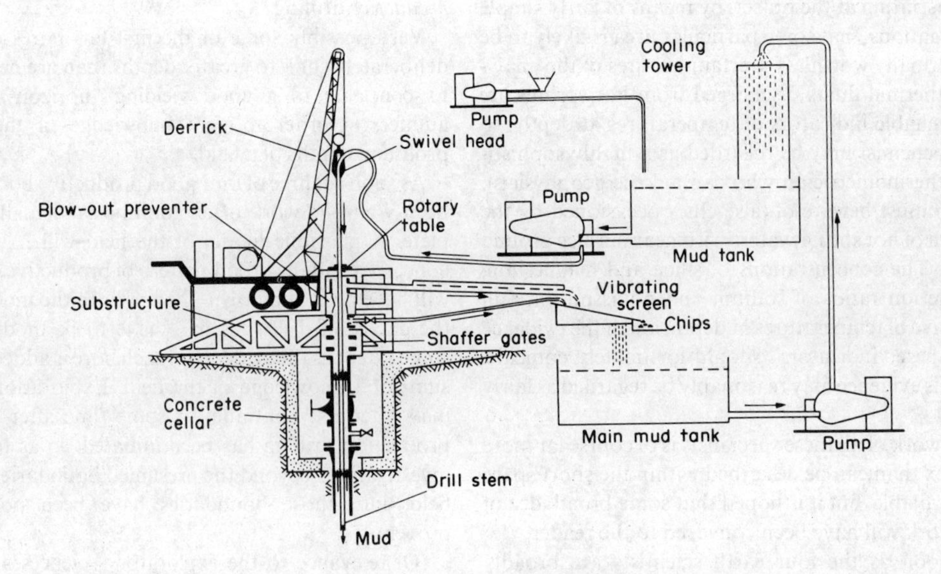

FIG. 10. Typical rotary drilling rig and mud circulation arrangements (after Craig, p. 123, ref. 46).

away from the bore walls, to act as a strainer for retaining rock particles which might otherwise be ejected up the bore with the rising fluids when the well is in service. The slotted casing should extend to the bottom of the bore. A typical arrangement of casings, as used at Wairakei, New Zealand, is illustrated in Fig. 11, which shows the nature of the formations there penetrated. Other fields, having other geological features will require a different arrangement of casings. Sometimes larger, and sometimes smaller, casings will be used than those shown in Fig. 11, depending upon the quality of the field.

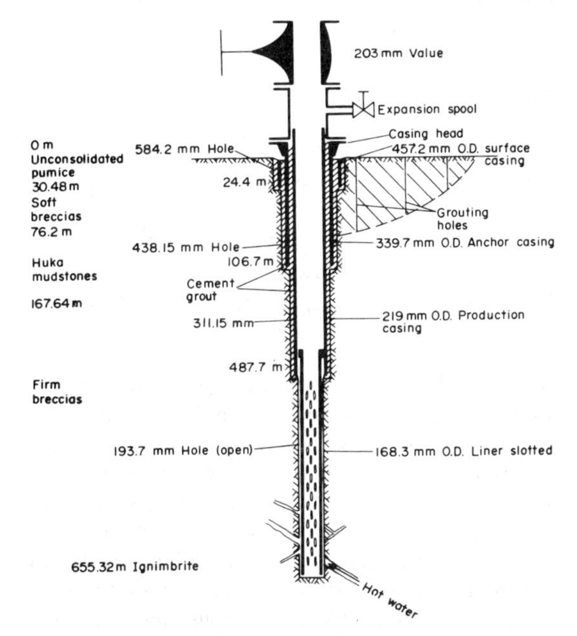

FIG. 11. Casing layout and geological formations at Wairakei, New Zealand (after Craig, p. 122, ref. 46).

It is of the utmost importance that casings are firmly and uniformly secured to one another and to the surrounding rock (except in the case of slotted liners) by adequate cementing. This is effected by forcing a cement slurry down each casing in turn by means of a plug, which drives the slurry round the bottom of the casing and up the annular surrounding space until it reaches the surface. Careless cementing can lead to various subsequent troubles such as collapsed casings or blow-outs. Casings must be carefully centred by means of spacers so that they are truly concentric with one another. Ordinary Portland cement is used for temperatures up to about 150°C, but certain additives are required in hotter zones.

The rate of drilling will depend upon the weight carried by the bit and upon the rotational speed, both factors being conducive to rapid penetration. But if either is too high, too much heat will be generated and the cooling mud or air will be unable to dissipate it, so that a blow-out may occur. Penetration rates will of course depend upon the hardness of the rock, the diameter of the bore, the ease of access to the site, the power of the rig, the quantities and sizes of the casings required, the lengths to be cored and the depth of penetration into the productive zone; but the following times[47] are fairly typical under favourable conditions:

depth	*actual drilling*	*finishing and testing*
500 m	15–30 days	10 days
1000 m	25–45 days	
1500 m	35–55 days	
2000 m	50–70 days	

In view of the possibility of blow-outs occurring during drilling, two precautionary devices are provided. First, just at the top of the surface casing, are fitted the "Shaffer gates" of rubber and of semi-circular profile sized so as to close tightly against a drill stem of casing, and capable of rapid closure by hand or by means of compressed air, so as to seal off the surrounding annular space. Above the Shaffer gates is a second device—the "blow-out preventor"—a flexible rubber jacket that can be pressed tightly against the drill string or casing under hydraulic pressure (Fig. 10).

The extraction of cores from a bore can provide a true record of the local geology, but as coring is rather a slow process it should only be done where the exploring geologist particularly needs the valuable information which it can provide for helping him to construct his field model.

Some thermal fluids are highly corrosive, and require the use of special casing steels. If they are too corrosive it may be necessary to abandon a hole, or even a whole area. Sand or small rock fragments can erode linings and wellhead equipment; but slotted liners are usually sufficient to restrict erosion to within acceptable limits.

Wells must not be so closely spaced as to interfere with one another's performance; nor must they be too widely spaced, as this would involve costly surface collection pipework and pressure drops. The practicable spacing of wells will depend upon the permeability of the underground formation, and will vary from field to field; but spacings ranging from about 100 to 300 m (equivalent to about 1–9 hectares respectively) for bore depths of 500–2000 m respectively are fairly typical. A method of economizing in surface pipework is to sink deviated, or curved (instead of straight vertical) bores, splaying outwards from a relatively small wellhead area (Fig. 12). In this way, a broad fluid collection base area may be tapped from a small surface area, thus requiring minimal surface collection pipework.

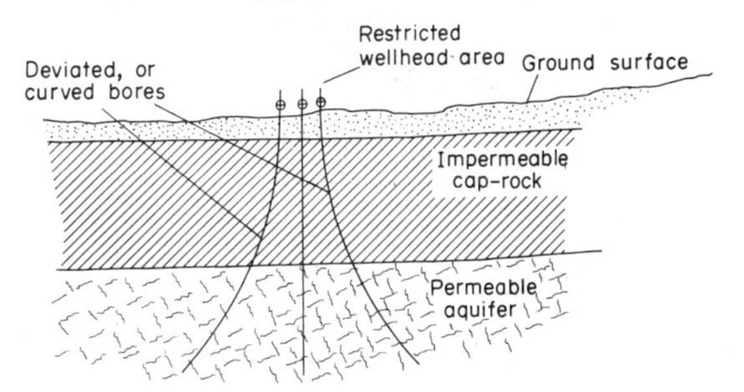

FIG. 12. Suggested method of using deviated, or curved bores to tap an extended producing zone from a restricted wellhead area.

Rare mishaps, such as collapsed casings and catastrophic blow-outs have been known to occur, and great ingenuity has been exercised in rectifying such incidents.[6] The taking of great care in cementing, the ensurance of adequate cooling during drilling, the selection of suitable bits and the avoidance of thermal shock by heating bores slowly when starting up and cooling them slowly when servicing, will generally suffice to prevent such mishaps—which are fortunately rare.

In this Section, only conventional rotary drilling methods have been briefly described. Entirely new techniques are now being developed which may perhaps revolutionize the methods of penetrating the Earth's crust to great depths. These methods will be briefly mentioned later.

(Of relevance to drilling are bibliography references 46–58.)

10. WELL CHARACTERISTICS AND THEIR MEASUREMENTS

After a newly drilled geothermal bore has been "blown" and shown to be productive it is necessary to take several important measurements in order to ascertain the physical properties and energy potential of the fluids discharged. (It will also be necessary to determine the chemical characteristics, but this will be dealt with later.) Most of the physical bore characteristics will be interdependent, so that much of the information collected will take the form of a series of curves rather than fixed figures.

The most important variable physical features of a productive bore are:

(1) the wellhead pressure;
(2) the wellhead temperature;
(3) the yield, or mass flow, of the steam (if any);
(4) the yield, or mass flow, of the hot water (if any);
(5) the fluid enthalpy;
(6) the fluid "quality"—i.e. the dryness or wetness factor.

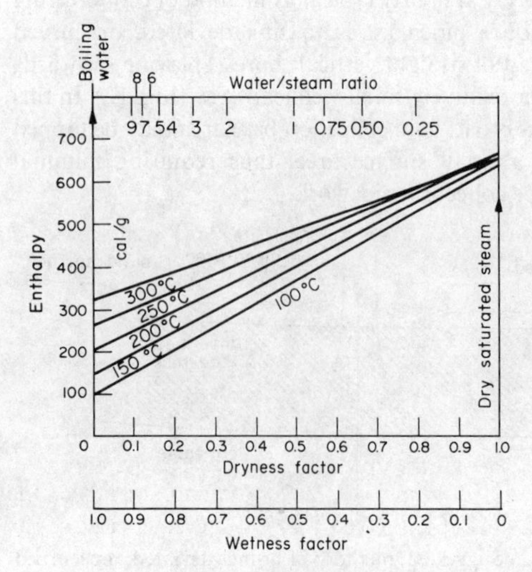

FIG. 13. Relationship between enthalpy, temperature and quality for saturated water/steam mixtures.

Wet Fields

With bores yielding water/steam mixtures, saturated conditions will of course prevail, so the variables (1), (2), (5) and (6) will be inter-dependent. The relationships between these four variables can be obtained from steam tables accurately, or from Figs. 13 and 14 approximately. The fluid yields—variables (3)

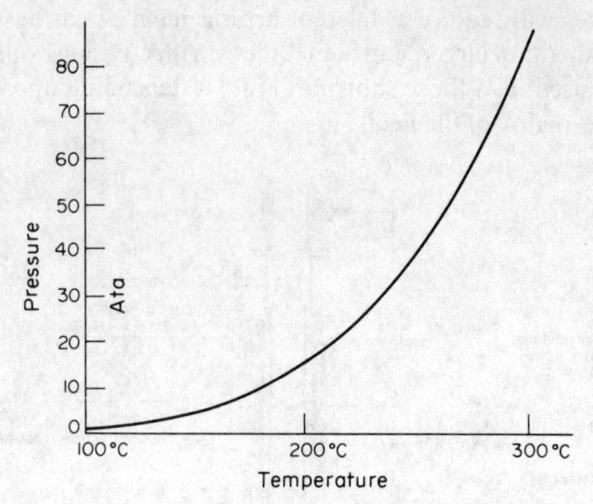

FIG. 14. Temperature/pressure relationship for dry saturated steam.

and (4)—will differ widely from bore to bore and from field to field, and must be determined by test measurements: they will vary with the wellhead pressure. These yields will be a function of the temperature at depth and of the flow resistances through the aquifer and up the bores; and since these factors will differ in every case, the flow characteristics of each individual bore will be unique to that bore. Figure 15 shows some typical bore characteristics at various exploited fields; but the word "typical" here requires qualification.

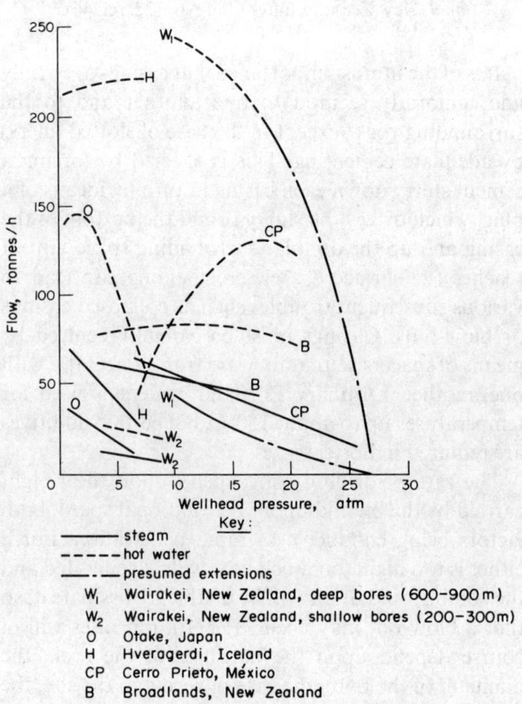

FIG. 15. Typical bore yield characteristics for "wet" geothermal fields.

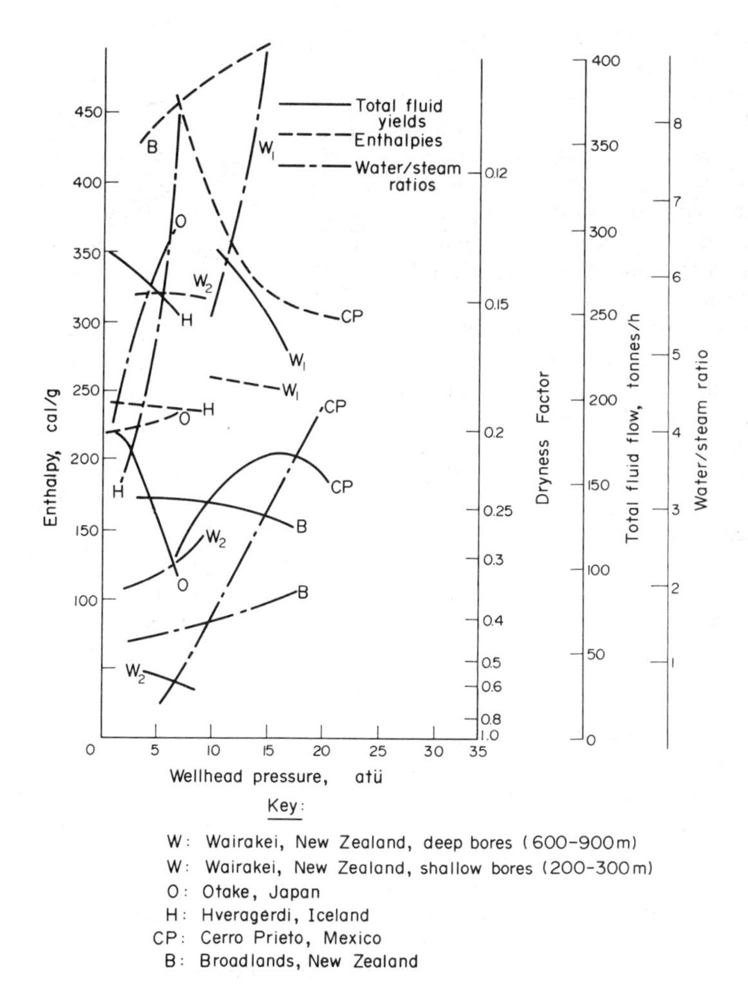

FIG. 16. Typical total bore yields, enthalpies and fluid qualities for "wet" geothermal fields.

First, for each field a random number of bores has been taken according to the available published data, and average yields have been deduced therefrom: the characteristics of any one of the selected bores may differ appreciably from the average. Secondly, the average figures do not all lie exactly upon the smoothed approximate curves here shown. Thirdly, the characteristics of a bore are not altogether constant: they often tend to become drier (i.e. the enthalpy increases) and rather less productive as time passes. The curves shown in Fig. 15 are all approximately representative of bores tested fairly early in the development of each field.

Obviously, when a well is closed, the flow of both water and steam will be zero; and the pressure instantaneously attained at the moment of closure is known as the "shut-in" pressure, which may later decline as the fluid in the bore cools off. In the case of the Wairakei deep bores (W_1 in Fig. 15) presumed extensions of the water and steam characteristics would lead one to expect a shut-in pressure of about 27 atü, though curvatures sometimes fall off rather abruptly as a well is gradually quenched by closure. Steam yield characteristics tend to approximate to straight lines, while hot water characteristics are usually curved.

From the steam and hot water characteristics it is a simple matter to deduce the total flow (by adding the two), the quality (by comparing the two) and the enthalpy (from steam tables or from Fig. 13). These further characteristics for the same fields are shown in Fig. 16, which reveals the following features:

(1) the very great variety of bore characteristics found from field to field;
(2) the relative constancy of the fluid enthalpy over a fair range of pressures at Wairakei (both shallow and deep bores), Hveragerdi and Otake;
(3) the *rising* enthalpy at Broadlands and the *falling* enthalpy at Cerro Prieto, with increased wellhead pressure;
(4) in every case, rising wellhead pressure causes increased wetness;
(5) the enthalpy of the shallow Wairakei bores is greater than that of the deep bores in the same field.

If a bore were to tap a zone of boiling water, the stagnant enthalpy of the flashing mixture (i.e. the enthalpy after allowing for the kinetic energy of the bore effluent) emitted by the bore would be expected to be constant, at the value of the enthalpy of the hot water tapped at depth. Thus the two classes of bore at

Wairakei and also the bores at Hveragerdi and (to a lesser extent) Otaka approximately conform with this expectation. The steeply rising and falling enthalpy curves for Broadlands and Cerro Prieto respectively suggest that the bores are fed from more than one "horizon" in each case, which contribute different proportions of the total heat with different pressure distributions down the bore. Increased wetness with rising pressure is only to be expected, especially if the enthalpy does not vary very greatly, since the higher pressure will suppress flashing. The relatively high enthalpy of the fluid emitted from the shallow Wairakei bores suggests the presence of free steam at the tapped zone, as the observed enthalpy is substantially higher than that of boiling water at the hydrostatic pressure corresponding with the depth at the base of the bore. It is presumed that these bores tap a rising plume of convected fluid which boils as it rises into zones of lower pressure just beneath the cap-rock, the observed enthalpy being that of more deeply seated hot water before it starts to boil on rising.

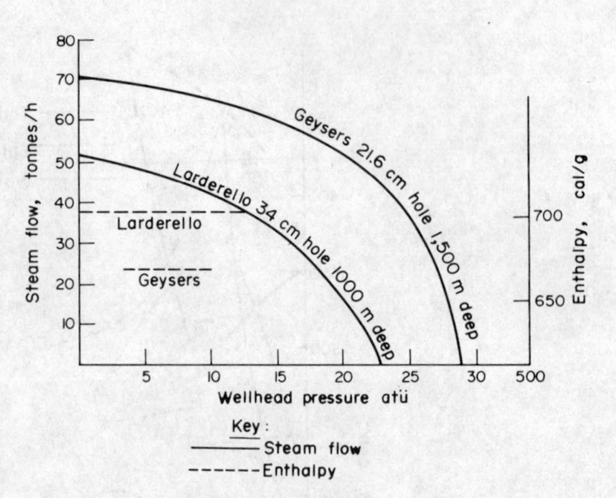

FIG. 17. Typical characteristics of bores in "dry" geothermal fields.

upon thermo-syphon action to lift the water from depth, to geo-pressurized fields which gush under partial lithostatic pressure.[25]

Dry Fields

As the steam discharged from dry fields is usually somewhat superheated, there can clearly be no direct relationship between the variables (1), (2) and (5), while (6) will have no meaning. At Larderello, in Italy, the degree of superheat ranges from about 83°C at a wellhead pressure of 6 ata to about 55°C at 16 ata. As low wellhead pressures are associated with high steam flows, there would be increased throttling action at low pressures as the steam passes at high velocity through the permeable formation and this will raise the degree of superheat, as would be expected with isenthalpic expansion. Expansion at Larderello does in fact approximate to the isenthalpic condition over the working range of pressures; but it is curious to note that the enthalpy is about 703 cal/g, which is more than the maximum heat that can be absorbed by water (669.8 cal/g). This seems to suggest that the steam derives its heat partly from hot rock above any boiling water reservoir which may exist at great depth. The rock must have been heated at some earlier time when the reservoir was filled with hotter water.

At the Geysers, California, the degree of superheat appears to decline with rising wellhead pressure, ranging from about 24°C at 5 ata to about 12°C at 9 ata. The enthalpy, however, is of the order of 669 cal/g only.

The steam flow and enthalpy characteristics, against a wellhead pressure base, are shown for Larderello and the Geysers—the two most famous "dry" fields in the world—in Fig. 17.

Hot Water Fields

Low enthalpy hot water fields are too idiosyncratic for generalizations to be made about their characteristics. Such fields range from deep wells initially at pressure equilibrium, that require pumping or rely

Measurements

The measurements of wellhead temperatures and pressures offer no problems: they can simply be made by thermometers and pressure gauges, and for wet bores only one of them need be measured. The measurements that require particular care are the mass flows. Once these have also been obtained, enthalpies and qualities can be deduced from steam tables or (for wet fields) approximately from Fig. 13.

Mass flow measurements of dry saturated or superheated steam can be made quite simply by means of sharp-edged orifices, after making sure that sufficient lengths of straight piping are placed upstream and downstream of the orifice to ensure non-turbulent flow (at least 25 diameters upstream and at least 10 diameters downstream). The mass flow will then be a function of the manometric pressure drop across the orifice and of the upstream pressure, and a calibration chart will be provided with each such orifice meter. Dry saturated steam may alternatively be measured by means of suitably proportioned cones discharging at sonic velocity into the atmosphere, and are therefore convenient for measuring steam flows at pressures of about 2 ata or more. The discharge from such a cone in kg/h/cm² of discharge area is approximately 54 times the absolute pressure atmospheres just upstream of the cone. (There are other slightly different empirical formulae which give slightly different results.)

The measurement of water/steam mixtures presents greater difficulties. The following are some of the methods used:

(1) *Calorimetry.*[59] The whole of the water/steam output of a bore is discharged for a measured time into a tank containing a known mass of cold water at a known temperature. When the well discharge has ceased, the additional weight of water in the tank is a measure of the mass flow during the period of the test; (the water content

is simply added while the steam content is condensed). From the rise in temperature of the water it is a simple matter to deduce the added heat, and by comparing this with the added mass the enthalpy can quite easily be deduced, and the dryness factor will be a derivative of this. Simple though this method is, it has its limitations. Unless a very large tank is used, the method is suited to bores of small output only; otherwise the duration of the test would have to be very short and the accuracy would be sensitive to the speed of operating the test valves. A special rocking arm has been devised in New Zealand[59] to reduce this difficulty; but where valves are used it is necessary to *start* opening them at the beginning of the measured time and to *start* closing them at the end of the measured time, so that the periods of valve operation will tend to cancel one another out. The method can be used for bores of larger output by splitting the discharge in two at a T-joint and measuring the discharge from one branch only and doubling it. If a valve on the *un*measured branch is adjusted to give the same back-pressure as on the measured branch, it is not unreasonable to suppose that the flow resistance in each branch is the same.

(2) *Sampler measurement.*[59] By means of a traversing sampler tube, the flows at different points on the diameter of a discharge pipe can be measured by calorimetry and integrated so as to arrive at the total discharge across the whole pipe. This is a very cheap method, as it requires only a small calorimeter and a light portable traversing device; but it is not very accurate. The rate of traverse is arranged so that the time in each position is proportional to the area of the annular ring being sampled. Ideally, the velocity of the fluid drawn off should equal the velocity of fluid in the main discharge pipe, so that the sampling tube causes a minimum of disturbance to the flow pattern. (Fig. 18.)

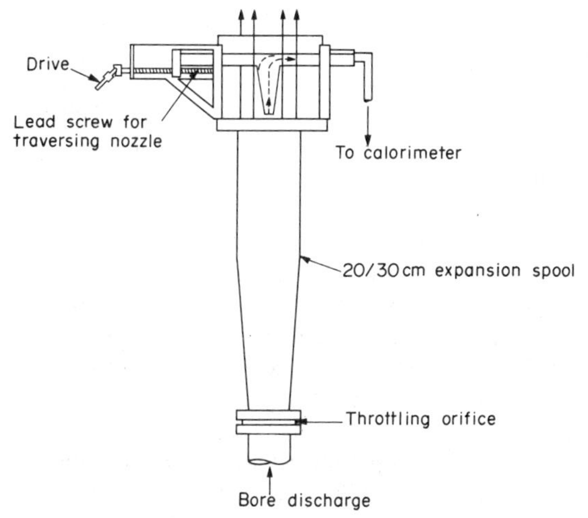

FIG. 18. Bore enthalpy measurement by steam sampler (after HUNT, p. 200, ref. 65).

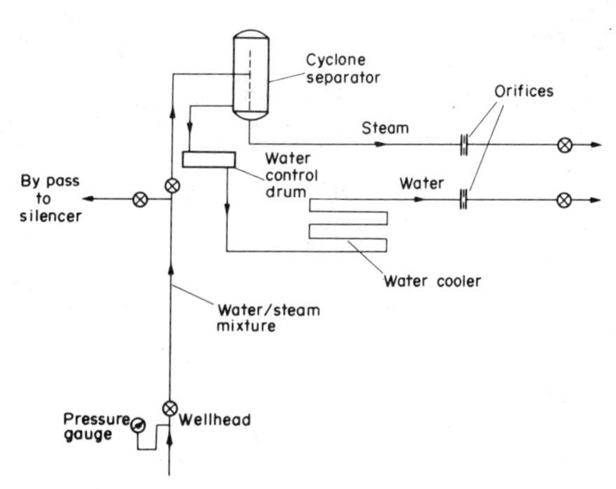

FIG. 19. Well yield measurement by phase separation and orifices (after Hunt, p. 197, ref. 59).

(3) *Phase separation and measurement of each phase separately.* This is the most accurate but most expensive method. The well fluid is passed through a cyclone separator (of about 99.9% efficiency); the steam is measured by orifice and manometer; and the hot water is measured in the same way, but it must first be cooled sufficiently to prevent boiling as it passes through the orifice. (Fig. 19.)

(4) *Critical lip pressure method.*[60] If hot water or a water/steam mixture is discharged at sonic velocity through an open-ended pipe, and if the pressure is measured at the discharge lip of the pipe, this pressure is a measure of the *heat flow* of the discharged fluid. The relevant formula is

$$G = \frac{38.455\, P^{0.96}}{h^{1.102}}$$

where G = flow in kg/sec/cm^2; P = critical lip pressure in ata; and h = fluid enthalpy in cal/g.

As an alternative to a V-notch or weir, this method can provide a cheap way of measuring the water discharged from the separator under method (3). For by measuring the temperature, the enthalpy of the water is known. With water/steam mixtures the enthalpy is unknown; but the combination of an upstream orifice and a lip pressure gauge can provide all the information required to deduce the enthalpy and mass flow, in the manner shown in Fig. 20. The orifice factor K is first determined by measurement, and the enthalpy then determined from the lower graph after substituting the value K in the ordinate and observing the upstream pressure and the manometer pressure. This is probably the cheapest method of measurement and is reasonably accurate within the idiosyncratic variations of a bore's performance from time to time.

Heat Capacity of a Bore

Once a bore characteristic is known, it is a simple matter to deduce its heat or power capacity, provided

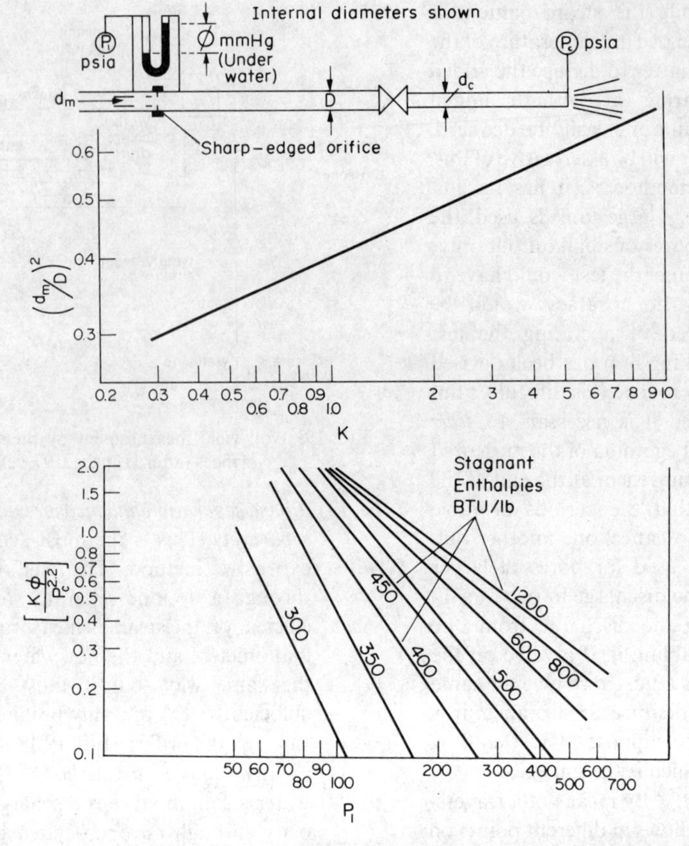

FIG. 20. Derivation of enthalpy by means of orifice and critical lip pressure at sonic velocity discharge (James, ref. 60).

Notes:

(1) d_c/D assumed to be 0.75, which gives convenient values of P_c. For other values of d_c, use correction formula:

$$\frac{P_{c_1}}{P_{c_2}} = \left(\frac{d_{c_2}}{d_{c_1}}\right)^{2.082}$$

(2) Recommended straight pipe lengths:

upstream of orifice	$\not< 25D$
orifice to valve	$\not< 10D$
valve to discharge	$\not< 25d_c$

(3) James's original chart, giving pressures and enthalpies in British units, is here reproduced. For conversions: 1 col/g = 1.8 BTU/lb; 1 psia = 0.06803 ata.

that its application is known. An example will illustrate this point. Figure 21 shows the flow characteristics of a hypothetical wet bore producing fluid at a constant enthalpy of 278 cal/g. It is an arbitrary, but not improbable, characteristic. It will be seen that the total heat emitted by the bore is fairly steady from atmospheric discharge conditions up to about 4 atü, or 150°C, and that it declines fairly rapidly after that, with rising wellhead pressure and temperature, to "shut-in" conditions of about 218°C/21 atü. With rising pressure and temperature at the wellhead, the steam contributes less and the hot water contributes more of the total heat. Figure 21 is only illustrative: other bores would behave differently.

The upper total heat curve in Fig. 21 represents a reckoning above 0°C, as taken from the steam tables. If at the field location the average ambient temperature of river or lake water were about 19°C, then the net total heat of the fluid at the bore reckoned above ambient conditions would be represented by the middle heat curve. If a 5% heat loss be assumed in transmission from the bore to the exploitation plant, then the usable heat would be as shown in the lowest heat curve.

Now let it be supposed that this particular bore is required for a timber-drying plant which, according to Fig. 2, needs heat at about 160°C; and let it further be assumed that a temperature loss of 3°C is incurred between the well and the plant. Figure 21 shows that the particular well represented could supply about 58 million net kg cal/h at the required temperature. A paper factory, on the other hand, requiring heat at 180°C (plus a 3°C drop from wellhead to plant) could obtain only about 52 million kg cal/h at the required temperature.

If a typical "Larderello" type of *dry* bore (as

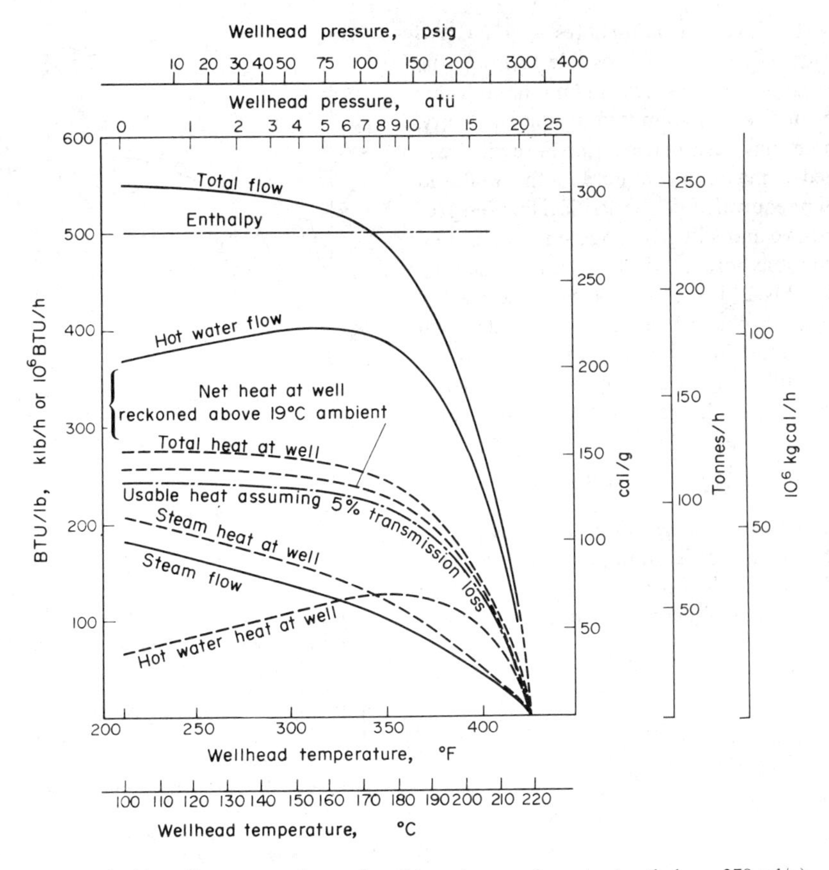

FIG. 21. Typical heat flow pattern from a "wet" bore (assumed constant enthalpy—278 cal/g).

illustrated in Fig. 22) were to be used, it will be seen that the grade of heat is higher at all wellhead pressures than required by either of the industries assumed in the above example for a wet bore. It would thus pay to operate the dry bore at the lowest practicable pressure (perhaps even subatmospheric) in order to obtain the maximum heat yield, and to degrade that heat to the temperature required for the particular industrial application contemplated. With a Geysers type of dry bore the temperatures are rather lower though the steam flows are higher than with a Larderello type. Except for processes requiring the highest grades of heat, where it might be necessary to operate at moderately high pressure in order to obtain the necessary temperature, it would again usually pay to operate the wells at very low pressures at which heat yields are still rising.

(It would be well here to mention a condition

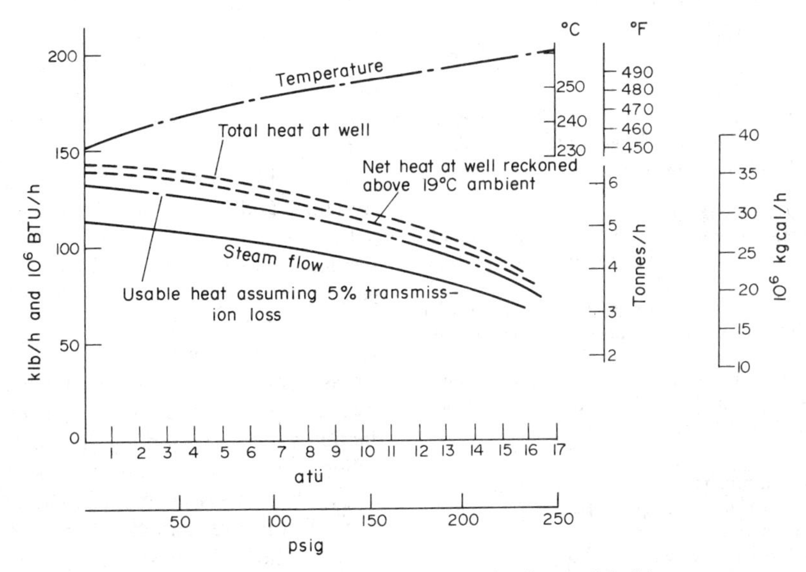

FIG. 22. Typical heat flow pattern from Larderello type "dry" bore.

sometimes met with very powerful bores—e.g. at Cerro Prieto, México—where it is impossible to lower the wellhead pressure below a certain minimum value because the fluid is being emitted at sonic velocity. Once this point has been reached, the bore yield will have attained a maximum value and the wellhead pressure will be equivalent to the "critical lip pressure" referred to above and will give a measure of the total heat flow from the bore. It will be seen that the bore represented by Fig. 21 just about reaches sonic velocity discharge at atmospheric pressure, because the total flow curve is almost horizontal. In Fig. 17, however, the yields are still rising as the pressure is lowered to atmospheric, so the discharge velocity is still subsonic.)

Power Capacity of a Bore

When it comes to the generation of electric power, however, the situation is quite different; for the efficiency of a power plant declines with temperature owing to inescapable thermodynamic constraints.

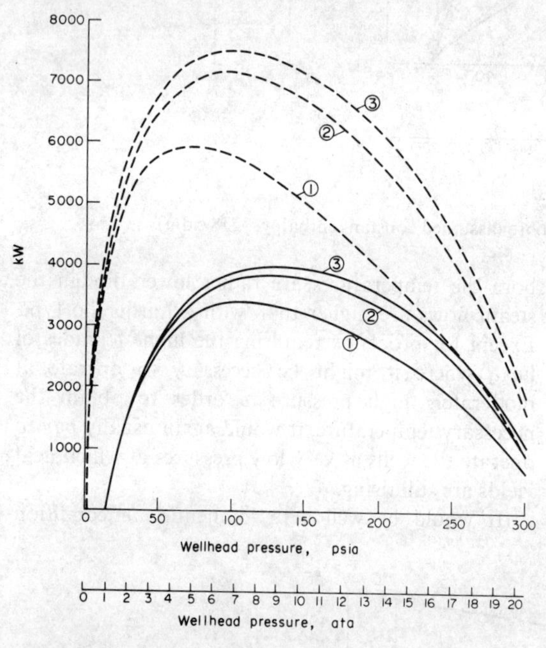

FIG. 23. Typical power potential from a "wet" bore (*N.B.* The powers shown above are at the generator terminals and exclude station auxiliary consumption).

Assumptions:
(1) Same bore as for Fig. 21.
(2) For condensing condition, 25% of the absolute wellhead pressure and 6% condensation lost in transmission.
(3) For non-condensing condition, 10% of the absolute wellhead pressure and 1% condensation lost in transmission.
(4) Condensing turbines sited at central power station fairly remote from bores. Non-condensing turbines sited fairly close to bores.

Key:
—— Exhaust to atmosphere.
--- Exhaust to 100 mm back-pressure.
(1) Steam phase alone.
(2) Total fluid with single flashing of hot water.
(3) Total fluid with double flashing of hot water.

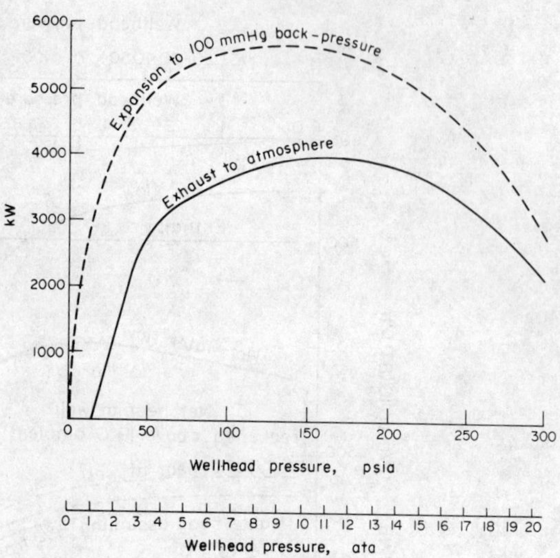

FIG. 24. Typical power potential pattern from a "Larderello" type of "dry" bore.

Assumptions:
(1) Same bore as for Fig. 22.
(2) For condensing condition, 25% of the absolute wellhead pressure and 5% of the heat content is lost in transmission.
(3) For non-condensing condition, 10% of the absolute wellhead pressure and 1% of the heat content is lost in transmission.
(4) Condensing turbines sited at central power station fairly remote from bores. Non-condensing turbines sited fairly close to bores.

(*N.B.* The powers shown above are at the generator terminals and exclude station auxiliary consumption.)

Thus the pressure of maximum power potential of a bore will not coincide with that of maximum heat potential. Rising wellhead pressure will both reduce the steam yield and increase the extractable energy per kg of steam. The result of these two opposing factors is a steam power potential curve that attains a maximum value at one particular wellhead pressure, falling off either with increasing or decreasing pressure on either side of that optimum. If consideration is also taken of the power extractable from the hot water phase of wet bores, a similar, though rather different, optimum pressure can be determined—despite the fact that over a certain pressure range the hot water yield may rise with increased wellhead pressure (as in Fig. 21).

Much will depend upon the cycle adopted, the fluid enthalpy, and the turbine back-pressure. A typical pattern of power potential for a *wet* bore, supplying straight condensing and non-condensing turbines, is illustrated in Fig. 23. The use of better vacua would improve the power potential of the bore when condensing turbines are used, but the gain would not necessarily be economically worthwhile because of the greater power required for gas exhaustion. By using the hot water and adopting double flashing after the wellhead, in the example chosen, a power gain of about 26% could be won by comparison with the use of steam alone. With lower enthalpy fluids (i.e. a higher proportion of hot water to steam) the proportional power gain could be appreciably more.

For a Larderello type of dry bore the pattern of power potential would be as illustrated in Fig. 24. The smaller steam yield, by comparison with the chosen example of wet bore, tends to be offset to a greater or lesser extent by the superheat, with the result that the steam phase alone of the chosen wet bore yields about 14% less power with non-condensing and about 5.5% more with condensing plants at 100 mm Hg back-pressure. If the hot water from the wet bore is also exploited for power, the wet bore of the example chosen could yield just about the same power as the dry bore when non-condensing, if double flashing be adopted. A Geysers type of dry bore (Fig. 17) could produce more power than either of the two bore types considered above.

It is emphasized that Figs. 23 and 24 are examples only: every individual bore will have a different power potential curve, depending upon its mass flow and enthalpy characteristics and upon how the bore is to be used. But in every case there will be an optimum pressure at which the power potential will reach a maximum value. This optimum pressure, however, will not necessarily be the same as the overall *economic* optimum pressure (see next sub-section).

Economic Optimum Wellhead Pressure

Operating a bore at the pressure at which the power yield is a maximum will clearly save capital expenditure in drilling costs and in wellhead equipment, etc., in that fewer bores will be needed to produce a desired quantity of power. On the other hand there are other economic criteria to be considered. Low pressures imply high specific steam volumes and therefore bulky pipework and vessels. Low pressures will also tend to shorten the life of a finite reservoir of heat by drawing down the stored fluid too quickly. High pressures, on the other hand, involve the use of short turbine blades which are more conducive to blade gap losses than long blades, and which are more vulnerable to the build-up of chemical deposits. High pressure valves and fittings tend to be costlier, and the associated higher temperatures require thicker thermal insulation. The interplay of all these considerations has been carefully studied by Russell James[61] who has concluded that for fluids of enthalpies exceeding 220 cal/g—and this covers most hyper-thermal fields—the overall economic wellhead pressure both for dry and wet fields seems to lie between 4.75 and 6 ata. This comparatively narrow range is somewhat surprising in view of the fact that the optimum power yield pressures in Figs. 23 and 24 range from about 5 to 11 atm. However, the curves in these figures are relatively flat-topped, and by choosing a wellhead pressure of about 6 ata the sacrifice in power from the theoretical optima would be within 3% for all condensing conditions. Only with atmospheric exhaust (rarely adopted for geothermal development) would the power sacrifice amount to about 11.5% when using steam alone with a Larderello type bore, and much less in other cases. Hence there is plenty of scope for the other economic considerations.

(Also of relevance to well characteristics and their measurement are bibliography references 62 and 63.)

11. FLUID COLLECTION, TRANSMISSION AND CONTROL

Wellhead Equipment

At each wellhead a certain amount of equipment is assembled to provide the means of controlling the fluids that emerge from the bores, of separating the water and/or dust from the steam (in hyper-thermal fields), of disposing of unwanted fluids, of silencing the bore fluids when discharged to the atmosphere and of protecting the equipment and pipelines against excessive pressures. A typical assembly of wellhead equipment at a wet field is shown in Fig. 25. The various components are connected together by means of pipework with conventional provision for thermal expansion—i.e. loops or bellows pieces—and most of them and their associated pipework will be lagged.

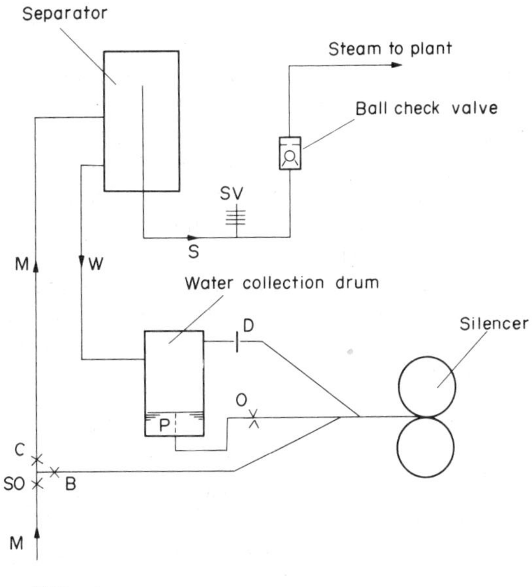

FIG. 25. Diagrammatic arrangement of wellhead equipment and safety devices at a "wet" bore (Armstead and Shaw, ref. 66).

Wellhead Valving

Figure 26 shows a typical wellhead valve arrangement as used at Wairakei. The valves are accommodated either in or just above the concrete cellar referred to in Section 9 and illustrated in Fig. 10. The

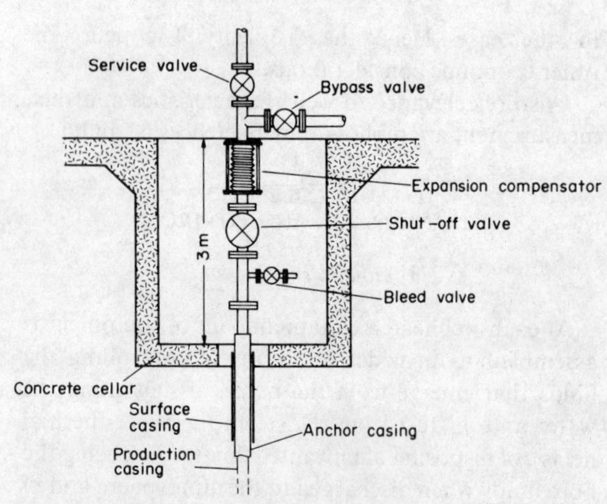

FIG. 26. Typical wellhead valving arrangement at Wairakei, New Zealand (after Haldane and Armstead, ref. 5).

service valve is used to regulate the flow and pressure of the emergent fluids during testing, while the shut-off valve enables the wellhead gear to be isolated for servicing. A bleed valve allows for the removal of accumulated non-condensable gases and a bypass valve permits the bore output to be discharged to waste. When a bore is taken out of service it is advisable always to bypass at least some of the fluid so as to keep the casings hot and thus avoid thermal shock due to alternating heating and cooling of the bore; and when a bore is in service, the control valve should normally be wide open so as to reduce wear: the control of flow and pressure is then exercised by the system control (see later).

Separators

In a wet field the water and steam are usually separated from one another at each wellhead (but see later remarks on two-phase fluid transmission). Various sophisticated separators have been devised, but one of the cheapest and most efficient is the Webre type[65] which is schematically illustrated in Fig. 27 and which can attain a separating efficiency of 99.9% or even more. The spiral inlet shown gives an improved performance over a simple tangential inlet that was tried earlier.

In dry fields the steam may contain a quantity of grit or dust, and this is usually removed by means of simple cyclone separators mounted co-axially in the steam take-off pipe; the dust being discharged continuously through a 6 or 7 mm orifice permanently open to the atmosphere. The loss of steam through this orifice is negligible. In wet fields, fine pieces of rock and dust are washed away with the water phase, but the occasional coarser fragments are trapped in the water collection drum by the perforated water take-off pipe (P in Fig. 25).

Hot Water Discharge

Often the hot water from a wet field is (regrettably) rejected to waste because the economics of extracting its inherent energy are not regarded as sufficiently attractive. Although the water/steam ratio is fairly steady at a constant pressure, it is nevertheless subject to short term fluctuations (gulping) and even to gradual long term variations. The water collection drum can take care of short term flow variations, but some cheap means of disposing of the hot water from this drum at the same average rate as that at which it is collected is necessary, if the water is to be discharged to waste. The use of float-controlled discharge valves would be costly and could sometimes be mechanically troublesome; and the problem has been solved in various wet fields simply by discharging the hot water through suitably sized bell-mouthed orifices. Such orifices, when passing boiling water, have the convenient property of enabling a wide range of water flows (about 3 : 1 typically) to be discharged without upstream flooding on the one hand and without loss of seal on the other hand.[66]

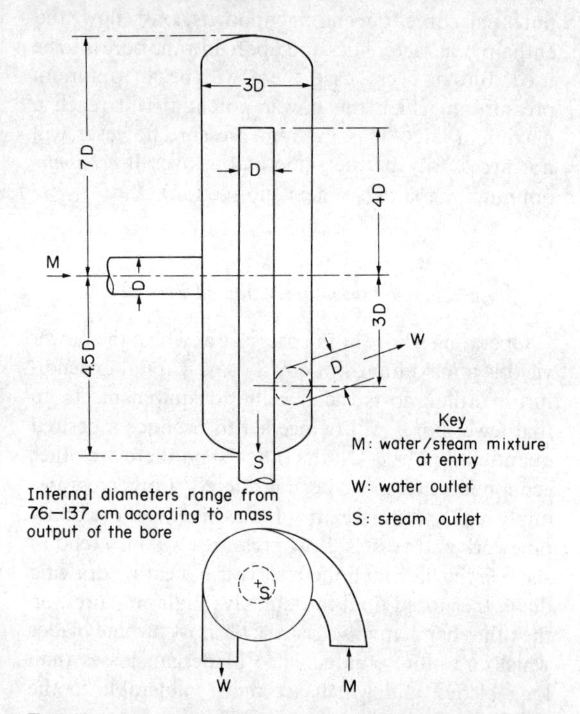

FIG. 27. Schematic arrangement and proportions of Webre type separator with spiral inlet (Bangma, ref. 65).

If the hot water is to be exploited for some useful purpose, it should be pumped—or perhaps removed by gravity if sufficient fall is available—from the water collection drum to the exploitation plant.

Silencers

When a bore is bypassed to waste, the noise can be ear-splitting. At Wairakei, even when the steam phase of a bore is being transmitted to the power plant the rejected hot water is usually discharged to waste, and the steam that flashes off from this hot water may sometimes be as much or more, in mass, as the useful steam. When the whole of a bore is being bypassed to waste, the volume of discharged fluid can be enormous. Thus there is a need for silencers at the Wairakei bores at all times, whether in service or bypassed. A simple design of silencer has been designed for use with the wet New Zealand field:[67] it is diagrammatically illustrated in Fig. 28. The fluid impinges upon a cusped steel dividing plate placed at the tangential junction of twin cylindrical concrete towers. The water swirls round the cylinders, losing its kinetic energy in friction, and is passed out over a measuring weir to a waste channel. The steam escapes to the air from the top of the cylinders. This type of silencer deflects most of the noise skywards and, more important, lowers the pitch of the noise from a high frequency scream to a tolerable deep-noted roar.

Steam Branch Pipes

From each well a steam branch pipe, sized to the output of the well and to allow for the pressure drop over its length, will conduct the steam from the wellhead to the nearest convenient point in the steam mains which carry the steam from the field to the power station or other utilization plant. These branches must be anchored at certain points by means of solid ground supports, and provision for thermal expansion and contraction between anchor points may be made by zig-zagging or "snaking" the pipes—usually in a horizontal plane, though sometimes by means of vertical loops where road access must be provided from one side of the branch pipe to the other.

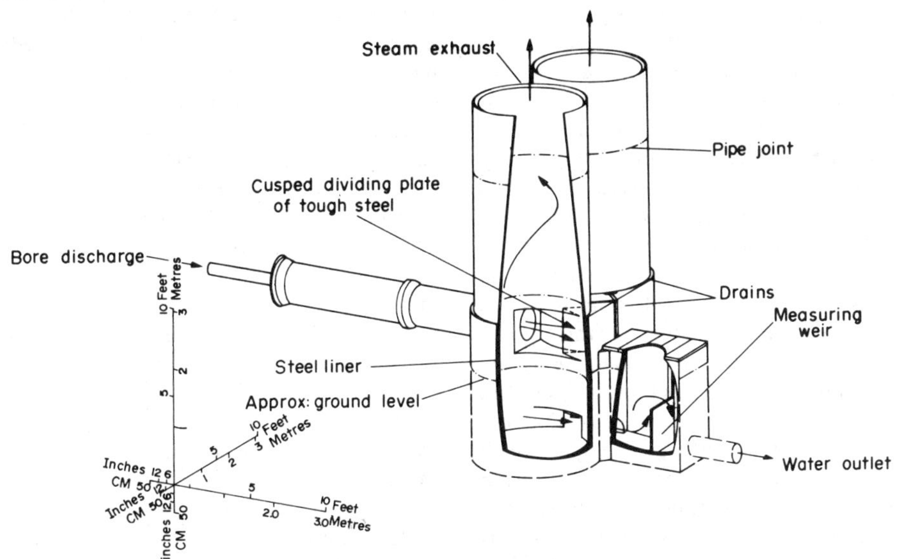

FIG. 28. Twin cyclone silencer, as used at Wairakei (Dench, p. 139, ref. 67).

Safety devices

The working pressure of a steam collection system will usually be a fraction of the shut-in pressure of the bore, and all equipment beyond the control valve at the wellhead will be designed for a lower pressure than the bore and its valving. A safety valve should therefore be provided at each wellhead as a statutory requirement; but as the lifting of a safety valve is an occurrence to be avoided except as a last resort (because of the erosion damage to seatings and because of the difficulty in getting them to re-seat without appreciably lowering the pressure) a cheaper safety device is also often provided at each wellhead in the form of a bursting disc. Such discs can be easily replaced and are relatively cheap, but as they are not usually accepted by Boiler Inspectors they have to be backed up by conventional spring or weight loaded safety valves. A further safety device is sometimes provided at wet wells in the form of a steel ball-float valve which lifts to shut off the steam supply in the event of a separator flooding or failing to perform its duty properly. The entry of a gulp of water into a steam main, and perhaps ultimately reaching a turbine, could be disastrous; hence the need for the ball check valve. The lifting of the ball would tend to induce the shut-in well pressure in all the wellhead equipment; hence, the bursting disc and safety valve should be placed between the top of the bore and the ball valve at the entry to the steam collection system (see Fig. 25).

Steam Mains

These will normally increase in carrying capacity as they progress from the remoter parts of the bore field towards the exploitation plant, gathering additional steam in the process. The increased capacity may simply be a matter of enlarged diameter, or it may be provided by adding more pipes in parallel. Wherever there is a change in the number of pipes, it is convenient to provide a valved manifold to enable individual pipes to be isolated for servicing. Such manifolds will be at anchor points. Junctions with well branch pipes should also be at anchor points, where pipe movement is zero, and anchors must of course also be provided wherever the pipelines change direction. Compensation for thermal expansion, as with branch pipes, may be provided either by zig-zagging or by bellows pieces. The latter can be of two kinds; *axial* bellows, which transmit large compressive or tensile forces to the anchor points which must be sufficiently robust to resist these forces; or *hinged* bellows which can be placed at loops, to form what are virtually three-pin arches. These arches, as adopted at Wairakei, transmit only minimal axial thrusts, so that some of their high intrinsic cost is saved by enabling lighter anchors to be used. The hinged expansion loops at Wairakei are in a vertical plane to provide road access beneath. Hinged bellows pieces can also be conveniently used to allow for expansion where pipelines descend steep slopes. Where axial bellows are used it is usual to take up half the pipe expansion movement when cold, so that the initial tension is more or less equal to the final compression when hot.

Lagging and Removal of Condensed Water from Pipelines

During its passage from the wellheads to the exploitation plant, the steam will lose some heat by conduction through the pipe walls. All pipes, both branches and mains, should therefore be lagged with thermal insulation. The cost of lagging is high, and it is unnecessary to restrict condensation to less than perhaps 5–7%, depending upon the length of the pipelines. Moreover, a certain amount of condensation is useful as a "scrubbing" device by diluting any small carry-over of saline water that may escape through the wellhead separators. At suitable intervals along the steam mains, downward pointing drain pots are provided, from which a large fraction of the pipe wall condensate is removed through traps, or even through small fixed orifices (which cost little in steam losses and save the cost of traps). By repeated dilution, partial removal, re-dilution and further partial removal it is possible to dilute the salinity of any small carry-over from the wellhead separators by a factor of many thousands, so that any small wetness factor in the steam entering the exploitation plant will consist of almost pure distilled water. Water detection devices may be provided in the drain-pots at strategic positions along the steam mains. These are arranged first to sound an alarm to alert the operating staff to the presence of excessive water far up the steam mains (possibly due to the failure of a wellhead ball check valve to operate) and progressively to trip off turbines or other exploitation plant units as any "slug" of water approaches nearer to the plant. This at first reduces the rate of steam flow in the mains and thus gives the drain pots a better chance of clearing the water from the line. Before the water reaches a point dangerously near to the plant, which could be damaged thereby, it may be necessary to trip off *all* the plant and bring the flow to a standstill until the trouble has been located and rectified.[66]

Hot Water Transmission

Hitherto, the transmission of very hot water over a considerable distance (about 1 mile) has only once been attempted, in an experimental plant at Wairakei. The problem to be solved in such a transmission scheme is to prevent actual boiling from occurring in the pipeline, as the formation of steam pockets and their subsequent collapse could give rise to very high and unpredictable stresses which might cause a burst pipe; and it is well to remember that boiling water is virtually an explosive fluid many times as dangerous as steam. Boiling in a pipeline can be prevented by assuring that the *hydraulic* pressure at all points exceeds the *vapour* pressure. The hydraulic pressure is a complex function of the initial vapour pressure at the point of collection, pipe friction, pipe gradients and dynamic heads arising from changes of water velocity due to valve movements. In the Wairakei scheme a safe margin between hydraulic and vapour pressure was assured by a combination of pumping, attemperation by injecting lower temperature water from second grade wells, and of carefully controlled speeds of valve operation. The problems of hot water transmission and its control are complex[5,66,68] but the Wairakei experimental scheme was absolutely successful technically. Unfortunately, the wells selected for the experiment, which for obvious reasons were those closest to the power station, proved to be near the edge of the field—a fact not suspected when the project was launched—and their output of hot water rapidly dried up. The scheme had therefore to be abandoned after about a year's successful operation at continuously reducing output, and the hot water pipeline was subsequently converted into a subsidiary steam line. Due to a fear that the whole field might ultimately yield dry steam, the hot water transmission scheme was not extended further into the heart of the bore field where there was (and still is) a large quantity of hot water available, with the result that several tens of megawatts of potential power from that hot water has been going to waste ever since the Wairakei installation was first initiated late in the 1950s. Thermal expansion of the hot water mains was taken up by means of axial bellows pieces so as to reduce pipe friction to a minimum, and the pipes were led beneath road crossings so as not to lose the static head inherent in tall vertical loops.

Something, however, was salvaged from the abandoned hot water transmission scheme at Wairakei, in the form of what is known as "*inter-flashing*". At Wairakei there are two distinct steam transmission systems—the high pressure system which delivers steam at the power station originally at not less than 13.5 ata but subsequently at a somewhat lower pressure owing to a change in the bore characteristics, and the intermediate pressure system which delivers steam at the power station at about 4.5 ata. Instead of transmitting hot water from the bore field to the power plant and there flashing it in two stages, as originally intended, the waste water from some of the high pressure bores is passed through separators in the bore field and partially flashed into steam which is then admitted into the intermediate pressure steam mains. This "inter-flashing" requires no separate transmission pipes except through short distances in the bore field; the existing IP steam mains carrying the inter-flashed steam from the field to the plant.[69] Although the system enables some small quantity of power to be recovered from the HP bore water, the system is not intrinsically very economic because the transmission of IP steam is costlier than that of HP steam and of hot water. Nevertheless it is physically simple and has produced a small bonus.

Two-Phase Fluid Transmission

The undesirability of the presence of steam in hot water mains, and of hot water in steam mains, blinded engineers for some time to the feasibility of transmitting water/steam mixtures all the way from the wells to the exploitation plant. The advantages of doing this, were it practicable, had been appreciated from the outset. No separators or hot water discharge facilities would be needed at the wellheads: valving and bypass silencers would suffice. Separation of the two phases could take place in large separators close to the plant, thus gaining the economic advantages of "scale effect", while the costly pumping and control features of hot water transmission could be avoided. But the fear remained that the transmission of water/steam mixtures over long distances would be dangerous on account of water-hammer. In the late 1960s experimental work on two-phase fluid transmission was undertaken in New Zealand[70] and it was found to be practicably feasible despite greater pressure drops than for the transmission of either phase by itself. At the Hatchobaru power plant in Japan, now under construction, two-phase fluid transmission is to be adopted, and the results of this enterprising project will be watched with great interest.[71]

System Control

When many bores, each having different flow/pressure characteristics, are connected to the same steam collection system, it is necessary to ensure that the exploitation plant—whether it be a power station, an industry or some other application—receives as much steam as it needs for its full and proper functioning with a minimum variation of temperature and pressure at the point of entry to the plant. Unlike a conventional combustion plant, in which control is effected by varying the rate of fuel feed and the quantity of combustion air, a geothermal plant must accept what nature gives it. The first requirement is to ensure that a sufficient number of bores are connected to the system, as otherwise the plant will be incapable of operating at its full capacity. The connection of too many bores, on the other hand, would give rise to excessive pressures in the steam collection and transmission system or to an extravagant waste of steam blown to the atmosphere.

Broadly, there are two principal requirements of a control system:

(1) either the delivery of a specified *quantity* of steam to the exploitation plant may be assured by allowing pressures to "float" between certain limits in the pipelines;

(2) or the maintenance of a certain specified *pressure* at the exploitation plant may be assured by allowing the flow quantity of delivered steam to vary.

To ensure either of these requirements it is necessary to provide vent valves, which may blow to waste varying quantities of steam. The setting of these vent valves may be automatically regulated in the first instance by the pressure at a chosen point in the pipeline, and in the second case by the pressure at the exploitation plant. The essential requirement of an efficient control system should be that under normal working conditions a minimum of steam shall be vented to waste, but that under emergency or transient conditions the vent valves shall be capable of temporarily releasing large quantities of steam until normal conditions have been restored. Space does not here permit a detailed description to be given of how these requirements can be met, but the problems and their satisfactory solution have been fully described elsewhere.[66]

The transmission of hot water is altogether more complicated. Some of the problems have been touched upon above, but they have been described in greater detail elsewhere.[5,66,68]

(Also of relevance to fluid collection, transmission and control are bibliography references nos. 64 and 72–75.)

12. GEOTHERMAL POWER GENERATION

After the emphasis that has been placed upon the value of geothermal heat for "non-power" applications, it may seem odd to single out for discussion electric power generation from all the possible uses for this heat, particularly as it is inherently an inefficient process. But clearly, in a review of this nature, it would be impossible to describe in detail even a few of the many other applications mentioned earlier, each of which has its own particular technology. It is also very probable that, despite the inefficiency, electricity production will for several years remain the principal form of geothermal development, partly for the reasons given earlier and partly because hyper-thermal fields, though limited in number and location, are likely to be exploited more rapidly than semi-thermal fields (or artificial fields) until all the more accessible ones have been developed. This is because hyper-thermal fields can produce far cheaper heat than low grade fields (and probably than very deep artificial fields), and this fact can commercially offset the inherent inefficiency of the power generation process. It is therefore considered justifiable to devote one section of this review to geothermal power development.

Cycles

In the early days of power development in the Larderello field in Italy, the corrosive nature of the steam was thought to be too aggressive for its direct admission to steam turbines. Accordingly, an "indirect" system was at first adopted, involving the use of a heat exchanger in which "clean" steam was raised from the "dirty" natural steam. This system, however, involved a loss of temperature and a sacrifice of about

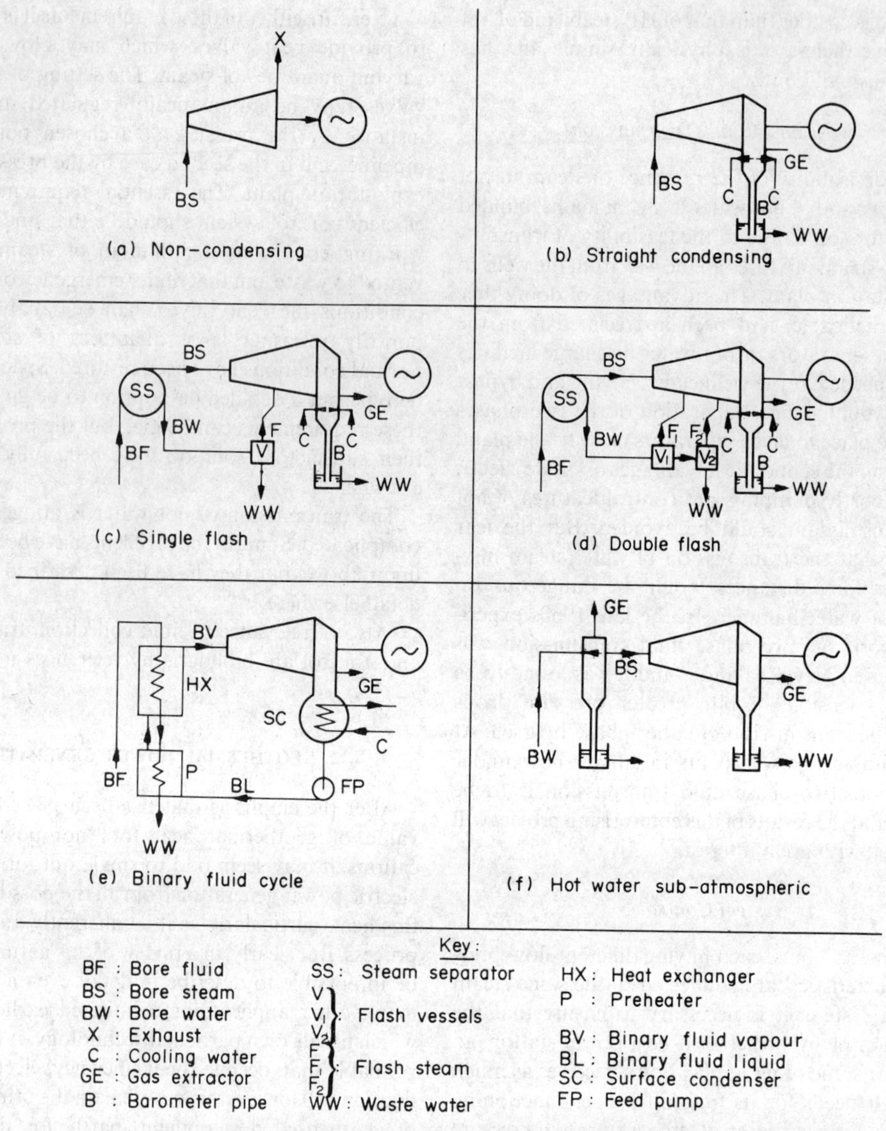

FIG. 29. Geothermal power generation cycles.

one-fifth of the power potential. Later, improved metallurgy enabled raw geothermal steam to be passed directly through turbines without this loss, and the indirect cycle may now be considered as virtually obsolete, although it could still be of use where valuable minerals (such as borax) are present in the geothermal fluids. The various practical cycles to be considered are illustrated diagrammatically in Fig. 29.

Non-Condensing

Fig. 29a shows the simplest and, in capital cost, the cheapest of all geothermal cycles. Bore steam, either direct from a "dry" bore or after separation from a "wet" bore, is simply passed through a turbine and exhausted to atmosphere. Such machines consume something like twice as much steam as condensing plants per kilowatt of output and are therefore wasteful of energy and costly in bores. Nevertheless, they have their uses as pilot plants, for small local supplies from isolated bores, as standby units and even perhaps (though to a very limited extent) for supplying peak loads.[76,77] Non-condensing machines must also be used if the content of non-condensable gases in the steam exceeds 15 or 20% by weight, because of the high power required to extract these gases from a condenser.

Straight Condensing

This cycle is shown in Fig. 29b. Since there is no need to recover the condensate for "feed" purposes, as in a conventional thermal plant, direct contact jet condensers with barometric discharge pipes can be used in place of costlier surface condensers.

Single Flash Cycle (Fig. 29c)

In wet fields it is possible to extract quite a lot of supplementary power from the hot water phase by passing it to a flash vessel operating at a lower pressure than that at which the main steam is admitted to the turbine(s). The flashed steam may then be used to pass through the lower pressure stages of the prime mover(s). Ideally, the maximum power yield from the hot water can be obtained if the flash vessel operates at a temperature midway between that of the bore water and that of the condenser. The unflashed fraction of the hot water is then rejected to waste. Although this cycle is here described as "single flash" some people prefer to call it "double flash", arguing that the first stage of flashing occurs in the bore itself. If the boiling water at depth in the aquifer is regarded as the original source of energy, this argument is sound; but it seems more logical to treat the wellhead fluid as the starting point of the thermal cycle from the engineering viewpoint. Otherwise cycle 29b would have to be called "single flash" although no surface flash vessels are involved.

Double Flash Cycle

Ideally, the maximum power would be extracted from the hot water of a wet field by using an infinite number of flash vessels connected in cascade—or at any rate by using one vessel for every turbine stage; but such a procedure would be economically absurd. It can, however, sometimes pay to provide two flash vessels as shown in Fig. 29d, operating ideally at temperatures of one-third and two-thirds way between the main admission temperature and the condenser temperature. This was the cycle adopted for the experimental hot water transmission scheme at Wairakei, described earlier. By analogy with what has just been said some people would prefer to describe this cycle as "treble flash". Approximately 68–70% of the supplementary power from the hot water is supplied by the first flash F_1: this shows that there would be diminishing returns from further stages of multiple flash which would render yet another stage of flashing uneconomic.

It is emphasized that in Fig. 29c and 29d, although a single turbine is shown with pass-in steam, it would be equally possible for the plant to take the form of separate turbines connected in cascade, with the exhaust from an upstream turbine mingling with flash steam and passing on to the entry of a downstream turbine.

Binary Fluid Cycle

Much thought has been and is being given to the use of refrigerant fluids having very low boiling points, such as freons and isobutane, in a closed turbine/feed/boiler cycle as shown in Fig. 29e. The theoretical advantages of the binary cycle are:

(1) it enables more heat to be extracted from geothermal fluids by rejecting them at lower temperature;
(2) it can make use of geothermal fluids that occur at much lower temperatures than would be economic for flash utilization;
(3) it uses higher vapour pressures that enable a very compact self-starting turbine to be used, and avoids the occurrence of sub-atmospheric pressures at any point of the cycle;
(4) it confines chemical problems to the heat-exchanger alone;
(5) it enables use to be made of geothermal fluids that are chemically hostile or that contain high proportions of non-condensable gases.

There are, however, the following disadvantages:

(1) it necessitates the use of heat exchangers which are costly, wasteful in temperature drop and can be the focus of scaling;
(2) it requires costly surface condensers instead of the cheaper jet type of condenser that can be used when steam is the working fluid;
(3) it needs a feed pump, which costs money and absorbs a substantial amount of the generated power;
(4) binary fluids are volatile and must be very carefully contained by sealing;
(5) makers are generally inexperienced, and high development costs are likely to be reflected in higher plant prices—at any rate unless and until binary cycles become commonly adopted in practice.

The only known operational example of a binary cycle geothermal plant, as distinct from laboratory prototypes, is the small 750 kW (nominal) unit at Paratunka in Kamchatka, USSR. This unit uses geothermal hot water at 81°C as the primary heat source, freon 12 as the binary fluid and cooling water at 6–8°C. High parasitic losses reduce the net output of the unit to 440 kW only.[78]

The binary cycle has its enthusiasts, but the case for it cannot yet be regarded as commercially proven for general use, though there could undoubtedly be conditions where it could be economic. Moreover, if the art of re-injection (see later) should ever become so perfected as to ensure that discharged hot geothermal waters can usefully serve as "boiler feed" to the underground heating cycle, the force of advantage (1) above would become somewhat weakened.

Hot Water Sub-atmospheric Cycle (Fig. 29f)

Only one case of this cycle is known to the author, namely at Kiabukwa in Zaire,[79] where a small British made 220 kW power plant was put into service as long ago as 1953. It is believed to be no longer in use. This plant used hot spring water at 91°C which was sprayed into a chamber at a low pressure of 0.3 ata, and the resulting flash steam passed through a three-stage low-pressure turbine exhausting to a condenser supplied with cooling water at 24°C. The plant was not self-starting, and the vacua in the flash chamber and condenser had first to be established by means of an ejector supplied with steam from an auxiliary wood-fired boiler. The natural occurrence of a hot spring enabled drilling costs to be avoided, but it is difficult to understand how such a plant could have been economic. Nevertheless, the plant was of great technical interest and is believed to have given good service for several years. It is doubtful whether the use of this cycle will ever be repeated.

Power Potential of Geothermal Fluids

The fact that geothermal fluids occur in nature at relatively low temperatures by comparison with those adopted in conventional thermal power plants means that the cycle efficiency of a geothermal power plant will always be low. Such plants are economic solely by virtue of the extreme cheapness of the heat. and are desirable from the aspects of energy conservation and environmental considerations.

The actual power potential of steam delivered to a power plant will not only depend upon its pressure, temperature and quality, but also upon the configurations of pipework, separators and valving within the power station, in all of which will be induced losses of pressure and energy. It will also depend upon the quality of the plant and upon the back-pressure at which the steam is finally exhausted. Considerable differences in power potential will therefore be experienced from plant to plant, but the curves shown in Fig. 30 may be regarded as fairly representative. They allow for reasonable turbine and alternator efficiencies and are expressed in terms of energy delivered at the alternator terminals; no allowance being made for the power consumed by gas exhausting equipment, cooling water pumping and miscellaneous station auxiliaries. These parasitic demands can vary widely according to the quantity of non-condensable gases present in the steam and the vacuum maintained in the condensers. They will vary from virtually zero for a non-condensing turbo-generating unit to as much as 47 or 48% for a condensing plant where the non-condensable gas content of the steam is as high as about 20%.

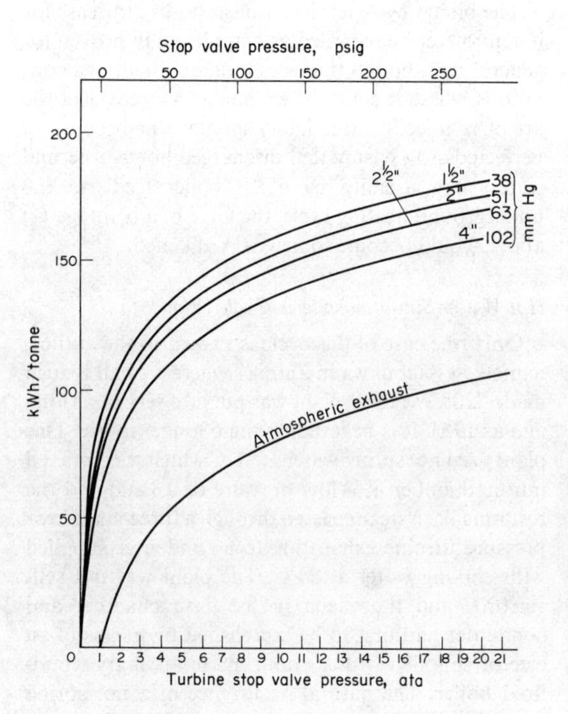

FIG. 30. Typical power potentials of dry saturated steam at various turbine admission pressures, exhausting to atmosphere or to different vacua.

Where dry superheated steam is available, its power potential cannot be represented by a simple series of curves as in Fig. 30 because much will depend upon the degree of superheat. To calculate the power potential in such cases, the available heat drop must be deduced from a Mollier chart and allowance made for the combined efficiencies of the turbine, the alternator and the station pipework, etc. These efficiencies would probably range from about 72.5 to 77.5% according to the admission pressure and the size and arrangement of plant units.

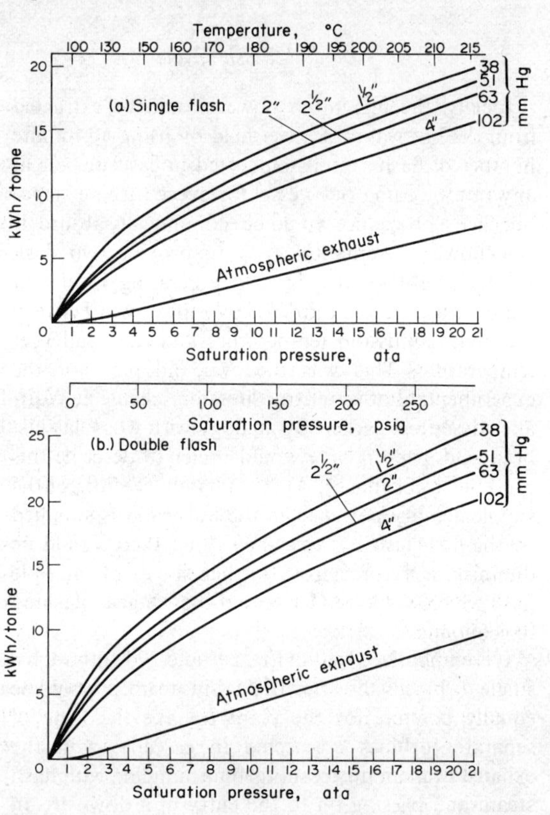

FIG. 31. Typical power potentials of boiling water at various pressures (and temperatures—see top scale) flashing at optimum pressures and exhausting to atmosphere or to different vacua.

The power potential of boiling water, as recoverable by the use of single or double flashing (within the meaning assigned earlier, and in accordance with Fig. 29c and d) is approximately as shown in Figs. 31a and 31b respectively. Although this potential, when expressed per tonne of fluid, is only a few percent of that of dry saturated steam at the same pressure, it should be remembered that the water/steam ratio is sometimes very high—perhaps 6 or 7 to 1—so that the total power potential of the hot water may form a substantial fraction of that of the steam in a wet field. At Wairakei, the use of all the hot water could have added about 50% to the power output when the field was first exploited. With the passage of the years the fluid has become less wet, and this proportion would now have dropped somewhat, but would still be substantial. The generation of power from hot water can mean a considerable saving in the cost of bores, wellhead equipment and steam pipework, which could outweigh the cost of transmitting and flashing the hot water. It is pointed out that although the flash potential for non-condensing turbines has been shown in Fig. 31, this is really of acadmic interest only: it would almost certainly not be worthwhile economically.

Choice of Condenser Vacuum

This will depend upon the availability and temperature of the cooling water (if any), upon the prevailing wet bulb temperature and upon the non-condensable gas content of the steam. Russell James[73] has attempted to show that a back-pressure of (127 mm Hg) is a more or less universal optimum, but his conclusion is not universally accepted. Except where large quantities of cold river, lake or sea water are available, it is generally not worth striving for too low a back-pressure, as this can sometimes necessitate costly cooling towers and high energy-consuming gas extraction equipment.

Cooling Towers

Where no adequate source of direct cooling water is available, cooling towers must be used. The heat dissipation of such towers per kilowatt of installed plant will be very much greater for a geothermal power installation than for a conventional power station owing to the low cycle efficiency of the former. The problem of cooling water make-up is simplified by the fact that the condensate from the turbines exceeds the evaporation loss, so that a permanent purge of surplus water can be provided, thus maintaining the concentration of impurities at a reasonably low level. This excess of condensate is due to the fact that the exhaust wetness from a geothermal power plant is much higher than for a conventional power plant owing to the low steam inlet conditions (saturated, or only moderately superheated).

Gas Extraction

Geothermal steam always contains non-condensable gases varying in amount from a fraction of 1% to as much as 25% by weight, or even more in rare cases. If the quantity exceeds about 15–20%, the energy required to extract the gases from a condenser would be too great for condensation to be economic. In such circumstances there is no alternative but to use non-condensing turbines despite their lower efficiency. For lower gas contents, the energy absorbed by the gas extraction equipment is more or less proportional to the gravimetric gas content and inversely proportional to the absolute pressure in the condenser. It is for this reason that the maintenance of very low back-pressures is seldom worthwhile unless the gas content is very low. The energy for gas extraction can be supplied either in the form of steam to ejectors, pumped water to ejectors or motive power to mechanical exhausters. For very low gas contents ejectors are the most economic; for higher gas contents mechanical exhausters (either rotary compressors or reciprocating air pumps) are to be preferred. The use of boiling water for ejectors has sometimes been mooted, but no published results of experiments appear to be available.

Evolutionary Turbine Units

In the Tuscan geothermal fields it is found that bores sunk in newly developed areas tend to produce steam with a very high gas content which gradually declines to a far lower level after 2 or 3 years of discharging. The Italians have devised a standard evolutionary turbine arrangement whereby a non-condensing turbo unit is first installed so that power can be generated from the high gas content steam. Later, when the gas content has fallen sufficiently, a second turbine with condenser and directly driven rotary exhauster is mounted as an extension to the shaft of the first unit, so that a two-cylinder condensing machine is thus obtained. The exhaust from the original non-condensing turbine is diverted to the inlet of the second unit—an operation that necessitates shutting down the plant for not more than 2 to 3 days. The steam consumption per kWh is then approximately halved.[80]

High and Low Level Condensing Plants

Opinions are divided as to the relative merits of what are known as the high level and low level turbine arrangements for condensing plants. With the former, the barometric jet condenser is mounted immediately below the turbine which, in consequence, must be placed about 9 or 10 m above ground level. This arrangement is adopted at Wairakei. With the low level arrangement the turbine is placed more or less at ground level and the condenser is mounted externally to the power house like a tower structure, with low pressure steam ducting connecting the turbine exhaust to the top of the condenser. The high level arrangement avoids the use of long exhaust ducts with their associated problems of drainage and corrosion, but it involves the use of larger buildings and of heavy structures to resist high level seismic forces; (hyperthermal fields usually occur in earthquake prone areas). The low level arrangement is used at the Geysers, California (units 1–4), Pauzhetka (USSR), Cerro Prieto (Mexico) and Matsukawa (Japan). There are also intermediate arrangements such as the medium level turbine placed over the condenser with the barometric pipe extending into a deep pit, as in Italy[80] and the medium level turbine placed over a shorter condenser provided with an extraction pump, as at the Geysers (units 5 and 6).

Fuelled Superheating

Some thought has been devoted to the idea of using supplementary fuel for superheating geothermal steam in order to achieve more efficient generation and also to reduce the wetness at the low pressure ends of the turbines. In 1970[76] it was shown that there could sometimes be a marginal advantage in superheating for the purpose of obtaining overload capacity from geothermal power plants during the hours of peak load; but owing to the greatly increased cost of fuels since then, it is almost certain that superheating in this way would never now be a paying proposition.

Load Contribution from Geothermal Plants

It is elementary knowledge that the cost of generating electricity in conventional thermal plants is made up of two components—a *fixed* cost per kilowatt installed plus a *variable* cost per kilowatt-hour generated. The variable component consists almost entirely of fuel costs. In a geothermal power plant there are no fuel costs and (except for a small element discussed later) all costs incurred are virtually fixed. By far the largest element is the capital charges on the investment. It is also elementary knowledge that an electricity supply system can be supplied most economically by a combination of base load plants, which are generally characterized by high fixed and low variable costs, and peak load plants, which are generally characterized by low fixed and high variable costs. Clearly, geothermal power is ideally suited to supplying base load; and practically every existing geothermal power plant in service is in fact operated at the highest attainable annual plant factor—sometimes exceeding 90%. In rare instances a case could be made for using geothermal power plants at medium load factors[76] or even for peak loads,[77] but as a general rule geothermal power plants will be used—especially when they form only one element of a large integrated power system—for supplying very cheap base load kilowatt-hours.

Byproduct Hydro Power

Where wet fields occur at sites high above sea level, it may sometimes be possible to win a "non-geothermal" bonus in the form of the hydro power potential from the hot water.[77] In some cases it may be the policy to re-inject the water (see later)—either very hot, as emitted from the wellheads, moderately hot after extracting flash steam power, or even fairly cool after using the heat for some industrial or other purpose. But if re-injection is thought to be inadvisable, the water (preferably after exploiting its heat content to the greatest practicable extent) may have to be rejected to waste. As this water by virtue of its height above sea level, has potential energy it could sometimes be used to generate hydro power rather than wastefully discharged to the sea through open water courses. There could sometimes be chemical problems—e.g. silica—

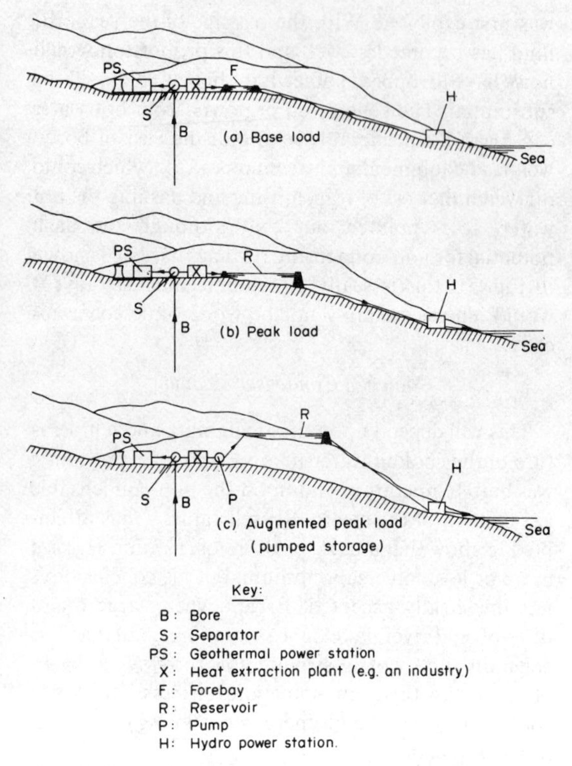

FIG. 32. By-product hydraulic power from a wet geothermal field (Armstead, ref. 77).

which could perhaps be overcome by means of settlement ponds or treatment. The discharged water could be used in three possible ways:

(1) for *base load*, by providing no storage other than some small forebay pondage (Fig. 32a);
(2) for *peak loads* by providing storage at the level of the geothermal field for, say 20 h flow, and by letting the water down through turbines for 4 h at peak load hours (Fig. 32b);
(3) for *augmented peak loads* (pumped storage) by pumping the water to a higher level reservoir during off-peak hours and generating a large output of peak load power when required (Fig. 32c).

As far as the author is aware, there is only one instance where hydro power is being generated as a byproduct from a geothermal development, namely at Wairakei where the waste bore water (regrettably, hot) is being discharged into the Waikato River and passes to the sea through a series of hydro plants, thereby generating about 2500 kW of continuous base load—a bonus of about 2% on the geothermal output of the Wairakei plant, and worth an annual saving of about 5000 tonnes of oil fuel.

Status of Geothermal Power Development in 1976

Table 1 gives a summary of the geothermal power plants now in service (1976), definitely planned, and contemplated. It is difficult to be accurate, partly because in some cases information is hard to obtain and partly because plans are constantly being revised.

It will be seen that nine countries are already generating power geothermally and that something approaching *1.4 million kW* of geothermal power plant are now in service. Most of the plants are capable of operating at very high annual plant factors—at least 90%—but some units are provided for standby purposes while a very few are incapable of producing their rated output owing to steam supply conditions. An average annual plant factor of something approaching 85%, reckoned on the total installed capacity, would not be an unreasonable assumption, so that the present annual generation of electricity by geothermal energy should be of the order of *10,000 GWh*, equivalent to the combustion of nearly 3 million tonnes of oil fuel per annum. Against the background of world oil consumption this is but a drop in the ocean—only a small fraction of one percent—and this serves to underline the fact that geothermal development is still in its infancy. Nevertheless it is something, and will doubtless increase rapidly.

Too much trust should not be placed on the figures in the second and third numerical columns of Table 1, for they may well greatly underrate the true position. It can be expected that there will be a continual shift from "contemplated" to "definitely planned"; and that as the extensive exploration work now in hand develops, there will be a steady growth in the figures of both columns. It may be noted that of the 2877 MW "definitely planned" no less than 652 MW are scheduled for 1977 and 1978, and are therefore already in the process of manufacture. This represents a 2-year average annual growth rate of 21.6% (compared with 14.2% p.a. for 1970/75). It seems highly probable that the actual growth rates in future years will be far more rapid than can be deduced from Table 2.

(Also of relevance to geothermal power generation are bibliography references nos. 4, 5, 8–10, 68, 71, 81–94.)

World Geothermal Power Generation Position in 1976

Most of the figures in the tables which follow have been directly communicated to the author or have been extracted from recent publications. With a few countries, either no response has been received to enquiries or plans are known to be under consideration but have not been declared: in these cases no figures have been shown in the tables, which may therefore give a slight understatement of the true position. Since the figures are closely known for all those countries at present in the lead in geothermal power development, the total MW capacities here shown will undoubtedly give a reasonably close estimate of the position of world geothermal power generation as in 1976, both of plants already in service and of those definitely or tentatively planned. Gross MW capacities are shown (i.e. as at the generator terminals without deduction for auxiliary consumption). The dates in brackets are the tentative programmed years of commissioning the definitely planned plants.

The key to "plant types" is as follows:

C—condensing;
N–C—non-condensing;
B–P—back-pressure;
MFC—mixed flow condensing;
B—binary cycle.

The details are set out in the following Table 2, and the figures are summarized above.

Table 1. Summary of world geothermal power generation position in 1976

	Gross MW capacities		
	In service	Definitely planned	Contemplated
United States of America	522	1621	
Italy	420.6	41	60
New Zealand	202.2	165	
Mexico	78.5	325	
Japan	70	100	
El Salvador	60	30	
USSR	5.75		
Iceland	2.8	70	
Turkey	0.5	15.5	
Philippines		443	
Nicaragua		65	
Indonesia			30
Chile			15
Santa Lucia		1.5	
Totals	1362.35	2877	105

In service	1362 MW	
In service and definitely planned	4239 MW	For details
In service, definitely planned and contemplated	4344 MW	see Table 2

In all the countries shown in the above summary, active exploration continues. In addition, exploration is also being vigorously pursued in the following countries, some of which are very likely to formulate plans for geothermal power generation in the near future:

Costa Rica	Guatemala
Ethiopia	India
French Antilles	Iran
Kenya	Taiwan
Panama	Thailand
Spain	

These exclude certain countries where exploration for low grade heat is being carried out.

The intentions of other countries, where geothermal energy could very probably be found, are not known.

TABLE 2. Details

Location	Unit nos.	Plant type	Number and size of units	In service (1976)	Planned: Definite	Planned: Tentative
Italy:			MW	MW	MW	MW
Larderello Area:						
(Total condensing plants 362.7 MW)						
Larderello 2		C	4 × 14.5	69		
		C	1 × 11			
Larderello 3		C	3 × 26	120		
		C	1 × 24			
		C	2 × 9			
Gabbro		C	1 × 15	15		
Castelnuovo		C	1 × 26	50		
		C	2 × 11			
		C	1 × 2			
Serrazzano		C	2 × 12.5	47		
		C	2 × 3.5			
		C	1 × 15			
Lago 2		C	1 × 14.5	33.5		
		C	1 × 12.5			
		C	1 × 6.5			
Monterotondo		C	1 × 12.5	12.5		
Sasso Pisano 2		C	1 × 12.5	15.7		
		C	1 × 3.2			
(Total non-condensing plants 57.9)						
Sasso Pisano 1		N–C	2 × 3.5	7		
Lagoni Rossi 1		N–C	1 × 3.5	3.5		
Lagoni Rossi 2		N–C	1 × 3	3		
Valonsordo		N–C	1 × 0.9	0.9		
Molinetto		N–C	1 × 3.5	3.5		
Travale-Radicondoli		N–C	1 × 15	18		
		N–C	1 × 3			
Monte Amiata Area:						
Bagnore 1		N–C	1 × 3.5	3.5		
Bagnore 2		N–C	1 × 3.5	3.5		
Piancastagnaio		N–C	1 × 15	15		
Monterotondo (1978)					10	
Travale (1978)					15	
Location Unspecified (1979)			2 × 8		16	
Torre Alfina						15
Efficiency improvements to existing plants						45
Totals for Italy				420.6	41	60
El Salvador:						
Ahuachapan	1 and 2	C	2 × 30	60		
	3 (1977)	C	1 × 30		30	
Totals for El Salvador				60	30	–
New Zealand:						
Wairakei	1, 4, 5 & 6	BP	4 × 11.15	44.6		
	7, 8, 9 & 10	C	4 × 11.15	44.6		
	2 & 3	BP	2 × 6.5	13		
	11, 12 & 13	MFC	3 × 30	90		
Total for Wairakei				192.2		

Table 2. (continued)

Location	Unit nos.	Plant type	Number and size of units	In service (1976)	Planned Definite	Planned Tentative
			MW	MW	MW	MW
Kawerau		BP		10		
Broadlands (1981)					165	
Totals for New Zealand				202.2	165	—
USA:						
The geysers	1	C	1 × 12	12		
	2	C	1 × 14	14		
	3 & 4	C	2 × 28	56		
	5, 6, 7, 8 9 & 10	C	6 × 55	330		
	11	C	1 × 110	110		
	12 (1978)	C	1 × 110		110	
	13 (1979)	C	1 × 140		140	
	14 (1978)		1 × 114		114	
	15 (1978)		1 × 57		57	
	16 & 17 (1980)		2 × 114		228	
	18 (1981)		1 × 114		114	
	19 & 20 (1982)		2 × 114		228	
	21 (1983)		1 × 114		114	
	22 & 23 (1984)		2 × 114		228	
	24 (1985)		1 × 114		114	
	25 (1986)		1 × 114		114	
Salton Sea and Brady Hot Springs (1978)					60	
Totals for USA				522	1621	—
Japan:						
Matsukawa		C	1 × 22	22		
Otake		C	1 × 13	13		
Onuma		C	1 × 10	10		
Onikobe		C	1 × 25	25		
Hatchobaru (1977)		MFC	1 × 50		50	
Kakkonda (1977)		C	1 × 50		50	
Totals for Japan				70	100	—
Nicaragua:						
Momotombo (1977)					10	
Momotombo (1978/79)					55	
Totals for Nicaragua				—	65	—
Chile:						
El Tatio						15
Total for Chile				—	—	15
México:						
Pathé		N–C	1 × 3.5	3.5		
Cerro Prieto	1 & 2	C	2 × 37.5	75		
Cerro Prieto Flash (1978)	3 & 4	C	2 × 37.5		75	
Cerro Prieto Flash (1979)	5	C	1 × 30		30	
Cerro Prieto Flash (1980)	6	C	1 × 55		55	
Cerro Prieto Flash (1981)	7	C	1 × 55		55	
Cerro Prieto Flash (1982)	8	C	1 × 55		55	
Cerro Prieto Flash (1983)	9	C	1 × 55		55	
Totals for México				78.5	325	—
Iceland:						
Namafjall		N–C	1 × 2.8	2.8		
Krafla (1977)		C	2 × 35		70	
Totals for Iceland				2.8	70	—
Turkey:						
Kizildere		N–C	1 × 0.5	0.5		
Kizildere (1980)					15.5	
Totals for Turkey				0.5	15.5	—

Table 2. (Continued)

Location	Unit nos.	Plant type	Number and size of units	In service (1976)	Planned Definite	Planned Tentative
			MW	MW	MW	MW
USSR:						
Pauzhetka		C	2 × 2.5	5		
Paratunka		B	1 × 0.75	0.75		
Total for USSR				5.75	—	—
Philippines:						
Tiwi		N–C	1 × 3		3	
Tiwi		C	4 × 55		220	
Los Baños		C	4 × 55		220	
Totals for Philippines				—	443	
Indonesia:						
Kamojang (W. Java)						30
Total for Indonesia				—	—	30
St. Lucia (Antilles):						
(1977)		N–C			$1\frac{1}{2}$	
Total for St. Lucia				—	$1\frac{1}{2}$	—

13. SOME ECONOMIC ASPECTS OF GEOTHERMAL ENERGY

At the outset it is well to recognize that the days of very cheap energy have probably gone for ever: those days ended in October 1973 when politically motivated swingeing increases in oil prices rocked the economies of many nations. With the alarming inflation that later afflicted so many countries, the goals of energy conservation and national self-sufficiency have largely overshadowed the question of costs. Frequent movements in the currency exchange markets have made it increasingly difficult for rational comparisons to be made between energy costs in different countries, or even between the costs of different forms of energy in the same country.

Unpredictability of Geothermal Costs

Even in stable times, so many uncertain factors can arise between the first moment of intent to develop an energy source and the moment when the energy starts to flow, that the precise forecasting of costs for any form of energy has always been a precarious business. But with geothermal energy, the task of predicting costs has always been rendered far more difficult by a number of other uncertainties peculiar to this form of energy. Exploration costs can seldom be foreseen with any degree of accuracy. A sum of money may be devoted to exploration, based on a rational plan of action, but until quite a lot of data have been collected and analyzed it is very difficult to predict how much money should be spent on each of the various exploration techniques available. The costs of exploratory drilling are particularly hard to foresee. Then it is hard to foretell the depths to which bores must be sunk to reach productive horizons, or the quality of rock to be penetrated by the drill. Again, the bore characteristics can never be known in advance, or the practical bore density per square kilometre, which will greatly affect the fluid transmission costs. There are other uncertainties such as the foundation conditions for a plant, the siting of which cannot be chosen until the pattern of productive bores is at least approximately known; or the ability or inability to discharge waste fluids into water courses without risk of pollution. If to all these difficulties be added those arising in the 1970s from unstable market prices for labour, equipment and financing, it will be appreciated that advance costing of geothermal energy must always be highly speculative. These uncertainties, however, should not be used as an argument against geothermal development; for somewhat paradoxically the very monetary troubles from which so much of the world is now suffering are tending to improve the economic prospects of geothermal energy. Geothermal exploration always involves some measure of risk, but experience has shown that where the risks have been taken it has nearly always paid handsome dividends. Moreover, exploration techniques are continually improving and the risks are steadily being reduced.

Past Experience of Geothermal Costs

Attempts have been made to rationalize geothermal costs, both in terms of heat and of electrical energy.[95] These attempts were reasonably valid at the time when they were made, but have since become outdated by inflation. However, comparative costs as between geothermal and alternative energy sources can be very significant even though absolute values may constantly be changing.

In the 1960s, when fossil fuels were cheap, it would have been approximately true to say that geothermal heat from exploited fields was of the order of US ¢12/million kg cal at the wellhead. After allowing for the cost of collecting and transmitting the geothermal fluids from the bore field, and of separating and (usually) rejecting the hot water at the wellheads, the cost of geothermal heat piped to a single delivery point for use in a power plant, a factory or a district heating scheme might have been approximately quadrupled to, say, US ¢50/million kg cal. In those same years, the cost of fuel heat ranged from about ¢100 to ¢250/million kg cal. If this heat was required in the form of steam or hot water, the price (after allowing for the capital charges, operation, repairs and maintenance incurred by the necessary boilers or heat exchangers) was raised to anything from about ¢150 to ¢400/million kg cal according to the location and scale of the enterprise. A fair average figure would have been about ¢275/million kg cal—about 5.5 times the cost of geothermal heat. In the case of small industries or micro power plants that could be served by a single bore without costly collection pipework, the cost advantage in favour of geothermal heat would have been very much greater. These approximate cost comparisons are based upon oil fuel as the alternative source of heat. Heat from coal or lignite could sometimes have been considerably cheaper, but not nearly cheap enough to cover this great basic advantage of about 5.5 times. In Reykjavik, Iceland, even after adding the costs of the extensive city heat reticulation system, the cost in 1961 of geothermal heat delivered to domestic consumers was only about 58% of the cost then prevailing of heat supplied from oil burned in small domestic boilers.

In the case of electric power, the advantage in the 1960s in favour of geothermal energy, though still positive, was less striking. This was because conventional power plants could operate at very much better thermal efficiency and could make use of power plant costing much less per kilowatt than geothermal power plants. Nevertheless, it was broadly true to say that geothermal power in those days cost only from 50 to 60% of conventionally generated thermal power at the same plant factor.

Effects of Inflation

From about 1970 world price inflation started to accelerate, and in late 1973 the price of oil fuel rocketed. By 1976 the price of oil fuel had at least quadrupled since the end of the 1960s, and other fuel prices had risen in sympathy, though not so steeply. But the fact that oil fuel prices have risen more rapidly than labour and material costs has rendered geothermal energy still more attractive relatively than conventional thermal energy, despite the higher interest rates on investment. This can be demonstrated by means of an arbitrary, but not unrealistic, example.

Let it be assumed that 10 years ago a geothermal power plant could generate electricity at 55% of the cost of that generated by a conventional thermal plant of the same capacity and serving the same duty, and that the proportional composition of the production costs were as follows:

Conventional		Geothermal	
Fuel	61%	Labour and materials	20%
Operation, repairs and maintenance	12%	Capital charges	80%
Capital charges	27%		
	100%		100%

Let it further be assumed that fuel prices have quadrupled, that labour and material costs have doubled, and that capital charges have risen from 9% to 14%. The relative production costs for new plants would now therefore be:

Conventional

Fuel	$61 \times 4 = 244$
Operation, repairs and maintenance	$12 \times 2 = 24$
Capital charges	$27 \times \frac{14}{9} \times 2 = 84$
	352%

Geothermal

Labour and materials	$20 \times 2 = 40$	
Capital charges	$80 \times \frac{14}{9} \times 2 = 249$	
	289	% of 55% = 159%

$$\text{Relative production costs } \frac{\text{geothermal}}{\text{conventional}} = \frac{159}{352} = 45\%$$

Thus, whereas geothermal energy was costing about 55% of conventional thermal energy about 10 years ago, it would now be costing only about 45%—equivalent to a relative improvement of (55 − 45)/55, or about 18%.

This comparison, while making no claim to accuracy, clearly illustrates the relatively improved economic position of geothermal energy resulting from the very high inflation of fuel costs. Insofar as industrial heat is concerned, the improved position of geothermal energy would be considerably greater owing to the higher fuel content of the cost of conventional heat.

Nature of Geothermal Costs

It has earlier been stated that geothermal costs are virtually fixed. Traditionally, it has been the custom to regard them all as fixed; but in the author's opinion it would be logical to assign a small variable cost component to the exploration costs. This is because the fruits of exploration are a finite quantity of energy, which may be expressed in MW-years or in teracalories. If the resources of a geothermal field were infinite, there could be no justification in regarding the exploration costs as anything but fixed; but it is generally believed that each field has a finite heat content, though that content may be difficult to assess with any degree of accuracy. This finite quantity of heat may be squandered quickly in a large plant, or it may be eked out slowly for a longer time in a smaller installation. It may be used wastefully in an inefficient plant, or economically in an efficient plant. It may be used "now" or, like fossil fuels, remain stored in the ground for future use. The general belief is that the infeed of magmatic and/or radioactive heat is small by comparison with the rate of draw-off in all exploited fields; so that the commercial development of a thermal field may be likened to the mining of a finite seam of fossil fuel. When we exploit a geothermal field, we are tapping a capital store of energy that has gradually been built up over thousands of years. The energy, except for a small fraction, is not strictly "renewable" like hydro power which is seasonally, or at least periodically, replenished.

If a geothermal installation is operated at a fairly constant annual plant factor, the assessment of a variable cost to cover exploration expenditure would be of little significance, once we have decided upon the method of exploitation. But when we come to consider different ways of exploiting a field, the existence or non-existence of a small variable cost component assumes some importance. For example, it becomes significant when we come to consider the relative merits of using condensing or non-condensing plants, or whether the latter could be used for carrying peak loads.

The actual evaluation of the variable cost component defies precision because of the inevitable uncertainty as to the thermal capacity of the field, and also because exploration is a process that may well be continued spasmodically for some time after a field is first exploited. But in theory the variable cost could be assessed by dividing the total costs incurred over the years of financing the exploration work (including interest and amortization of borrowed money) by the estimated total exploitable energy capacity of the field. In terms of generated electrical energy, the variable cost assessed in this manner is likely to be small—probably not more than, say, 0.5 US mil/kWh for condensing plants and perhaps 0.85 US mil/k Wh for non-condensing plants (which would consume about 70% more steam than condensing plants per kWh after allowing for differences of admission pressure and condensation losses arising from the fact that condensing plants would be sited at a central station at some distance from the bore field while non-condensing plants would only be used if they were sited close to the bores—see later). The corresponding variable cost component for heat at the wellhead would be of the order of 4.4 US ¢/million kg cal.

When it comes to considering the use of non-condensing plants for peaking purposes, about 10% should be added to the variable cost component to allow for scavenging steam to exclude air from the turbines when not in service.

These figures are intended to be no more than very rough illustrative examples, and probably err on the high side. By comparison with the variable cost component for conventional thermal power plants, the geothermal variable costs would be extremely small.

Relative Economics of Using Condensing and Non-Condensing Turbines

The normal method of exploiting a thermal field for power generation is to pipe the steam from a number of bores to a central power station, where electricity is generated by means of condensing turbines. (The question as to whether the hot water from a wet field is used or wasted need not affect the argument which here follows.) This practice is because of the relatively greater efficiency of condensing plants by comparison with non-condensing plants. An alternative method of exploiting the field would be to install a group of relatively small non-condensing plants close to the wellheads and to interlink them electrically. Such plants would require minimal pipework, the simplest of weather protection and foundations, no condensers and no cooling water system. Moreover, by siting the turbines close to the bores, the heat losses inherent in long steam mains can be reduced and higher turbine admission pressures can be adopted. It is true that their higher steam consumption would mean sinking about 70% more bores in order to achieve the same output, but this can be more than offset by the economies effected. It would in fact not be unreasonable to expect that the capital cost per kW of the non-condensing plant arrangement might perhaps be from 12 to 14% lower than for the central condensing plant alternative, and that the production cost per k Wh (despite the higher variable cost, as deduced above) could be about 10% lower.

This would seem to pose a paradox; for if power can be generated more cheaply by means of small non-condensing plants scattered around the bore field than by means of a central piped-up condensing plant, how is it that all major existing geothermal power plants make use of condensing turbines? The answer is that production costs per kWh, though important, are not the only criterion on which the economic choice of the type of a geothermal power plant should be based. Another criterion of even greater importance is energy conservation.

Let it be assumed that the assessed thermal capacity of a field, based on the use of condensing turbines, be estimated at 3400 MW-years. Since non-condensing plants would consume about 70% more steam, then the capacity of the field based on the use of non-condensing plants would be only about 2000 MW-years. In other words, by using condensing plants the life of the field would be about 70% longer for a given installed plant capacity. Would it be right to squander about 1400 MW-years of energy for the sake of perhaps a 10% saving in production cost/kWh? There is a difference between the *cost* and the *value* of a kWh. One way of looking at the problem is to compare the geothermal production costs with those of the most nearly competitive alternative source of base load energy. To take a hypothetical example, let it be supposed that the production costs of the alternative source are 17 US mils/kWh, while those of condensing and non-condensing geothermal plants are 10 and

9 US mils/kWh respectively. Over the life of the field, the savings effected by using geothermal energy rather than the alternative source would be:

for a central condensing plant—

$$3400 \times 1000 \times 8760 \times \frac{(17-10)}{1000} = \text{US } \$208.49 \text{ millions}$$

for non-condensing plants—

$$2000 \times 1000 \times 8760 \times \frac{(17-9)}{1000} = \text{US } \$140.16 \text{ millions}$$

Although the latter would save 8 mils/kWh, as against 7 mils for the former, by comparison with the alternative source of base load, the larger saving could be applied to a much smaller number of kWh, so that the long term profitability of the costlier condensing plant would be US $68.33 millions greater than if the cheaper non-condensing plants were used. This, of course, is an over-simplification, for it assumes a constant money value over the life of the field. In an inflationary market the savings in favour of the condensing plant would probably be far greater if the alternative energy source were fuel-consuming. The example serves to show, however, that the normal considerations of "merit order" generally applied to thermal plants in terms of their variable cost component per kWh would usually have to be overruled if any non-condensing geothermal plants were connected to the system; otherwise such plants would tend to be offered base load. There would, however, be two exceptions to this overruling—pilot plants and non-condensing geothermal plants using high gas content steam.

The Economics of Geothermal Pilot Plants

In the early stages of developing a geothermal field it is customary for the first few successful bores to be blown to waste at high output for a considerable time in order to study the bore behaviour and the field characteristics. A pilot plant could put this steam to good use. Since the steam thus used would otherwise be wasted, it could be of greatest value if consumed by a plant in almost continuous operation, so that it could make a small contribution towards the system base load. Pilot plants are of an experimental nature, installed to foster confidence and to gain experience. They should be cheap, as their function is comparatively ephemeral, so non-condensing units would be suitable for the purpose. They can be quickly installed, and as they make use of a single bore, or very few bores, they will be of fairly small capacity—say, 2–5 MW; and since they only use waste steam they should not be expected to bear any share of the costs of the bores or wellheads, these being primarily provided to serve a later and larger more permanent plant. Nor, for the same reason, need they be burdened with any variable (exploration) costs. However, if a pilot plant is to enjoy these exemptions, it would be unfair to compare its production costs (as computed on the basis of these exemptions) with the full production costs of other plants. It would be more logical to compare them only with the direct incremental savings earned by the generated output of the pilot plant. For practical purposes, these savings may be taken as the value of fuel saved in any conventional thermal plants that may be contributing energy to the integrated electricity system. Every kW of pilot plant capacity, if supplying base load, can save about 2 tonnes of oil fuel per annum (or equivalent for other fuels) that would otherwise have to be burned elsewhere in the system. Even under 1976 conditions of inflated costs it should be possible to operate a non-condensing geothermal pilot plant at an annual cost of perhaps US $70/kW or thereabouts. If this plant can save two tonnes of oil fuel per annum it would not only pay for itself but would also contribute some modest excess revenue if the cost of fuel oil exceeded US $35/tonne—a condition likely to be found in non-oil producing countries.

The fact that a pilot plant may have served its original purpose within 2 or 3 years of its installation does not mean that its life is finished after that period. It could be shifted to a new field for further pioneering work, retained *in situ* as an emergency generator, or even used for peak load purposes.

The Economics of Geothermal Power for Carrying Peak Loads

This is a rather complicated subject that has been analyzed elsewhere,[77] but it may here be said that a case could sometimes be made for very limited use of non-condensing geothermal plants for supplying "secondary peak loads"—i.e. thin flat-topped slices of load in the duration curve just below the extreme peak. If a use can be found for off-peak steam, possibly with the help of thermal storage, for industrial or other profitable application, much larger quantities of secondary peak load could be supplied geothermally.

Scale Effect

With conventional thermal power plants the capital cost per kW installed is sensitive to what is generally known as the "scale effect": that is to say, a very large plant will generally cost less per kW than a small plant of similar type. An approximate formula for this effect, which broadly applies to conventional thermal plants, is:

$$\text{Total capital cost} \propto \text{kW}^{0.815} \text{ (for conventional plants)}$$

Thus, if plant A is ten times the capacity of plant B, its total cost would be about 6.5 times that of plant B, and the cost per kW of plant A would be about 65% of that of plant B. This advantage in favour of the larger plant is partly due to the spread of overheads over a greater number of kW, and other general economies of large scale manufacture, and partly due to the fact that the larger plants use higher pressures and temperatures which are conducive to higher efficiencies and to more compact turbine blading and the accommodation of a larger number of stages on a single shaft.

With geothermal plants the scale effect is much less pronounced, for the following reasons:

(1) drilling costs are more or less directly proportional to the installed plant capacity (except that since an integral number of bores must always be used the proportion of a fractionally unused bore will tend to be higher for a small plant);
(2) turbo-generator costs are less subject to scale effect, partly because there is a more limited choice as to the maximum plant unit size, and partly because admission pressures and temperatures are usually modest, being influenced by factors other than mere plant size.

The net result is that for geothermal power installations the scale effect formula would be more like:

$$\text{Total capital cost} \propto \text{kW}^{0.875} \text{ (for geothermal plants)}$$

Thus if plant A is ten times the capacity of plant B its total cost would probably be about 7.5 times as great, and its cost per kW would be about 75% that of plant B. The advantage of size is therefore less marked with geothermal than with conventional plants. However, this is true only if the costing theory outlined above—that of assigning a variable cost component per kWh to cover the exploration costs—is accepted. If the explorations costs were treated simply as capital expenditure, the scale effect of a geothermal plant would be *greater* than for a conventional thermal plant, and the formula would become something like:

$$\text{Total capital cost} \propto \text{kW}^{0.7}$$
(for geothermal plants if exploration costs are treated as capital)

That is to say, if plant A were ten times the size of plant B, its total cost (including exploration costs) would be about five times as great, and its capital cost per kW would be about 50% that of plant B. This formula, however, should be treated as indicative only, since exploration costs can vary so widely from project to project.

It is clearly more logical to treat exploration costs as a variable cost per kWh so that the scale effect is less marked. The alternative of treating the exploration costs as a capital item means that if only a small installation were first adopted, the exploration costs would have to be fully recovered on that installation and that a large quantity of residual "free heat" would remain in the ground for use in future developments which would quite illogically be exempted from any share of the exploration costs.

Scope for Cost Improvements

In what way can we seek to reduce the costs of exploiting geothermal energy? There is possible scope in several directions.

Reduced Exploration Costs

In the early days of geothermal exploration, when knowledge concerning the nature and performance of geothermal fields was rudimentary, the search for exploitable geothermal energy was largely a "hit-or-miss" operation. There was much "wild cat" drilling; that is to say, bores were sunk in the vicinity of surface thermal manifestations without any proper understanding of underground processes. Drilling for steam or hot water was largely a gamble, which sometimes did, and sometimes did not, pay off. During the last quarter century exploration methods have become greatly refined and improved. The skills of the geologist, the geochemist, the geophysicist, the geohydrologist and the drilling engineer have all been coordinated so as to enable exploration to become much more systematic than in former years. Methods still continue to improve, and the risk element is steadily being reduced in the task of choosing good drilling sites. It cannot yet be claimed that the risk element has been entirely removed, but the chances of finding productive drilling sites have become greatly improved so that the costly sinking of abortive bores may now form a much smaller part of the total exploration costs than formerly.

Two-Phase Fluid Transmission

The technical feasibility of two-phase fluid transmission has already been mentioned. If this should become standard practice for wet fields the economic savings could be very great, not only in fluid transmission costs but also from the fact that by bringing the hot water right up to the plant the exploitation of its flash potential would be so greatly simplified. Moreover, the scale effect of supplying and maintaining a very few very large separators at the plant, by comparison with having to provide very many small separators scattered all over the bore field, would also be considerable.

Binary Fluids

As already shown, the use of binary fluids has both advantages and disadvantages, and it is too early to draw positive conclusions concerning its economic aspects. It is possible, however, that in certain circumstances appreciable economic gains might be achieved. It is important to understand that the use of binary fluids does not imply any improvement in cycle efficiency as such, but that it enables use to be made of heat that would otherwise never enter the turbine cycle at all, being rejected to waste.

Dual and Multi-Purpose Plants

The potentially great economic advantages of such plants are self-evident.

By-Product Hydro Power from Wet Fields

Generating byproduct hydro power from wet fields, could, where practicable bring economic advantages, though these would be too individual to particular projects for any generalizations to be made. The question would arise as to whether a value (variable cost) should be assigned to the water. This could be quite a logical allocation, as indeed it would be if the heat content of the bore water were first usefully exploited. In any case, any variable cost assigned to the steam should be appropriately reduced, since the charges on the exploration costs could be spread over both fluids—logically in proportion to the utilized energy content of each.

Novel Drilling Methods

Herein lies another possible source of economy in future geothermal development, as will later be shown.

Heat Pumps

The use of heat pumps is a subject by no means exclusively relevant to geothermal development. Nevertheless, where low grade waste heat must unavoidably be discarded from a geothermal exploitation plant it might sometimes be possible to upgrade the heat at a high performance ratio so that the natural fluid is rejected at a lower temperature and more heat extracted therefrom, while at the same time providing a reasonably large quantity of higher grade heat that could perhaps be used for some other practical purpose. Too much optimism should not, however, be placed upon this possibility, as heat pumps are only serviceable if the vapour pressure of the refrigerant is not too high—i.e. only at rather tepid temperatures. It is important not to fall into the error of proposing the use of heat pumps to up-grade the heat from the warm condensing water from a dry steam field power plant; for it never can pay to degrade heat to a low level and then pump it "uphill" again. In such cases it would be better either to use back-pressure or pass-out turbines, the exhaust temperature of which is sufficiently high for direct application to space heating or some other purpose.

Actual Costs

In view of the present instability of prices, the author is reluctant to quote actual geothermal costs to any large extent; but certain published data are here reproduced as a matter of interest (Table 3). An examination of the figures will clearly show how difficult it is to draw any positive conclusions from them. It is emphasized that projects constructed some years ago were of course far cheaper than what they would now cost. Some rough idea of trends in geothermal power plant costs may be gleaned from the figures quoted in the table which follows below and which show the actual and estimated capital investment by the Pacific Gas and Electric Company in geothermal power plants in the Geysers field, California.[7] The costs shown include step-up transformers, substations and electrical transmission, but they exclude the steam supply system. The estimated costs include an allowance for expected inflation. The figures were published in 1975 and amended in 1976. They are of limited significance only, partly because of the exclusion of the steam supply system (part of which would be accounted for by exploration and drilling), and partly because of the wide variety of unit capacities. The sudden drop in the cost of units nos. 7 and 8 by comparison with units nos. 5 and 6 is presumably explained by the gains from a repeat order; so it would be wiser to treat these four units together at an average cost/kW of $112.0. It would be difficult to attempt to allow for scale factor because some of the quoted costs are for single units and others are for pairs. It is of interest to note that the estimated cost of a single 55 MW unit in 1977 is $319.1 per kW whereas the cost of 53 MW units in 1971/72 was only $112 per kW. It would be unwise, however, to conclude that this implies a 2.85% inflationary factor, because the former figure was for a single machine while the latter was for four units of similar size. Nevertheless, the effects of inflation in those 5 or 6 years must have been very great.

No figures are available (for California) of the capital investment in exploration, drilling and the steam supply system (which is owned by a different enterprise); but it is of interest to note that at Wairakei the heat supply system accounted for approximately

TABLE 3. Actual and estimated costs (as in 1976) of geothermal power plants in Geysers field.

Unit no.	Year of actual or expected commissioning	Net rating in MW	Investment US dollars	Incremental cost per net kW US dollars		
1	1960	11	4,010,000 (units 1–2)	167.1		Actual
2	1963	13				
3	1967	27	7,610,000 (units 3–4)	140.9		
4	1968	27				
5	1971	53	12,756,000 (units 5–6)	120.3	112.0 average (units 5–8)	
6	1971	53				
7	1972	53	10,982,000 (units 7–8)	103.6		
8	1972	53				
9	1973	53	14,280,000 (units 9–10)	134.7		Estimated
10	1973	53				
11	1974	106	16,560,000	156.2		
12	1976	106	21,710,000	204.8		
14	1976	110	28,560,000	259.6		
15	1977	55	17,550,000	319.1		
13	1977	135	29,380,000	217.6		
Totals		908	163,400,000	180.0 (average)		

half of the total costs excluding the electrical transmission. However, the Wairakei heat supply system costs included the experimental hot water transmission, and the distance from the bore field to the power plant was much greater than at the Geysers; so the gross Geysers capital costs, including exploration, drilling and steam supply would probably be considerably less than double the figures quoted above. No idea of the Geysers steam supply investment can be gleaned from the price charged to the Pacific Gas and Electric Co. for steam in bulk, because this price which is directly proportional to the quantity of generated energy) is a flexible function of alternative heat costs and are not directly related to the cost of supplying the steam. Thus, if the price of oil rises, the price of steam rises. This flexible interlinking of steam costs with fuel costs is not without its critics; but the steam supply concessionnaires argue that it is a fair reward for the risks they have taken, that it provides them with funds for further geothermal exploration and that it offers encouragement to future concessionnaires.

Total production costs of electricity delivered to the Pacific Gas and Electric Co. system from the Geysers field have been estimated as follows (in US mils/kWh delivered to the system):

	Annual capacity factor		
	70%	80%	90%
1973 estimate based on delivery from unit No. 14 (110 MW net), with steam supply @ 4.8 mils/kWh plus re-injection costs (see later) @ 0.5 mil/kWh[96]	9.7	9.2	8.8
1978 estimate based on delivery from unit No. 15 (55 MW), with steam supply @ 12.5 mil/kWh plus re-injection costs @ 0.5 mil/kWh[7]	—	23.4	—

Thus a 154% increase is foreseen in the cost/kWh at 80% capacity factor during the 5 years 1973/78 and part of this will be accounted for by the less advantageous scale factor of the later unit. A direct comparison is therefore difficult. Nevertheless, the increase represents about 20.5% p.a. cost growth. It is of interest to note that in 1960, when the first Geysers plant was commissioned, the steam price was only 2 mil/kWh. The expected increase over 18 years is thus 6.25 times, equivalent to an annual average steam price increase of about 10.7% (disregarding reinjection costs). The average steam price growth rate from 1973 to 1978, as revealed by the above figures, is about 21% p.a.

Geothermal power continues to be the cheapest source of energy in the Pacific Gas and Electric Co.'s system. The 14 year average cost of geothermal energy in California (1961–1974) has been 5.612 mil/kWh as compared with 8.256 mil/kWh for fuel-fired plants.[97] When adjusted to the same annual plant factor the geothermal costs would still only be 89% of the conventional thermal power costs, despite the small unit capacities of the earlier geothermal plants.

For the 75 MW plant at Cerro Prieto, México, first commissioned in 1973, the total investment was US $14,400,000, equivalent to $192/kW (gross) at the generator terminals, which would probably work out at rather more than $200/kW (net) after allowing for auxiliary consumption. This figure compares very favourably with the Geysers figures, considering that it includes exploration, drilling, steam supply and plant, and that two units of only 37.5 MW each are installed. No production cost figures have yet been published for Cerro Prieto since an estimate was prepared in 1974 of 12.56 US mil/kWh generated, based on 80% annual plant factor, a 30 year life for the plant and a 10 year life for the bores.

There is a paucity of other updated geothermal cost data.

(Also of relevance to geothermal economics are bibliography references nos. 32, 62, 98, 99.)

14. CHEMICAL AND METALLURGICAL PROBLEMS

Non-Condensable Gases

Geothermal steam is invariably accompanied by non-condensable gases in widely differing proportions from field to field. At Hveragerdi in Iceland the gas content is very low, being only about 0.1% by weight, while in the Monte Amiata district of Italy it may vary from rather less than 10% to at least 25%. Typical intermediate figures are: Wairakei, from 0.35 to 0.5%; The Geysers, from 0.6 to 1.0%; Cerro Prieto, about 1.25%; Larderello, from 4.5 to 5%. Of these gases CO_2 forms by far the largest component (from 80 to 97%), with H_2S as the next largest ingredient (from $<1\%$ to 19%). The remaining ingredients consist of small quantities of CH_4, N_2, H_2 and sometimes traces of NH_3, H_3BO_3 and the rare gases.

Geothermal Waters

Geothermal hot waters will usually contain chlorides, silica and perhaps metaboric acid, sulphates and traces of fluorides. Sometimes they are alkaline—e.g. Wairakei, pH 8.6—and sometimes acid—e.g. Matsukawa, pH about 5. Rare cases have been known—e.g. the Tatun area in Taiwan—of pH values as low as 1.5, where the acidity has destroyed bore casings in a very short time and the field has been unusable.

Chemical Effects on Plant Components

These chemical ingredients are sometimes apt to cause corrosion or chemical deposition. Strict attention must be paid to the metallurgy of the exploitation equipment taking due consideration of the chemical and physical conditions of the fluids where they come into contact with the various plant components. Wet fields tend to be more troublesome than dry, in this respect. In the early days of geothermal development the usual practice, before manufacturing the plant, was to submit stressed and unstressed test pieces for long periods to environments of bore steam, air/steam mixtures and bore water, so that careful measurements could be made of corrosion rates and samples could be subjected to microphotographic examination for subsurface changes of metallic structure. In this way, vulnerable materials could be avoided and resistant materials used for the construction of the plant. So much is now known of the effects of bore fluids upon construction materials that much less empirical testing is now necessary.

Chemical Deposition

This sometimes occurs at the following points:

(1) *Calcite deposits in wet bores.* Fortunately this is not a very common form of trouble, but with certain waters it is apt to occur at the point where the uprising hot water starts to flash into steam. Hard calcium deposits then build up, first reducing the fluid output and ultimately almost choking the bore altogether. There is little that can be done about this except periodically reaming the bore to clear the deposits. If this is necessary only once in a year, or thereabouts (depending upon the fluid yield when the bore is clear), the bore may be economically serviceable; but if calciting builds up too quickly it may be necessary to abandon the bore. Calciting can also sometimes occur in the rock formation if the permeability is not very good, as flashing may take place before the fluid reaches the bore.

(2) *Silica deposits from wet bores.* At Wairakei and other wet fields, silica may be present in the bore water to saturation. As the temperature falls from the original value at depth, the silica will tend to come out of solution. There are different forms of silica, and that most commonly encountered is fortunately capable of remaining in solution in a state of super-saturation for sufficient time to prevent precipitation until the hot water is clear of the plant. Thus at Wairakei, no silica deposition occurs in the bores, wellhead separators or flash tanks; but in open discharge channels carrying cooling waste hot water, very large deposits are built up which have to be periodically and laboriously chipped away. In the course of time, the underground waters feeding the base of the bores through rock fissures may slowly deposit silica if flashing occurs in the formation where fluid velocities are low. This can gradually reduce the fluid yield from a bore and, perhaps after a few years, lead to its abandonment.

(3) *Turbine blade deposits.* Sometimes, through imperfect separation at the wellheads, and where bores are close to the generating plant (and the scrubbing effect of steam lines is therefore not very effective), traces of solids reach the turbine with the wet content of the steam and are deposited on the blades as the temperature falls during expansion. This has caused trouble at Cerro Prieto, where the station annual plant factor has been only about 70% during the first 2 or 3 years of operation. Blade deposits there have caused a loss of power of 4 or 5 MW (out of 75 MW nominal) and turbines have had to be taken out of service periodically for the deposits to be sandblasted away. Steam washing with oil has been found to be effective in Italy[100] in reducing the quantity of blade deposits and in facilitating their removal. The risk of blade deposits is a factor that favours the adoption of rather low working pressures, which are conducive to larger steam passages.

(4) *Iron sulphide deposits.* The action of H_2S on mild steel is to form a skin of iron sulphide, which fortunately protects the metal from further attack. Thus, although a black deposit will usually be found on the walls of all wellhead pressure vessels and pipes, this is quite harmless. It was therefore possible to use cheap mild steel for all these plant components.

Corrosion and Stress-Corrosion

Although the chemical impurities in geothermal fluids will vary from field to field, the broad characteristics are similar in quality though they may differ in degree. Table 4 may be taken as a broad guide to the resistance or vulnerability of various materials likely to be used in the component parts of a geothermal exploitation plant.

Turbine Blade Materials

As shown in Table 4, 13% chrome iron of the type normally used for turbine blades is, in the hardened or martensitic state, susceptible to stress-corrosion cracking in the presence of geothermal steam, but enjoys immunity if in the soft or ferritic state provided that chlorides are not present except in minute quantities. Turbine blades of softened and tempered stainless iron are therefore commonly used. But softened metal is liable to erosion, and owing to the higher wetness of the steam at the lower pressure stages of a geothermal turbine, this could be a hazard. It is inadvisable to protect the blades by brazing on erosion shields on account of the risks of local hardening and the vulnerability of spelter. It is therefore customary to limit blade tip speeds to a moderate figure (say, 275 m/sec) to reduce the erosion risks, which are proportional to the square of the tip speed.

Condensers and Gas Extraction Equipment

The most arduous conditions from the corrosion aspect occur in the condenser and gas extracting equipment, where bore gas, water vapour and oxygen are all present. The internal surfaces of jet condensers and ejector condensers should therefore be protected with various proprietory resins or with aluminium spraying. Austenitic stainless steels should be used for

TABLE 4. Resistance or vulnerability of materials to chemical attack in the presence of geothermal fluids

Material	Remarks
Category 1. Resistant to Corrosion	
Good quality close-grained cast iron	
Ordinary mild steel or low alloy steel (in the unhardened, fully pearlitic condition).	H_2S forms a protective coat of FeS, but below 30°C this is not so, and the material becomes vulnerable. Also, the presence of oxygen makes the material vulnerable.
13% chromium iron in the soft, or ferritic state (but not in the martensitic state). Also the same material if hardened and tempered at 750°C.	This material is suitable for turbine blades. It is essential, however, that chlorides are either absent or present to a minute degree only.
Ordinary 60/40 brass Admiralty brass (70% Cu, 29% Zn, 1% Sn) Aluminium brass	Resistance is poor in the presence of oxygen.
Pure aluminium	Vulnerable only at very high temperatures
Hard chromium plate	
Stellites Nos. 1 and 6	
18/8/3 (austenitic) chrome–nickel–molybdenum steel	
18/8 (austenitic) chrome–nickel steel, columbium or titanium stabilized; (non-precipitation hardening).	
Category 2. Vulnerable to Corrosion	
Copper and copper-based alloys, other than the brasses referred to above.	
Nickel and nickel-based alloys.	
Brazing spleter	Potentially dangerous
Zinc	Unreliable at higher temperatures
Aluminium alloys	*Pure* aluminium is reliable (see Category 1 above).
Category 3. Vulnerable to Stress-Corrosion	
Aluminium bronze	
13% chromium iron in the hardened or martensitic state.	
ATV (an austenitic steel)	11% Cr, 35% Ni, 1.3% Mn, 0.5% Si, 0.3% C
1.5/0.3 manganese–molybdenum steel	
Hardened carbon and low alloy steels (incorporating martensitic or quasi-martensitic structure).	
Most, if not all stainless steels if chlorides are present (aggravated if oxygen is also present).	External stress corrosion has occurred on stainless steel bellows where bore water has leaked through flanges onto them.
13% chromium iron, even in soft state, if chlorides present except minimally.	Hence the need for flash steam scrubbers to protect turbine blades.

gas exhauster rotors where the gases are wet, and intercooler tubes are usually made of 99.5% pure aluminium, which is resistant to geothermal steam except at high temperatures.

Attacks from Polluted Atmosphere

In the neighbourhood of geothermal fields, even before exploitation, there is usually rather a strong smell of H_2S, and while natural selection may have ensured that local vegetation is unharmed by this gas, this immunity will not extend to the works of Man after exploitation, as this will usually result in a great increase in local H_2S concentration unless very special precautions are taken.

Aluminium conductors in substations and transmission lines will usually take on a protective coating of sulphide, which inhibits further attack; but instruments and relay contacts will almost certainly suffer if they feature exposed copper, as sealing is seldom perfect. Silver contacts and bare connectors are advisable and, as an additional precaution, control rooms are sometimes sealed with air locks and ventilated with purified air. Exciter commutators of copper can be very troublesome, not only because the copper itself is attacked but also because the sulphide film causes sparking at the brushes which wear away at an alarming rate. Self-excited generators are now therefore generally used.

Sulphate Attack on Cooling Towers

Carry-over of bore gases into the condensate and cooling water can cause sulphate attack on the concrete structures of cooling towers. At Otake, Japan, alkaline river water has been introduced into the cooling system as a means of countering this, while at the Geysers, California, a coal tar epoxy protection is provided on concrete surfaces.

(Also of relevance to chemical and metallurgical problems are bibliography reference nos. 101–105.)

15. ENVIRONMENTAL FACTORS

Since the latter part of the 1960s there has been a growing public consciousness—long overdue—of the dangers of pollution. Geothermal enthusiasts have generally been rather too ready to claim that the exploitation of earth heat is a "clean" process. Although far less guilty of pollution than fuel combustion, it must nevertheless be frankly recognized that geothermal development is not altogether blameless unless certain precautions are taken. Fortunately, these precautions can be, and in some cases are being, taken with remarkable efficiency, though at some (but not exorbitant) cost. It is necessary to be aware of the possible sources of geothermal pollution and to avoid the *laissez-faire* attitude that was prevalent in the early days of geothermal development.

The USA have enacted very stringent anti-pollution laws which are now being rigidly enforced in the Geysers field at all the newer installations. Although these laws are in some ways to be welcomed, it is arguable that they have been too suddenly and too rigidly applied; for full compliance with the law has had the effect of delaying construction, and delay can cost a lot of money. It has already been pointed out that every kW of geothermal base load can save about 2 tonnes of oil fuel annually in an integrated system. With oil prices at the 1976 level, the cost of delaying the construction of a geothermal power plant by 1 year has approximately the same effect as doubling the true construction cost. Perhaps a more gradual introduction of the anti-pollution laws, with some permissible relaxation over a few years, might have been more in the national interest, especially as the marginal effects of geothermal development in California upon pollution would probably be small by comparison with the pollution caused by the combustion of fuel that could have been saved by avoiding the delays consequent upon the rigid and immediate enforcement of the new laws. As a *reductio ad absurdum* it may be mentioned that the natural thermal phenomena in Yellowstone Park are themselves "breaking the law"!

It is convenient to consider the problems of geothermal pollution one by one; but as the solution to several of these problems can lie in 're-injection' it is well first to consider this subject.

Re-Injection

For many years, the possibility of re-injecting into the ground the hot bore water emitted from wet fields was mooted. It was recognized that re-injection would avoid discharging huge quantities of heat into rivers, with consequent hazards to fisheries, and would also avoid infecting rivers and streams with toxic substances emitted from the bores which could endanger downstream drinking water supplies, fisheries or farming activities. Furthermore, since the removal of huge quantities of underground waters is known sometimes to cause land subsidence, re-injection could at least mitigate this. But for a long time there was a certain timidity of approach towards re-injection. Would the introduction of cooler waters into the permeable substrata interfere with the useful output of heat from producing wells? Would excessive power be absorbed in forcing the unwanted fluids into the ground? Would the permeability of the aquifer be destroyed by chemical deposition? Would re-injected waters outcrop elsewhere, thus simply transferring the pollution problem from one place to another? Would re-injection trigger off seismic shocks?

At the Pisa UN Geothermal Symposium of 1970 some discussion was devoted to re-injection. By the time the next UN Geothermal Symposium was held in San Francisco in 1975 much useful empirical data had been gained; and although it is too early to be dogmatic about the *pros* and *cons* of re-injection, practical experience now offers promising evidence that it could often not only be practicable but perhaps even thermally advantageous. For by re-injecting at some strategic point in a field, well removed from the producing bores, it may perhaps be possible to impose a warm barrier that delays the ultimate ingress of cold water from beyond the confines of a field. Re-injection could thus perhaps become a useful tool in field management, by re-cycling both water and heat, and could provide the equivalent of "boiler feed" as an alternative to the wasteful rejection of heat at the surface. Even if the heat of the hot water can be usefully exploited before re-injection, the return of the cooled liquid to the substrata could avoid toxic pollution and could perhaps provide a medium for picking up the heat of the hot rock left behind after the extraction of the original hot water contained in an aquifer. More experience is still needed before proper judgement can be given, but it is now known that re-injection can be achieved by gravity alone (i.e. without pumping) and it is already being practised in El Salvador,[106] California,[107] France,[108] Japan[109] and Italy.[110] There could undoubtedly be occasions where local conditions, such as excessive silica or low subterranean permeability, would preclude re-injection—at least without prior chemical treatment and/or settlement facilities; but it seems probable that the practice will become increasingly adopted and will thus greatly assist in preventing pollution.

Hydrogen Sulphide

H_2S is almost always present in geothermal fields. This noxious gas, in moderate and harmless concentrations, has a characteristic and rather unpleasant smell; but when more strongly concentrated it paralyses the olefactory nerves and thus becomes odourless. Therein lies its danger, for when it is present in lethal quantities it gives no warning of its presence. It is also 17.5% heavier than air at the same temperature, and is therefore apt to collect in low-lying pockets. Rare fatalities have occurred in the vicinity of fumaroles, but no incidents have yet been reported from this hazard in geothermal exploitation plants. This is probably attributable to the provision of adequate ventilation in cellars and basements. H_2S also attacks electrical equipment and it may have adverse effects on crops and on river life.

At power plants H_2S occurs in high concentration at the gas ejector discharge points. Until recently it has been deemed sufficient to place these points at high level and to rely on the high temperature of the discharged gases to provide the necessary buoyancy and thus ensure their wide dispersal. Apart from this precaution, and the provision of good ventilation in buildings, the past tendency has been more or less to ignore H_2S on the argument that in any case it occurs naturally. But the scale of development has now grown so rapidly in certain fields that greater precautions are now thought to be necessary. It has been estimated that at Cerro Prieto about 55 tonnes of H_2S are being discharged daily into the air,[111] and that in 1975 (if suitable precautions had not been taken) the corresponding figure in the Geysers area would have been about 28 tonnes/day.[112] Such quantities cannot be ignored indefinitely, and Californian legislation now insists on the reduction of the H_2S content of the air to within the threshold of odour; so that if it can be smelled the law is being broken. At the Geysers, all regular escape paths of H_2S are being trapped and the gas is being burned to form SO_2, which is then scrubbed with cooling tower water. Elemental sulphur is thus precipitated, which can be filtered out, while the contaminated water is re-injected into the ground.[113] The only "untamed" escape paths at the Geysers would be from "rogue" bores which, thanks to good drilling practice, are unlikely to occur, and through escaping steam from wells under test, or from piped steam at times when the plant cannot absorb all of it so that some has to be vented to the air—both comparatively infrequent occurrences. Where river cooling is adopted, some H_2S would escape in solution in the cooling water. With rivers of large flow (e.g. at Wairakei) this is probably unimportant (though this is arguable), but with small streams the only solution could be conversion to cooling towers. Where bore water from wet fields is discharged into rivers, the dissolved H_2S could affect fisheries and weed growth: re-injection could then be the only practicable solution.

Carbon Dioxide

By far the greater part of the incondensable gases that accompany the bore fluids consist of CO_2. This can escape into the air or local water courses. The fact that fuel combustion usually produces far greater quantities of this gas than geothermal exploitation on the same thermal scale has generally been regarded as an excuse for inaction, particularly as the gas is not toxic; but in certain fields such as Monte Amiata in Italy the CO_2 discharged to the air may be far greater than from fuel-fired plants of comparable size and duty. It is believed that the growing CO_2 content of the atmosphere, mainly due to fuel combustion, may be having a gradual adverse effect upon the world climate, while high CO_2 content in waters discharged into rivers can aggravate weed growth. It is undesirable that geothermal development should contribute towards these ill effects, and ideas are being considered for the commercial extraction of CO_2 from geothermal effluents. Meanwhile the emission of large quantities of CO_2 from geothermal installations seems inevitable. The problem is not yet one of urgency but if geothermal development grows dramatically, as well it soon may, it will have to be tackled seriously before long.

Land Erosion

At the Geysers field where heavy rains and steep slopes of incompetent rock often cause natural landslides and high erosion rates, the artificial levelling of ground for the accommodation of field works, roads and power plants has sometimes aggravated these conditions by creating very steep local gradients and by the removal of vegetation which normally acts as a deterrent by the binding action of its roots. Closer control, replanting of shrubs and trees, more careful site selection and improved construction methods are helping to solve this problem.[112] The close spacing of several wells within a single levelled area, combined with directional drilling (Fig. 12) could help in this respect.

Water-Borne Poisons

The water phase in wet fields sometimes contains toxic ingredients—notably boron, arsenic, ammonia and mercury—which, if discharged into water courses could contaminate downstream waters used for farming, fisheries or drinking water.[111,114,115] Although not strictly "poisonous", highly saline bore waters too can be harmful. Possible solutions to this problem include re-injection, disposal into the sea (if not too remote) through ducts and channels, the use of evaporator ponds as at Cerro Prieto,[10,116] storage during the dry season with subsequent release into rivers in spate during the wet season when dilution would render the offending substances more or less harmless, or even chemical treatment.[114]

Air-Borne Poisons

From ejector exhausts, from the upward effluents from cooling towers, from silencers, drains and traps, from discharging bores under test, from "wild" bores and also from control vent valves, various harmful substances (besides H_2S) may sometimes escape into the air at thermal sites. These may include mercury and arsenic compounds and radioactive elements. Certain noxious, though not poisonous, emissions such as rock dust and silica-laden spray may also be air-borne. At Wairakei a certain amount of forest trees were damaged in this way, while in El Salvador the contamination of coffee plantations was a problem. At Cerro Prieto salt deposition on buildings and agricultural lands occurred.[116] Horizontal well discharge in a controlled direction could sometimes offer a solution to this problem. Generally, air-borne toxicants are seldom present in harmful concentrations, but systematic monitoring is advisable to enable a careful watch to be kept on possible future dangers.

Noise

The nuisance caused by noise can cause a serious health hazard. Workers on new well sites have to wear ear-plugs or muffs lest their hearing be damaged. Though the noise from regular escape paths of fluids can be reduced by means of silencers, there are occasions when noisy discharges cannot be avoided; e.g. newly "blown" bores undergoing test. The erection of temporary sound barriers can sometimes be helpful in such cases. Re-injection could eliminate the noise due to flashing steam where hot water is now being discharged to waste (as at Wairakei). Drilling operations can be somewhat noisy, but do not persist for very long. Noise also occurs in and near power plant buildings from machinery. This is difficult to control, and is generally no worse than in conventional power plants. Control rooms and offices can be soundproofed. Legislative action against noise has been taken in the USA and other countries, and though its strict enforcement may sometimes be difficult it should act as a powerful incentive to designers to overcome the nuisance.

Heat Pollution

The necessary adoption of relatively low temperatures for geothermal power production results in low efficiencies and the emission of huge quantities of waste heat. Where cooling towers are used this waste heat escapes into the air by way of the surplus condensate. Where direct river cooling is adopted it is mostly spent in raising the temperature of the river water. In wet fields another enormous source of heat wastage can arise from the rejection of very hot unwanted bore water into rivers and streams (as at Wairakei) or into storage ponds and thence into the atmosphere (as at Cerro Prieto). Heat pollution of river water can damage fisheries and encourage the growth of unwanted water weeds. Possible remedies include re-injection of surplus cooling tower water and unwanted hot bore water; the generation of additional power by means of binary fluid cycles; or the establishment of dual or multi-purpose projects which usefully extract low grade heat from turbine exhausts and from rejected bore waters. The escape of heat and moisture from cooling towers may affect local climate, to a greater extent than with highly efficient fuel-fired plants, particularly in the formation of fog and ice;[117] on the other hand the increased humidity could sometimes have a beneficial effect.

Silica

Silica can be particularly troublesome with district heating systems.[118] Re-injection of silica-laden waters could foul up the permeability of the substrata, thus necessitating the constant changing of re-injection bore sites. Settlement ponds could mitigate this trouble, but various physical and chemical methods of treatment may offer more promising solutions to the problem.[111,114]

Subsidence

The withdrawal of huge quantities of subterranean water from a wet field can cause substantial ground subsidence, which could cause tilting and stressing of pipelines and surface structures, perhaps with serious or even disastrous results. Vertical movements have been observed at certain points at Wairakei of up to 4.5 m in 10 years,[119] together with appreciable horizontal movements also. Re-injection is really the best solution to this problem. Although it would not replace the whole of the withdrawn fluid (owing to the extraction of the steam phase) natural recharge could well make up the difference.

Seismicity

Fears have sometimes been expressed that prolonged geothermal exploitation could trigger off earthquakes, especially if re-injection is practised in zones of high shear stress where fairly large temperature differentials could occur. These fears arise because all existing geothermal exploitations of hyper-thermal fields are in naturally seismic areas, and it could be that any artificial interference with nature could precipitate seismic shocks.[117,120] At the present state of development the risks on this account do not seem to be great, but if very large scale geothermal exploitation should be undertaken in future the problem could perhaps become a serious one. Fortunately it is already the subject of study, so that reliable information should be available in good time.

Escaping Steam

The formation of fog from cooling tower effluents has already been mentioned. Of more serious impact can be the huge volumes of flash steam escaping from the hot water rejected in the field, as at Wairakei, or bore steam released through control vent valves. Dense fogs can occur from these discharges, which may drift across nearby roads and cause traffic hazards. Traffic warning signs and diversionary routes can of course mitigate the trouble, but the best palliatives are to use the hot water productively or to re-inject it into the ground, though that would not nullify the effects from control venting. This is unavoidable at times, as also are fogs formed from blowing bores newly opened or under test; but in a well exploited field the proportion of escaping steam should be small, and the hazard not serious.

Scenery Spoilation

It is difficult to be objective on matters of aesthetics, but it is a fact that thermal areas often occur in natural beauty spots, highly prized by the local population and frequented by tourists. Conservationists will sometimes oppose geothermal development on the grounds that scenic amenities are being destroyed. Though it cannot be denied that man-made engineering works can seldom compete with natural scenic beauty, it must be admitted that the power plants in the Geysers field, California, have been most tastefully camouflaged. The pipelines have been coloured to blend with the background; scarcely a puff of steam is visible; the power plants are inconspicuous; and the dry climate quickly absorbs the plumes of vapour arising from the cooling towers. In New Zealand, where the scarred ground surface has been rehabilitated by careful "landscaping" and damaged trees have been removed so that only the healthy forest is visible, visitors flock to see the geothermal development in greater numbers than those who frequented the area before exploitation. The Wairakei scene can even claim to a certain majesty of its own, where the billowing steam from the wellhead silencers—itself a form of "pollution"—contributes a dramatic aesthetic quality.

It is true that geothermal exploitation can interfere with natural surface thermal manifestations. In New Zealand, for instance, the activity of the Geysers and hot springs in the once famous Geyser Valley, close to the exploited area, has virtually ceased. The question of scenic amenities is one that can only be judged subjectively on a balance of considerations, by weighing the value of the energy won against that of touristic attractions and national heritage—a balance that must largely be emotional and cannot be strictly quantitative. The declaration of an area such as Yellowstone Park as a zone of outstanding beauty and interest, not to be exploited for energy development, would be a value judgement having its protagonists and its opponents; there can be no absolute standards in such matters. It is only fair, however, to point out that we cannot do without energy, and that a geothermal power plant, which produces no smoke, has no unsightly chimney stacks, no ungainly coal- or ash-handling equipment, no coal storage yards or oil storage tanks and no boiler-house can be far less displeasing to the eye than a fuel-fired plant of the same capacity. For similar reasons, industrial establishments using geothermal heat are likely to be far less obtrusive than those that rely upon fuels. On balance, it may justly be claimed that geothermal exploitation is far less guilty of scenery spoilation than fuel combustion.

Ecology

This is a subject that has received but little attention as yet. The discharge of chemicals into the air and into streams and rivers and thence into the ground water; small but appreciable changes of local temperature and humidity; noise; a degree of deforestation: all these factors could, and possibly do, disturb the natural balance of nature prevailing in a thermal area before exploitation. The effects of geothermal exploitation upon local fauna and flora has already been studied in New Zealand, California and possibly elsewhere. Protectors of wild life and of fisheries would do well to encourage more intensive research in this direction, though it is only fair to say that at present there does not appear to be any evidence that gives cause for alarm.

Conclusion

In recent years a change of mood can be discerned in geothermal circles from one of unreasoning optimism to one of sober realization of the environmental aspects of geothermal exploitation. Gone is the pious belief that geothermal exploitation is entirely "clean" and wholly guiltless of infecting the environment. Nevertheless, it is an undoubted fact that geothermal exploitation is far less culpable than fuel combustion of fouling the human nest, and that an antidote can probably be found to almost every source of geothermal pollution. Timely legislation in certain countries has enforced attention to this very important matter, even though it could perhaps have been a little less drastic in its pace of enforcement. The advances made in environmental studies in the 1970s have been impressive, and the good work may be expected to continue.

(Also of relevance to environmental problems are bibliography references nos. 82, 121–124.)

16. THE FUTURE

Nature has strictly rationed the supply of earth heat at present accessible to us, by confining it to some hyper-thermal fields within the Seismic Belt and some semi-thermal fields which may occur elsewhere. The combined area of thermal fields of either class probably form only a tiny fraction of the Earth's surface. Yet we know that at all points in the world there are immense quantities of heat below ground. If only we could gain access to this heat at places other than where natural fields occur, we should go a long way towards solving Man's energy problems. If natural thermal fields are a rarity, then our task must be to create artificial fields. The author is confident that we shall succeed in this task, but success will not come overnight; so while we are tackling the problem of creating artificial fields we must press ahead with renewed efforts to exploit as many as possible of the natural thermal fields that nature has rather grudgingly provided. First we must examine all available worldwide records, that have been collected by the earth scientists, and at the same time extend our inventory of relevant knowledge by undertaking in likely zones as much fieldwork as we can afford, in order to locate promising areas where hyper- or semi-thermal fields could perhaps be found. Then we must attack these areas systematically.

Systematic Approach to Geothermal Development

So often in the past, countries have squandered their limited resources in looking for geothermal fields in an unsystematic way. They may have flitted from area to area, sinking a few holes here and a few holes there, solely on the superficial evidence of surface thermal manifestations, in the optimistic hope that they might be lucky and strike steam somewhere. They may have invited international "experts" to pay short visits to their thermal areas and have expected these experts to assess geothermal potentialities from a superficial examination lasting only a day or two. Such unsystematic activities are not necessarily entirely wasted, for every hole sunk adds to the cumulative knowledge of an area, and every visiting expert may contribute some useful suggestions as to the next steps to be taken. Nevertheless, if exploration costs are to be kept to a minimum, and if good results are to be obtained quickly, it is essential that a systematic procedure be adopted from the outset. A suggested methodology[125] is illustrated in Fig. 33, which is more or less self-explanatory. Its main purpose is to concentrate first on the less expensive investigations and to postpone the costlier drilling operations until a reasonably probable field model has been conceived.

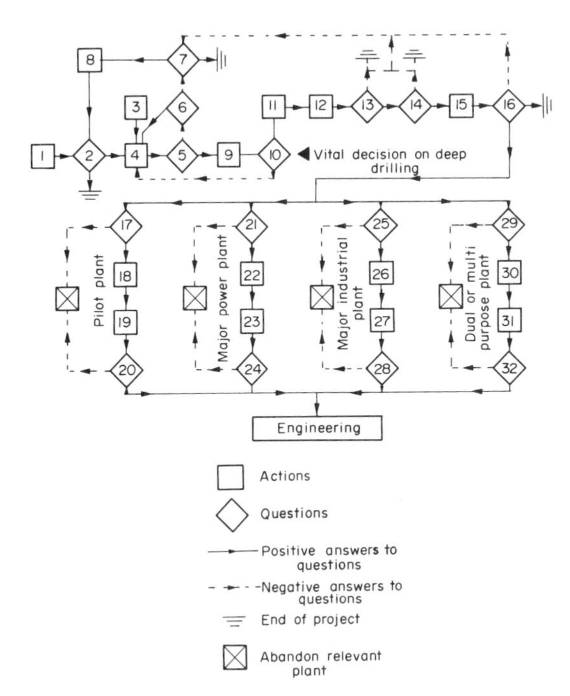

FIG. 33. Systematic approach to geothermal development: process diagram (Armstead, Gorhan and Müller, p. 50, ref. 125).

Key: 1. Inventory of alternative energy costs; 2. Is geothermal energy likely to be competitive? 3. Collection and review of preliminary field data; 4. Collection of more extensive field data; 5. Are data enough for conjectural field model? 6. Are data sufficiently promising to proceed? 7. Is there an alternative area worth investigating? 8. Selection of alternative area; 9. Conjecture of tentative working model; 10. Are data sufficient to justify deep exploration drilling? 11. Minimum economic yield of bores; 12. Deep exploratory drilling and borehole measurements; 13. Are production bores likely to yield economic minimum? 14. Are corrosion problems soluble? 15. Estimation of energy potential of field; 16. Are other engineering problems soluble? 17. Is a pilot plant required? 18. Preliminary design of pilot plant, with estimates; 19. Re-appraisal of economics of pilot plant; 20. Is pilot plant still economic? 21. Is field potential enough for major power plant (>20 MW)? 22. Preliminary design of major power plant, with estimates; 23. Re-appraisal of economics of major power plant; 24. Is major power plant still economic? 25. Is field potential enough for major industrial plant? 26. Preliminary design of major industrial plant, with estimates; 27. Re-appraisal of economics of major industrial plant; 28. Is major industrial plant still economic? 29. Is there a demand for waste heat from power plant? 30. Preliminary design of dual or multi purpose plant, with estimates; 31. Economic study of dual or multi purpose plant; 32. Is dual or multi purpose plant economic?

Artificial Permeability (Rock-Fracturing)

It has already been emphasized that the fundamental difference between a thermal area and a thermal field is one of rock permeability. If we wish to create an artificial thermal field it is first necessary to create permeability, and this can only be done by rock-fracturing. Rocks are poor thermal conductors; so to obtain a worthwhile yield of heat an extensive labyrinth of voids and fissures must be created in the hot rock through which water can be circulated and heat extracted continuously until the fractured zone becomes thermally depleted. The labyrinth must have a large rock-to-water interface area, and must be capable of retaining a fairly substantial volume of water so as to ensure low water velocities. These features would tend to compensate for the low thermal conductivity of the rock and would also tend to reduce flow resistance and parasitic pumping losses. The problem is to find some method of shattering hot rocks in such a way as to produce a labyrinth having these characteristics.

Once the underground labyrinth has been created in hot rock, heat could be extracted by means of an open cycle with cold feed (Fig. 34a), or by a closed cycle with re-injection and a heat-exchanger (Fig. 34b) or, if a

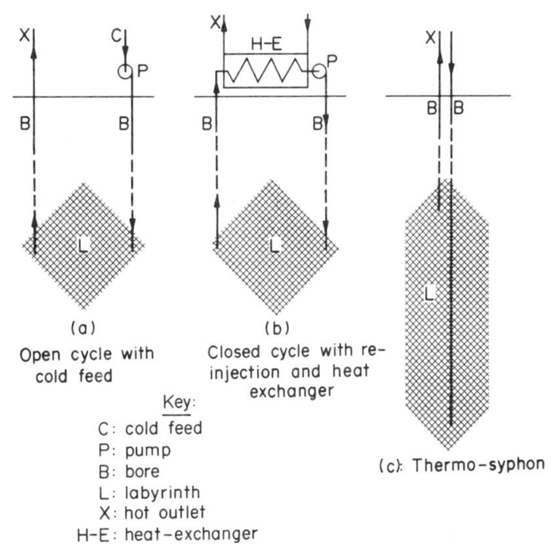

FIG. 34. Extraction of earth heat by circulating water through artificially created underground labyrinths.

suitable vertically orientated labyrinth could be formed, by thermo-syphon (Fig. 34c). This last method would not only dispense with pumping, but it could be used either with the open or closed cycle and could be formed either by two closely adjacent bores or even concentrically through a single bore. The closed cycle would be less vulnerable to chemical deposition than the open cycle.

How are we to create the necessary labyrinth? Some years ago it was proposed to do this by means of underground nuclear explosions[126] which would form a rubble-filled chamber having all the necessary characteristics of the required labyrinth. The energy of the explosion would be added to that of the rock heat, thereby increasing it by about 20%. Although it was expected that most of the radioactive matter would be trapped in a layer of vitrified melted rock at the bottom of the chamber, there was nevertheless an alleged risk of radiation contamination of the circulated fluid. Moreover, the seismic shock of the explosion would restrict this method to very sparsely populated, or even desert, areas. Nothing has yet come of this proposal, though it is understood that the Ploughshare Division of the US Atomic Energy Commission have by no means abandoned the idea for possible adoption in some of the desert areas of the Western States where semi-thermal gradients are known to exist. Certainly nuclear explosions offer a fairly cheap form of "potted" energy, so maybe the last of this has not yet been heard. It has also been suggested that nuclear explosions could safely be released beneath the continental shelf, in some parts of which (e.g. the North Sea) fairly high temperature gradients are known to occur.[127]

Shortly after the oil crisis of late 1973 the author gave some thought to the possibility of rock-fracturing by hydraulic impact. He contended that such a method could be used not only in hot rocks at fairly shallow or moderate depth, as may be found in hyper- and semi-thermal areas respectively, but also in non-thermal areas by shattering rocks at much greater depth. In this way he hoped to find the key to geothermal exploitation in all parts of the world, and not only where nature dictated. His proposed method is illustrated in Fig. 35. A deep bore would be provided with a hydraulic accumulator and a high-pressure pump at surface level. The proposal was to fill the bore with water and to raise the accumulator weight by pumping to a pressure equal to the lithostatic pressure at the bottom of the bore. After closing the valve to isolate the pump a dynamic blow was then to be administered to the water column by means of a device resembling a pile-driver, with the expected result of shattering the rock at the base of the bore by providing an instantaneous pressure sufficient to overcome both the weight of the over-burden and the tensile strength of the rock. By repeating the operation at another bore at a suitable distance away, it was hoped that the two shattered zones would intersect, thus opening up a path for water circulation. It was expected that the circulating system would have to be permanently pressurized, to an extent equal to the difference

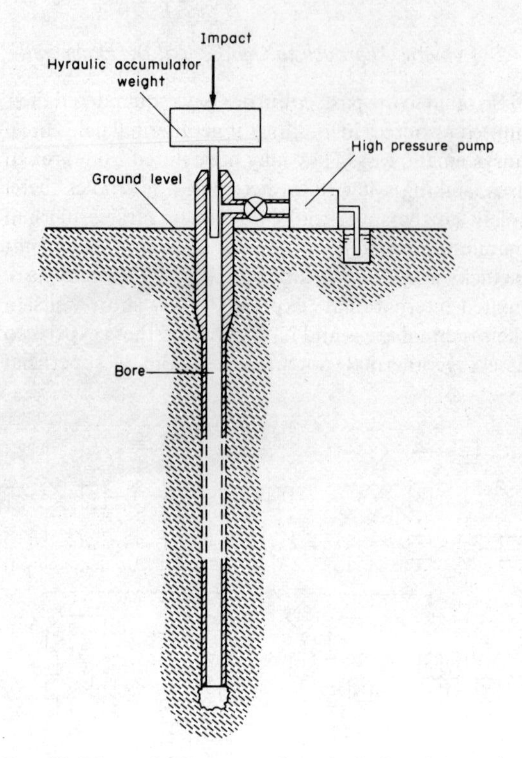

FIG. 35. The author's proposed method of rock-fracturing by hydraulic impact.

between the lithostatic and hydrostatic pressures, in order to hold the fissures open after their formation. It was suggested that the first aim should be to recover low grade heat suitable for space-heating, industry or farming, from depths of the order of 3000 m, and later to aim at recovering high grade heat for power generation at depths of about 6000 m or more. These depths would apply to typical non-thermal areas: in hyper- or semi-thermal areas it would be unnecessary to penetrate nearly so deeply. It so happened, however, that the Americans were in fact already conducting rock fracturing experiments in New Mexico, but by using straight hydraulic pressure, without dynamic impact. After studying the problem in greater detail, the author realized that his first proposals had been rather crude and that he had made two fundamental errors of assumption. First, whereas he had expected the direction of fissuring to be more or less horizontal, the tendency is in fact to form more or less vertical fractures. Secondly, he greatly overrated the pressures required to form and to hold open the fissures: in fact, vertical fissures, once formed, can be held open by pressures only slightly above the hydrostatic (and sometimes perhaps by hydrostatic pressure alone). It is understood that the dynamic impact method has not yet been attempted; but it is to be hoped that this method may still be tried, as it could give rise to very high instantaneous pressures while using pumps designed for more moderate duties.

Great work has been, and is being, done by the Los Alamos Scientific Laboratory of the University of California in rock-fracturing experiments. It remains to be seen whether serviceable labyrinths in hot rocks can in fact be formed at reasonable cost. If so, the way will be opened up to the exploitation of large hyper- and semi-thermal areas in various parts of the world, and perhaps later extended to non-thermal areas. The fracturing process could be aided by the thermal stresses resulting from the contact of relatively cool water with hot rock surfaces, so that fissure formation may be a self-propagating process to some extent; it is also possible that thermal stresses may cause horizontal cracks to ramify from vertical fissures, thus forming a honeycomb structure. Chemical leaching could also sometimes aid the process.

Already the Los Alamos Scientific Laboratory experiments have attained some measure of success. A hole penetrating 2927 m into hot granite encountered a temperature of 196.8°C at the bottom. By the application of hydraulic pressure (at a level not yet specified) a near-vertical crack was formed, believed to be of nearly 250 m diameter. A second hole was then sunk, directionally, in a spiral track to a depth of 3065 m, where a temperature of 205.5°C was encountered, and contact was established with the artificial fracture so that circulation between the two holes was established. Unfortunately the impedance to flow is still unsatisfactorily high, and efforts are now being made to determine the location and nature of the points of high impedance, and to reduce them by various means such as propping, leaching and the use of explosives. The place chosen for the experiments is typical of a "semi-thermal area", as defined earlier; the temperature gradient being about 62°C/km (assuming a surface ambient temperature of 15°C). Circulation between the two experimental holes is maintained by means of pressurized water, in preference to water/steam, so as to improve the rock-to-fluid conductivity and the rate of heat extraction from the exit hole. Although the life of the original fracture is expected to be about 10–15 years, it is hoped that this will be substantially extended by the action of thermal stresses which are expected to cause the tips of the crack to advance in the course of time.

Novel Drilling Techniques

Conventional drilling methods could enable us to penetrate to the hot rocks in hyper- and semi-thermal areas without difficulty. Technically, it would even be possible to reach the hot rocks in non-thermal areas also by the same means, but the costs would be formidable—especially if high grade heat is to be won. As a corollary to rock-fracturing, some cheaper method of ground penetration is clearly desirable if earth heat is to be won from non-thermal areas. A cheaper method, of course, would also benefit the exploitation of natural fields and of hyper- and semi-thermal areas.

In parallel with their rock-fracturing activities, the Los Alamos Scientific Laboratory of the University of California has for some time been engaged in some very interesting work on entirely new methods of ground penetration, the most promising of which is probably "melt-drilling".[128,129] The application of very intense localized heat not only melts the rock and thus enables the penetration tool to advance, but also results in the formation of a hard glass-like self-supporting bore wall of great mechanical strength, intimately bonded with the surrounding rock. This dispenses with the need to remove the débris, by forcing the melted displaced material into the walls of the hole: it also renders unnecessary the provision of steel casings. Rates of penetration are satisfactory, and although the energy consumption is higher per metre of penetration for a hole of a given size than for conventional drillings, this is more than offset by savings arising from the absence of steel casings, the avoidance of cementation and the freedom from delays that arise from bit renewals, stringing and circulation losses. Moreover, penetration rates are more or less independent of the composition, hardness and structure of the rock; and as hotter ground is reached with increased depth, less heat needs to be added, or alternatively penetration rates should rise. Early experiments with electrically heated penetrators have shown promise, and the application of nuclear heating devices is under active study. Confidence has been expressed that the "nuclear subterrene", which could be used not only for drilling but also for tunnelling, should be capable of producing holes up to several metres in diameter and several tens of kilometers long or deep. Hope has even been expressed that by about the mid-1980s it will be possible to use this technique to penetrate the Earth's crust and enter the outer mantle. Clearly the potentialities of melt-drilling for geothermal development could be immense, and could open the door to the extraction of earth heat from non-thermal areas. Other novel drilling techniques are also under investigation by the LASL.

Direct Tapping of Volcanoes

The direct tapping of earth heat from active volcanoes, as distinct from hydro-thermal fields, has in recent years been the subject of much discussion; but not very much seems yet to have been achieved in the way of tangible results. Between 300 and 400 active volcanoes are known to exist in the world, and since many inactive ones may merely be dormant the total number of volcanoes that could become active may perhaps be numbered in thousands. Volcanoes have the advantage over hydro-thermal fields in that their existence and location are self-evident: they do not have to be "discovered" by exploration, even though their hidden structure may have to be divined by exploration methods. A joint US–Japanese seminar was held in Hawaii in February, 1974, when the possibilities of direct volcanic exploitation was discussed. It has also been reported that the Russians plan to develop about 300 MW of power by boring nearly 3000 m into the heart of Avachinskaya Spoka volcano in Kamchatka. A paper was also presented on this subject at the UN Geothermal Symposium held in San Francisco in 1975.[130]

It is well to draw attention to two other good reasons for intensifying our efforts to exploit earth heat. First, the threat to a nation's energy hunger may not always come from abroad: it may be an internal threat from a small section of the community engaged in providing indigenous energy (e.g. coal) demanding an unrealistically high reward for their labour. Herein lies a strong inventive to seek an energy source that is not labour-intensive; and earth heat, once won, is such a source. Secondly, by comparison with an off-shore oil rig, geothermal bores are far less vulnerable to sabotage—a risk to be treated seriously in these lawless days.

In conclusion, the author would like to recall what he stated publicly in 1967 in Washington, DC, when he suggested that Humanity might have reaped a far greater reward had the vast expenditure hitherto lavished upon space exploration been devoted instead to activities in a downward direction. He still stands by that opinion.

(Also of relevance to future developments are bibliography references nos. 131–134.)

REFERENCES

The task of choosing a bibliography for a review of this nature is a difficult one. To quote more than a small fraction of the many thousands of publications relevant to geothermal development would be confusing to the reader. A very large number of excellent articles and papers has therefore of necessity had to be omitted from the references which follow: their exclusion is not intended to be any reflection upon their excellence, and the author would like to extend his apologies to those many writers of valuable works which do not appear in the following list. His aims have been primarily to include some typical references that have direct bearing upon points brought out in the text and to quote one or two documents of diversified interest.

The following United Nations publications have been printed as composite documents, and contain a wealth of specialized papers and articles, many of which are included in this bibliography. These publications, to avoid tedious repetition, are referred to by the following abbreviations:

Rome: Proceedings of the United Nations Conference on New Sources of Energy, held in Rome, August, 1961. Volumes 2 and 3. Publication ref. E/CONF. 35/3 and 4. Document Sales No. 63.I.36 and 37.

Pisa: Proceedings of the United Nations Symposium on the Development and Utilisation of Geothermal Resources, held in Pisa, Sept./Oct., 1970. Published in *Geothermics* Special Issue 2, Volumes 1 and 2.

UNESCO: *Geothermal Energy*, Review of research and development. Earth Sciences 12. ISBN 92-3-101063-8.

San Francisco: Proceedings of the 2nd United Nations Symposium on the development and use of geothermal resources, held in San Francisco, May, 1975. (*N.B.* At the time of going to press these proceedings had not yet been published. The author is therefore only able to quote from those papers which he has personally seen in his capacity as Rapporteur to Sections IV, VII and VIII of the Symposium, but he cannot quote page or paper numbers and cannot include papers from other sections which may well be of relevance.

Another abbreviation used in the bibliography is:

LASL: The Los Alamos Scientific Laboratory of the University of California.

1. WORLD ENERGY CONFERENCE, Survey of energy resources, Table I-2 (1974).
2. UNITED NATIONS STATISTICAL PAPERS, World Energy Supplies, Series J-17 (Ref. ST/STAT/SER J/17) (1969–1972).
3. CHIOSTRI, E. and BALSAMO, A., Geothermal resources for heat treatment, San Francisco, Sec. VIII.
4. ENTE NAZIONALE PER L'ENERGIA ELECTRICA (ENEL), Larderello and Monte Amiata: electric power by endogenous steam, Official brochure.
5. HALDANE, T. G. N. and ARMSTEAD, H. C. H., The geothermal power development at Wairakei, New Zealand, *Proc. Inst. Mech. Eng.* **176** (1962).
6. NEW ZEALAND GOVERNMENT, Power from the Earth, Official brochure on geothermal development in New Zealand, particularly at Wairakei.
7. WORTHINGTON, J. D., Geothermal development, Reprint from Chapter 9, Status report—Energy resources and technology, a report of the *ad hoc* committee on energy resources and technology, Atomic Industrial Form, Inc. 1975 (modified by personal communication with the author).
8. JAPAN GEOTHERMAL ENERGY ASSOCIATION, Geothermal energy utilisation in Japan, 1975 brochure.
9. NAYMANOV, O. S., A pilot geothermoelectric power station in Pauzhetka, Kamchatka, Vol. 2, Pisa.
10. GUIZA, J., Power generation at Cerro Prieto geothermal field, México, San Francisco, Sec. VII.
11. EINARSSON, S. S., Geothermal space heating and cooling, San Francisco, Sec. VIII.
12. BOLDISZAR, T., Geothermal energy production from porous sediments in Hungary, Vol. 2, Pisa.
13. REYNOLDS, G., Cooling with geothermal heat, Vol. 2, Pisa.
14. LÍNDAL, B., Industrial and other applications of geothermal energy, pp. 135–148, UNESCO.
15. KUNZE, J. F., RICHARDSON, A. S., HOLLENBAUGH, K. M., NICHOLS, C. R. and MINK, L. L., Non-electric utilisation project, San Francisco, Sec. VIII.
16. REISTAD, G. M., The potential for non-electric applications of geothermal energy and their place in the national economy, San Francisco, Sec. VIII.
17. ARMSTEAD, H. C. H. and RHODES, C., Desalination by geothermal means, Proceedings of the 3rd International Symposium on Fresh Water from the Sea, pp. 451–459, Vol. 3, Dubrovnik (1970).
18. VALFELLS, A., Heavy water production with geothermal steam, Vol. 2, Pisa.
19. EINARSSON, S. S., Rapporteur's report, pp. 112–121, Vol. 1, Pisa, Sec. X.

20. EINARSSON, S. S., Geothermal district heating, pp. 123–134, UNESCO.
21. BULLARD, SIR E., Basic theories, pp. 19–29, UNESCO.
22. BULLARD, SIR E., Reversals of the Earth's magnetic field, *Phil. Trans. Roy. Soc. A* **263**, 481 (1968).
23. BULLARD, SIR E., The origin of the oceans, *Sci. Am.* **221**, 66 (1969).
24. TAMRAZYAN, G. P., Continental drift and thermal fields, Vol. 2, Pisa.
25. HOUSE, P. A., Potential power generation and gas production from Gulf Coast geopressure reservoirs, San Francisco, Sec. VII.
26. FACCA, G. and TONANI, F., Self-sealing geothermal fields, *Bulletin Volcanologique* **XXX** (1967).
27. FACCA, G., The structure and behaviour of geothermal fields, pp. 61–69, UNESCO.
28. FACCA, G. and TONANI, F., Theory and technology of a geothermal field, *Bulletin Volcanologique* **XXVII** (1964).
29. GOGUEL, J. Le rôle de la convection dans la formation des gisements géotermiques, Vol. 2, Pisa.
30. JAMES, R., Rapporteur's Report, pp. 91–98, Pisa, Sec. VII.
31. TONGIORGI, E., Rapporteur's report, pp. 1–7, Vol. 1, Pisa, Sec. I.
32. BANWELL, C. J. Geothermal drill-holes: physical investigations, paper G/53, Vol. 2, Rome.
33. BANWELL, C. J., Rapporteur's report, pp. 32–57, Vol. 1, Pisa, Sec. IV.
34. BANWELL, C. J., Geophysical methods in geothermal exploration, pp. 41–48, UNESCO.
35. ELLIS, A. J., Quantitative interpretation of chemical characteristics of hydrothermal systems, Vol. 2, Pisa.
36. FOURNIER, R. O. and ROWE, J. J., Estimation of underground temperatures from the silica content of water from hot springs and wet steam wells, *Am. J. Sci.* **264** (1966).
37. MCNITT, J., Rapporteur's report, pp. 24–31, Vol. 1, Pisa, Sec. III.
38. MCNITT, J., The rôle of geology and hydrology in geothermal exploration, pp. 33–40, UNESCO.
39. MEIDAV, T., Application of electrical resistivity and gravimetry in deep geothermal exploration, Vol. 2, Pisa.
40. SIGVALDASON, G. E., Geochemical methods in geothermal exploration, pp. 49–59, UNESCO.
41. STRANGWAY, D. W., Geophysical exploration through geologic cover, Vol. 2, Pisa.
42. THOMPSON, G. E. K., BANWELL, C. J., DAWSON, G. B. and DICKINSON, D. J., Prospecting for hydrothermal areas by surface thermal surveys, paper G/54, Vol. 2, Rome.
43. TONANI, F., Geochemical methods of exploration for geothermal energy, Vol. 2, Pisa.
44. WHITE, D. E., Preliminary evaluation of geothermal areas by geochemistry, geology and shallow drilling, paper G/2, Vol. 2, Rome.
45. WHITE, D. E., Rapporteur's report, pp. 58–80, Pisa, Sec. V.
46. CRAIG, S. B., Geothermal drilling practices at Wairakei, New Zealand, paper G/14, Vol. 3, Rome.
47. MATSUO, K., Drilling for geothermal steam and hot water, pp. 73–83, UNESCO.
48. BOLTON, R. S., Blowout prevention and other aspects of safety in geothermal drilling, paper G/43, Vol. 3, Rome.
49. BRUNETTI, V. and MEZETTI, E., On some troubles most frequently occurring in geothermal drilling, Vol. 2, Pisa.
50. CIGNI, U., Machinery and equipment for harnessing endogenous fluid, Vol. 2, Pisa.
51. CIGNI, U. and GIOVANNONI, A., Planning methods in geothermal drilling, Vol. 2, Pisa.
52. CIGNI, U., GIOVANNONI, A., LUSCHI, E. and VIDALI, M., Completion of producing geothermal wells, Vol. 2, Pisa.
53. CONTINI, R. and CIGNI, U., Air drilling in geothermal bores, paper G/70, Vol. 3, Rome.
54. DENCH, N. D., Casing string design for geothermal wells, Vol. 2, Pisa.
55. FABBRI, F. and GIOVANNONI, A., Cements and cementation in geothermal well drilling, Vol. 2, Pisa.
56. FABBRI, F. and VIDALI, M., Drilling mud in geothermal wells, Vol. 2, Pisa.
57. GIOVANNONI, A., Rapporteur's report, pp. 81–90, Vol. 1, Pisa, Sec. VI.
58. WOODS, D. I., Drilling mud in geothermal drilling, paper G/21, Vol. 3, Rome.
59. HUNT, A. M., The measurement of borehole discharges, downhole temperatures and surface heat flows at Wairakei, paper G/19, Vol. 3, Rome.
60. JAMES, R., Factors controlling borehole performance, Vol. 2, Pisa.
61. JAMES, R., Optimum wellhead pressure for geothermal power, *New Zealand Engineering* **22**(6), (1967).
62. JAMES, R., Power life of a hydro-thermal system, 2nd Australasian Conference on Hydraulics and Fluid Mechanics (1965).
63. JAMES, R., Metering of steam/water two-phase flow by sharp-edged orifices, *Proc. Inst. Mech. Eng.* **180**, Pt. I (1965/66).
64. JAMES, R., MCDOWELL, G. D. and ALLEN, M. D., Flow of steam/water mixtures through a 12″ diameter pipeline: test results, Vol. 2, Pisa.
65. BANGMA, P., The development and performance of a steam/water separator for use on geothermal bores, paper G/13, Vol. 3, Rome.
66. ARMSTEAD, H. C. H. and SHAW, J. R., The control and safety of geothermal installations, Vol. 2, Pisa.
67. DENCH, N. D., Silencers for geothermal bore discharge, paper G/18, Vol. 3, Rome.
68. ARMSTEAD, H. C. H., The extraction of power from hot water, Proceedings of the World Power Conference, Moscow. Paper 174, Sec. C-4 (1968).
69. WIGLEY, D. M., Recovery of flash steam from hot bore water, Vol. 2, Pisa.
70. JAMES, R., Pipeline transmission of steam/water mixtures for geothermal power, *New Zealand Engineering* **23**(2) (1968).
71. AIKAWA, K. and SODA, M., Advanced design in Hatchobaru geothermal power station, San Francisco, Sec. VIII.
72. JAMES, R., Rapporteur's report, pp. 99–105, Sec. VIII, Pisa.
73. JAMES, R., Power station strategy, Vol. 2, Pisa.
74. SMITH, J. H., Collection and transmission of geothermal fluids, pp. 97–106, UNESCO.
75. TAKAHASHI, Y., HAYASHIDA, T., SOEZIMA, S., ARAMAKI, S. and SODA, M., An experiment on pipeline transportation of steam-water mixtures at Otake geothermal fields, Vol. 2, Pisa.
76. ARMSTEAD, H. C. H., Geothermal power for non-base load purposes, Vol. 2, Pisa.
77. ARMSTEAD, H. C. H., Some unusual ways of developing power from a geothermal field, San Francisco, Sec. VII.
78. MOSKVICHEVA, V. N., Geothermal power plant on the Paratunka River, Vol. 2, Pisa.
79. ROLLET, A., Centrale géotermique de Kiabukwa: leçons tirées de quatre années d'exploitation. Bulletin des Séances, Académie Royale des Sciences Coloniales, Brussels (1957).
80. CIAPICA, I., Present development of turbines for geothermal application, Vol. 2, Pisa.
81. ARMSTEAD, H. C. H., Rapporteur's report, pp. 106/111, Vol. 1, Pisa, Sec. IX.
82. ARMSTEAD, H. C. H., Rapporteur's reports, San Francisco, Secs. IV, VII and VIII.
83. COMISIÓN FEDERAL DE ELECTRICIDAD, MÉXICO, Energía del subsuelo, Official brochure (1971).
84. DAL SECCO, A. G., Turbo-compressors for geothermal plants, Vol. 2, Pisa.
85. DAL SECCO, A. G., Geothermal plant gas removal for jet condensers, San Francisco, Sec. VII.
86. FACCA, G., Rapporteur's report, pp. 8–23, Vol. 1, Pisa, Sec. II.

87. JAMES, R., The applicability of the binary cycle, San Francisco, Sec. VII.
88. PESSINA, S., RUMI, O., SILVESTRI, M. and SOTGIA, G., Gravimetric loop for the generation of electric power from low temperature water, Vol. 2, Pisa.
89. SATO, H., On Matsukawa geothermal power plant, Vol. 2, Pisa.
90. USUI, T. and AIKAWA, K., Engineering and design features of Otake geothermal power plant, Vol. 2, Pisa.
91. WOOD, B., Geothermal power, pp. 109–121, UNESCO.
92. BUDD, C. F., Producing geothermal steam at the Geysers field. Reprint from *Geothermal Energy Resources, Production and Stimulation*, KRUGER and OTTE (eds.), Stanford University Press (1975).
93. JONSSON, V. K., TAYLOR, A. J. and CARMICHAEL, A. D., Optimisation of geothermal power plant by use of freon vapour cycle, *Tímarit VFÍ*, Iceland (1969).
94. LAWRENCE LIVERMORE LABORATORY, Geothermal development of the Salton Trough: California and Mexico, Brochure UCRL-51775 (1975).
95. ARMSTEAD, H. C. H., Geothermal economics, pp. 161–174, UNESCO.
96. DAN, F. J., HERSHAM, D. E., KHO, S. K. and KRUMLAND, L. R., Development of a typical generating unit at the Geysers geothermal project: a case study, San Francisco, Sec. VII.
97. MATTHEW, P., Geothermal operating experience at Geysers power plant, San Francisco, Sec. VII.
98. BRADBURY, J. J. C., Rapporteur's report, pp. 122–131, Vol. 1, Pisa, Sec. XI.
99. KREMNJOV, O. A., ZHURAVLENKO, V. J. and SHURTSKOV, A. V., Technical-economic estimation of geothermal resources, Vol. 2, Pisa.
100. RICCI, G. and VIVIANI, G., Maintenance operations in geothermal power plants, Vol. 2, Pisa.
101. ALLEGRINI, G. and BENVENUTI, G., Corrosion characteristics and geothermal power plant protection, Vol. 2, Pisa.
102. HERMANNSSON, S., Corrosion of metals and the forming of a protective coating on the inside of pipes in the thermal water used by the Reykjavik Municipal District Heating Service. Vol. 2, Pisa.
103. MARSHALL, T. and BRAITHWAITE, W. R., Corrosion control in geothermal systems, pp. 151–160, UNESCO.
104. OZAWA, T. and FUJII, Y., A phenomenon of scaling in production wells: the geothermal power and plant in the Matsukawa area, Vol. 2, Pisa
105. TOLIVIA, E. M., Corrosion measurements in a geothermal environment, Vol. 2, Pisa.
106. EINARSSON, S. S., VIDES, R. A. and CUÉLLAR, G., Disposal of geothermal waste water by re-injection, San Francisco, Sec. IV.
107. CHASTEEN, A. J., Geothermal steam condensate re-injection, San Francisco, Sec. IV.
108. GRINGARTEN, A. C. and SAUTY, J. P., Recovery of heat energy from acquifers,
109. EJIMA, Y., Re-injection of geothermal hot water in the Otake geothermal field, San Francisco, Sec. IV.
110. TONGIORGI, E., Re-injection and recharge of a geothermal field as from experience at Larderello, San Francisco, Sec. IV.
111. AXTMANN, R. C., Chemical aspects of the environmental impact of geothermal development, San Francisco, Sec. IV.
112. REED, M. J. and CAMPBELL, G. E., Environmental impact of development in the Geysers geothermal fields, San Francisco, Sec. IV.
113. ALLEN, G. W. and MCCLUER, H. K., Hydrogen sulphide emissions abatement for the Geysers power plant, San Francisco, Sec. IV.
114. ROTHBAUM, H. P. and ANDERTON, B. H., Removal of silica and arsenic from geothermal discharge waters by precipitation of useful calcium silicates, San Francisco, Sec. IV.
115. ANDERSON, S. O., Environmental impact of geothermal resource development on commercial agriculture: a case study of land use conflict, San Francisco, Sec. IV.
116. MERCADO, S., Pollution and basic protection: geothermal electric power plant at Cerro Prieto, México. San Francisco, Sec. IV.
117. SWANBERG, C. A., Physical aspects of pollution related to geothermal energy development, San Francisco, Sec. IV.
118. THÓRHALLSSON, S., RAGNARS, K., ARNÓRSSON, S. and KRISTMANNSDÓTTIR, H., Rapid scaling of silica in two district heating systems, San Francisco, Sec. IV.
119. STILWELL, W. B., HALL, W. K. and TAWHAI, J., Ground movement in New Zealand fields, San Francisco, Sec. IV.
120. CAMELI, G. M., Seismic control during a reinjection experiment in the Viterbo region, Italy, San Francisco, Sec. IV.
121. AXTMANN, R. C. Environmental impact of a geothermal power plant, *Science* **187**, 4179, March (1975).
122. BOLTON, R. S., Management of a geothermal field, pp. 175–184, UNESCO.
123. CUÉLLAR, G., Behaviour of silica in geothermal waste waters, San Francisco, Sec. IV.
124. HATTON, J. W., Ground subsidence of a geothermal field during exploitation, Vol. 2, Pisa.
125. ARMSTEAD, H. C. H., GORHAN, H. L. and MÜLLER, H., Systematic approach to geothermal development, *Geothermics* 3, no. 2, June (1974).
126. KENNEDY, G. C., A proposal for a nuclear power programme, Third Ploughshare Symposium—Engineering with nuclear explosives. University of California (1964).
127. PARKER, K., Personal communication.
128. LASL The subterrene programme, LASL Mini-Review 75-2 (1975).
129. ROBINSON, E. S., ROWLEY, J. C., POTTER, R. M., ARMSTRONG, D. E., MCINTEER, B. B., MILLS, R. L. and SMITH, M. C. (eds.), A preliminary study of the nuclear subterrene, LASL, Ref. LA-4547. UC-38 (1971)
130. HAYASHIDA, T., Volcanic power plant with artificial fractured zone, San Francisco, Sec. VII.
131. HARLOW, F. H. and PRACHT, W. E., A theoretical study of geothermal energy extraction, *J. Geophysical Res.* **77**, 35 (1972).
132. HARRISON, E., KIESCHNICK, W. F. and MCGUIRE, W. J., The mechanics of fracture induction and extension, *Petroleum Transactions AIME* (1953).
133. SMITH, M. C., Geothermal energy, LASL, Ref. LA-5289-MS. Informal Report UC-34 (1973).
134. SMITH, M. C., BROWN, D. W. and PETTITT, R. A., Los Alamos dry geothermal source demonstration project, LASL Mini-Review 75-1 (1975).

Manuscript received May 1976

Prog. Energy Combust. Sci., Vol. 2, pp. 239–255, 1977. Pergamon Press. Printed in Great Britain

COMBUSTION FUNDAMENTALS RELEVANT TO THE BURNING OF NATURAL GAS

P. F. JESSEN
Watson House, British Gas Corporation, Peterborough Road, London SW6, UK

and

A. MELVIN
London Research Station, British Gas Corporation, Michael Road, London SW6 2AD, UK

1. AVAILABILITY OF NATURAL GAS

Natural gas is one of the most highly-prized fuels in world-wide use[1] at present and is likely to become even more important in fossil fuel-based energy economies over the next 25 years. More than two-thirds of the World's proved reserves of natural gas are located in North America, the USSR and the oil states of the Middle East. Both Western Europe and North Africa have proved reserves in excess of 10% of the world total. It is generally believed that most natural gases arise, like crude oil, from the degradation of marine organisms in particular geological conditions which occurred after the drying up of shallow seas more than 60 million years ago. The formation of natural reservoirs usually requires the presence of an impermeable basement rock supporting a porous rock layer, usually sandstone, and a cap which may again be impermeable rock or a salt dome formed by sea evaporation and subsequent plastic deformation. The sandstone is the reservoir which holds the gas or oil, but in general one cannot assume that the hydrocarbons were actually formed in the locations in which they are stored. Migration from source rock is an important process in the concentration of hydrocarbons in particularly favourable geological structures and little is known in detail of such mechanisms of migration. Although natural gas sometimes occurs in close proximity with coal (as for example in the Dutch Slochteren gas field where the Rotliegendes sandstone reservoir rests on the coal-bearing Westphalian strata), it is generally believed that most natural gases do not originate from coal degradation.[1]

Natural gases from various parts of the world can be substantially different in composition and such differences can markedly affect their combustion characteristics. United Kingdom natural gas from the Leman Bank in the southern North Sea contains 94.8% methane. Most North American gases contain more than 85% methane but with somewhat greater concentrations of higher hydrocarbons than in North Sea gases. Helium is an important constituent of some North American gases, particularly in fields in Colorado and the Texas Panhandle. In the gases from which it is extracted commercially, the helium is present in concentrations of 1–8%. Leman Bank gas contains only 0.03% helium, a level not economically worth recovering. Some Hungarian natural gases contain as much as 40% of carbon dioxide and in fact in the untreated state do not support combustion. The French natural gas field at Lacq contains 15% hydrogen sulphide and is a commercially viable source of sulphur.

At present, proved natural gas reserves are growing world-wide. However, energy demand has risen so rapidly since the Second World War, that a modest rate of increase in energy use, even when controlled by energy conservation policies, would be likely to lead to rapid depletion of both oil and natural gas reserves at some period in the first quarter of the next century. One can therefore see naturally-occurring crude oil and gas fulfilling an intermediate role in the next half century before the large scale advent of new coal technologies and the commercial proving of the fast breeder reactor.[2] Since the major capital asset of a natural gas system is the investment in the transmission system for transporting the gas at high pressure (6.9–10.4×10^6 Pa) across country and in the systems for local distribution at low pressure (2–5×10^3 Pa) in conurbations, one can see new coal technologies leaning heavily towards the production of substitute natural gas from coal in the next century.

Substitute natural gas from oil and light petroleum distillate is already a reality in several parts of the world. The use of natural gas for base load heating is at present most fully developed commercially in the United States. Most of the existing plants operated by US utilities are based on the "Catalytic Rich Gas" process developed by the British Gas Corporation. This is based on the steam reforming of naphtha over a nickel on alumina catalyst as the first stage with subsequent methanation of the product gases in one or two further stages over a similar catalyst. Hydrogenation processes for heavier petroleum fractions are also available. There is also considerable interest in the United States, where coal is still relatively cheap, in the Lurgi process for the gasification of coal in association with subsequent methanation of the product gases. In the United Kingdom the main use envisaged for substitute natural gas is in peak load "shaving" in periods of excessive demand. Base load substitute natural gas may well not be needed here until the early part of the next century. Catalytic steam reforming processes give a gas, when the carbon dioxide has been removed, which is quite similar to natural gas in properties. Non-catalytic hy-

TABLE 1.

Component	Composition (volume %)				
	North Sea: Leman Bank	SNG : CRG + double methanation	Fluidised bed hydrogenator/ crude oil feed	Gas recycle hydrogenator/ naphtha feed	Algerian LNG
Nitrogen	1.22	—	0.1	trace	0.36
Hydrogen	—	0.60	10.0	23.9	—
Helium	0.03	—	—	—	—
Carbon monoxide	—	0.05	1.5	3.3	—
Carbon dioxide	0.04	0.50	0.2	0.7	—
Methane	94.81	98.85	79.5	44.4	87.2
Ethane	3.00	—	8.7	25.3	8.61
Propane	0.55	—	trace	2.4	2.74
Butanes	0.19	—	—	—	1.07
Higher hydrocarbons	0.16	—	—	—	0.02
Gross calorific value (MJ m^{-3})	38.84	36.86	37.21	39.15	42.62

drogenation processes, however, although they give high concentrations of methane in the product gases, also invariably contain some hydrogen. Relatively small amounts of hydrogen change the combustion properties drastically and the problems with these processes depend on the ease, or otherwise, with which the product gases can be blended in with other natural gases or substitute natural gases to give a composite gas of acceptable combustion characteristics.

A further variable in natural gas classification is provided by liquefied natural gas, which has a composition somewhat different from the original natural gas because of cryogenic processing. Major Japanese gas utilities are basing their conversion to natural gas distribution largely on imported liquefied natural gas from Brunei. The United Kingdom has been importing LNG from Algeria for well over a decade. Here the application of LNG is in peak load "shaving" and supplemental base load. Extensive base load applications of LNG are unlikely to be important in the United Kingdom until the end of the century and then only if the economic and political climates are capable of justifying the importation of LNG on a large scale.

Table 1 summarizes the compositions of selected gases, both natural and substitute, together with an Algerian LNG.[3] Also given is the calorific value of each gas. In general, in the United Kingdom, where the calorific value of the substitute gas falls below 38.8 MJ m^{-3}, it is necessary either to blend the gas in with a richer gas or enrich directly with liquefied petroleum gas (LPG) to ensure that the legal requirement to maintain the declared calorific value is maintained.

2. PREDICTION OF COMBUSTION CHARACTERISTICS—INTERCHANGEABILITY OF GASES

Since the composition of natural gas and its manufactured substitutes can vary significantly, it is important that gas utilities possess accurate methods of assessing the suitability of proposed new supplies before distribution. Not only would this help to avoid appliance operating problems but also the needless pretreatment of gases.

Mixtures of fuel gases distributed solely on the basis of maintaining the declared calorific value are unlikely to be satisfactory in use even though gas-burning equipment is designed to tolerate some variations in gas quality. The important criterion that must apply to a potential substitute gas, whether natural or synthetic, is that it should burn satisfactorily on all types of existing burners with negligible change in performance and without the need for special adjustments. Two gases are said to be interchangeable if they can ensure satisfactory operation of appliances so that:

(a) the flame is stable, with no tendency to lift or light back;
(b) there is no significant formation of carbon monoxide or soot;
(c) the heat output remains reasonably constant, and ignition is satisfactory.

Several methods of predicting the combustion characteristics of gas mixtures have been developed but they are all based on empirical relationships and require a large number of laboratory measurements to ensure their validity. This restricts their application to the limited range of gases and appliances for which they have been tested. Generally, the approach has been to adopt a small number of indices, calculated from the composition of the gas, which are then used to assess its combustion performance.[4a]

The most important, internationally recognized, index is the Wobbe number, defined as the gross calorific value of the gas divided by the square root of its specific gravity with respect to air. It is proportional, with reasonable precision, to the heat input of a burner at a fixed pressure of fuel gas. With some approximation, it is also inversely proportional to the primary aeration* of an aerated burner.[4b]

Other indices have been proposed to take into account variations in the burning characteristics of different fuel gases. In the UK, for example, the so-called Weaver flame speed factor is used as the second

* The primary aeration is the amount of air pre-mixed with the fuel prior to combustion, usually expressed as a percentage of the stoichiometric requirement.

axis in a two-dimensional graphical representation of gas interchangeability. The Weaver flame speed factor is an empirical parameter relating the maximum burning velocity of the gas and its stoichiometric air requirement.[4a] It is defined as being the burning velocity of a stoichiometric fuel/air mixture expressed as a percentage of the burning velocity of the same mixture of hydrogen/air. Values of the burning velocities of pure gases given by Weaver's formula are not in good agreement with the best literature values of maximum burning velocity. Weaver suggested that this is partly because, for aerated burners, the burning velocity is that of the primary air/gas mixture in the burner, and not the maximum.

We wish to point out that any method of assessing gas interchangeability based on Wobbe number and burning velocity—however accurately the latter may be determined—can be of limited applicability only. In practice, the factors influencing flame stability and completeness of combustion are considerably more complex than indicated by premixed flame considerations alone. Most aerated burners for domestic appliances burn a richer than stoichiometric mixture of gas and air to form a premixed flame, the "inner cone", and a diffusion flame, the "outer cone" or secondary flame.

That the outer diffusion flame plays an important role in determining the overall flame structure is easily demonstrated in the case of natural gas burners producing a primary aeration of 60% or less. Contrary to simple prediction, such flames exhibit a well-defined inner cone, even though the mixture composition issuing from the burner ports lies outside the normal flammability limits and should, therefore, have burning velocity equal to zero.

The stability of laminar diffusion flames and aerated flames is considered in greater detail in Sections 4 and 5 respectively. Fundamental burning velocity data is briefly reviewed in the next section, for, despite the limitations referred to above, it provides a useful insight into premixed flame behaviour and may help as a means of classifying gaseous fuels.

3. BURNING VELOCITY

The normal burning velocity, sometimes called the fundamental, adiabatic or laminar burning velocity, and usually given the symbol S_u, is a useful combustion parameter characterizing the overall reaction rate in a premixed flame. It is generally regarded as a fundamental constant, depending only on the initial temperature, pressure and composition of the gas mixture. If the coupling between the chemical kinetics and molecular transport properties is known in detail, it is possible to calculate the burning velocity because S_u represents the eigenvalue of the time-independent, one-dimensional flame equations.[5] Examples of burning velocity prediction on this basis are available for flames of hydrogen and air,[6,7] hydrogen and bromine[8] and hydrazine.[9] Unfortunately the combustion kinetics of hydrocarbon mixtures are not known in sufficient detail to enable their burning velocities to be determined from first principles and, therefore, experimental methods are used.

The apparent simplicity of the theoretical definition of burning velocity, i.e. the velocity relative to the cold unburnt gas with which a plane flame front travels along the normal to its surface—belies the experimental difficulties in measuring this quantity for a given fuel/oxidant mixture. To have any physical significance, such a definition can only refer to a large, flat flame isolated from all heat sinks. In practice, a flame holder acts as an energy sink, curvature of the flame causes variations in velocity across the flame and a vessel confining the gas mixture to a limited volume causes aerodynamic interactions, all of which are difficult to correct. It is not surprising, therefore, that wide discrepancies exist in the published data on burning velocities and experimental values appear to be partially dependent on the method of measurement.

The maximum burning velocity of methane/air, e.g. is today regarded to be 45 ± 2 cm/s, whereas for many years it had been thought to be about 36 cm/s.[10]

Andrews and Bradley[11] have published a critical view of the experimental methods for measuring burning velocity and derived correction factors to compensate for errors inherent in the different techniques. The methods of measurement fall into two broad groups (a) study of burner flames, in which the flame is held stationary by a contrary flow of gas, and (b) measurement of velocity of a flame travelling through an initially quiescent gas mixture. Of the stationary flame methods, those involving a water-cooled, "constant velocity" nozzle seem preferable because a straight-sided conical flame front is produced even though it is oriented at an oblique angle to the gas flow. In order to obtain the highest accuracy it is necessary to identify the flame front and to measure the magnitude and direction of the gas velocity relative to it. This is usually achieved by a combination of the particle tracking technique to map the flow distribution through the flame and schlieren photography or thermocouple measurements to locate the position of the flame front. Professor Guenther's group at Karlruhe University have successfully employed this method to investigate flame stabilization mechanisms on burners by following the increase in local burning velocity as the flow streamlines pass through the flame zone. Full details of the experimental procedures have been published.[12,13,14] Laser–doppler anemometry offers an attractive, modern, alternative approach to the measurement of local gas velocities. It is quicker and easier to apply than particle tracking methods, and appears to be capable of yielding accurate burning velocity results.[15]

Of the experimental methods involving non-stationary flames, the double kernel method of Raezer and Olsen[16] is considered to be one of the most accurate. The principle of this technique is that when two flame kernels are propagating towards each other, the gas velocity at the midpoint of the line between their centres is zero and the flame speed of each of the kernels as they meet must become equal to the burning velocity. Experimental difficulties are principally associated with synchronization of the two ignition sparks leading to flame kernels of unequal size.[17] If the experiments are conducted in a closed vessel it is possible to determine burning velocities of mixtures at initial pressures other than 1 atm.

A selection of maximum burning velocities of various fuel–oxidant mixtures at 1 atm and room temperature is given in Table. 2.

TABLE 2. Maximum burning velocities of gases at 1 atm and 298 K

Gas mixture	Maximum burning velocity (cm/s)	Reference
Methane–air	45	11
Ethane–air	48	19
Propane–air	46	19
n-Heptane–air	43	19
Acetylene–air	160	19
Ethylene–air	79	15
Benzene–air	45	19
Hexane–air	40	19
Hydrogen–air	350	17
Methane–oxygen	450	19
Acetylene–oxygen	1140	19
Hydrogen–oxygen	1400	19

Figures 1 and 2 show the burning velocities of methane and other hydrocarbon–air mixtures as a function of stoichiometry. The dependence of burning velocity on mixture composition roughly follows that of flame temperature—in each case the maximum values are to be found on the rich side of stoichiometric. Hydrogen and carbon monoxide are exceptional in that their maximum burning velocities occur at about twice the stoichiometric mixture compared with 1.05–1.1 stoichiometric for hydrocarbon fuels.[18]

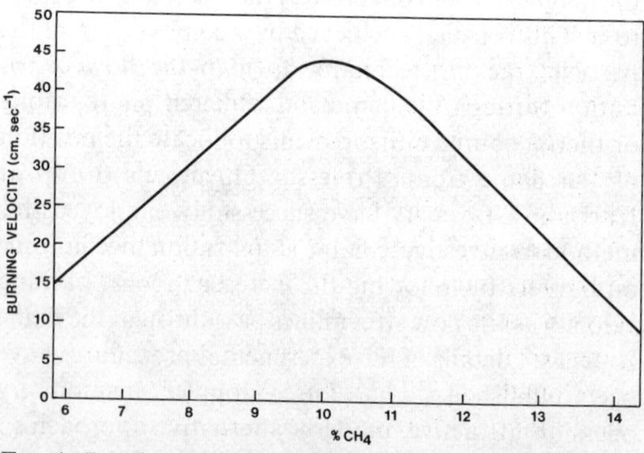

FIG. 1. Burning velocities of methane–air at one atmosphere pressure and 293 K (after reference 17).

The Temperature and Pressure Dependence of S_u

For a given pressure and mixture composition, preheating the reactants prior to combustion raises the overall burning velocity and widens the flammability limits enabling leaner mixtures to be burned without loss of flame stability. In general, we may write

$$S_u = A + BT^n$$

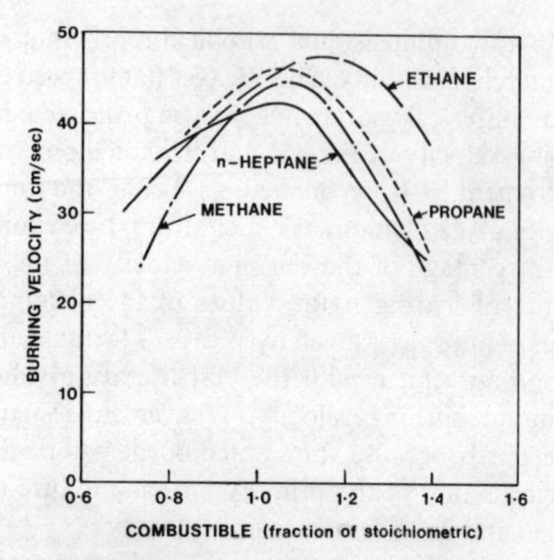

FIG. 2. Burning velocities of methane, ethane, propane and n-heptane–air mixtures at atmospheric pressure and room temperature (after Zabetakis, reference 19).

where A and B are constants, T is the temperature and n is a constant for a given mixture composition. Stoichiometric methane–air mixtures closely follow the formula[10]

$$S_u = 10 + 0.000\,371\, T_u^2 \text{ cm/s}$$

where T_u is the unburnt gas temperature in K. A similar relation holds approximately for other paraffin hydrocarbons.[20]

The pressure effect on burning velocity is less consistent. In general, there is a decrease of burning velocity with increase of pressure for fuels whose burning velocity at atmospheric pressure is below about 50 cm/s and an increase for mixtures whose burning velocity at atmospheric pressure is higher than this value. Why this should be so is not clear.

A useful approximate formula is

$$S_{u1}/S_{u2} = (P_1/P_2)^n$$

where S_{u1}, S_{u2} are the respective burning velocities at pressures P_1 and P_2. Lewis has reported measurements of n for a large number of flames using the spherical bomb technique and his graph is reproduced in the book of Gaydon and Wolfhard.[21] For methane–air mixtures over a pressure range 0.1–100 atm, a value of $n = -0.5$ has been recommended.[10]

Burning Velocities of Fuel Mixtures

Compared with the many experimental and theoretical studies that have been made of laminar flame propagation with mixtures of a single fuel and oxidant, relatively few reports are available of the burning velocity of mixtures containing several fuel gases. Scholte and Vaags[22] have published experimental data on the burning velocities of carbon monoxide–hydrogen and carbon monoxide–hydrogen–methane mixtures and Leason[23] has investigated the effect of the addition of several gaseous fuels to propane–air mixtures. These results are of limited use to today's Natural Gas Industry, however, which is more interested in the combustion characteristics of methane–air mixtures containing various amounts of higher hydrocarbons, inerts and hydrogen.

A simple empirical mixing rule, giving the maximum burning velocity of a mixture in terms of the maximum burning velocities of the component fuels has been proposed by Payman and Wheeler.[24]

$$S_{um} = \frac{aS_{ua} + bS_{ub} + \ldots}{a + b + \ldots}$$

where S_{um} = burning velocity of the mixture.

S_{ua}, S_{ub}, etc. = burning velocities of the components a, b etc. and a, b etc. are the concentrations by volume of the component combustible mixtures.

This formula corresponds to no conceivable physical system and its use is restricted to fuels whose burning velocities and flame temperatures do not differ substantially from each other.

Spalding[25] has suggested a mixing rule which avoids this difficulty. He assumed that no chemical interaction occurs, but that the final mixture may have properties which are different, in a thermal sense, from those possessed by the components.

Thus it is recognized that the final flame temperature of the mixture is different from that of any component if it alone were present. If A represents air, B and C represent two different fuels, and the mass ratio of AB and AC in the ABC mixture is b/c where $b + c = 1$, the burning velocity of the ABC mixture is given by the expression

$$\frac{I_{BC}}{k_{BC}} = b\frac{I_B}{k_B} + c\frac{I_C}{k_C}$$

Here $I = (\bar{c}\Delta T \rho_u S_u)^2$, $\bar{c}$ being the specific heat at constant pressure, ΔT the temperature rise of the adiabatic flame, ρ the density and k the thermal conductivity. The data required for the evaluation of the rule are (a) the flame temperature of the mixture of AB and AC; (b) the variation of the burning velocity of each component with final flame temperature. This information is generally available from the literature.

Martin[26] has compared the predictions of the mixing rule against experimental results. His paper includes data of the burning velocities of mixtures of methane, hydrogen and air at various equivalence ratios and with hydrogen concentrations varying from 0 to 10% of the methane volume. Good agreement was obtained between predicted and measured burning velocities for these and other mixtures of fuel gases. The only evidence of any unusual interaction effects between two fuel components was provided by carbon monoxide–hydrogen mixtures where the predicted value of burning velocity was always well below that found experimentally.

On the other hand, the experimental results of Corbeels and Van Tiggelen[27] showed a weak inhibition of hydrogen combustion by methane and a weak promotion of methane combustion by hydrogen. Support for the occurrence of a chemical interaction between hydrogen and methane also comes from the recent work of Borie.[28] It is difficult to judge whether the effects reported by these authors are real or apparent because of the limited accuracy of their burning velocity measurements.

Inert gases such as nitrogen and carbon dioxide act as straightforward diluents. Their effects on burning velocities are purely thermal, causing a reduction in flame temperature. Reed *et al.*[29] have published extensive experimental data on the burning velocity of methane/nitrogen/air, methane/carbon dioxide/air, and methane/water vapour/air mixtures. They used the Semenov equation to extrapolate the measured burning velocity data to obtain burning velocities at higher diluent concentrations and for leaner mixture compositions than could be obtained, with good accuracy, experimentally, and to predict also the burning velocities of methane/combustion product/air mixtures. The latter data is useful for the prediction of the effect on burning velocities of vitiated air.

4. DIFFUSION FLAME STRUCTURE

In the United Kingdom, in the latter days of the town gas era and particularly in the 1950s, diffusion flame burners played an important part in the development of appliances for space-heating. A well known example is the radiant-convective gas fire which played an important part in the expansion of the use of gas in the period just before rapid expansion in central heating installations began. Most of these radiant convectors were based on the Bray jet burner. This gave in effect a batswing type of flame on town's gas. A typical town's gas contained about 50% of hydrogen and for such hydrogen-rich gases, the Bray jet diffusion flame was very stable. There was obviously no light-back problem and considerably less susceptibility than with natural gas to blow-off. The lack of any need to entrain air in the piping to the burner obviated the entrainment of dust and fluff into the burner, the so-called "linting" problem which can lead to partial blocking of the mixing plenum in a natural gas–air premixed system. Noise emission from such a burner was so low that a radiant convector based on a battery of such burners could have a noise rating in a living room as low as 15 pndb.

The advent of natural gas in the 1960s in the United Kingdom meant a decline in use of such diffusion flame burners and the United Kingdom turned to the premixed natural gas–air systems in common use in the United States since the beginning of the Second World War. Such systems have, of course, the stability problems which have been described earlier. Diffusion flames are, however, still important in the natural gas era. Most appliance pilot flames for burner ignition are pure diffusion flames. In recent years a number of British burner manufacturers have produced what are in effect pilot flame-stabilized diffusion flame burners of high heat output and designed for natural gas operation: the so-called "pin-hole" burners. In such a burner, natural gas at high velocity issues from a bank of holes in a gas distribution plenum which are no more than 3×10^{-4} m in diameter.

Close to and parallel to this row of pinholes is a further bank of larger holes, from which the gas issues at very low velocity. On ignition, the low velocity gas stream is stable on natural gas and maintains a continuous source of ignition and preheat at the base of the high velocity jet to provide a counter against blow-off. For UK natural gases, such a system works reasonably well and provides a relatively simple method of avoiding some of the problems inherent in premixed burner operation. The interaction between the main flames and the stabilizing flames involves, however, a delicate balance among a number of factors and, in particular, a change of natural gas composition to, for example, a substitute natural gas, could disturb this balance and give rise to both sooting and stability problems. Furthermore, the premixed flames burnt in domestic appliances in the UK are invariably considerably richer than stoichiometric and a major part of the volume of the practical flame is in effect the secondary diffusion flame which can have a considerable effect on the stability of the whole. In consequence, it is important not to ignore the properties of the diffusion flame in practical application since, in terms of stability at least, these are quite different from the properties of the classical premixed flame.

Considerations of burning velocity, so important to the premixed flame, do not apply to the diffusion flame. In fact, in the most efficiently burning steady-state diffusion flames the fuel and oxidant do not co-exist. The lack of any burning velocity criteria has the profound implication that diffusion flame stability cannot be discussed in the simple burning velocity terms described above in their application to premixed flames. The properties of diffusion flames have been neglected as a subject of study until recently. In view of this, it is worth outlining typical diffusion flame structure in some detail. Figure 3 shows schematically, the structure of a diffusion flame sheet. A typical diffusion flame may consist of two flat flame sheets (e.g. the Bray jet flame) or a single sheet (axisymmetric flame on a cylindrical port, e.g. the pinhole burner flame).

One can identify four important regions in a diffusion flame as follows:

(1) The outer convective/diffusive region which forms the bulk of the flame volume.
(2) The inner region in the upper part of the reaction zone where chemical reactions are equilibrated.
(3) The inner region at some distance from the burner where the flame structure is controlled by the competition between the reaction kinetics and diffusive processes into and out of the flame sheet.
(4) The mixed inner/outer region at the base of the flame where strong fluid dynamic effects exist and where the chemical reaction probably goes at a very slow rate.

Naturally enough, these regions are not separate entities but shade into each other gradually. The regions having the greatest effect on flame stability are

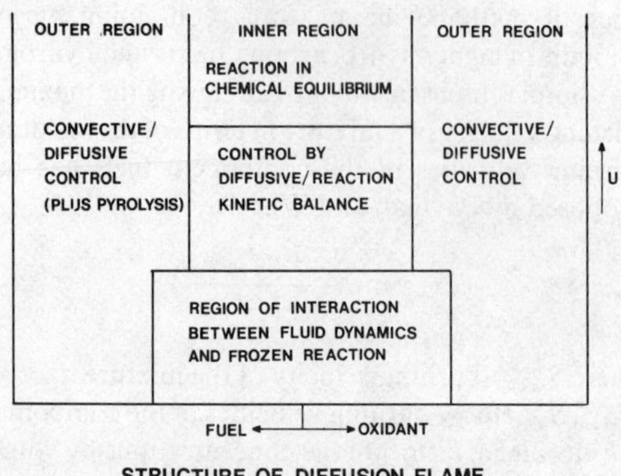

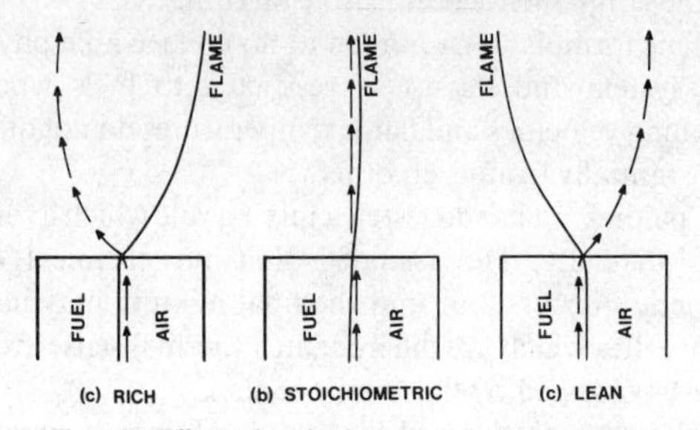

FIG. 3. Effect of stoichiometry on flame shape.

(3) and (4). That is, the regions where chemical kinetic processes compete with diffusion and where there is strong interaction between the chemical kinetic processes and the gross convective structure at the base of the flame.

In considering methane diffusion flames, Melvin and Moss[30] were able to carry over much of the information derived by Clarke, Moss and Melvin,[31] and Moss and Clarke[32] on stability parameters based on both chemical kinetic criteria and predicted fluid dynamic structure for hydrogen diffusion flames. In particular, Melvin and Moss showed that the kinetic features important in a methane diffusion flame could be adequately represented by adding a further four reactions to a conventional hydrogen–oxygen carbon monoxide kinetic scheme.

$$OH + CH_4 \rightleftarrows CH_3 + H_2O$$
$$OH + HCHO \rightleftarrows CO + H_2 + H$$
$$CH_3 + O_2 + M \rightleftarrows HCHO + OH + M$$
$$CH_3 + OH \rightleftarrows HCHO + H_2$$

This mechanism, which is the simplest one can postulate for methane, has elements of reaction competition which can be expressed in terms of a chemical kinetic extinction condition for the diffusion flame.[30] However, the fluid dynamic processes at the flame base also have to be taken into account here. It is possible to show that the velocity field at the base of the flame is controlled by the stoichiometry of the diffusion flame. Stoichiometry for diffusion flames is perhaps an unfamiliar concept, particularly as the reactants are not mixed. In the analysis of diffusion flame structure, it does, however, become a necessary condition that fuel and oxidant diffuse into the flame sheet in stoichiometric proportions. If we write C_α = mass fraction of a species α with a molecular weight W_α and D_α

= diffusion coefficient of species into products at the flame sheet in the direction normal to the sheet, we may express the stoichiometry condition as follows:

$$\frac{D_{O_2}}{W_{O_2}} \cdot \frac{\partial C_{O_2}}{\partial Y} = -2 \cdot \frac{D_{CH_4}}{W_{CH_4}} \cdot \frac{\partial C_{CH_4}}{\partial Y}$$

In general, one can then say that a "stoichiometric" diffusion flame will burn with its flame sheet at the fuel–air interface. A "rich" flame will have its flame sheet position on the oxidant side and a "lean" flame on the fuel side. The shape of the flame is thus determined by the stoichiometry. Figure 3 shows the three cases for a simple flat flame and also shows the velocity field associated with each case. Neat methane jets burning in air are always rich flames and for this case it can be seen that streamlines from the oxidant side at the base of the flame are refracted to the fuel side. The analysis and experimental evidence for this phenomenon are presented in detail in reference 33. The process of refraction provides a plausible fluid dynamic mechanism for extinction and blow-off in that the transfer of oxidant across the base of the flame sheet destroys the diffusion structure and creates a region of frozen reaction which separates the flame from the burner rim on which it was anchored.

5. THE LAMINAR AERATED FLAME

Problems of stability for premixed flames of natural gas and air occupied the US Bureau of Mines in the 1930s and a good summary of this work appears in the book by Lewis and von Elbe.[34] Most organizations distributing natural gas have leant heavily on this work as a systematization of their stability problems for blow-off and lightback in particular. Nevertheless, as Lewis and von Elbe make very clear in their discussions of boundary velocity gradient criteria, the resulting correlations of boundary velocity gradient with stoichiometry are purely empirical guides. Attempts to improve on these correlations have been made without much success. It is therefore fair to say that the present position, is, surprisingly, that we have no adequate scientific treatment of flame stability problems which would allow us to go further than merely correlate existing experimental data.

In this section, we shall deal with the blow-off of premixed flames as an illustration of one stability problem and as a reminder to the reader of the patois in this area. The earliest theory of blow-off which needs serious consideration is, as indicated above, that proposed by Lewis and von Elbe as involving boundary velocity gradient criteria. Over the diameter of the burner port the velocity profile can assume various shapes, depending on the length to diameter ratio of the port. Close to the inner port surface one has a range of flow velocities, to which one is trying to match a single burning velocity characteristic of the fuel–air mixture being burnt. Such a situation would not be stable and Lewis and von Elbe suggested that interdiffusion with the surrounding air reduces the burning velocity of the fuel–air mixture in the region of steep

(a)

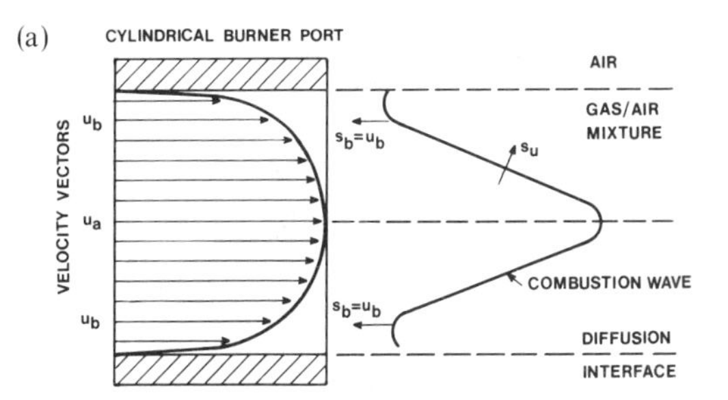

(b)

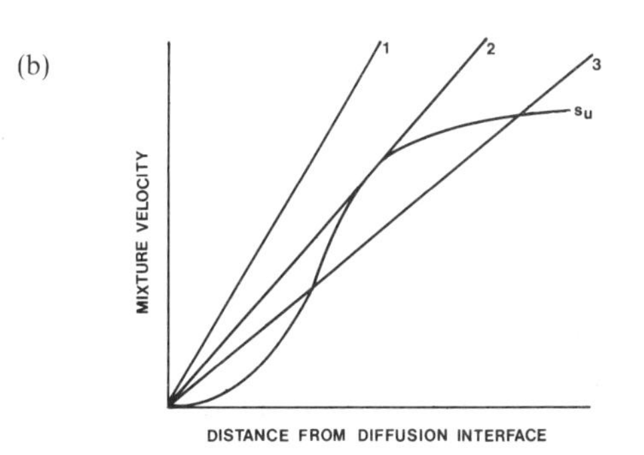

FIG. 4. (a) Matching burning velocity to gas velocity profile; (b) Condition 1: blow-off; 2: stable flame; 3: lightback.

velocity gradient. Assuming that the gas velocity increases freely with distance from the port surface we may illustrate the processes of blow-off, stable combustion and light-back in the region of the port surface by the three cases shown in Fig. 4.[34] Although attractive as a means of visualizing flame stability, the boundary velocity gradient theory does not fulfil the function of a comprehensive theory, which should ideally allow us to predict flame behaviour from knowledge of fuel composition.

It is worth, however, at this stage giving an illustration of the use which is generally made of it quantitatively. Since boundary velocity gradients for cylindrical ports are normally required, the usual first assumption is that the velocity profile is described by fully-developed Poiseuille flow. The boundary velocity gradient can then be defined by

$$g = \frac{4V}{\pi r^3}$$

where V is the total volumetric flow rate through the port and r is the port radius. One can then define g_f and g_b as boundary velocity gradients calculated from experimentally determined flow rates at flashback and blow-off respectively. Information on these critical boundary velocity gradients is then normally presented as a plot of g_f or g_b against fuel concentration, expressed as a fraction of stoichiometric. Such a plot for blow-off for methane is shown in Fig. 5 and illustrates the purely empirical nature of the boundary velocity gradient correlation. Its claim to utility is that it allows one to predict blow-off rates for burners of different sizes given the experimental blow-off data for one particular burner size. Its weakness is that it gives no insight into the factors affecting flame stability. There is no indication, for instance, in the correlation for the methane case that the leanest flame shown is

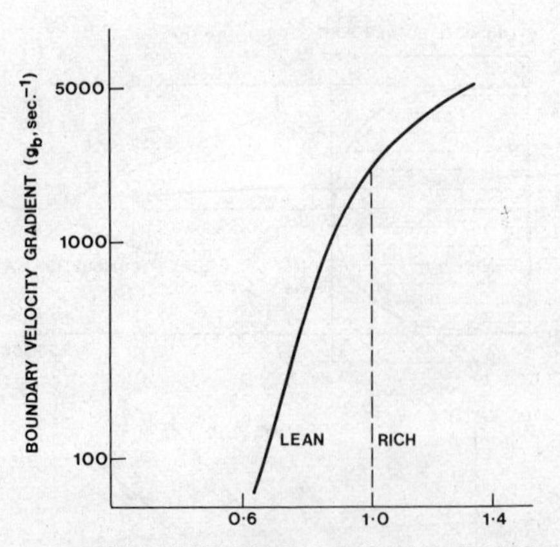

FIG. 5. Critical boundary velocity gradient curve for methane–air premixed flames.

qualitatively different from the richest flame, which is dominated by the outer diffusion flame structure with quite different criteria for stability. One may regard the boundary velocity gradient correlation largely as a burner size-scaling correlation and it is here that it finds most application in gas industry applications.

A more ambitious theory based on boundary velocity gradient considerations was developed by Karlovitz[35] as the "flame stretch" theory and was later applied by Reed[36] to aerated flames on cylindrical ports. Here it is suggested that the curvature of the flame surface which occurs in the region of changing flow velocity near the port boundary may lead to an energy loss from the flame reaction zone which is not matched by that gained in the pre-heat zone.

One can then define a non-dimensional number, the Karlovitz flame stretch factor, which is a measure of the degree to which the flame surface area increases as a result of the velocity gradient. Lewis and von Elbe believed that, in some circumstances, excessive stretching of the flame surface could lead to local quenching of the flame reaction, although they were unable to establish a correlation for premixed flames on burner ports. Reed, however, claimed to have established such a correlation and it is therefore worth briefly considering recent developments in this area.

For flows substantially in excess of the burning velocity where the flame front is nearly parallel to the direction of flow, the Karlovitz factor may be approximated by

$$K = \frac{\eta_0}{U} \cdot \frac{dU}{dy}$$

where η_0 is a length characteristic of the "thickness" of the pre-heat zone and is defined by

$$\eta_0 = \frac{K}{\rho C_p} \frac{1}{S_u}$$

where K is the thermal conductivity, ρ the density, C_p the specific heat at constant pressure and S_u the burning velocity for the gas–air mixture. In the treatment by Reed, the flame stretch factor is further approximated to

$$K = \frac{\eta_0 g_b}{S_u}$$

and the theory proposes that K should be constant ($= 0.23$) and that a plot of S_u vs g_b should give a single line correlation for all fuel–air mixtures at least leaner than stoichiometric. In practice, Reed had to exclude hydrogen–air mixtures and mixtures of fuels with oxygen from his correlation. Edmondson and Heap[37] suggested that Reed did not have data of sufficient accuracy to test the correlation and themselves produced data which seemed internally more self-consistent. They concluded, largely from a study of blow-off of methane–air flames, that the flame stretch correlation could not hold.

Melvin and Moss[38] showed that the apparent success of the correlation for Reed's data was due to the dominance in the data of his own results for ethylene–air mixtures at elevated temperatures, which were shown to fit a quite different but trivial correlation where g_b varies linearly with S_u. A detailed study of local velocity and temperature fields in methane–air flames also led Günther and Janisch[39] to the conclusion that the flame stretch correlation did not hold. Indeed, the flame stretch equation appears to yield a spurious correlation and Lewis and von Elbe were justified in their original claim that it did not apply to premixed flames on cylindrical ports.

Of recent work in the area of premixed flame stability that of Günther and Janisch[39,40] has influenced present thinking in both the German and UK gas industries. Günther and Janisch mapped velocity and temperature fields for three stoichiometric methane–air flames near the blow-off limit, near flashback and in the middle of the stable region. One interesting conclusion to arise from this work is that these authors reject as in conflict with the experimental evidence any suggestion that flame stabilization correlates with the normal burning velocity as defined above for the cold boundary of the flame front. Instead they point out that the local balance between velocity and gas flow only occurs in the middle of the luminous part of the reaction zone (about 1500–1600 K). Here it is necessary to define local burning velocity S_1 as

$$S_1 = u_1 \sin \beta$$

where u_1 is the local gas velocity and β is the angle between the flame front and the flow direction. The relation between S_u and S_1 is then

$$S_1 = S_u \rho_u / \rho_1 \cdot A_u / A_1$$

where ρ is the density, A is the cross-sectional area of the stream-tube, and u and 1 designate cold boundary and local "hot" values respectively. Günther and Janisch suggest that, in the stabilizing region, both S_u and S_1 are reduced by intermixing with the surrounding atmosphere and that blow-off occurs when the composition in this region reaches about 5%, the lower flammability limit for methane–air ignition.

In taking the ideas contained in this work on stoichiometric flames further and applying them to the "rich" flames in common use in gas appliances, one must necessarily replace this simple flammability limit criterion for ambient diffusion effects by considerations of the diffusion flame structure and the associated modifications to the aerodynamics which the discussion of Section 4 shows to be necessary.

6. CATALYTIC COMBUSTION

In previous sections we have concentrated on the kind of free burning flame in general use in most domestic appliances. In such applications, no advantage is taken of the unique high temperature properties of the flame. On a cooker hotplate which is being used to boil water one is merely degrading high temperature heat to its low temperature equivalent by flame impingement. In fact, if the flame could be made to burn at 600 K one would still transfer the same heat, i.e. the product of density flux and enthalpy. Catalytic combustion provides a means whereby methane can be burnt at such temperatures and its feasibility opens up a range of novel possibilities in application: e.g. flameless cooker hot-plates and integrated catalytic burner/heat exchangers for central heating boilers among others. A further important advantage of catalytic combustion at such low temperatures is that emissions of NO_x can be reduced to zero and carbon monoxide emissions to negligible levels.

It is possible to operate a catalytic combustor in either the premixed or diffusion flame modes. At present the diffusion flame system is by far the most popular in commercial applications. Figure 6 shows schematically a typical commercially-available catalytic heater. The catalyst pad consists of about 1% platinum on asbestos. Gas is forced through the pad

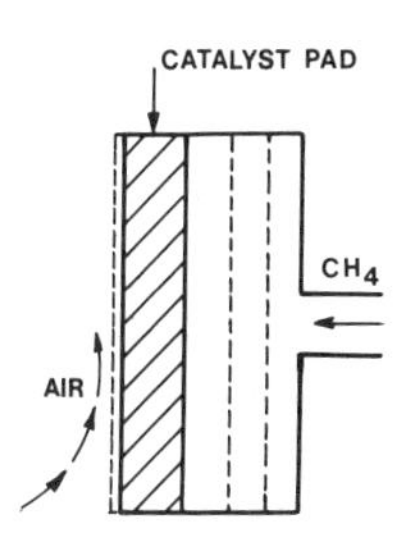

FIG. 6. Non-premixed catalytic combustion.

via a distribution system and catalytic combustion occurs ideally on the face of the pad, to which air is supplied by free convection. Heat emission is then by free convection and radiation from the pad face. Such catalysts operate at 600–700 K and so it is necessary to heat the catalyst before combustion on the face can occur. This can be done in one of two ways. Either the pad is heated electrically to the required temperature or a small free-burning pilot flame is allowed to impinge on an area of the catalyst pad surface until it reaches the operating temperature and the combustion spreads over the whole surface. Both methods are relatively unsatisfactory since in the first case electric power and a relatively complex control system are required and in the second considerable amounts of unburnt gas could well be emitted into the room during the start-up period.

Radcliffe and Hickman[41] have described the practical performance in the combustion of natural gas of sixteen commercially available catalytic pads operated in the diffusive mode. Of these, only three gave maximum combustion efficiencies in excess of 90%. In all cases, some methane was observed to be discharged from the pad unburnt. Furthermore, the combustion efficiency of pads tested fell off with time at the rate of about 5%/1000 h of operation. Clearly, despite the particular advantages of catalytic combustion, considerable improvement would be required in these aspects of performance before the technique could be considered for large scale application. Dongworth and Melvin[42] have considered the role of the catalyst in such combustors from the point of view of assessing both scope for development in the technique and the limitations which the catalyst itself might place on improvements. The conclusions were that, under the conditions of use of combustion catalysts in practice the catalytic reaction would be diffusion-controlled and that methane emission from a combustor could only be reduced by increasing the bed depth by an unacceptable amount. Under these conditions, attempts to improve the catalyst by increasing its activity would be irrelevant and the only hope for improvement would be to increase the rates of mass transfer occurring not only within the catalyst bed but also within the individual catalyst particles. Clearly, with the present laminar flow diffusive combustor designs available commercially, such a route for improvement in performance would be impracticable. One could envisage specialized applications in the gas industry where it would be feasible to make purpose-built catalytic burners which operate under turbulent flame conditions, but even here the rates of mass transfer within the catalyst particles themselves may not be greatly increased. Such specialized applications would, in any case, be few and far between. Dongworth and Melvin also showed that the long term loss in efficiency of combustion catalysts was probably due to fuel-cracking in the combustion catalyst and the consequent build-up of carbon to reduce access of fuel and oxidant to the interior of the catalyst particles.

Catalytic combustion of premixed methane and air is still a diffusion-controlled process. As a possible alternative to diffusive combustion, it has a number of serious disadvantages. Since the combustion temperature is higher (*ca.* 1000 K), great care must be taken to ensure that the fuel–air mixture never becomes richer than stoichiometric; otherwise, large quantities of carbon monoxide are emitted. Premixed catalytic combustors are not easy to make intrinsically safe since the high temperature on the inlet face of the catalyst pad can give rise to ignition of the incoming mixture in the gas phase followed by propagation of a flame against the flow into the mixing lines. The high temperature of the inlet face is a further disadvantage since the high level of radiation from this face cannot easily be used for heating and, unless complex precautions are taken, represents a considerable loss in useful heat from the burner.

One may therefore conclude that there does not seem to be a future within the gas industries of the world for large scale use of catalytic combustion. Present small scale specialized applications are likely to continue but in the most widespread of these, e.g. the heating of caravans and chicken-houses, natural gas has no particular advantage over liquefied petroleum gas which is clearly more convenient when a mobile, self-contained heating source is required.

7. TURBULENT FLAMES

In the preceding sections, the emphasis has been on laminar flames, with the consequent field of application being largely in the use of natural gas in domestic and small-scale commercial circumstances. Although laminar combustion has some industrial application, turbulent combustion in large-scale furnaces and in process heating is of considerably greater importance. As a result, there is considerable practical expertise available throughout the world on the design of systems for large-scale heating. It is important to note, however, that this know-how is largely based on hard-won experience and the empirical formulation of relatively crude design principles. Basic research on the principles of turbulent combustion has had little or no effect on the principles of design of turbulent burners. Here, of course, one finds practice outstripping science since, although there has long been a requirement for the large turbulent flame in industrial application, there is still only a rudimentary knowledge of the science of turbulent combustion. This is not to say that there has not been a long-felt need among furnace designers for more scientific procedures in design. Such a need was clearly the motivation in Europe for the establishment of the Ijmuiden Furnace Laboratory of the International Flame Research Foundation. Nearly 30 years later, however, we are still not at the stage where reliable predictive procedures for turbulent combustion are available to the furnace designer. With the furtherance of energy conservation policies over the next 25 years, it will become even more important to improve the efficiency of use of turbulent combustion in industry. Although, in general in gas utilities worldwide, the emphasis on energy conservation will probably mean that the domestic sector will expand in relation to the industrial sector, there will still be a heavy and increasing demand for natural gas from industry.

The consequence will almost certainly be a relative rise in natural gas fuel costs to industry in the long-term and there will hence be a strong economic incentive to improve the efficiency of industrial combustion. That a serious interest and involvement still exists in the 1970s in the modelling of turbulent flames is illustrated by the symposium organized by the Institute of Fuel and the British Flame Research Committee in 1972 on "Predictive methods for industrial flames".[43]

At this stage, it is worthwhile briefly summarizing some current approaches to the problem of describing scientifically the physical processes occurring in the turbulent flame. Given that no single approach can universally be regarded as adequate, such a review can hardly result in a recommendation for a particular approach. Rather, it should serve to indicate some measure of the gap which exists between the present state of the art and the position which would have to be reached before such techniques could be of practical application in design for the combustion of natural gas in industry. For such combustion of natural gas, one requires ideally a predictive method which could permit calculation of the rate of mixing of fuel and oxidant, the shape and size of the turbulent flame brush, the flame stability characteristics and the generation and emission of undesirable components such as carbon monoxide and NO_x. Since the heat transfer from natural gas flames is usually convective, with a relatively small radiant contribution, the problem of prediction of characteristics should be simpler than for the corresponding oil flame where one often tries to optimize the radiant component and has an extremely complex two—or even three—phase combustion system.

One can divide turbulent flames into two classes: (a) premixed; and (b) jet flames (or "turbulent diffusion flames"). Characterization of the gross physical properties of such flames was carried out by Wohl, Gazley and Kapp[44] and Wohl, Shore, Rosenberg and Weil.[45] Early attempts to predict the properties of premixed flames were made by Karlovitz[35] in the development of the flame stretch concept and of fuel jet flames by Hawthorne, Weddell and Hottel.[46] The latter work was particularly interesting since it stressed the importance of "unmixedness" in the flame and defined a measure of it in terms of the root mean square of the fluctuating concentration. Here the model of the instantaneous combustion process was given as pockets of oxidant plus products separated from pockets of fuel plus products by a thin fluctuating flame sheet. The concomitant temperature fluctuations imply that individual pockets may be cold or hot. Similarly, turbulent premixed flames are envisaged as consisting of pockets of hot, burnt gas separated from pockets of cold, unburnt gas by thin reaction zones. Earlier theoretical treatments of the mixing process assumed an infinitely thin flame sheet. Although this may be a reasonable approximation for the instantaneous state, models based on this assumption do not describe the time averaged case and provide an inadequate description of the turbulent flame brush.

Nevertheless, if the practical furnace designer takes any account of the flame at all, he will base his picture of the turbulent flame on the descriptions given in these early publications. The newer work on turbulent flame modelling has had little or no influence even on the designer's picture of the processes occurring in the flame.

It is difficult to classify recent developments other than as semi-analytic approaches and purely numerical techniques. Clearly, the objective of the semi-analytic approach is to identify and rank in terms of importance the physical processes controlling the detailed characteristics of the flame. Such approaches do not as yet lead to direct prediction but rather indicate that there has to be a better identification of controlling processes before prediction is possible. Among such approaches one may cite the treatment by Chung[47] of the turbulent diffusion flame, where the turbulent fluctuations are treated in a manner analogous to "classical treatments of fluctuation in Brownian motion". This is a purely physical model with the emphasis on modelling by analogy and probably would not lead to the kind of prediction required by the furnace designer.

Another such approach is that of Toor[48] and of Bush and Fendell[49] for turbulent diffusion flames where scalars are approximated by probability density functions. The options in this kind of approach lie in the choice of an appropriate form for the probability density function. The appeal ultimately is to experiment and the choice of a form for the probability density function which gives predictive agreement with experiment, e.g. O'Brien.[50] More familiar and perhaps more easily applied by the designer are the purely numerical techniques developed by Spalding.[51] One such model is the so-called "*k-w-g*" model of the turbulent diffusion flame where closure in the conservation equations is achieved by defining fluctuation equations for turbulent kinetic energy, k, vorticity, w, and concentration, g.

Programs for handling systems of equations in boundary layer form are available (Patankar and Spalding)[52] and it is possible to solve the equations of the *k-w-g* model to predict the structure of a turbulent diffusion flame. For axi-symmetric turbulent diffusion flames, experimental information is available from the work of Kent and Bilger[53] to enable an estimate to be made of the degree of success of the prediction. In making such a comparison, Kent and Bilger concluded that the *k-w-g* model overestimated by a factor of 2 the mixing rate in the diffusion flame. The overprediction of mixing rate has the necessary consequence that predicted fluctuations in velocity and concentration are higher than the measured values in the flame. One may therefore conclude that, in this case at least, although the *k-w-g* model provides an acceptable semi-quantitative prediction, it does fall short as an accurate quantitative prediction, although it may be of use to the furnace designer within these limits of accuracy.

A clearer view of the limitations of simple boundary layer-type approaches has been recently provided by Bray.[54,55,56] He points out that although there is a great attraction in reducing the turbulent flame equations to boundary layer form because of the ease of subsequent computation, the assumptions necessary to force them into this form mean that many important terms, comparable in magnitude to those retained, are omitted.

Because of the length and complexity of Bray's treatment, it is not possible to give an adequate exposition here. As a sample of the degree of complication required to provide adequate *approximate* equations, one may compare the time-averaged incompressible flow equation in two dimensions

$$\frac{\partial \bar{U}}{\partial X} + \frac{\partial \bar{V}}{\partial Y} = 0$$

which is the simplest equation appearing in boundary layer treatments, with the simplest equation possible in the flame case, evaluated from an order of magnitude analysis

$$\frac{\partial \bar{U}}{\partial x} + \frac{\partial \bar{V}}{\partial y} = \frac{-\alpha - 1}{\alpha} \cdot \frac{1}{\bar{p}} \sum_{i=1}^{n} \frac{H_i \bar{w}_i}{W_i}$$

where α = specific heat ratio, $\bar{p}$ = mean pressure, H_i, W_i = enthalpy and molecular weight of the ith species and w_i = time averaged rate of generation of species i.

Comparison of these two equations illustrates the severe problems in modelling turbulent flames. Although in incompressible laminar flow, the velocity divergence (i.e. the left hand side of the above equations) is zero, in turbulent flow it interacts strongly with the chemical rate of production.

At present, the results of Bray's analysis do not allow us to proceed to more accurate predictions of turbulent flame properties. The complexity of the equations probably means that an adequate predictive method will not be available for a long time. One can, however, make some estimate from Bray's work of the range of applicability of conventional boundary layer treatments. It is reasonably clear that these may provide good approximate prediction for small fluctuations in turbulence parameters, i.e. for flames of low relative intensity of turbulence. For large fluctuations of the magnitude present in industrial turbulent flames, one might expect only semi-quantitative prediction from boundary layer treatments. One can draw the same conclusions from the work of Chung referred to earlier,[47] which shows that the Boussinesq approximation essential to the boundary layer formulation, is only a valid approximation for small fluctuations.

Despite the progress which has been made in recent years, the way forward in turbulent flame prediction does not seem too clear. Old ideas are being rediscussed, e.g. the quasi-flame stretch hypothesis of Klimov[57] and more careful attempts are being made to describe the details of turbulent flame structure observed experimentally, e.g. Ballal and Lefebvre.[58] Since most treatments ultimately involve an appeal in some form to experiment, one can envisage future developments as involving a judicious combination of prediction and experiment, with complex interaction terms which appear in the equations being replaced by experimental information obtained in model turbulent flames closely related to the flame for which the properties are being predicted. The development of the laser Doppler technique for flow measurement in recent years has been useful in both laboratory flame characterization and in industrial study of practical burner performance.[59,60] The use of the laser Raman spectroscopy technique for species concentration measurement in turbulent flames has long been problematical, but with the advent of tandem argon ion lasers of 40 W output, it is likely that some progress will be made in the next few years. We can clearly expect progress in turbulent flame research over the next few years, but in the present state of the art, it is difficult to say how soon it will be before modern techniques are well enough developed to be easily applied by the furnace designer.

8. ENVIRONMENTAL POLLUTION

Over the past 6 years a great deal has been written about the mechanisms of pollutant formation in flames, consequently there is no shortage of articles reviewing the basic combustion chemistry and various aspects of the control technology. Rather than repeat this material here, we refer the reader to recent issues of *Combustion Science & Technology* in which the present state of knowledge is comprehensively reviewed. We confine ourselves, in this section, to some general remarks relevant to the burning of natural gas.

All fossil fuels contribute towards the pollution of the environment. Neglecting carbon dioxide and water vapour, the major combustion-generated pollutants are particulates, oxides of sulphur, oxides of nitrogen, carbon monoxide, hydrocarbons including partially oxidized species and soot. The contribution of gas to the total emission of each pollutant can be estimated from known emission factors and total fuel usage patterns.[61,62] Overall, less than 1% of the total man-made production of particulates, sulphur dioxide and carbon monoxide is produced by natural gas-fired installations. NO_x emissions can be significant although generally below those produced by burning thermally equivalent amounts of coal or oil. There is no doubt that the relative cleanliness of natural gas has enhanced its value as a premium fuel in the industrially developed countries of the world.

Average emission factors for NO_x from various sources are given in Table 3. Some improvements in control techniques have been made since these figures were compiled but they serve to indicate trends amongst different fuel types. The lower NO_x emissions from gas-fired equipment is largely because there is no contribution from the oxidation of chemically bound nitrogen in the fuel. Whatever the fuel used, the formation of oxides of nitrogen varies markedly with combustion conditions and usually increases with the heat input of the installation. This is illustrated in Fig. 7 which represents the average trend observed from many hundreds of data points.[63]

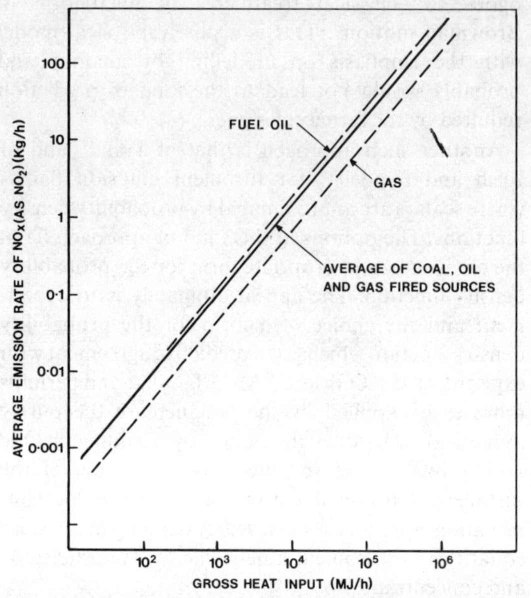

FIG. 7. Average emission rate of NO_x as function of gross heat input to combustion source (after Saltzman, reference 63).

Sulphur content of natural gas is generally very low. Most British natural gases as received contain less than 0.2 ppm sulphur compounds by volume. The majority of the sulphur content in distributed gases has been added as an odorant largely in the form of diethyl sulphide, methyl ethyl sulphide and various mercaptans totalling between 4 and 10 ppm, i.e. an equivalent sulphur content of 6–14 mg m^{-3}.[64] In contrast, manufactured town gases contained much larger amounts of sulphur, typically up to 500 mg m^{-3}.

TABLE 3. Emissions of oxides of Nitrogen from stationary sources in the UK (after ref. 62).

Appliance or source	Fuel	Capacity	Range of (NO) in flue gases (parts 10^{-6})	Value used (parts 10^{-6})	Rate of emissions of NO_x (g/MJ)
Space heater	Gas	Domestic	34–62	50	0.040
Water heater	Gas	Domestic	4–96	50	0.077
Cooker	Gas	Domestic	8–33	15	0.024
Refrigerator	Gas	Domestic	1–8	5	0.012
Grate or cooker	Coal	Domestic	4–30	11	0.069
Grate or cooker	Smokeless fuel	Domestic	6–11	8	0.047
Central heating	Oil	Domestic	20–400	70	0.086
Central heating, commercial, industrial	Oil	5 MW	20–400	150	0.172
General industrial and commercial	Solid fuel	5 MW	161–1204	470	0.198
General industrial and commercial	Gas	5 MW	45–216	90	0.061
General industrial and commercial	Oil	5 MW	180–400	300	0.21
Incinerators	—	Municipal	24–92	60	0.09
Electricity generating stations	Oil	—	180–900	400	0.16
Electricity generating stations	Coal	—	100–1460	400	0.22

After carbon dioxide, carbon monoxide is the most abundant pollutant in the atmosphere, and generally results from the incomplete combustion of carbon-based fuels. Although the overall contribution of gas to pollution by CO is negligible (the bulk being due to vehicular emissions), high local concentrations can be generated under adverse conditions. This is obviously undesirable, particularly for unflued appliances. Careful attention must, therefore, be paid to the proper design of burners to ensure that flame stability is good, flame quenching does not occur, and the air supplied for combustion is both sufficient and properly distributed. In the UK and elsewhere, regulations exist to ensure that these conditions are met in commercially available appliances by checking their performance against nationally agreed standards.[65] Most countries have adopted combustion performance criteria which limit the maximum permissible CO concentration to 0.1–0.2% v/v, measured in the burnt gas on a dry, air-free basis.

Excessive aerodynamic quenching or chilling by flame impingement can sometimes give rise to the emission of small concentrations of partially oxidized hydrocarbon species, principally formaldehyde and acrolein. Aldehydes are well-known eye irritants and readily detectable in trace amounts by their characteristic odour.

The chemistry of formaldehyde formation and destruction in methane flames is now fairly well understood but only fragmentary data are available on aldehyde emission factors from different combustion sources. A preliminary survey in the United States indicates that 1–2 lb of aldehydes are generated per million cubic feet of natural gas combusted.[66] This is thought to be insignificant compared with the contribution from transportation and open burning sources.

9. IGNITION

When a flammable mixture is heated to an elevated temperature, chemical reaction is initiated that may proceed with sufficient rapidity to ignite the whole mixture. It is found that ignition may occur spontaneously under certain conditions of temperature and pressure but under other conditions must be forced to occur by use of an external energy source.

Spontaneous ignition is usually studied in closed vessels, the walls of which are maintained at a controlled temperature. There is a fairly definite temperature called the minimum spontaneous-ignition or auto-ignition temperature above which ignition occurs. Below this temperature slow combustion may occur, and around this ignition temperature there is usually an appreciable delay, the induction period before ignition occurs, this induction period usually decreasing as the temperature is raised.[67] Auto-ignition temperatures are mainly of interest from the standpoint of safety and adiabatic compression in internal combustion engines. A list of minimum auto-ignition temperatures together with flammability limits for some common gases and vapours is given in Table 4. Unlike flammability limits, where "mixing rules" are well established, very little is known about the auto-ignition temperatures of multi-component fuel mixtures. It is usual practice, therefore, to assume that the auto-ignition temperature of a mixture of fuels corresponds to the component with the lowest ignition temperature.

TABLE 4. Minimum auto-ignition temperatures and flammability limits of some individual gases in air at 25°C and 1 atm.[68]

Combustible	Lower flammability limit (volume percent)	Upper flammability limit (volume percent)	Auto-ignition temperature (°C)
Methane	5.0	15.0	540
Ethane	3.0	12.4	515
Propane	2.1	9.5	450
n-Butane	1.8	8.4	405
Propylene	2.4	11.0	460
Ethylene	2.7	36.0	490
Hydrogen	4.0	75.0	400

The lighting of a flowing gas mixture is most conveniently achieved using an external energy source, e.g. electric spark, hot wire or pilot flame. In such cases it is necessary to supply sufficient energy to heat the gas locally to a temperature well above the minimum spontaneous ignition temperature and, in the case of sparks, greatly in excess of the adiabatic flame temperature.

A fundamental description of the development of steady-state combustion from a non-steady state initiated by a local source of ignition requires solution of the time-dependent conservation equations governing flame propagation. So far, this has been attempted only for fuel-rich mixtures of hydrogen–air because of insufficient knowledge about the detailed combustion chemistry of hydrocarbon fuels.[69] Most models of ignition are therefore based on energy considerations since it appears that a certain minimum energy is required to ignite a flammable mixture.

The concepts of "minimum ignition energy", "critical flame diameter" and "quenching distance" have been particularly valuable in characterizing instantaneous, point ignition of which the high voltage electric spark is a good approximation. Experimentally it is found that for a given gas mixture and physical state there is an absolute minimum amount of energy, E_{min}, which will just cause ignition. This minimum ignition energy appears to be necessary for a small spherical flame (kernel) generated by a point-like spark to attain a critical diameter, d_0, beyond which the flame is able to propagate unaided.

In a typical capacitive spark discharge, a small volume of hot, ionized gas is formed that, initially at least, is in thermal disequilibrium with the free electrons present and attains peak temperatures at thermal equilibrium in excess of 10^4 K. Thus, the processes that lead to ignition may be thermal or non-thermal, the primary distinction being whether or not the initial excitation decays to thermal equilibrium with its surroundings before significant chemical reaction occurs.

Simple models of spark ignition fall into two main groups, depending on the assumptions made about the nature of energy transfer to the unburnt gas. For example, so-called activation or chain theories[70] consider the spark as a source of chemical energy in the form of atoms and free radical dissociation products, while the presence of a high initial gas temperature (spark plasma) and its effect on the ignition process is ignored or taken to be of secondary importance. Thermal theories on the other hand, assume that the heating effect of the spark predominates and treats the phenomenon strictly as a problem of transient heat conduction, neglecting all effects due to mass diffusion.

Lewis and von Elbe[34] have suggested two criteria which relate minimum ignition energies to measurable or calculable flame parameters. The relationships concerned are based on separate thermal concepts of ignition and provide at least order of magnitude estimates of minimum spark energy requirements. The first of these relationships is derived simply on the assumption that $E_{\min}$ is equal to the excess thermal energy contained in a flame kernel of minimum critical diameter d_0, which Lewis and von Elbe equate to the quenching distance. The diameter d_0 may be measured experimentally using glass-flanged spark electrodes, when it is found that for flange separations greater than a critical value (d_0), the ignition energy remains roughly constant, while for separations less than d_0, the energy rises steeply with decreasing flange separation. The excess thermal energy per unit area of a plane flame front is given by the expression $\lambda_u(T_b - T_u)/S_u$, so for a kernel of area πd_0^2, the total excess thermal energy is given by

$$E_{\min} = \frac{\pi d_0^2 \lambda_u}{S_u}(T_b - T_u)$$

Where λ, T and S_u are the thermal conductivity, temperature and normal burning velocity of the gas respectively, the subscripts u and b denoting its state, i.e. unburnt or burnt.

Experimentally it is found that for hydrocarbon–air mixtures $E_{\min} \simeq \text{constant} \times d_0^2$, a dependence that is not obvious from the above equation. The physical significance of this, coupled with the observation noted earlier that there exists a critical size to which an incipient flame kernel must grow in order to propagate unaided can be appreciated in terms of flame stretch theory—a concept discussed in Section 5.

Using this concept Lewis and von Elbe showed that for a spherical flame to expand through a distance equal to its preheat zone thickness, a certain minimum amount of energy is involved. In the case of an incipient flame kernel having a critical quenching diameter d_0, the minimum energy required to support its expansion through a distance equal to the preheat zone thickness can be related to the minimum energy expression and appears to be directly proportional to d_0^2, as found experimentally.

It would seem reasonable, therefore, that for a given flammable mixture there exists some minimum size of flame kernel below which the diverging flame front cannot supply sufficient heat to the surrounding preheat zone for self-propagation to be sustained. When the energy supplied by a spark is equal to the minimum ignition energy for the gas mixture, the incipient flame kernel just succeeds in attaining the critical size required for it to propagate unaided thereafter.

Considering the assumptions made in Lewis and von Elbe's approach, the order of magnitude agreement between experimental and predicted minimum ignition energies is remarkably good. One serious limitation of all the purely thermal approaches to spark ignition, however, is their inability to predict the stoichiometry of the most readily ignitable fuel–air mixture. From the elementary theory one would expect the minimum ignition energy to occur at the maximum burning velocity of the particular mixture, corresponding to a stoichiometric fraction of about 1.1 for hydrocarbon fuels (see Section 3). Figure 8 shows that there are important differences in the mixture ratios at which minimum ignition energies occur and, in general, there is a shift away from stoichiometric towards richer mixtures as the molecular weight of the fuel increases. These differences are thought to be due to the effects of preferential diffusion which manifest themselves in highly curved flame fronts.[71,72] Thus methane, being lighter than oxygen, has its minimum ignition energy at a stoichiometric fraction of 0.8, while for those fuels whose diffusivity is less than that of oxygen, the minima appear on the fuel-rich of stoichiometric.

In addition to neglecting mass diffusion, there is no sound basis for assuming that the temperature of incipient flame kernels are in practice either spatially uniform or equal to the appropriate adiabatic flame temperature. Similarly, the use of one-dimensional laminar flame parameters for deriving minimum ig-

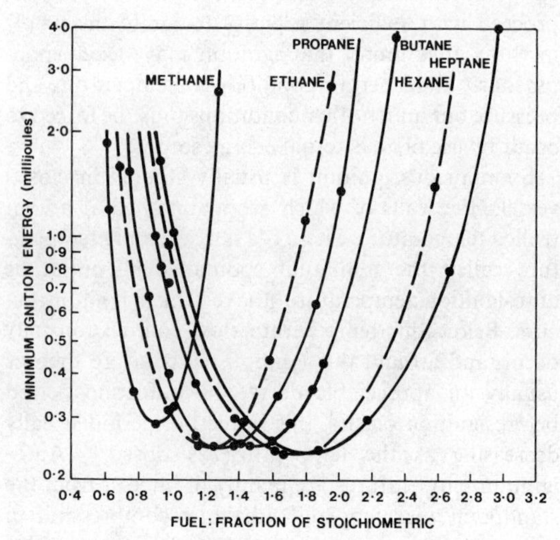

FIG. 8. Minimum ignition energies plotted against stoichiometric fraction for a number of hydrocarbons (after Lewis and von Elbe).

nition energies must be considered to be largely speculative.

When a source possesses mass, and energy release is not practically instantaneous (i.e. > 1 μs), the concept of minimum source energy is not adequate for characterizing ignition. For a source releasing energy at a comparatively slow rate, the total energy required for effecting ignition depends on the rate and duration of energy supply to the mixture. Under such conditions a more useful parameter of characterization is the source temperature. Since the rate of heat transfer between the source and surrounding gas is proportional to the source temperature, a lower source temperature will require a longer time, which is usually called the contact time or critical heating period, to impart sufficient energy for ignition.

Hot bodies such as heated wires and pilot flames have been used extensively in the past as slow ignition sources. The electrically heated filament igniter usually takes the form of a coiled wire supported between two metallic electrodes. Power may be supplied continuously by mains electricity or intermittently by batteries, although battery operation is only practical for fine wires exhibiting catalytic activity. Owing to the large number of independent variables required to define even the most idealized hot wire ignition process, most investigations have been concerned with establishing empirical relationships between coil parameters and gas parameters. Two important sets of results of this type are those of the American Gas Association[73] and the more recent studies carried out by British Gas Corporation.[74]

As a first step towards obtaining a more basic understanding of hot wire ignition, Wills and Nolan[78] have carried out an investigation of the heat transfer from non-catalytic wires with a view to optimizing coil design. By measuring the temperature distribution along the wire they were able to estimate graphically the loss of heat by radiation and conduction to the end supports, and so deduce the convective heat transfer to the gas. The energy transferred increased roughly linearly with coil pitch and was slightly higher in rich mixtures of methane–air than in lean mixtures. For a typical optimized coil it was found that, of the total electrical energy supplied to the igniter, 30% went into heating the coil, 50% was lost as radiation and 5% conducted to the supports leaving only 15% to heat the gas. This demonstrates the importance of radiation and of the thermal capacity of the coil and underlines the gross inefficiency of the process compared to spark ignition.

The effect of coil pitch on the minimum ignition temperature was studied in some detail. It was found that the ignition temperature passes through a minimum at a pitch corresponding to about 2.5 times the close-wound value (see Figs. 9 and 10). This confirms fairly closely to the findings of reference 73 which recommends a pitch of about twice the wire diameter. A qualitative explanation for the existence of an optimum pitch for ignition was given in terms of a quenching effect inside the coil. Schlieren photographs of the ignition process revealed that for large diameter coils (7 mm ID), ignition was initiated inside the coil if the pitch was 1 mm or below, and in the wake of the coil for pitches greater than this. For a small diameter coil (2 mm ID), ignition was found to be initiated inside the coil for all pitches investigated (up to 2 mm).

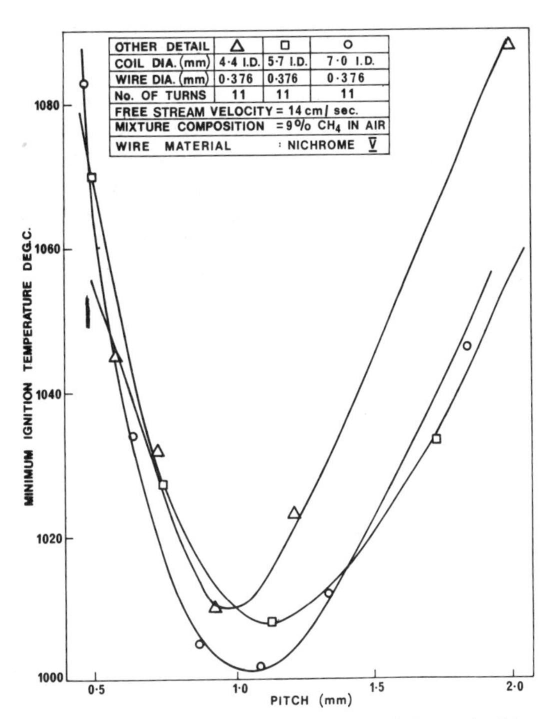

FIG. 9. The effect of coil pitch on the minimum ignition temperature.

Similar to the behaviour of other slow ignition sources, the ability of a flame to induce ignition of a

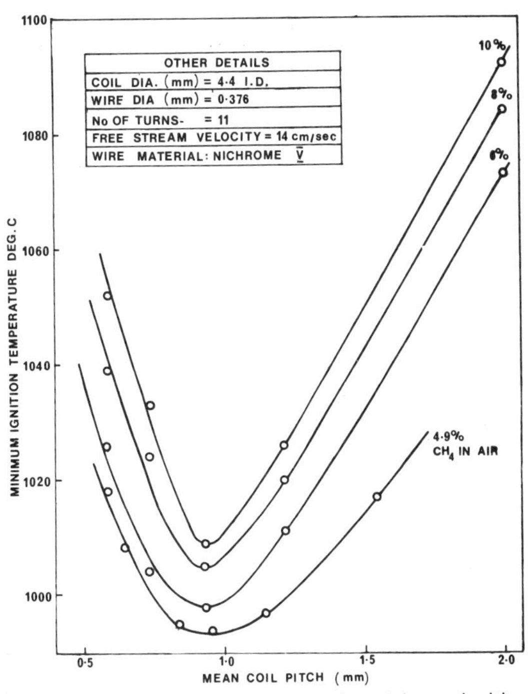

FIG. 10. The effect of coil pitch on the minimum ignition temperature for different mixture compositions.

flammable gas mixture may also be characterized by three main parameters: the temperature and size of the flame and the time of contact between flame and gas. The variation of the critical contact time with fuel concentration follows the same pattern as that of minimum ignition energy in spark ignition. Generally, however, the problems associated with pilot flames are more concerned with practical difficulties such as flame stability and jet blockage than with a lack of knowledge about the fundamental mechanism of ignition.

The trend in recent years has been to move away from forms of ignition involving permanent pilot lights and heated wires, to spark ignition which is inherently more efficient, robust and versatile. For those readers interested in the practical applications of spark ignition to both domestic and industrial burners we recommend a number of review papers published by the Institution of Gas Engineers.[74–77]

REFERENCES

1. TIRATSOO, E. N., *Natural Gas*, Scientific Press Ltd., Beaconsfield (1972).
2. ENERGY R. AND D. IN THE UNITED KINGDOM—a discussion document, Advisory Council on Research and Development for Fuel and Power, Department of Energy (1976).
3. BRITISH GAS DATA BOOK, Volume 1. *Properties of Natural Gas: Treatment, Transmission, Distribution and Storage*, British Gas Corporation, London Research Station (1974).

4a. WEAVER, E. R., Formulas and graphs for representing the interchangeability of fuel gases. *J. natn. Bur. Standards* **46**, 213 (1951).

4b. GILBERT, M. G. and PRIGG, J. A., The prediction of the combustion characteristics of town gas. *Trans. Inst. Gas. Eng. (Lond)* **106**, 530 (1956).

5. FRISTROM, R. M. and WESTENBERG, A. A., *Flame Structure*, McGraw Hill, New York (1965).
6. DIXON-LEWIS, G., Burning velocities in hydrogen–air mixtures. *Combust. Flame* **15**, 197 (1970).
7. STEPHENSON, P. L. and TAYLOR, R. G., Laminar flame propagation in hydrogen, oxygen, nitrogen mixtures. *Combust. Flame* **20**, 231 (1973).
8. SPALDING, D. B. and STEPHENSON, P. L., Laminar flame propagation in hydrogen and bromine mixtures. *Proc. Roy. Soc.* **A324**, 315 (1971).
9. SPALDING, D. B., STEPHENSON, P. L. and TAYLOR, R. G., A calculation procedure for the prediction of laminar flame speeds. *Combust. Flame* **17**, 55 (1971).
10. ANDREWS, G. E. and BRADLEY, D., The burning velocity of methane–air mixtures. *Combust. Flame* **19**, 275 (1972).
11. ANDREWS, G. E. and BRADLEY, D., Determination of burning velocities: a critical review. *Combust. Flame* **18**, 133 (1972).
12. LINDOW, R., Zur Bestimmung der laminaren Flammen–geschwindigkeit, Thesis, Universität Karlsruhe (T.H.) (1966).
13. GÜNTHER, R. and JANISCH, G., Measurement of burning velocity in a flat flame front. *Combust. Flame* **19**, 49 (1972).
14. JANISCH, G., Zur stabilisierung laminarer Vormischflammen, Thesis Universität Karlsruhe (T.H.) (1972).
15. FRANCE, D. H. and PRITCHARD, R., Laminar burning velocity measurements using a laser-doppler anemometer. *J. Inst. Fuel* **49**, 81 (1976).
16. RAEZER, S. D. and OLSEN, H. L., Measurement of laminar flame speeds of ethylene–air and propane–air mixtures by the double kernel method. *Combust. Flame* **6**, 227 (1962).
17. ANDREW, G. E. and BRADLEY, D., Determination of burning velocity by double ignition in a closed vessel. *Combust. Flame* **20**, 77 (1973).
18. GIBBS, G. J. and CALCOTE, H. F., Effect of molecular structure on burning velocity. *J. Chem. Eng. Data* **4**, 226 (1959).
19. ZABETAKIS, M. G., Flammability characteristics of combustible gases and vapors. U.S. Bureau of Mines, Bulletin 627, p. 49 (1965).
20. See, for example ref. 19.
21. GAYDON, A. G. and WOLFHARD, M. G., *Flames: their structure, radiation and temperature*, 3rd edn, Chapman and Hall (1970).
22. SCHOLTE, T. G. and VAAGS, P. B., Burning velocities of hydrogen, carbon monoxide and methane with air. *Combust. Flame* **3**, 511 (1959).
23. LEASON, D. B., The effect of gaseous additions on the burning velocity of propane–air mixtures. *Fourth (International) Symposium on Combustion*, p. 369, Williams and Wilkins, Baltimore (1953).
24. PAYMAN, W. and WHEELER, R. V., *Fuel (Lond.)* **1**, 185 (1922).
25. SPALDING, D. B., A mixing rule for laminar flame speed. *Fuel (Lond.)* **35**, 347 (1956).
26. MARTIN, D. G., Flame speeds of mixtures containing several combustible components or a known quantity of diluent. *Fuel (Lond.)* **35**, 352 (1956).
27. CORBEELS, R. and VAN TIGGELEN, A., Flame propagation velocity methane–hydrogen–oxygen mixtures. *Bull. Soc. Chim. Belg.* **68**, 613 (1959).
28. BORIE, V., Influence du methane sur les vitesses de propagation et des limites d'inflammabilité des melanges $H_2/O_2/N_2$. *2nd European Combustion Symposium*, p. 115, The Combustion Institute (1975).
29. REED, S. B., MINEUR, J. and MCNAUGHTON, J. P., The effect on the burning velocity of methane of vitiation of combustion air. *J. Inst. Fuel* **44**, 149 (1971).
30. MELVIN, A. and MOSS, J. B., Structure in methane–oxygen diffusion flames. *Fifteenth Symposium (International) on Combustion*, p. 625, The Combustion Institute (1974).
31. MELVIN, A., MOSS, J. B. and CLARKE, J. F., The structure of a reaction-broadened diffusion flame. *Combust. Sci. Tech.* **4**, 17 (1971).
32. CLARKE, J. F. and MOSS, J. B., On the structure of a spherical H_2–O_2 diffusion flame. *Combust. Sci. Tech.* **2**, 115 (1970).
33. MELVIN, A., MOSS, J. B. and CLARKE, J. F., Streamline deflection by a diffusion flame. *Combust. Sci. Tech.* **6**, 135.
34. LEWIS, B. and VON ELBE, G., *Combustion Flames and Explosions of Gases*, 2nd edn, Academic Press, New York.
35. KARLOVITZ, B., DENNISTON, D. W., KNAPSCHAEFER, D. H. and WELLS, F. E., Studies on turbulent flames. *Fourth Symposium (International) on Combustion*, p. 613, Williams and Wilkins, New York.
36. REED, S. B., Flame stretch—a connecting principle for blow-off data. *Combust. Flame* **11**, 177 (1967).
37. EDMONDSON, H. and HEAP, M. P., The correlation of burning velocity and blow-off data by the flame stretch concept. *Combust. Flame* **15**, 179 (1970).
38. MELVIN, A. and MOSS, J. B., Evidence for the failure of the flame stretch concept for premixed flames. *Combust. Sci. Tech.* **7**, 189 (1973).
39. GÜNTHER, R. and JANISCH, G., The stabilising region of premixed flames. *Combustion Institute European Symposium*, p. 689, Academic Press, New York.
40. GÜNTHER, R. and JANISCH, G., On the stability of premixing flames. *Gaswärme International* **24**, 489 (1975).
41. RADCLIFFE, S. W. and HICKMAN, R. G., Diffusive catalytic combustors. *J. Inst. Fuel*, 208, December, 1975.

42. DONGWORTH, M. R. and MELVIN, A., Diffusive catalytic combustion. *Sixteenth Symposium (International) on Combustion*, The Combustion Institute (in the press).
43. Predictive Methods for Industrial Flames, *J. Inst. Fuel* (several papers) (1972).
44. WOHL, K., KAPP, N. M. and GAZLEY, C., The stability of open flames. *Third Symposium (International) on Combustion*, p. 3, Williams and Wilkins, New York (1949).
45. WOHL, K., SHORE, L., VON ROSENBERG, H. and WEIL, C. W., The burning velocity of turbulent flames. *Fourth Symposium (International) on Combustion*, p. 620, Williams and Wilkins, New York.
46. HAWTHORNE, W. R., WEDDELL, D. S. and HOTTEL, H. C., Mixing and combustion in turbulent gas jets. *Third Symposium (International) on Combustion*, p. 266, Williams and Wilkins, New York (1949).
47. CHUNG, P. M., Diffusion flame in homologous turbulent shear flows. *Phys. Fluids* **15**, 1735 (1972).
48. TOOR, H. L., Turbulent mixing of two species with and without chemical reaction. *Ind. Eng. Chem. Fundamentals* **8**, 655 (1969).
49. BUSH, W. B. and FENDELL, F. E., On diffusion flames in turbulent shear flows—the two step symmetrical chain reaction. *Combust. Sci. Tech.* **11**, 35 (1975).
50. O'BRIEN, E. E., Turbulent mixing of two rapidly reacting chemical species. *Phys. Fluids* **14**, 1326 (1971).
51. SPALDING, D. B., Mixing and chemical reaction in steady, confined turbulent flames. *Thirteenth Symposium (International) on Combustion*, p. 649, The Combustion Institute (1971).
52. PATANKAR, S. V. and SPALDING, D. B., *Heat and Mass Transfer in Boundary Layers*, 2nd edn, Intertext Books, New York (1970).
53. KENT, J. M. and BILGER, R. W., The prediction of turbulent jet diffusion flames, *Fourteenth Symposium (International) on Combustion*, p. 615, The Combustion Institute (1973).
54. BRAY, K. N. C., Equations of turbulent combustion. I Fundamental equations of reacting turbulent flow, University of Southampton, AASU Report No. 330.
55. BRAY, K. N. C., Equations of turbulent combustion. II Boundary layer approximation, University of Southampton, AASU Report No. 331 (1973).
56. BRAY, K. N. C., Kinetic energy of turbulence in flames. University of Southampton, AASU Report No. 332 (1974).
57. KLIMOV, A. M., Flame propagation under conditions of strong turbulence. *Dokl. Akad, Nauk SSSR* **221**, (1) 56 (translated as *Sov. Phys. Dokl.* **20** (3), 168) (1975).
58. BALLAL, D. R. and LEFEBVRE, A. H., The structure and propagation of turbulent flames. *Proc. Roy. Soc.* **A344**, 217 (1975).
59. DURST, F., MELLING, A. and WHITELAW, J. H., The application of optical anemometry to measurements in combustion systems. *Combust. Flame* **18**, 197 (1972).
60. BAKER, R. J., HUTCHINSON, P. and WHITELAW, J. H., Detailed measurement of flow in the recirculation region of an industrial burner by laser anemometry. *Combustion Institute European Symposium*, p. 583, Academic Press, New York. (1973).
61. ROBINSON, E. and ROBBINS, R. C., Sources, abundance and fate of gaseous atmospheric pollutants. Project PR-6755. Menlo Park, California. Stanford Research Institute.
62. DERWENT, R. G. and STEWART, H. N. M., Air pollution from oxides of nitrogen in the United Kingdom. *Atmos. Env.* **7**, 385 (1973).
63. SALTZMAN, B. E., Method of measuring and control of oxides of nitrogen in the atmosphere, and their derivatives. World Health Organisation Publication, WHO/AP/68-31.
64. DENSHAM, A. B. and GIBBONS, R. M., The properties of British Natural Gas. Gas Council Research Communication GC 187 (1971).
65. LAKE, F. and PURKIS, C. H., The international harmonisation of domestic gas appliance standards. Gas Council Communication GC 821 (1970).
66. STAHL, Q. R., Preliminary air pollution survey of aldehydes: a literature review. National Air Pollution Control Administration Publication No. APTD 69-24 Raleigh N.C. (1969).
67. BRADLEY, J. N., *Flame and Combustion Phenomena*, Chapter 2, Methuen, London (1969).
68. ZABETAKIS, M. G., Flammability characteristics of combustible gases and vapors. U.S. Bureau of Mines, Bulletin 627.
69. DIXON-LEWIS, G. and SHEPHERD, I. G., Some aspects of ignition by localised sources, and of cylindrical and spherical flames. *Fifteenth Symposium (International) on Combustion*, p. 1483, The Combustion Institute (1975).
70. LANDAU, H. G., The ignition of gases by local sources. *Chem. Rev.* **21**, 245 (1937).
71. STREHLOW, R. A., *Fundamentals of Combustion*, p. 240, International Textbook Company, New York.
72. STREHLOW, R. A., The effect of curvature on flame structure, *2nd European Combustion Symposium*, p. 496, The Combustion Institute (1975).
73. AMERICAN GAS ASSOCIATION, Research Bulletin No. 28.
74. ELKINS, B. and BROWN, A. M., Gas Council Research Communication GC 146 (1967).
75. ATKINSON, P. G., MARSHALL, M. R. and MOPPETT, D. J., The ignition of industrial burners. Gas Council Research Communication GC 147 (1967).
76. SAYERS, J. F., TEWARI, G. P., WILSON, J. R. and JESSEN, P. F., Spark ignition of natural gas—theory and practice. Gas Council Research Communication GC 171.
77. EKINS, B. and WILSON, J. R., Some characteristics of practical spark-ignition systems. Gas Council Research Communication GC 185 (1971).
78. WILLS, B. J., An investigation of the hot wire ignition of methane–air mixtures. M. Phil. Thesis to be submitted to the Polytechnic of the South Bank, London.

Manuscript received 26 October 1976

AUTHOR INDEX

SUBJECT INDEX